Theory of Nonlinear Operators

SCHRIFTENREIHE DES ZENTRALINSTITUTS FÜR MATHEMATIK UND MECHANIK

BEI DER AKADEMIE DER WISSENSCHAFTEN DER DDR

Herausgeber: Heinz Ahrens, Helmut Koch, Klaus Matthes, Günther Schmidt, Kurt Schröder

Heft 20

THEORY OF NONLINEAR OPERATORS

AKADEMIE-VERLAG · BERLIN
1974

THEORY OF NONLINEAR OPERATORS
PROCEEDINGS OF A SUMMER-SCHOOL

held in October 1972 at Neuendorf (Hiddensee), GDR

Herausgegeben von

Professor Dr. J. Nečas und Professor Dr. J. Kolomý, Prag
Professor Dr. R. Kluge und Professor Dr. A. Langenbach, Berlin

Mit 3 Abbildungen

AKADEMIE-VERLAG · BERLIN
1974

Zentralinstitut für Mathematik und Mechanik, 108 Berlin, Mohrenstr. 39
Erschienen im Akademie-Verlag, 108 Berlin, Leipziger Straße 3–4

Lizenznummer: 202 · 100/418/74
Gesamtherstellung: VEB Druckerei „Thomas Müntzer", 582 Bad Langensalza
Bestellnummer: 761 947 6 (2069/20) · LSV 1065
Printed in GDR
EVP 68,–

Inhaltsverzeichnis

S. Burýšek: Eigenvalue problem, bifurcations and equations with analytic operators in Banach spaces . . . 1

J. Daneš: On densifying and related mappings and their application in nonlinear functional analysis . . . 15

S. Fučík, M. Kučera, J. Souček, V. Souček: Topics in nonlinear spectral theory . . . 57

H. Gajewski: Zur Fehlerabschätzung bei Gleichungen mit nichtlinearen Potentialoperatoren . . . 75

K. Gröger: Zum Galerkin-Verfahren für Evolutionsgleichungen . . . 85

R. Hünlich: Positive Fixpunkte parameterabhängiger Operatoren . . . 105

O. John, J. Stará: Regularity results for non linear integro-differential operators in two dimensions . . . 117

J. Kačur: The Rothe method and nonlinear parabolic equations of arbitrary order . . . 125

R. Kluge: Näherungsverfahren für einige nichtlineare Probleme . . . 133

R. Kluge:, G. Bruckner: Gewöhnliche nichtlineare Differentialgleichungen und Differentialungleichungen im Hilbertraum . . . 147

J. Kolomý: Normal solvability and solvability of nonlinear equation . . . 155

E. Krauß: Zur Steuerung mit Operatorgleichungen . . . 169

A. Langenbach, B. Wegner, K. Wiedemann: Differenzierbare Zweige von Eigenfunktionen im Bifurkationselement . . . 177

M. Müller: Über die Differenzierbarkeit der Lösungen gewisser quasilinearer elliptischer Differentialgleichungen am Rande des Gebiets . . . 199

J. Naumann: On parabolic regularization of an undamped problem in nonlinear plate vibration . . . 211

J. Nečas: Mathematical models of elastic-inelastic materials . . . 227

R. Nürnberg: Zur starken Konvergenz des Galerkin-Verfahrens für Evolutions-Gleichungen . . . 233

G. Stoyan: On the stability of twodimensional Janenko-scheme with respect to boundary values . . . 241

K. Zacharias: Zum Galerkin-Verfahren bei einigen Klassen nichtlinearer Differentialgleichungen im Hilbertraum . . . 247

E. Zeidler: Existenz, Eindeutigkeit, Eigenschaften und Anwendungen des Abbildungsgrades im R^n . . . 259

R. Zezula: On a class of nonlinear eigenvalue problems, arising in reactor physics . . . 313

Eigenvalue problem, bifurcations and equations with analytic operators in Banach spaces

SLAVOMÍR BURÝŠEK, Prag

Introduction. In this note we want to show an elementary approach to the non-linear eigenvalue problem. Further we briefly sketch the theory of analytic operators with an application to the study of non-linear operators equations.

According to standard definition, a complex number λ is said to be an eigenvalue of a non-linear operator $F: X \to X$ in a Banach space X if there is an element $x \in X$, $x \neq 0$ such that $F(x) = \lambda x$. This definition is transferred from the linear analysis. But there are great differences between the linear and non-linear case. In the linear case the eigenvalues and eigenvectors are somehow "exceptional" points. The cause of this exceptionality is the homogeneity of the same degree of a linear and the identity operator. For non-linear operators this homogeneity is no longer true. For example, the operator F on the space $C(-\pi/2, \pi/2)$ defined by

$$F(x) \equiv y(s) = \int_{-\pi/2}^{+\pi/2} [\sin s \cdot \cos t \cdot x(t) + \cos s \cdot \sin t \cdot x^2(t)] \, dt$$

has eigenvectors of the form $x_\lambda(t) = (3\lambda/4) \left(\sin t + (2\lambda/\pi) \cos t\right)$. Thus, each λ is an eigenvalue of the operator F. It is natural to consider the nonlinear eigenvalue problem for the couple (F, G) of homogeneous or (roughly speaking) near to homogeneous operators F, G with the same degree of homogeneity as the problem of non-zero solutions of the equation $F(x) = \lambda G(x)$. Some results in this way were reached by J. Nečas [17], S. Fučík and J. Nečas [18] and others. Unlike these deep studies using, among others, the Ljusternik-Schnirelmann theory of categories, we shall try to present here a different, rather an elementary access to this difficult problem avoiding the assumption of potentiality.

1. Definition of a spectrum of a nonlinear operator

Let X, Y be Banach spaces and let C be the space of complex numbers. Denote $K_r(a) = \{x \in X / \|x - a\| < r\}$ the open ball in X, $\overline{K_r(a)}$ its closure and $S_r(a) = \{x \in X / \|x - a\| = r\}$ the sphere in X. If X, Y are Hilbert spaces then $(.,.)$ denotes the scalar product. Consider an operator $G: X \times C \to Y$ such that $G(0, 0) = 0$ and let $M \subset X$ be a given nonempty set.

Definition 1.1. We shall say that $\lambda \in C$ is a point of a spectrum of the operator G with respect to the set M if there is a sequence $\{x_n\} \subset M$, $x_n \neq 0$ such that

$\lim_{n\to\infty} G(x_n, \lambda) = 0$. The set of all points of a spectrum of the operator G with respect to M we denote by $\mathscr{S}_G(M)$ and call the spectrum of the operator G with respect to M. A point $\lambda_0 \in \mathscr{S}_G(M)$ is said to be an eigenvalue of the operator G with respect to M if there is a point $x_0 \in M$, $x_0 \neq 0$ such that $G(x_0, \lambda_0) = 0$. The point x_0 is called the eigenvector of the operator G with respect to M corresponding to the eigenvalue λ_0.

Proposition 1.1. *Let $M \subset X$ be a bounded set and let $G: X \times C \to Y$ be bounded operator satisfying the Lipschitz's condition with respect to the varible $\lambda \in C$ (i.e., there is a bounded functional $f(x)$ on X such that $||G(x, \lambda) - G(x, \mu)|| \leqq f(x) |\lambda - \mu|$ for any $x \in X$, $\lambda, \mu \in C$). Then $\mathscr{S}_G(M)$ is a closed set in C.*

Proof. Let $\{\lambda_k\} \subset \mathscr{S}_G(M)$ be a sequence such that $\lim \lambda_k = \lambda_0 \in C$. Then there are sequences $\{x_n^k\} \subset M$ such that $\lim_{n\to\infty} ||G(x_n^k, \lambda_k)|| = 0$, $k = 1, 2, \ldots$. Denote $a_n^k = ||G(x_n^k, \lambda_k)||$ and choose for a given $k = 1, 2, \ldots, a_{n_k}^k$ from the sequence $\{a_n^k\}$ such that $a_{n_k}^k \leqq 1/k$. Putting $y_k = x_{n_k}^k$, $k = 1, 2, \ldots$, we obtain a sequence $\{y_k\} \in M$ such that $\lim_{k\to\infty} ||G(y_k, \lambda_k)|| = 0$. Hence, $||G(y_k, \lambda_0)|| \leqq ||G(y_k, \lambda_k)|| + ||G(y_k, \lambda_k) - G(y_k, \lambda_0)|| \leqq ||G(y_k, \lambda_k)|| + f(y_k) |\lambda_k - \lambda_0|$, so that $\lim_{k\to\infty} ||G(y_k, \lambda_0|| = 0$ and thus $\lambda_0 \in \mathscr{S}_G(M)$.

Definition 1.2. Let $G(0, \lambda) = 0$ for any $\lambda \in C$. We shall say that a point $\lambda_0 \in C$ is a bifurcation point of the operator G if for any $\varepsilon > 0$ and any neighbourhood U_0 of the point zero in X there is an eigenvalue $\lambda \in \mathscr{S}_G(U_0)$ such that $|\lambda - \lambda_0| < \varepsilon$. The set of all bifurcation points of the operator G we denote be $\mathscr{B}_G$.

Proposition 1.2. *Let us suppose that G satisfies the assumptions of the proposition 1.1 and, further, let $G(0, \lambda) = 0$ for any $\lambda \in C$. Then $\mathscr{B}_G \subseteq \mathscr{S}_G(U_0)$ for any neighbourhood U_0 of the point zero in X.*

Proof: Let U_0 be a neighbourhood of the point $0 \in X$. Obviously, we can assume that U_0 is bounded. According to definition 1.2 there is a sequence $\lambda_n \in \mathscr{S}_G(U_0)$ of eigenvalues of the operator G with eigenvectors $x_n \in K_{1/n}(0)$ such that $\lim_{n\to\infty} \lambda_n = \lambda_0$ and $\mathscr{S}_G(K_{1/n}(0)) \subset \mathscr{S}_G(U_0)$ for sufficiently large n. Then, according to proposition 1.1, $\lambda_0 \in \mathscr{S}_G(U_0)$.

Proposition 1.3. *Let $M \subset X$ be a compact and closed (resp. weakly compact and weakly closed) set in X and let $0 \notin M$. Suppose $G: X \times C \to Y$ is a continuous (resp. strongly continuous) operator in the variable $x \in X$ for any $\lambda \in C$. Then any non-zero point from $\mathscr{S}_G(M)$ ist an eigenvalue of the operator G with respect to M.*

Proof: Let $\lambda_0 \in \mathscr{S}_G(M)$, $\lambda_0 \neq 0$. Then there is a sequence $\{x_n\} \in M$, $x_n \neq 0$ such that $||G(x_n, \lambda_0)|| \to 0$ as $n \to \infty$. Now it is possible to choose a subsequence $\{x_{n_k}\}$ which converges (resp. weakly converges) to a non-zero point $x_0 \in M$. Using the continuity (resp. strong continuity) of the operator G we obtain $\lim_{k\to\infty} ||G(x_{n_k}, \lambda_0)|| = ||G(x_0, \lambda_0)|| = 0$ and thus, λ_0 is an eigenvalue with eigenvector x_0.

2. The spectrum of a couple of operators

Definition 2.1. Let $S, T\colon X \to Y$ be a couple of operators and put $G(x, \lambda) = S(x) - \lambda T(x)$ for any $x \in X$, $\lambda \in C$. Let $M \subset X$ be a nonempty set. Then the spectrum $\mathcal{S}_G(M) \equiv S_{S,T}(M)$ is called the spectrum of the couple (S, T) with respect to M. Analogously we define eigenvalues and eigenvectors of the couple (S, T) with respect to M. In case that $X = Y$ and $T = I$ is the identity operator then the spectrum $\mathcal{S}_{S,I}(M) \equiv \mathcal{S}_G(M)$ is called the spectrum of the operator S with respect to M. Eigenvalues and eigenvectors of the operator S with respect to M are defined analogously.

Proposition 2.1. *Let $M \subset X$ be a bounded set, $S, T\colon X \to Y$ be a couple of operators and let S be a bounded operator (i.e., it maps bounded sets from X onto bounded sets in Y). Suppose, further, that* $\operatorname{dist}(T(M), \{0\}) = d > 0$. *Then $\mathcal{S}_{S,T}(M)$ is a bounded and closed set in G.*

Proof: According to proposition 1.1 $\mathcal{S}_{S,T}(M)$ is closed. Suppose, on the contrary, that $\mathcal{S}_{S,T}(M)$ is not bounded. Then for any $K > 0$ there is a $\lambda \in \mathcal{S}_{S,T}(M)$ such that $|\lambda| > K$. Denote $||S||_M = \sup_{x\in M} ||S(x)||$ and put $K = (||S||_M + 1)/d$. According to definition 1.1, there is a sequence $\{x_n\} \in M$ such that $\lim_{n\to\infty} ||S(x_n) - \lambda T(x_n)|| = 0$. But $||S(x_n) - \lambda T(x_n)|| \geqq |\lambda| \cdot ||T(x_n)|| - ||S(x_n)|| > K.\, d - ||S||_M = 1$ and we come to a contradiction. Thus, $\mathcal{S}_{S,T}(M)$ ist a bounded set.

The following theorem shows a sufficient condition for the existence of a non-empty spectrum of a couple of bounded operators.

Theorem 2.1. *Let X be a Banach space, Y a Hilbert space and let $S, T\colon X \to Y$ be a couple of bounded operators. Suppose $M \subset X$ is a nonempty bounded set and let, further, the following condition (p_1) hold.*

$$(p_1) \qquad 0 < \sup_{x\in M} |(S(x), T(x))| = ||S||_M \cdot ||T||_M ,$$

where

$$||S||_M = \sup_{x\in M} ||S(x)||, \quad ||T||_M = \sup_{x\in M} ||T(x)|| .$$

Then $\mathcal{S}_{S,T}(M)$ is a closed set in C containing at least one non-zero point. If, in addition, $\operatorname{dist}(T(M), \{0\}) = d > 0$ *then $\mathcal{S}_{S,T}(M)$ is a bounded closed set contained in the circle $|\lambda| \leqq ||S||_M \cdot ||T||_M/d^2$.*

Proof: Let $\varepsilon > 0$ be an arbitrary positive real number. Then there is a point $x_0 \in M$, $x_0 \neq 0$ such that

$$|(S(x_0), T(x_0))| > ||S||_M \cdot ||T||_M - \frac{\varepsilon}{2} ||T||_M/||S||_M .$$

Denote $\lambda_0 = e^{i\theta}||S||_M/||T||_M$, where θ is the argument of the complex number $(S(x_0), T(x_0))$ or $\lambda_0 = \operatorname{sign} S((x_0), T(x_0)) \, ||S||_M/||T||_M$ if Y is a real Hilbert space. Then it holds:

$$||S(x_0) - \lambda_0 T(x_0)||^2 = ||S(x_0)||^2 - 2 \operatorname{Re} [\lambda_0(T(x_0), S(x_0))] + |\lambda_0|^2 ||T(x_0)||^2$$
$$\leqq ||S||_M^2 - 2|(S(x_0), T(x_0))| \cdot ||S||_M/||T||_M + ||S||_M^2 < 2||S||_M^2 + 2(\frac{\varepsilon}{2} ||T||_M/||S||_M - ||S||_M \, ||T||_M) \cdot ||S||_M/||T||_M < \varepsilon.$$

Putting $\varepsilon = 1/n$, $n = 1, 2, \ldots$, we obtain a sequence $\{x_n\} \in M$, $x_n \neq 0$ and a sequence $\lambda_n \in C$, $|\lambda_n| = ||S||_M/||T||_M$ such that $\lim\limits_{n\to\infty} ||S(x_n) - \lambda_n T(x_n)|| = 0$. At the same time we can assume that $\lim\limits_{n\to\infty} \lambda_n = \lambda_0 \neq 0$ (choosing a convergent subsequence, if necessary). By the triangular inequality we have

$$||S(x_n) - \lambda_0 T(x_n)|| \leqq ||S(x_n) - \lambda_n T(x_n)|| + |\lambda_n \quad \lambda_0| \cdot ||T(x_n)|| \,.$$

Hence, $\lim\limits_{n\to\infty} ||S(x_n) - \lambda_0 T(x_n)|| = 0$ and thus, $\lambda_0 \in \mathscr{S}_{S,T}(M)$. According to proposition 1.1 $\mathscr{S}_{S,T}(M)$ is closed. If, in addition, $\operatorname{dist}\big(T(M), \{0\}\big) = d > 0$ then, according to proposition 2.1, $\mathscr{S}_{S,T}(M)$ is bounded. Finally, for any $\lambda \in \mathscr{S}_{S,T}(M)$ there is a sequence $\{x_n\} \in M$ such that $\lim\limits_{n\to\infty} ||S(x_n) - \lambda T(x_n)|| = 0$. Denote $R(x_n) = \big(S(x_n) - \lambda T(x_n)\big)/||T(x_n)||^2$. Then $||R(x_n)|| \leqq ||S(x_n) - \lambda T(x_n)||/d^2$, so that $\lim\limits_{n\to\infty} ||R(x_n)|| = 0$. It is obvious that $|\lambda| \leqq |\big(S(x_n), T(x_n)\big)|/||T(x_n)||^2 + ||R(x_n)|| \cdot ||T(x_n)||$. Hence $|\lambda| \leqq ||S||_M \cdot ||T||_M/d^2$ and the proof is finished.

Proposition 2.2. *Let $S, T: X \to Y$ be bounded operators, Y be a Hilbert space and let $M \subset X$ be a bounded nonempty set. Suppose, further, that for any $x \in M$ it holds*

$$\{y \in Y / ||y|| = ||T(x)||\} \subset T(M) \,.$$

Then the condition (p_1) from theorem 2.1 is equivalent with the following condition (p_2)

$$(p_2) \qquad 0 < \sup_{x \in M} |\big(S(x), T(x)\big)| = \sup_{\substack{x \in M \\ y \in M}} |\big(S(x), T(y)\big)| \,.$$

Proof: Let the condition (p_2) be satisfied. Then for an arbitrary positive real number $\varepsilon > 0$ there are points $x_0, y_0 \in M$ such that $||S(x_0)|| > ||S||_M - \varepsilon$, $||T(y_0)|| > ||T||_M - \varepsilon$. According to the hypothesis of the proposition we can choose a point $z_0 \in M$ such that $T(z_0) = S(x_0) \cdot ||T(y_0)||/||S(x_0)||$. Thus $\big(S(x_0), T(z_0)\big) = ||S(x_0)|| \cdot ||T(y_0)|| > (||S||_M - \varepsilon) \cdot (||T_M|| - \varepsilon)$, so that $\sup\limits_{\substack{x \in M \\ y \in M}} |\big(S(x), T(y)\big) \geqq ||S||_M \cdot ||T||_M$. On the other hand, $\sup\limits_{\substack{x \in M \\ y \in M}} |\big(S(x), T(y)\big)| \leqq \sup\limits_{x \in M} ||S(x)|| \times \sup\limits_{y \in M} ||T(y)|| = ||S||_M \cdot ||T||_M$. Hence, $\sup\limits_{x, y \in M} |\big(S(x), T(y)\big)| = ||S||_M \cdot ||T||_M = \sup\limits_{x \in M} |\big(S(x), T(x)\big)|$ and the condition (p_1) is also satisfied. To complete the proof we must show that (p_1) implies (p_2). But this is evident because $\sup\limits_{x \in M} |\big(S(x), T(x)\big)| \leqq \sup\limits_{x, y \in M} |\big(S(x), T(y)\big)|$ and the opposite inequality follows immediately from the condition (p_1).

Theorem 2.2. *Let $M \subset X$ be a bounded closed nonempty set such that $0 \notin M$. Let $S: X \to Y$ be a completely continuous operator, $T: X \to Y$ be a continuous operator with the Lipschitzian inverse operator T^{-1}. Then any non-zero point from the spectrum $\mathscr{S}_{S,T}(M)$ is an eigenvalue of the couple (S, T) with respect to M.*

Proof: Let $\lambda \in \mathscr{S}_{S,T}(M)$, $\lambda \neq 0$. According to definition 1.1, there is a sequence $x_n \in M$ such that $\lim_{n\to\infty} ||S(x_n) - \lambda T(x_n)|| = 0$ and we can assume that $\{S(x_n)\}$ is a convergent sequence choosing a convergent subsequence if necessary. Denote $z_n = T(x_n)$, so that $x_n = T^{-1}(z_n)$, $n = 1, 2\ldots$. Let m, n be arbitrary natural numbers. Then $||x_n - x_m|| = ||T^{-1}(z_n) - T^{-1}(z_m)|| \leqq K||z_n - z_m|| \leqq 1/|\lambda| \cdot K||S(x_n) - S(x_m)|| + 1/|\lambda| \cdot K||S(x_n) - \lambda T(x_n)|| + 1/|\lambda| K||S(x_m) - \lambda T(x_m)||$, where $K > 0$ is a constant. Now it is evident that $\{x_n\}$ is a fundamental sequence in the Banach space X, so that there is a point $x_0 \in M$, $x_0 \neq 0$ such that $\lim_{n\to\infty} x_n = x_0$. Further,

$$||S(x_0) - \lambda T(x_0)|| \leqq ||S(x_0) - S(x_n)|| + ||S(x_n) - \lambda T(x_n)|| + |\lambda| \cdot ||T(x_n) - T(x_0)|| .$$

Using the continuity of the operators, S, T we obtain that $S(x_0) - \lambda T(x_0) = 0$. Hence, λ is an eigenvalue of the couple (S, T) with respect to M.

Corollary 2.1. *Let $M \subset X$ be a bounded closed nonempty set such that $0 \notin M$. Let X be a Hilbert space. Suppose $S: X \to X$ is a completely continuous operator and $T: X \to X$ is a continuous strongly monotone operator on M. Suppose, further, that the condition (p_1) from theorem 2.1 is satisfied. Then there is at least one eigenvalue of the couple (S, T) with respect to M.*

Corollary 2.2. *Let X be a Hilbert space. Suppose $S: X \to X$ is a completely continuous operator and $T: X \to X$ is a continuous strongly monotone operator. Suppose, further, that there is $r_0 > 0$ such that for any $r \in (0, r_0)$, the following condition is satisfied:*

$$0 < \sup_{||x||=r} |(S(x), T(x))| = \sup_{||x||=r} ||S(x)|| \cdot \sup_{||x||=r} ||T(x)|| .$$

Then the point $0 \in C$ is a bifurcation point of the couple (S, T).

Remark 2.1. It would be very useful to know some classes of operators satisfying the condition (p_1) or (p_2). If $T = I$ is the identity operator and S is a continuous selfadjoint linear operator on a Hilbert space then it is well-known that the couple (S, I) satisfies the condition (p_1) on every ball $K_r(0)$. But there is an interesting class of nonlinear operators satisfying (p_1) on every ball $K_r(0)$ and on every sphere $S_r(0)$, the class of symmetric continuous homogeneous polynomial operators in a Hilbert space. An example of such an operator is the operator P: $L_2(0, 1) \to L_2(0, 1)$ defined by

$$P(x) \equiv y(s) = \int_0^1 \cdots \int_0^1 K(s, t_1, t_2, \ldots, t_k)\, x(t_1)\, x(t_2) \cdots x(t_k)\, dt_1 dt_2 \cdots dt_k ,$$

where $K(s, t_1, t_2, \ldots, t_k)$ is a symmetric, in all variables quadratically integrable function on the unit cube in E_{k+1}. This operator belongs to the well-known Ljapunov-Lichtenstein potential operators ([8], [9]). But there exist operators satisfying (p_1) which are not potential as the following simple example shows. Consider an operator F on $L_2(0, 1)$ defined by

$$F(x) \equiv y(s) = \int_0^1 s \cdot x^2(t)\, dt .$$

Then $\sup_{||x||=r} ||F(x)|| \cdot r = \sup_{||x||=r} |(F(x), x)| = r^3/3$ for any $r > 0$.

As an example of an operator which does not satisfy the condition (p_1) can serve the following

$$F(x) = y(s) = \int_0^{\pi}\int_0^{\pi} \cos s \cdot \sin t_1 \cdot \cos t_2 x(t_1)\, x(t_2)\, dt_1 dt_2 \,.$$

Its spectrum with respect to any sphere $S_r(0)$ does not contain nonzero points.

3. Spectrum of a positive homogeneous operator with respect to a sphere

Let us consider $S, T: X \to Y$ positive homogeneous operators of the degree $k > 0$, $m > 0$ respectively, (i.e., $S(t \cdot x) = t^k S(x)$, $T(t \cdot x) = t^m T(x)$ for any $x \in X$ and any $t \in C$). Let $M \subset X$ be a nonempty set. It is obvious that $t^{k-m}\mathscr{S}_{S,T}(M) = \mathscr{S}_{S,T}(t \cdot M)$ for any $t \neq 0$ so that, if $k = m$ we obtain the situation analogous to the linear case: If the couple (S, T) has an eigenvalue with respect to the unit sphere then it has an eigenvalue with respect to any sphere. Hence, every eigenvalue of the couple (S, T) with respect to a sphere is a bifurcation point.

Remark 3.1. If k is not an integer we take the main value of the power t^k in case that X, Y are complex spaces and suppose $t \in R^+$ (nonnegative real numbers) in case that X, Y are real spaces.

Proposition 3.1. *Let X, Y be complex (resp. real) Banach spaces. Consider $S, T: X \to Y$ and let T be homogeneous of the order k. Then the following assertions hold.*

a) *If T has an inverse operator T^{-1} then T^{-1} is homogeneous of the order $1/k$ and a $\lambda \in C$, $\lambda \neq 0$ (resp. $\lambda \in R^+$, $\lambda > 0$) is an eigenvalue in $\mathscr{S}_{S,T}(M)$ if and only if λ is an eigenvalue in $\mathscr{S}_{T^{-1}S}(M)$.*

b) *If T has a Lipschitzian inverse operator then $\mathscr{S}^{S,T}(M) \subset \mathscr{S}_{T^{-1}S}(M)$ (resp. $\mathscr{S}_{S,T}(M) \cap R^+ \subset \mathscr{S}_{T^{-1}S}(M)$).*

Proof: The assertion a) ist obvious. To prove b) let us suppose $\lambda \in \mathscr{S}_{S,T}(M)$. Then there is a sequence $\{x_n\} \in M$ such that $\lim_{n\to\infty} ||S(x_n) - \lambda T(x_n)|| = 0$. Let us put $R(x_n) = T^{-1}(S(x_n)) = \lambda^{1/k} x_n$, $n = 1, 2, \dots$. Then $||R(x_n)|| = ||T^{-1}(S(x_n)) - T^{-1}(\lambda T(x_n))|| \leq K||S(x_n) - \lambda T(x_n)||$, so that $\lim_{n\to\infty} ||R(x_n)|| = 0$ and thus $\lambda \in \mathscr{S}_{T^{-1}S}(M)$.

Proposition 3.2. *Let X be a complex (resp. a real) Hilbert space and let S, T: $X \to X$ be bounded homogeneous operators of the order k. Suppose S is completely continuous, T is a continuous strongly monotone operator and let the following condition hold:*

$$0 < \sup_{||x||=1} |(S(x), T(x)| = \sup_{||x||=1} ||S(x)|| \cdot \sup_{||x||=1} ||T(x)|| \,.$$

Then it holds:

a) *The couple (S, T) has an eigenvalue in $\mathscr{S}_{S,T}(S_1(0))$ and every nonzero point in $\mathscr{S}_{S,T}(S_1(0))$ is a bifurcation point.*

b) *Every bifurcation point is in $\mathscr{S}_{S,T}(S_1(0))$.*

c) *$\lambda \in \mathcal{S}_{S,T}(S_1(0))$ if and only if $\lambda^{1/k} \in \mathcal{S}_F(S_1(0))$ (resp. $\lambda \in \mathcal{S}_{S,T}(S_1(0)) \cap R^+$ if and only if $\lambda^{1/k} \in \mathcal{S}_F(S_1(0)) \cap R^+$), where $F = T^{-1}S$.*

d) *If F has continuous Fréchet derivative at the point zero then $\mathcal{S}_{S,T}(S_1(0))$ is at most a countable set and the unique limit point of $\mathcal{S}_{S,T}(S_1(0))$ can be only the point zero.*

Proof: According to theorem 2.1, there is a nonzero point in $\mathcal{S}_{S,T}(S_1(0))$. Being strongly monotone T has a Lipschitzian inverse operator and thus, according to theorem 2.2, every nonzero point in $\mathcal{S}_{S,T}(S_1(0))$ is an eigenvalue of the couple (S, T) with respect to $S_1(0)$. The homogeneity implies that any eigenvalue is a bifurcation point. The assertion b): For a bifurcation point λ_0 there is a sequence $\lambda_n \to \lambda_0$ and a sequence $\{x_n\}$, $x_n \neq 0$ such that $S(x_n) - \lambda_n T(x_n) = 0$. Put $y_n = x_n/||x_n||$. Then $||S(y_n) - \lambda_0 T(y_n)|| \leqq ||S(y_n) - \lambda_n T(y_n)|| + |\lambda_n - \lambda_0| \cdot ||T(y_n)||$, so that $\lambda_0 \in \mathcal{S}_{S,T}(S_1(0))$. The assertion c) follows from proposition 3.1 and theorem 2.2. To prove d) we recall that the derivative $F'(0)$ is a completely continuous linear operator. Moreover, from the homogeneity follows that $F'(0)(h) = F(h)$ for any $h \in X$. Thus, the spectrum of the operator F is at most a countable set with at most one limit point zero and the same holds for $\mathcal{S}_{S,T}(S_1(0))$.

For a bounded homogeneous operator F on a Hilbert space we define the norm $||F||$ and the absolute norm as follows.

$$||F|| = \sup_{||x||=1} ||F(x)|| , \qquad |||F||| = \sup_{||x||=1} |(F(x), x)| .$$

Theorem 3.2. *Let X be a Hilbert space and let $F: X \to X$ be a bounded positive homogeneous operator of the order $n > 0$. Let $||F|| = |||F|||$. Then the following assertions hold.*

a) *For any $r > 0$ there is a $\lambda_r \in \mathcal{S}_F(S_r(0))$ such that $|\lambda_r| = r^{n-1}||F||$.*

b) *If F is completely continuous then every non-zero point of $\mathcal{S}_F(S_r(0))$ is an eigenvalue with respect to $S_r(0)$ and the point zero is a bifurcation point of the operator F.*

c) *If $n > 1$ then the point zero is the unique bifurcation point of F.*

d) *If F has Gateaux's differential $VF(x, h)$ which is symmetric on $S_r(0)$ (i.e., $(VF(x, h), k) = (h, VF(x, k))$ for any $x \in S_r(0)$, $h, k \in X$) then $\mathcal{S}_F(S_r(0))$ is a compact set contained in the interval $[r^{n-1}m, r^{n-1}M]$, where $m = \inf_{||x||=1} (F(x), x)$, $M = \sup_{||x||=1} (F(x), x)$ and both $r^{n-1}m$, $r^{n-1}M$ belong to the spectrum $\mathcal{S}_F(S_r(0))$.*

Proof: The assertions a) and b) follow from theorems 2.1 and 2.2. If $n > 1$ then $|(F(x), x)|/||x||^2 \leqq ||F|| \cdot ||x||^{n+1}/||x||^2 = ||F|| \cdot ||x||^{n-1}$ so that $\lim_{||x|| \to 0} (F(x), x)/||x||^2 = 0$ and thus, zero is the unique bifurcation point of the operator F. To prove d) we recall that $VF(x, x) = n \cdot F(x)$. Hence, $n \cdot (F(x), x) = (VF(x, x), x) = (x, VF(x, x)) = \overline{(VF(x, x), x)}$ and thus, $(F(x), x)$ is a real number for any $x \in S_r(0)$. Consider $\lambda = a + i \cdot b \in C$ such that $b \neq 0$. Then for $x \in S_r(0)$, $y = F(x) - \lambda x$ we obtain $(y, x) = (F(x), x) - \lambda(x, x)$, $(x, y) = \overline{(y, x)} = (F(x), x) - \bar{\lambda}(x, x)$. Therefore $(x, y) - (y, x) = (\lambda - \bar{\lambda})(x, x) = 2ib\,(x, x) = 2ibr^2$. From this follows that $2|b|\, r^2 = |(x, y) - (y, x)| \leqq 2||y||\, r$. Hence $||y|| = ||F(x) - \lambda x|| \geqq |b| \cdot r > 0$. Thus, $\lambda \notin \mathcal{S}_F(S_r(0))$ and the spectrum with respect to $S_r(0)$ must be real. Assume, further, that $\lambda = M \cdot r^{n-1} + d$, where $d > 0$. Then $(F(x) - \lambda x, x) \leqq M \cdot r^{n+1} - (M \cdot r^{n-1} + d)\, r^2 = -dr^2 < 0$. Thus $|(F(x) - \lambda x, x)| \geqq d \cdot r^2$, so that $||F(x) - \lambda x|| \cdot r \geqq |(F(x) - \lambda x, x)| \geqq d \cdot r^2$ and thus, $||F(x) - \lambda x|| \geqq d \cdot r > 0$. Therefore $\lambda \notin \mathscr{S}_F(S_r(0))$. The case $\lambda < m \cdot r^{n-1}$ can be examined analogously. The rest of the assertion d) follows from the proof of theorem 2.1.

The last proposition shows a kind of a perturbation theorem.

Proposition 3.3. *Let $S, T: X \to Y$ be homogeneous operators of the order $k > 0$ and let T be a bounded operator. Consider the following operator $G(x, \lambda) = S(x) - \lambda T(x) - R(x, \lambda)$, where $R(x, \lambda)$ is an operator satisfying*

(i) $R(0, \lambda) = 0$ *for any* $\lambda \in C$

(ii) $\lim\limits_{||x|| \to 0} R(x, \lambda)/||x||^k = 0$ *locally uniformly with respect to* $\lambda \in C$.

Then $\mathscr{B}_G \subseteq \mathscr{S}_{S,T}(S_1(0))$.

Proof: Let $\lambda \in \mathscr{B}_G$. Then there are sequences $\{x_n\} \in X$, $x_n \neq 0$ and $\{\lambda_n\} \in C$ such that $\lim\limits_{n\to\infty} ||x_n|| = 0$ and $G(x_n, \lambda_n) = 0$. Therefore, using the homogeneity of S and T, we obtain

$S(x_n/||x_n||) - \lambda_n T(x_n/||x_n||) = R(x_n, \lambda_n)/||x_n||^k$. Putting $y_n = x_n/||x_n||$

we have $\lim\limits_{n\to\infty} ||S(y_n) - \lambda_n T(y_n)|| = 0$ and thus $\lim\limits_{n\to\infty} ||S(y_n) - \lambda T(y_n)|| = 0$, because $||S(y_n) - \lambda T(y_n)|| \leqq ||S(y_n) - \lambda_n T(y_n)|| + |\lambda_n - \lambda| \cdot ||T(y_n)||$. Hence, $\lambda \in \mathscr{S}_{S,T}(S_1(0))$.

As examples of the operator G from proposition 3.3 can serve some holomorphic operators, especially Ljapunov-Lichtenstein operators, polynomial operators, etc. [13, 8, 19].

4. Analytic operators on Banach spaces

In this part of our note we want to draw attention to the ideas of analyticity on general Banach spaces. It is evident that the theory of analytic operators on both real and complex Banach spaces can be useful, for example, for the treatment of nonlinear operator equations. Let us recall problems concerning the existence and the number of solutions, branching of solutions, continuation, etc. [14, 15, 22, 23, 24]. We shall start from the definition and some basic properties of holomorphic operators [1, 2, 4].

Definition 4.1. Let $D \subset X$ be an open set in a Banach space X. We shall say that an operator $F: D \to Y$ is a holomorphic operator on D if the following assumptions hold.

(i) There exists Fréchet's differential $d^nF(x, h^n)$ of the order n at each point $x \in D$ for any natural number n.

(ii) For any (but fixed) $x \in D$ there exists $r > 0$ such that

$$F(x + h) = \sum_{n=0}^{\infty} 1/n!\, d^nF(x, h^n) \quad \text{for any} \quad h \in K_r(0)\,,$$

where $d^0F(x, h^0) = F(x)$ and the convergence is considered in the norm topology of the Banach space Y.

Remark 4.1. In the case that X, Y are complex Banach spaces it is sufficient to assume in the definition 4.1 that F has only the Baire's property on D and possesses the Gateaux's differential of the first order on D in order to the assumptions (i) and (ii) are satisfied. In this complex case, the well-known Cauchy's formula

$$d^nF(x_0, h^n) = n!/2\pi i \oint_{|z|=\varrho} F(x_0 + zh) \cdot z^{-n-1}dz ,$$

holds for a given $x_0 \in D$ and any $h \in X$, $||h|| \leqq 1$, where $\varrho > 0$ is a given (sufficiently small) constant and z is a complex variable. If, in addition, there is a constant $M > 0$ such that $||F(x)|| \leqq M$ for any $x \in \overline{K_\varrho(x_0)}$ then

$$||d^nF(x_0, h^n)|| \leqq (M \cdot n!/\varrho^n)\, ||h||^n \quad \text{for any} \quad h \in X,\ n = 0, 1, 2, \ldots$$

Proposition 4.1. *An operator $F: D \subset X \to Y$ is holomorphic on an open set D of a complex Banach space X if and only if the function $f(z) = \langle F(x + zh), g\rangle$ of the complex variable z is a holomorphic function in some neighbourhood of the origin for any (but fixed) $x \in D$, $h \in X$, $g \in Y'$, where Y' is the dual space to Y.*

Theorem 4.1. (*Theorem on the uniqueness*, [4]). *Let $F: D \subset X \to Y$ be a holomorphic operator on an open connected set D and suppose that there is an open ball $K_r(a) \subset D$ such that $F(x) = 0$ for any $x \in K_r(a)$. Then $F(x) = 0$ for any $x \in D$.*

We recall the notion of a polynomial operator in order to give the definition of abstract power series in Banach spaces [2], [13].

Definition 4.2. We shall say that an operator $P: X \to Y$ is the homogeneous polynomial operator (briefly h.p.-operator) of the order k if there is a continuous k-linear operator $P^*(x_1, \ldots, x_k)$ from the product space $X^k = X \times X \times \cdots \times X$ into Y which is symmetric in all variables (i.e., its value does not change under arbitrary permutation of all variables) and such that $P^*(x, \ldots, x) = P(x)$ for any $x \in X$. The operator P^* is called the polar operator to P. Further we define the norms $||P||$, $||P^*||$ as follows.

$$||P|| = \sup_{||x||=1} ||P(x)|| , \qquad ||P^*|| = \sup_{||x_i||=1,\ i=1,2,\ldots,k} ||P^*(x_1, \ldots , x_k)|| .$$

Proposition 4.2. *Let $P: X \to Y$ be h.p.-operator of the order k. Then*

$$||P|| \leqq ||P^*|| \leqq (k^k/k!)\, ||P|| \quad (\textit{see } [2]) .$$

If, in addition, $X = Y$ is a Hilbert space and P is a symmetric h.p.-operator (i.e., the $(k + 1)$-linear functional $p(x_1, \ldots, x_{k+1}) = (P(x_1, \ldots, x_k), x_{k+1})$ is symmetric in all $k + 1$ variables) then

$$||P|| = ||P^*|| \ (\textit{see } [13]) .$$

Definition 4.3. Let $P_k: X \to Y$ be h.p.-operators of the order $k = 1, 2, \ldots$; $P_0 \in Y$ be a constant and let $x_0 \in X$ be a given point. Then the series

$$P(x, x_0) = \sum_{k=0}^{\infty} P_k(x - x_0) ,$$

converging in some neighbourhood of the point x_0 is called the F-power series (briefly F-p.s.) with the centre x_0. Let us denote $C(P)$ the set of all points in X for which the F-p.s. converges in the norm topology and put

$$\varrho = 1/\limsup_{k\to\infty} ||P_k||^{1/k}\,, \qquad \varrho^* = 1/\limsup_{k\to\infty} ||P_k^*||^{1/k}\,.$$

The number $\varrho(\varrho^*)$ is called the radius (the polar radius) of the uniform convergence. The F-p.s. is called symmetric if all P_k are symmetric h.p.-operators.

Proposition 4.3. ([4]). *The F-p.s. $P(x, x_0)$ with $\varrho > 0$ converges uniformly on every closed ball $K_r(x_0) \subset K_\varrho(x_0)$ and the convergence is not uniform on any sphere $S_r(x_0)$ with $r > \varrho$. Moreover, the following assertions are equivalent.*

(i) $\varrho > 0$;
(ii) $\varrho^* > 0$;
(iii) $P(x, x_0)$

is a holomorphic operator on the interior of the set $C(P)$.

Remark 4.2. It is easy to prove from the proposition 4.2 and the well-known Stirling's formulae that $\varrho^* \leqq \varrho \leqq \varrho^* \cdot e$ ($e = 2{,}718\ldots$).

Theorem 4.2. (*The maximum modulus principle*, [4]) *Let $F: X \to Y$ be holomorphic operator on an open connected set $D \subset X$. Further, let $||F(x)||$ be nonconstant on D. Then $||F(x)||$ does not attain its maximum in D.*

Theorem 4.3. *Let $\{F_k\}$ be a sequence of holomorphic and locally uniformly bounded operators on a domain D. If there is a ball $K_r(a) \subset D$ such that $\lim_{k\to\infty} F_k(x) = F(x)$ for any $x \in K_r(a)$ then $\lim_{k\to\infty} F_k(x)$ exists for any $x \in D$ and F is a holomorphic operator on D. Moreover, $\lim_{k\to\infty} d^nF_k(x, h^n) = d^nF(x, h^n)$ for any n and any $x \in D$.*

Starting from the notion of F-p.s. we can follow the classical method of analytical continuation and define analytic (multivalued) operators on Banach spaces. Number of properties analogous to the properties of classical analytic function can be proved, among others, the monodromy theorem, some properties of the line integral and primitive function, etc. We shall not deal with this theory here but we pay attention to operator equations with holomorphic operators.

The following theorem can be useful in the theory of nonlinear operator equations with holomorphic operators.

Theorem 4.4. ([6]). *Let X be a complex Banach space, $D \subset X$ be a bounded connected open set in X. If $F: D \to D$ is a holomorphic operator on D and $F(D)$ has a positive distance from the complement of D then F has a unique fixed point in D.*

Definition 4.4. We shall say that a h.p.-operator $P: X \to X'$ of the order $k \geqq 1$ is positive (resp. positive defined) on a set $M \subset X$ if $\langle P^*(x^{k-1}, h), h\rangle > 0$ for any $x \in M$, $x \neq 0$ and any $h \in X$, $h \neq 0$ (resp. there is a constant $c > 0$ such that $\langle P^*(x^{k-1}, h), h\rangle \geqq c||x||^{k-1}\,||h||^2$ for any $x \in M$, $h \in X$). We shall say that a F-p.s. is positive (positive defined) on M if all P_n are positive (all P_n are positive and there is a $k \geqq 1$ such that P_k is positive defined) on M and the series converges on M.

Lemma 4.1. *Let $P: X \to X'$ be a positive symmetric h.p.-operator (i.e., the functional $\langle P^*(x_1, \dots, x_k), x_{k+1}\rangle$ is symmetric in all $k+1$ variables) of the order $k \geqq 1$ on an open convex set $M \subset X$. Then P is a monotone operator, the functional $f(x) = \langle P(x), x\rangle$ is a convex functional on M (thus, f is weakly lower semi-continuous on M) and* grad $f(x) = P(x)$ *for any $x \in M$.*

Proof: It is evident that P has Fréchet's differential of arbitrary order. Hence, f is also F-differentiable and $df(x, h) = \langle dP(x, h), x\rangle + \langle P(x), h\rangle$. But $dP(x, h) = k \cdot P^*(x^{k-1}, h)$, where P^* is the polar operator to P. So that, using symmetricity of P, we obtain $\langle dP(x, h), x\rangle = \langle k \cdot P^*(x^{k-1}, h), x\rangle = k\langle P(x), h\rangle$. Hence, $df(x, h) = (k+1)\, \langle P^*(x^{k-1}, h), x\rangle = (k+1)\, \langle P(x), h\rangle$. By the same way we obtain $d^2f(x, h^2) = k \cdot (k+1)\, \langle P^*(x^{k-1}, h), h\rangle > 0$ and thus, P is monotone operator and f is convex continuous and therefore weakly lower semi-continuous functional on M.

As an example of a positive (resp. positive defined) h.p.-operator can serve the following operator P defined on $L_2(G)$ by

$$P(x) \equiv y(s) = x(s) \int_G K(s, t)\, x^2(t)\, dt\,,$$

where $G \subset E_n$ is a bounded domain and $K(s, t)$ is a continuous and positive (resp. $K(s, t) \geqq k_0 > 0$) kernel on G G. ([19]). If, in addition, $K(s, t)$ is a symmetric kernel then P is symmetric h.p.-operator of the order $k = 3$. A generalization on higher order h.p.-operators is evident.

Lemma 4.2. *Let $F: X \to X'$ be a strongly continuous operator. Then the functional $f(x) = \langle F(x), x\rangle$ is weakly continuous. If, in addition, F is a homogeneous potential operator of the order k then the functional $(1/(k+1))\, f(x)$ is the potential of F.*

Proof: Let $\{x_n\} \in X$ be a weakly convergent sequence to a point $x_0 \in X$. Then

$$|f(x_n) - f(x_0)| = |\langle F(x_n), x_n\rangle - \langle F(x_0), x_0\rangle| \leqq |\langle F(x_n) - F(x_0), x_n\rangle|$$
$$+ |\langle F(x_0), x_n - x_0\rangle| \leqq ||F(x_n) - F(x_0)||\, ||x_n|| + |\langle F(x_0), x_n - x_0\rangle|\,.$$

Then $\lim\limits_{n\to\infty} ||F(x_n) - F(x_0)|| = 0$ and $||x_n||$ is a bounded sequence. Thus, $\lim\limits_{n\to\infty} |f(x_n) - f(x_0)| = 0$ and f is weakly continuous at the point x_0.

Proposition 4.4. *Let $F = \sum\limits_{n=0}^{\infty} P_n: X \to X'$ be a symmetric positive F-p.s. with radius of uniform convergence $\varrho > 0$. Then F is a potential operator and its potential*

$$f(x) = \sum_{n=0}^{\infty} \frac{1}{n+1} \langle P_n(x), x\rangle$$

is a holomorphic weakly lower semi-continuous functional on $K_\varrho(0)$. If, in addition, P_n are strongly continuous then f is a wekly continuous holomorphic functional on $K_\varrho(0)$.

The proof follows easily from lemma 4.1 (resp. lemma 4.2), proposition 4.3 and theorem 4.3.

Theorem 4.5. *Let X be a reflexive Banach space and let $F = \sum P_n\colon X \to X'$ be a positive defined F-p.s. with radius of uniform convergence $\varrho > 0$. Suppose $0 < r < \varrho$. Then there is a positive real number $c_0 > 0$ and a natural number $k \geqq 1$ such that the equation*

$$F(x) = y$$

has a unique solution for any $y \in X'$ satisfying $\|y\| < c_0 \cdot r^k$.

Proof: According to proposition 4.3, the F-p.s. converges uniformly on $K_r(0)$ and, according to proposition 4.4, F is a potential operator with its potential $f(x) = \sum_{n=0}^{\infty} \frac{1}{n+1} \langle P_n(x), x\rangle$. The definition 4.4 implies that there is a constant $c > 0$ and a natural number k such that $\langle P_k(x), x\rangle \geqq c\|x\|^{k+1}$ for any $x \in \overline{K_r(0)}$. Denote $c_0 = c/(k+1)$ and let $\|y\| < c_0 \cdot r^k$. Then

$$g(x) = f(x) - \langle x, y\rangle \geqq c_0\|x\|^{k+1} - \|x\|\,\|y\| .$$

Hence, for any $x \in S_r(0)$ we have

$$g(x) \geqq c_0 \cdot r^{k+1} - r \cdot \|y\| = r(c_0 r^k - \|y\|) > 0 .$$

Thus $g(x) > g(0)$ for any $x \in S_r(0)$, so that the functional g has the "m-property" (see [9]). Further, g is weakly lower semi-continuous according to proposition 4.4 and thus, g attains its infimum at a point $x_0 \in K_r(0)$. Then it is obvious that grad $g(x_0) = F(x_0) - y = 0$. Suppose that there is another point $x_1 \neq x_0$ such that $F(x_1) = y$. Then, using the mean value theorem, we can find $t \in (0, 1)$ such that $\langle F(x_1) - F(x_0), x_1 - x_0\rangle = \langle dF(x_0 + t(x_1 - x_0), x_1 - x_0), x_1 - x_0\rangle \geqq k \cdot c \cdot \|x_0 + t(x_1 - x_0)\|^{k-1} \cdot \|x_1 - x_0\|$. If $x_0 + t(x_1 - x_0) \neq 0$ then we come to a contradiction. If $x_0 + t(x_1 - x_0) = 0$ then $dF(0, x_1 - x_0) = P_1(x_1 - x_0)$ and we obtain a contradiction again.

Similarly it is possible to prove the following

Proposition 4.5. *Let $F = \sum_{n=0}^{\infty} P_n$, $G = \sum_{n=0}^{\infty} Q_n\colon X \to X'$ be symmetric F-p.s. with radius of uniform convergence $\varrho > 0$. Let F be positive defined on $K_\varrho(0)$ and let $0 < r < \varrho$. Then there is a $c_0 > 0$ and a natural number k such that the equation*

$$F(x) - \lambda G(x) = 0$$

has a unique solution $x(\lambda) \in K_r(0)$ for any $\lambda \in C$ satisfying $|\lambda| < c_0 \cdot r^k / \sum_{n=0}^{\infty} \frac{1}{n+1} \times \|Q_n\|\, r^n$.

An answer to eigenvalue problem for holomorphic operators gives

Proposition 4.6. *Let $F, G\colon X \to X$ be holomorphic operators on a bounded domain D of a Hilbert space X. Let $F(0) = G(0) = 0$ and $0 \notin \partial D$. Suppose, further, that the couple (F, G) satisfies the condition (p_1) from theorem 2.1 on $\overline{D}$. Then there is a nonzero point in $\mathscr{S}_{F,G}(\partial D)$, where ∂D is the boundary of D.*

The proof follows immediately from theorem 2.1 and theorem 4.2.

Let us remark that it is possible to define the resolvent of a holomorphic operator F (as a solution of the equation $x - \lambda F(x) = y$) and to give conditions under which the resolvent exists and is a holomorphic operator.

Unsolved problems. Number of authors deal with branching theory, bifurcations, small solutions [22, 25, 26, 27, 28] and many results were obtained in the case that the equation $G(x, \lambda) = 0$ can be linearized i.e. (*)

$$G(x, \lambda) = \lambda x - Ax - R(x, \lambda)\,, \tag{*}$$

where A is a linear (mostly bounded, selfadjoint or, at least, closed) operator with finitedimensional null-space of $\lambda_0 I - A$ and $R(x, \lambda)$ is a holomorphic operator. But the situation is still rather not clearly mapped out. Let us consider a simple example of such an equation

$$x(s) = \lambda \int_0^1 3\, s \cdot t \cdot x(t)\, dt + 2 \sin \left[\int_0^1 t \cdot x(t)\, dt\right] - 4\, s^2\,.$$

It is evident that there is a unique solution for λ near to zero of the form $x(s) = 3\lambda\xi s + 2 \sin \xi - 4s^2$, where ξ satisfies the "branching equation" $\xi = \lambda\xi + \sin \xi - 1$. The number of solutions grows as $\lambda \to 1$ and for $\lambda = 1$ we obtain the infinite number of solutions.

Problem 1. The study of the equation (*) in case that the nullspace of $\lambda_0 I - A$ is infinite.

Problem 2. The study of the equation $S(x) - \lambda T(x) = R(x, \lambda)$, where S, T are homogeneous operators of the order $k > 1$ and $R(x, \lambda)$ is a term of higher order.

The second problem seems to be much more complicated then the first one.

References

[1] Alexiewicz, A., Orlicz, W., Analytic operators in real Banach spaces, Studia Math. **14** (1953) 57–78.

[2] Mazur, S. W., und Orlicz, W., Grundlegende Eigenschaften der polynomischen Operationen, Studia Math. **5** (1935) 50–68, 179–189.

[3] Kurzweil, J., On approximation in real Banach spaces by analytic operations, Studia Math. **16** (1957) 124–127.

[4] Hille, E., and Phillips, R. S., Functional analysis and semi-groups, Providence (1957).

[5] Bochnak, J., Analytic functions in Banach spaces, Studia Math. **35** (1970) 273–292.

[6] Earle, Clifford, J., Hamilton, R. S., A fixed point theorem for holomorphic mappings, Global Analysis (Proc. Symp. Pure Math. Vol. XVI Berkeley, Calif. 1968, p.p. 61–65, Providence, R. I. 1970.

[7] Nemyzki, W. W., Структура спектра нелинейных вполне непрерывных операторов, Матем. Сборник Т **33** [75] (1953) 555—558.

[8] Krasnoselski, M. M., Топологические методы в теории нелинейных интегральных уравнений, Москва (1956).

[9] Wainberg, M. M., Вариационные методы исследования нелинейных операторов, Москва 1956.

[10] Wainberg, M. M., Trenogin, W. A. Теория ветвления решений нелинейных уравнений, Москва 1969, (Deutsche Übersetzung: Theorie der Lösungsverzweigungen bei nichtlinearen Gleichungen, Akademie-Verlag Berlin, 1973).
[11] Katschurowski, R. I., О монотонных операторах и выпуклых функционалах, Успехи Мат. Наук **15** (1960) 213—215.
[12] Katschurowski, R. I., Регулярные точки, спектр и собственные функции нелинейных операторов. ДАН СССР **188** (1969) 274—277.
[13] Burýšek, S., Some remarks on polynomial operators, Comment. Math. Univ. Carolinae **10** (1969) 285—306.
[14] Langenbach, A., Über Lösungsverzweigungen bei Potentialoperatoren, Math. Nachr. **42**, 1—3, (1969) 61—67.
[15] Browder, F. E., Nonlinear mapping of analytic type in Banach spaces, Math. Annalen **185** (1970) 259—278.
[16] Browder, F. E., Nonlinear eigenvalue problems and Galerkin approximations, Bull. Amer. Math. Soc. **74** (4) (1968) 651—656.
[17] Nečas, J., Remark on the Fredholm alternative for non-linear operators with application to non-linear integral equation of generalized Hammerstein type, Comment. Math. Univ. Carolinae **13**, 2 (1972) 297—310.
[18] Fučík, S., and Nečas, J., Ljusternik-Schnirelmann theorem and non-linear eigenvalue problems, Math. Nachr. **53**, 1—6, (1972) 277—289.
[19] Bazley, N., Zwahlen, B., A branch positive solutions of nonlinear eigenvalue-problems, Manuscripta Math. **2** (1970) 365—374.
[20] Berger, M. S., Bifurcation theory and the type numbers of Marston Morse, Proc. Nat. Acad. Sci. U. S. A. **69**, 7, (1972) 1737—1738.
[21] Berger, M. S., Perspectives in nonlinearity, W. A. Benjamin 1968.
[22] Sather, D., Branching of solutions of an equation in Hilbert space, Arch. Rat. Mech. Anal. V **36**, 1 (1970) 47—64.
[23] Shimizu, T., Analytic operators and analytic operational equations, Math. Japan **1** (1948) 36—40.
[24] Achmedow, K. T., Об аналитическом продолжении решений нелинейных операторных уравнений в пространствах Банаха. Учен. Зап. Азерб. Инст. **4** (1958) 3—15.
[25] Dancer, E. N., Bifurcation theory in real Banach spaces, Proc. London Math. Soc. XXIII, 4 (1971) 699—734.
[26] Burýšek, S., On spectra of nonlinear operators, Comment. Math. Univ. Carolinae **11**, 4 (1970) 727—743.
[27] Keller, J. R., Bifurcation theory and nonlinear eigenvalue problems, New York (1968).
[28] Westreich, D., Banach space bifurcation theory, Trans. Amer. Math. Soc.

Anschrift des Autors:

Technical university of Prague
Department of mathematics

Praha 2, Karlovo náměsti 13.

On densifying and related mappings and their application in nonlinear functional analysis

JOSEF DANEŠ, Prag

To my daughter Markéta

Introduction. The role of completely continuous mappings as well as contractions in nonlinear functional analysis, especially for fixed point and existence theorems, is well-known. Schauder and Banach fixed point theorems and Leray-Schauder degree theory play an important role in applications.

First extension of both Banach and Schauder fixed point theorems to mappings being the sum of a completely continuous mapping and a contraction was given by G. Darbo [21] in 1955 using the notion of the Kuratowski measure of noncompactness [50, 51] α (for a related but weaker result see M. A. Krasnosjelskij [46].) For X a metric space and $M \subseteq X$, $\alpha(M)$ is the infimum of all $\varepsilon > 0$ such that M can be divided into a finite number of sets having diameters $< \varepsilon$; clearly, $\alpha(M) = 0$ iff M is precompact — hence the name. Darbo proved that every (k,α,α)-bounded continuous selfmapping F of a bounded closed convex (nonempty) subset C of a Banach space has a fixed point provided $k < 1$. The (k,α,α)-boundedness of F means that $\alpha(F(M)) \leqq k\alpha(M)$ for all $M \subseteq C$. Let us remember that every mapping of the form $T+K$, where T is a contraction (with constant $k < 1$) and K is a completely continuous mapping, is (k,α,α)-bounded and Darbo's theorem applies.

Second paper in this direction was published by B. N. Sadowski [86] in 1967. He used the Hausdorff measure of noncompactness introduced by L. S. Goldenštein, I. C. Gochberg and A. A. Markus [26, 27]. For X a metric space and $M \subseteq X$, the Hausdorff measure of non-compactness of M, $\chi(M)$, is the infimum of all $\varepsilon > 0$ such that M has a finite ε-net in X; as above, $\chi(M) = 0$ iff M is precompact. Sadowski proved that every χ-densifying (that is, diminishing the Hausdorff measure of noncompactness of non-precompact sets) continuous mapping of a bounded closed convex subset of a Banach space into itself has a fixed point.

In [88] — 1968, B. N. Sadowski introduced a quite general definition of a measure of noncompactness in a locally convex space and the notion of densifying mappings in such spaces. He extended the Leray-Schauder theory (in the language of the rotation) to densifying (and asymptotically compact) mappings; this extension heavily relies on a paper by Yu. G. Borisovič ["On the relative rotation of compact vector fields in linear spaces", Trudy seminara po funkc. anal., Vyp. 12, Voronež 1969, pp. 3—27].

Another extension of Leray-Schauder theory was made by R. D. Nussbaum [60] in 1969 using the work of F. E. Browder [6] and F. E. Browder and R. D.

Nussbaum [8]. He studied (k,α,α)-bounded mappings (that is, k-set-contractions in his definition) mainly.

An extension of the notion of topological degree for multivalued densifying mappings was given by V. V. Obuchowski [72] in 1971.

At present time, there is a lot of papers devoted to the theory of measures of noncompactness, densifying and related mappings and their applications to integral, ordinary differential, hyperbolic, functional-differential equations etc. An excellent introduction to the theory of densifying mappings and its application is the paper by B. N. Sadowski [93]. It contains almost all related literature up to 1971 (126 items).

In this paper, we accepted Sadowski's approach rather than Nussbaum's one because it seems to be simpler and, at the same time, it is developed in the setting of locally convex spaces.

The present paper contains results obtained by the author mainly during 1971 and the first half of 1972. These results were announced in [17], on Equadiff III (Czechoslovak Conference on Differential Equations and Their Applications, Brno, Aug. 28—Sept. 1, 1972) and on the Conference on Monotone Operators held Oct. 2—Oct. 6, 1972, in Neuendorf on Hiddensee, German Democratic Republic.

In Section 1 we remember the notions of measure of noncompactness, asymptotically compact, densifying and (k,μ,ν)-bounded mappings [88, 93] and introduce notations. We study the relation between α and χ in normed linear spaces and the behavior of (k,μ,ν)-bounded and related mappings under (ordinary and asymptotical) Fréchet differentiation. Some results can be considered as examples of densifying and (k,μ,ν)-bounded mappings.

In Section 2 we recall Sadowski's extension of Leray-Schauder theory to asymptotically compact and densifying mappings in locally convex spaces. We derive a number of fixed point and surjectivity theorems generalizing many well-known classical results. We are also interested in Fredholm alternative for nonlinear mappings.

Section 3 contains existence theorems for systems of equations, the definition of the index of an isolated fixed point of an asymptotically compact mapping, its application to computing the rotation of a (k,μ,μ)-bounded mapping $[k < 1]$ by means of ordinary and asymptotical Fréchet derivatives. Continuous dependence of fixed points and eigenvectors on parameters and the invariance of domain is studied. (Compare with M. A. Krasnosjelski [47] and R. D. Nussbaum [60] mainly.)

References contain works either used in our paper, or related to densifying mappings and not contained in Sadowski's review [93]. Let us note that some papers appeared recently (or became known to the author recently) and related to our paper are contained in References but have not carresponding reflection in the text, because the paper was finished already.

Acknowledgements. The author would like to thank to Professor R. I. Kačurowski, Professor M. A. Krasnosjelski and, especially, to Professor B. N. Sadowski for very useful discussions and the possibility of reading some of their preprints during his stay in Moscow and Voronež, Oct. 14, 1971—Jan. 12, 1972.

§ 1. Measures of noncompactness, asymptotically compact and densifying mappings

First of all we introduce some notations and definitions.

Let (X, p) be a pseudometric space and C a subset of it. Then:

Bo (C) denotes the set of all bounded subsets of C;
Cl (C) denotes the set of all closed subsets of C (in the relative topology);
Cpt (C) the set of all compact subsets of C;
RelCpt (C) the set of all relatively compact subsets of C;
PreCpt (C) the set of all precompact subsets of C;
BoCl (C) $=$ Bo $(C) \cap$ Cl (C) etc.
$B_X(x, r) = B_p(x, r) = \{y \in X: p(y, x) \leqq r\}$ $(r \in R^+ = [0, +\infty), x \in X)$.

If X is a linear space, then Co (C) denotes the set of all convex subsets of a set C in X. For X a set, $2^X = \exp(X)$ denotes the set of all subsets of X.

Let Λ be a set, X a LCS (i.e. a locally convex space), C a subset of X and $F: \Lambda \times C \to X$ a multivalued mapping. Let us define a transfinite sequence of sets as follows:

$$C_0 = \operatorname{cocl} F(\Lambda \times C),$$
$$C_\alpha = \operatorname{cocl} F(\Lambda \times (C \cap C_{\alpha-1})) \quad \text{if} \quad \alpha - 1 \text{ exists},$$
$$C_\alpha = \bigcap_{\beta<\alpha C_\beta} \quad \text{if} \quad \alpha - 1 \text{ does not exist};$$

here cocl (M) denotes the convex closed hull of a subset M of X. It is easy to see that the sequence $\{C_\alpha\}$ stabilizes, i.e. $C_\alpha =$ const. for sufficiently large α. This limit is denoted by $F^\infty(\Lambda \times C)$ and is called the asymptotic range of F on $\Lambda \times C$. F is said to be *asymptotically compact* and we shall write $F \in$ AsCpt $(\Lambda \times C)$ if the set $F(\Lambda \times (C \cap F^\infty(\Lambda \times C)))$ is relatively compact. Then $F^\infty(\Lambda \times C) = \operatorname{cocl} F(\Lambda \times (C \cap F^\infty(\Lambda \times C)))$, F^∞ is monotone in $\Lambda \times C$ etc.

Let $\mathfrak{M}$ be a subset of 2^X which is invariant under the operation cocl, i.e. $M \in \mathfrak{M}$ implies cocl $M \in \mathfrak{M}$. A mapping $\mu: \mathfrak{M} \to A$, where A is a partially ordered set, is said to be a measure of noncompactness (shortly, a mnc) in X iff $\mu(\operatorname{cocl} M) = \mu(M)$ for each $M \in \mathfrak{M}$. In this case we shall write $\mathfrak{M}_\mu = \mathfrak{M}$ and $A_\mu = A$.

μ is said to be monotone iff $M \subseteq N$, $M, N \in \mathfrak{M}$ imply $\mu(M) \leqq \mu(N)$. Let $F: \Lambda \times C \to X$ be a multivalued mapping. It is said to be *μ-densifying on C* iff the following three conditions are satisfied:

1) $\mathfrak{M}_\mu \supseteq 2^C$;
2) $F(\Lambda \times M) \in \mathfrak{M}_\mu$ for each $M \in 2^C$;
3) $M \in 2^C$ and $\mu(F(\Lambda \times M)) \geqq \mu(M)$ imply $M \in$ RelCpt (X).
   In this case we shall write $F \in$ Dens $(\Lambda \times C, \mu)$ or $F \in$ Dens (μ). If the condition 3) is replaced by

3') $M \in 2^C$, $M \in$ RelCpt (X) imply $\mu(F(\Lambda \times M)) < \mu(M)$, then, clearly, $F \in$ Dens $(\Lambda \times C, \mu)$.

In what follows we shall consider only single-valued mappings. If $F: U \to V$ and $G: U \to V$ are (single-valued) mappings then $F \cup G$ is the multivalued mapping defined as follows: $(F \cup G)(x) = \{Fx, Gx\}$, $x \in U$.

The following two theorems give relations between the classes Dens and AsCpt, and are due to B. N. Sadowski [93].

Theorem 1 *If* $C \in \mathrm{Cl}(X)$, *where* X *is a* LCS, Λ *is a compact topological space and* F *is a continuous mapping lying in* $\mathrm{Dens}(\Lambda \times C, \mu)$, *then* $F \in \mathrm{AsCpt}(\Lambda \times C)$.

Theorem 2. *Let* X *be a* LCS, $C \in \mathrm{CoCl}(X)$, Λ *a compact topological space,* μ *a* mnc *in* X, F *a continuous mapping satisfying* $F \in \mathrm{Dens}(\Lambda \times C, \mu)$ *and* $F(\Lambda \times C) \subseteq C$. *Then* $F \in \mathrm{AsCpt}(\Lambda \times C)$.

Let us remember some classical measures of noncompactness. Let (X, p) be a pseudometric space and $M \in 2^X$. The Kuratowski's mnc $\alpha_X(M) = \alpha_p(M)$ of M is defined by $\alpha_X(M) = \inf\{d > 0$: there is a finite d-p-decomposition of M, i.e. M can be divided into a finite number of subsets with p-diameters less than $d\}$; the Hausdorff mnc $\chi_X(M) = \chi_p(M)$ of M is given by the formula $\chi_X(M) = \inf\{d > 0$: there is a finite d-p-net of M in X, i.e. M can be covered by a finite number of d-p-balls centered at points of $X\}$. It should be remarked that these two measures are not mnc's in the sense just introduced. (Here we take $\inf \emptyset = +\infty$.)

Now, let X be a LCS and P a system of continuous pseudonorms on X. The Kuratowski's mnc α_P (with respect to P) is defined as follows: $\mathfrak{M}_{\alpha_P} = \mathrm{Bo}(X)$, $A_{\alpha_P} =$ the set of all nonnegative functions on P with the uniformity of simple convergence and the natural order, $\alpha_P(M)(p) = \alpha_p(M)$ for $M \in \mathrm{Bo}(X)$. If $C \in 2^X$, then the Hausdorff mnc $\chi_{P,C}$ is defined as follows: $\mathfrak{M}_{\chi_{P,C}} = \mathrm{Bo}(X)$, $A_{\chi_{P,C}} A = A_{\alpha_P}$, and $\chi_{P,C}(M)(p) = \chi_{(C,p|C)}(M)$ for $M \in \mathrm{Bo}(X)$. The mappings α_P and $\chi_{P,C}$ for $C \in \mathrm{Co}(X)$ are mnc's in the sense introduced above, i.e. they are invariant with respect to the operation cocl (see [88, 93] and [21, 60]).

A mnc μ is said to be:

1) subadditive iff for each pair of elements of A_μ has a supremum in A_μ and $M, N, M \cup N \in \mathfrak{M}_\mu$ implies $\mu(M \cup N) = \sup\{\mu(M), \mu(N)\}$;
2) nonsingular iff A_μ is a cone (i.e. a closed convex cone in a LCS with $A_\mu \cap (-A_\mu) = \{O\}$); and
   $M \in \mathfrak{M}_\mu$, M a singleton implies $\mu(M) = 0$;
3) of type (O, PC) iff A_μ is a cone and
   $M \in \mathfrak{M}_\mu$, $\mu(M) = O$ implies M precompact;
4) of type (PC, O) iff A_μ is a cone and
   $M \in \mathfrak{M}_\mu$, M precompact implies $\mu(M) = 0$;
5) quasiregular iff it is of type (PC, O) and of type (O, PC);

3′) of type (O, RC): as in 3) with "precompact" replaced by "relatively compact";

4′) of type (RC, O): as in 4) with "precompact" replaced by "relatively compact";

5′) regular iff it is of type (O, RC) and of type (RC, O);

6) continuous iff A_μ is a topological space and for each M from $\mathfrak{M}_\mu$ and each neighbourhood W of $\mu(M)$ in A_μ there is a neighbourhood V of O in X such that
   $$N \in \mathfrak{M}_\mu, \quad M \subseteq N + V, \quad N \subseteq M + V \text{ implies } \mu(N) \in W;$$
7) uniformly continuous iff A_μ is a uniform space and for each vicinity W in A_μ there is a neighbourhood V of O in X such that
   $M, N \in \mathfrak{M}_\mu$, $M \subseteq N + V$, $N \subseteq M + V$ implies $(\mu(M), \mu(N)) \in W$;
8) algebraically subadditive iff A_μ is a cone and
   $M, N, M + N \in \mathfrak{M}_\mu$ imply $\mu(M + N) \leq \mu(M) + \mu(N)$;

9) positively homogeneous iff A_μ is a cone and
$t > 0$, M and tM in $\mathfrak{M}_\mu$ imply $\mu(tM) = t\,\mu(M)$;
10) homogeneous iff A_μ is a cone and
$t \in R^1$, M and tM in $\mathfrak{M}_\mu$ imply $\mu(tM) = |t|\,\mu(M)$;
11) translation invariant iff
$x \in X$, M and $x + M$ in $\mathfrak{M}_\mu$ imply $\mu(x + M) = \mu(M)$.

For example, α_P is monotone and has properties 1), 2), 4), 6)—11); and 4′); if P is defining then α_P has properties 3) and 5); if X is, in addition, quasi-complete, then α_P has properties 3′) and 5′). Similarly for $\chi_{P,C}$, but here we have to suppose in many cases that $\mathrm{cl}\,(C) = X$, i.e. that C is dense in X. It is an easy exercise that the mnc's α_P and $\chi_P = \chi_{X,P}$ are equivalent in the sense: $\chi_P \leqq \alpha_P \leqq 2\chi_P$.

A trivial mnc is the following: $\mathfrak{M}_\mu = 2^X$ (X a LCS), $A_\mu = \{0, 1\}$,

$$\mu(M) = \begin{cases} 0 & \text{iff } M \text{ is precompact,} \\ 1 & \text{iff } M \text{ is not precompact.} \end{cases}$$

For other "natural mnc's" in Banach spaces with basis, $C[a, b]$, $L_p(a, b)$ etc. see [88, 93]. These two references also justify the study of general mnc's (applications in the theory of differential equations etc.).

Let us consider the relation between α and χ once more. We restrict ourselves to the case of normed linear spaces.

Lemma 1. *Let X be a NLS (normed linear space) and* $\delta(\varepsilon) = \sup \left\{ \left\| \frac{x+y}{2} \right\| : x, y \in B_X(0, 1), \|x - y\| \geqq \varepsilon \right\}$, $\varepsilon \in [0, 2]$. *Then*

$$\frac{1}{2}\alpha_X \leqq \chi_X \leqq \delta(1) \cdot \chi_X \leqq \chi_X .$$

Proof. Let $M \in \mathrm{Bo}\,(X)$, $d = \mathrm{diam}\,(M)$ and $\varepsilon \in (0, 1)$. Then there exists $t \in X$ such that

$$M \subseteq B_X\big(t, d\,\delta(1 - \varepsilon)\big) . \tag{*}$$

Indeed, let $p, q \in M$ be taken so that $\|p - q\| \geqq d - \varepsilon d$. Let $s \in M$ be arbitrary and set $x = \frac{p-s}{d}$, $y = \frac{q-s}{d}$ and $t = \frac{p+q}{d}$. Then we have $x, y \in B_X(0, 1)$, $\|x - y\| = \frac{\|p-q\|}{d} \geqq \frac{d - \varepsilon d}{d} = 1 - \varepsilon$. Therefore, by the definition of $\delta(\cdot)$, $\left\| \frac{x+y}{2} \right\| \leqq \delta\,(1 - \varepsilon)$, i.e. $\|s - t\| = d \cdot \frac{1}{d} \left\| \frac{p+q}{2} - s \right\| = d \left\| \frac{x+y}{2} \right\| \leqq d \cdot \delta(1 - \varepsilon)$, and (*) is proved.

The inequality (*) easily implies the following one:

$$\chi_X(M) \leqq \delta(1 - \varepsilon)\,\alpha_X(M)\,, \qquad M \in \mathrm{Bo}\,(X)\,.$$

Since the function δ is continuous on $[0, 2)$ (see V. I. Gurarij [29]), we can conclude that

$$\chi_X(M) \leqq \delta(1) \cdot \alpha_X(M)\,.$$

Other inequalities of the lemma are known and trivial.

Remark. If, for example, X is an inner product space, then $\delta(1) = \sqrt{3}/2 = 0{,}866 < 1$.

Let X and Y be locally convex spaces, A a cone, $\mu\colon \mathfrak{M}_\mu \to A$ and $\nu\colon \mathfrak{M}_\nu \to A$ two measures of noncompactness in X and Y, respectively, C a subset of X, Λ a set and $k \in R^+$. A multivalued mapping $F\colon \Lambda \times C \to Y$ is said to be (k,μ,ν)-*bounded* iff for each M from 2^C the following holds:

1) $M \in \mathfrak{M}_\mu$,
2) $F(\Lambda \times M) \in \mathfrak{M}_\nu$,
3) $\nu(F(\Lambda \times M)) \leqq k\mu(M)$.

In this case we shall write $F \in \mathrm{Bo}\,(C, k, \mu, \nu) = \mathrm{Bo}\,(k, \mu, \nu)$.

Some simple properties are contained in the following

Lemma 2. a) *$F \in \mathrm{Bo}\,(C, k, \mu, \nu)$ and $G \in \mathrm{Bo}\,(D, l, \nu, \pi)$ imply $G \circ F \in \mathrm{Bo}\,(C, kl, \mu, \pi)$.*

b) *If ν is monotone and algebraically subadditive, $F_i \in \mathrm{Bo}\,(C, k_i, \mu, \nu)$ for $i = 1, 2$, then $F_1 + F_2 \in \mathrm{Bo}\,(C, k_1 + k_2, \mu, \nu)$, provided that $F_1(M) + F_2(M) \in \mathfrak{M}_\nu$ for each $M \in 2^C$.*

c) *$F \in \mathrm{Dens}\,(C, \mu, \nu)$, $G \in \mathrm{Dens}\,(D, \nu, \pi)$, A ordered $(= A_\mu = A_\nu = A_\pi)$, μ of type (O, RC), π of type (RC, O) and the condition "G maps relatively compact sets into relatively compact sets" imply that $G \circ F \in \mathrm{Dens}\,(C, \mu, \pi)$. The condition under quotation marks is satisfied if, for example, G is continuous and the domain of G is closed.*

Let us show how densifying mappings can arise. The first proposition below is due to B. N. Sadowski [93]. (Strictly speaking he assumes that H ranges in X only, but this is not sufficient; it has to be supposed that F ranges in D.)

Proposition 1. *Let X be a* LCS, *$C \in \mathrm{Bo}\,(X)$, Λ a compact topological space, $C \subseteq D \in \mathrm{Co}(X)$, P a defining system of pseudonorms on X, $\mathrm{cl}(D)$ quasicomplete (it is sufficient that $\mathrm{PreCpt}(\mathrm{cl}(D)) = \mathrm{RelCpt}(\mathrm{cl}(D))$), $\varphi_p\colon R^+ \to R^+$ a nondecreasing function such that $\varphi_p(t^+) < t$ for each $t > 0$ and $p \in P$. Let $H\colon \Lambda \times C \times D \to D$ be a mapping such that:*

a) *for each $y \in D$, H is continuous in $\lambda \in \Lambda$, uniformly with respect to $x \in C$;*
b) *for each $(\lambda, y) \in \Lambda \times D$, $H(\lambda, C, y)$ is precompact;*
c) *for each $(\lambda, x, y, z, p) \in \Lambda \times C \times D \times D \times P$,*

$$p\big(H(\lambda, x, y) - H(\lambda, x, z)\big) \leqq \varphi_p\big(p(y - z)\big)\,.$$

Then the mapping $F\colon \Lambda \times C \to D$ defined by $F(\lambda, x) = H(\lambda, x, x)$ is $\chi_{D,P}$-densifying on $\Lambda \times C$, i.e. $F \in \mathrm{Dens}\,(\Lambda \times C, \chi_{D,P})$.

The following proposition is a generalization of a result of R. D. Nussbaum (see the Remark below). Its proof is an extension of the corresponding result of R. D. Nussbaum.

Proposition 2. *Let (X, ϱ), (Y, σ) be metric spaces, A cone such that each pair of elements of A has a supremum (in the order given by A), $\mu\colon \mathrm{Bo}(X) \to A$ and $\nu\colon \mathrm{Bo}\,(Y) \to A$ two mappings such that:*

a) *ν is subadditive, i.e. $M, N \in \mathrm{Bo}\,(X)$ implies $\nu(M \cup N) = \sup\,\{\nu(M), \nu(N)\}$;*
b) *there is a mapping $\psi\colon [0, +\infty) \to A$ such that $\lim_{\varepsilon \to 0^+} \psi(\varepsilon) = 0 \in A$ and $\nu(B_Y(N, \varepsilon)) \leqq \nu(N) + \psi(\varepsilon)$ for all $\varepsilon > 0$ and $N \in \mathrm{Bo}\,(X)$.*

Let $\varphi: A \to A$ and $H: X \times X \to Y$ be two mappings such that:

(1) $\nu(H_y(M)) \leqq \varphi(\mu(M))$ *for all* $y \in X$, $M \in \mathrm{Bo}(X)$, *where* $H_y = H(.,y)$ *(this implies that each* H_y *is bounded, i.e. maps bounded sets into bounded sets):*

(2) *for each* $M \in \mathrm{Bo}(X)$, *the set* $\{H_{y|M}\}_{y\in M}$ *is precompact in* $C_b(M, Y)$, *the space of all bounded continuous mappings from* M *into* Y *with the supremum metric (this implies that each* H_y *is continuous).*

Then the mapping $F: X \to Y$ *defined by* $F(x) = H(x, x)$, $x \in X$, *satisfies the inequality:*

$$\nu(F(M)) \leqq \varphi(\mu(M)) \text{ for each } M \in \mathrm{Bo}(X).$$

Proof. It is easy to see that ν is monotone ($\mathrm{Bo}(Y)$ partially ordered by inclusion), since it is subadditive. Let $M \in \mathrm{Bo}(X)$ and $\varepsilon > 0$ be given. Then there is a finite set π in M such that $\{H_{z|M}\}_{z\in\pi}$ is an ε-net for the set $\{H_{y|M}\}_{y\in M}$ in the space $C_b(M, Y)$. This implies that

$$F(M) \subseteq H(M \times M) \subseteq \bigcup_{z\in\pi} B_Y(H_z(M), \varepsilon).$$

Hence we have (see the properties of ν):

$$\begin{aligned}\nu(F(M)) &\leqq \nu\Big(\bigcup_{z\in\pi} B_Y(H_z(M), \varepsilon)\Big)\\ &= \sup_{z\in\pi} \nu(B_Y(H_z(M), \varepsilon))\\ &\leqq \sup_{z\in\pi} [\nu(H_z(M)) + \psi(\varepsilon)]\\ &\leqq \sup_{z\in\pi} [\varphi(\mu(M)) + \psi(\varepsilon)]\\ &= \varphi(\mu(M)) + \psi(\varepsilon).\end{aligned}$$

Tending ε to 0, we conclude that

$$\nu(F(M)) \leqq \varphi(\mu(M)).$$

Remark. Setting $A = [0, +\infty)$, $\mu = \alpha_X$, $\nu = \alpha_Y$, $\psi(\varepsilon) = 2\varepsilon$, $\varphi(t) = kt$, we obtain Proposition I.A.8 of R. D. Nussbaum [60].

Corollary 1. *Assume, in addition to the hypotheses of proposition 2, that* $\mu(M) \nleqq \varphi(\mu(M))$ *for each non-precompact* $M \in \mathrm{Bo}(X)$. *Then* F *is "densifying with respect to* μ *and* ν*" in the following sense:* $M \in \mathrm{Bo}(X)$ *and* $\nu(F(M)) \geqq \mu(M)$ *imply that* M *is precompact.*

Corollary 2. *Let* C *be a complete bounded subset of a* NLS X, $\varphi: R^+ \to R^+$ *a function such that* $\varphi(t) < t$ *for each* $t > 0$ *and* $H: C \times C \to X$ *a mapping satisfying the following conditions:*

(1) $\alpha_X(H_y(M)) \leqq \varphi(\alpha_X(M))$ *for each* $M \in 2^C$, $y \in C$;

(2) $\{H_y\}_{y\in C}$ *is precompact in* $C_b(C, X)$.

Then the mapping $F: C \to X$ *defined by* $F(x) = H(x, x)$ *is densifying with respect to* α_X, *i.e.* $F \in \mathrm{Dens}(C, \alpha_X)$.

Proof. In this case, our condition (2) is equivalent to the condition (2) of Proposition 2. Then use Corollary 1 and the fact that C is complete and bounded.

A mapping F of a subset C of a metric space (X, d) into X is said to be a *φ-contraction* iff $\varphi: R^+ \to R^+$ is a nondecreasing right hand continuous function such that $\varphi(t) < t$ for each $t > 0$ and $d(Fx, Fy) \leqq \varphi(d(x, y))$ for all $x, y \in C$.

Proposition 3. *Let C be a complete bounded subset of a* NLS X, $T: C \to X$ *and* $S: C \to X$ *two mappings. Let T be a φ-contraction and $S(C)$ precompact. Then* $F = T + S \in \operatorname{Dens}(C, \alpha_X)$.

Proof. Let us define $H(x, y) = Tx + Sy$ for $x, y \in C$. Then H satisfies conditions (1) and (2) of corollary 2 above (easy).

Similarly, proposition 1 implies.

Proposition 4. *Let D be a complete convex subset of a* NLS X, *C a closed subset of D, $T: D \to D$ a φ-contraction and $S: C \to D$ a mapping with precompact image. Then* $F = T + S \in \operatorname{Dens}(C, \chi_D)$.

Remark. The last two propositions can be proved directly. Indeed, T is easily seen to be densifying and hence, so is $F = T + S$ because S has precompact image.

Proposition 5. *Let (X, ϱ), (Y, σ), A, μ and ν be as in proposition 2. Let Y be also a (multiplicative and, in general, noncommutative) group, $\varphi: A \to A$, $B: X \to Y$ and $K: X \to Y$ three mappings satisfying the conditions:*

(1) *B is continuous and $\nu(B(M) \cdot Ky) \leqq \varphi(\mu(M))$ for all $y \in X$, $M \in \operatorname{Bo}(X)$;*

(2) *for each $M \in \operatorname{Bo}(X)$, $B(M)$ is bounded and $K(M)$ is precompact:*

(3) *$\sigma(z \cdot Ky, z \cdot Ky') \leqq \varkappa_{M,N}(\sigma(Ky, Ky'))$ for all $M \in \operatorname{Bo}(X)$, $N \in \operatorname{Bo}(Y)$ $y, y' \in M$ and $z \in N$, where, for each $M \in \operatorname{Bo}(X)$ and $N \in \operatorname{Bo}(Y)$, $\varkappa_{M,N}: R^+ \to R^+$ is a function continuous at 0 with $\varkappa_{M,N}(0) = 0$ (this last condition is satisfied if, for example, σ is multiplication invariant).*

Then the mapping $F: X \to Y$ defined by $F(x) = (Bx) \cdot (Kx)$, $x \in X$, satisfies the inequality:

$$\nu(F(M)) \leqq \varphi(\mu(M)) \text{ for each } M \in \operatorname{Bo}(X) .$$

Proof. Follows from proposition 2 and it is left to the reader.

Remark. A similar result can be derived from proposition 1.

Two propositions below show that the property to belong to a class $\operatorname{Bo}(C, k, \mu, \nu)$ is invariant with respect to Fréchet differentiations (in some cases). Both results are very important in applications and contain as special cases the well-known results on Fréchet differentiations of completely continuous mappings. The first of these two propositions extends corresponding results of many authors (see, for example, [16, 23, 63, 93]).

Proposition 6. *Let X and Y be* NLS *s, x a point in X, $r > 0$, $k \in R^+$, A a cone, $\mu: \operatorname{Bo}(X) \to A$ and $\nu: \operatorname{Bo}(Y) \to A$ two* mnc*'s. Assume that ν is algebraically subadditive monotone and both μ and ν are positively homogeneous, and $\mu(x + M) = \mu(M)$ for each subset M of $B_X(0, r)$. Let $F: B_X(x, r) \to Y$ be a mapping which is Fréchet differentiable at x. Assume that $\nu(\{-Fx\}) = 0$. If $F \in \operatorname{Bo}(B_X(x, r), k, \mu, \nu)$, then $F'(x) \in \operatorname{Bo}(X, k, \mu, \nu)$.*

Proof. Let M be a subset of $B_X(0, r)$, $0 < t \leqq 1$, $d = \sup\{||x||: x \in M\}$. Then $x + Mt$ lies in $B_X(x, r)$ and

$$\begin{aligned}\nu(F'(x)(tM)) &\leqq \nu(F(x + tM)) + \nu(\{-Fx\}) + \nu(-r(x, tM)) \\ &\leqq k\mu(x + tM) + \nu(B_Y(0, \varphi(td))) \leqq k\mu(tM) + \varphi(td)\, \nu(B_Y(0, 1)) ,\end{aligned}$$

where $r(x, h) = F(x + h) - F(x) - F'(x)\, h$, $\varphi(s) = \sup\{||r(x, h)|| : ||h|| \leqq s\}$.

Hence

$$\nu(F'(x)(M)) \leqq k\mu(M) + \inf_{0<t\leqq 1} \frac{\varphi(td)}{t} \nu(B_Y(0, 1)) = k\mu(M) . \qquad (*)$$

Given M in Bo (X), there is $t > 0$ such that tM is contained in $B_X(0, r)$. Then $(*)$ implies that

$$\nu(F'(x)(M)) = t^{-1}\nu(F'(x)(tM)) \leqq t^{-1}k\mu(tM) = k\mu(M) .$$

Proposition 7. *Let X, Y, r, and k be as in proposition* 6, *A a cone, μ:* Bo $(X) \to A$ *and ν:* Bo $(Y) \to A$ *two mnc's. Assume that ν is algebraically subadditive monotone and both μ and ν are positively homogeneous and translation invariant. Let $C \supseteq X \backslash B_X(0, r)$ be a subset of X and $F: C \to Y$ an asymptotically Fréchet differentiable mapping with $F \in$* Bo (C, k, μ, ν). *Then $F'_{as} \in$* Bo (X, k, μ, ν).

Remark. Let us remember that the asymptotic Fréchet differentiability of F means that there is a (unique) linear bounded mapping F'_{as} from X into Y such that the mapping $R = F - F'_{as}$ satisfying the condition: $\lim_{||x|| \to +\infty} ||Rx||/||x|| = 0$.

Proof. Let $R = F - F'_{as}$ and $\psi(s) = \sup \{ ||Rx|| \cdot ||x||^{-1} : ||x|| \geqq s \}$; hence $\psi(s) \to 0$ as $s \to +\infty$. Let $M \in$ Bo (X) be a subset of $X \backslash B_X(0, r)$. Then, for each $t \geqq 1$,

$$F'_{as}(M) \subseteq \frac{F(tM)}{t} + \left\{ -\frac{R(tx)}{t} : x \in M \right\},$$

and therefore

$$F'_{as}(M) \subseteq \frac{k\mu(tM)}{t} + \nu(\psi(tr) \cdot \sup ||M|| \cdot B_Y(0, 1))$$

$$\leqq k\mu(M) + \sup ||M|| \cdot \psi(tr) \cdot \nu(B_Y(0, 1)) \to k\mu(M) \text{ as } t \to +\infty ,$$

i.e. $\nu(F'_{as}(M)) \leqq k\mu(M)$. Now, let $M \in$ Bo (X) be arbitrary. Then $x + M \subseteq X \backslash B_X(0, r)$ for some $x \in X$, and

$$\nu(F'_{as}(M)) = \nu(F'_{as}(x + M)) \leqq k\mu(x + M) = k\mu(M) .$$

At the end of this section, we shall present here some considerations on mappings of the form $T = \sum_{i=1}^{n} F_i + R$, where F_i are positively p_i-homogeneous and R is small (near 0 or ∞); this can be considered as generalizations of propositions 6 and 7. Such mappings were studied by M. A. Krasnosjelski [47], V. B. Melamed [55], V. B. Melamed and A. I. Perow [56] and P. P. Zabreiko and M. A. Krasnosjelski [109] and applied there to the computation of the (topological) rotation of some completely continuous fields. The results below are extensions of theorem 11 from J. Daneš, "Surjectivity and fixed point theorems and existence theorems for equations of Hammerstein type", Math. Inst. of Charles Univ., Prague 1971 (this was the first version of [17, 18]). For related results see also R. H. Moore and M. Z. Nashed [59]. Many results of [59] are special cases of considerations made below (we do not consider families of mappings, but the extension to this case is easy).

Remarks. 1. Let X and Y be linear spaces, $C \subseteq X$ starshaped with respect to ∞ (i.e., x in C and $t \geqq 1$ implies that tx lies in C), $\mathfrak{M}_\mu \subseteq 2^C$ and $\mathfrak{M}_\nu$ $\mathfrak{M}$ 2^Y two systems such that:

$$M \in \mathfrak{M}_\mu \quad \text{and } t \geqq 1 \text{ implies} \quad tM \in \mathfrak{M}_\mu ;$$

$$M, N \in \mathfrak{M}_\nu \quad \text{and } t \geqq 1 \text{ implies} \quad M + N,\ -M,\ tM \in \mathfrak{M}_\nu ;$$

let A be a cone (i.e. a closed convex cone in a LCS with vertex at 0), $\mu \colon \mathfrak{M}_\mu \to A$ and $\nu \colon \mathfrak{M}_\nu \to A$ two mappings such that μ is positively homogeneous and ν is positively homogeneous, algebraically subadditive and monotone (μ, ν need not be mnc's, i.e. need not be invariant under the operation "cocl"). Further, let $p > 0$, $\varphi, \psi \colon A \to A$ be two mappings such that $t^{-p}\varphi(ta) \to \psi(a)$ as $t \to +\infty$ for each a in A. Let $F, R \colon C \to Y$ be two mappings transforming sets of $\mathfrak{M}_\mu$ into sets of $\mathfrak{M}_\nu$ such that:

(1) F is positively p-homogeneous on C (i.e. x in C and $t \geqq 1$ implies $F(tx) = t^p Fx$);

(2) $t^{-p}\nu\big((-R(tM))\big) \to 0$ as $t \to +\infty$, for each $M \in \mathfrak{M}_\mu$. If the mapping $T = F + R$ satisfies the inequality

$$\nu\big(T(M)\big) \leqq \varphi\big(\mu(M)\big)\ , \text{ for each } M \in \mathfrak{M}_\mu\ , \qquad (*)$$

then

$$\nu\big(F(M)\big) \leqq \psi\big(\mu(M)\big)\ , \text{ for each } M \in M_\mu\ . \qquad (**)$$

Proof. For each M in $\mathfrak{M}_\mu$ and $t \geqq 1$, we have:

$$t^p F(M) = F(tM) \subseteq T(tM) + \big(-R(tM)\big)\ .$$

Hence:

$$t^p \nu\big(F(M)\big) = \nu\big(t^p F(M)\big) \leqq \nu\big(T(tM)\big) + \nu\big(-R(tM)\big)$$
$$\leqq \varphi\big(\mu(tM)\big) + \nu\big(-R(tM)\big) = \varphi\big(t\mu(M)\big) + \nu\big(-R(tM)\big)\ .$$

The rest of the proof is not difficult and it is left to the reader.

2. The preceding remark easily extends to the case $T = \sum_{i=1}^{n} F_i + R$, where F_i is positively p_i-homogeneous, $0 < p_1 < p_2 < \dots < p_n$, if the condition (2) is replaced by:

$$(2^n) \qquad t^{-p_{k+1}}\nu\left(\left(-\sum_{i=1}^{k} F_i - R\right)(tM)\right) \to 0 \text{ as } t \to +\infty\ ,$$

for each $M \in \mathfrak{M}_\mu$ and $k = 0, 1, \dots, n-1$

$\left(\text{where } \sum_{i=1}^{0} F_i = 0\right)$. The conclusion is that $\nu\big(F_i(M)\big) = 0$ and $\nu\big(F_i(M)\big) \leqq \psi\big(\mu(M)\big)$ for all $i = 2, \dots, n$ and $M \in \mathfrak{M}_\mu$. The proof consists in the inductive use of the preceding remark.

3. Example. X, Y normed linear spaces, $C \subseteq X$ starshaped with respect to ∞ with *dist* $(0, C) > 0$, $\mathfrak{M}_\mu = \mathrm{Bo}\,(C)$, $\mathfrak{M}_\nu = \mathrm{Bo}\,(Y)$, A a cone, $\mu \colon \mathrm{Bo}\,(C) \to A$ positively homogeneous, $\nu \colon \mathrm{Bo}\,(Y) \to A$ positively homogeneous, algebraically subadditive, monotone, $p > 0$, $F_i \colon C \to Y$ positively p_i-homogeneous (on C), $R \colon C \to Y$ satisfying $\left\|\sum_{i=1}^{k} F_i(x) + Rx\right\| \cdot \|x\|^{-p_{k+1}} \to 0$ as $\|x\| \to +\infty$, $x \in C$

for each $k = 0, 1, \ldots, n-1$. If $T = \sum_{i=1}^{n} F_i + R$ satisfies the inequality (*), then the conclusions of the preceding remark hold for each $M \subseteq \mathrm{Bo}(C)$.

4. The three preceding remarks can be dualized: replace "C starshaped *at* ∞", "$t \geqq 1$", "$t \to \infty$" and "$0 < p_1 < \cdots < p_n$" by "C starshaped with respect to 0", "$0 \leqq t \leqq 1$", "$t \to 0^+$" and "$0 < p_n < \cdots < p_1$" respectively. Clearly, the condition $dist(O, C) > 0$ of Remark 3 is omitted here. The explicit formulation and the proof is left to the reader.

§ 2. The rotation of asymptotically compact and densifying mappings. Fixed point and surjectivity theorems

In [88] and [93] B. N. Sadowski defined the rotation of asymptotically compact (and densifying) vector fields. Here we list the main properties of the rotation. Let X be a LCS, $D \in \mathrm{CoCl}(X)$, C a nonempty (relatively) open subset of D, and $F\colon \mathrm{cl}(C) \to D$ a continuous mapping lying in the class $\mathrm{AsCpt}(\mathrm{cl}(C))$ and satisfying the condition: $Fx \neq x$ for all $x \in \partial_D C$. (Here $\partial_D C$ denotes the boundary of C in the space D.) Then there can be defined an integer $\gamma_D(I - F, \mathrm{cl}(C))$, the rotation of $I - F$ over $\mathrm{cl}(C)$ (with respect to D), having following properties:

(i) *If $S \in \mathrm{CoCl}(X)$, $T = F^\infty(\mathrm{cl}(C)) \subseteq S \subseteq D$, $F(C \cap S)$ is relatively compact and $F(\mathrm{cl}(S \cap C)) \subseteq S$, then there is defined*

$$\gamma_S((I-F)|_{\mathrm{cl}(S\cap C)}, \mathrm{cl}(S \cap C)), \quad \text{and} \quad \gamma_D(I - F, \mathrm{cl}(C))$$
$$= \gamma_S((I-F)|_{\mathrm{cl}(S\cap C)}, \mathrm{cl}(S \cap C)) = \gamma_T((I-F)|_{\mathrm{cl}(T\cap C)}, \mathrm{cl}(T \cap C))$$

(the stabilization property and the definition).

(ii) *If $F_0, F_1\colon \mathrm{cl}(C) \to D$ are continuous mappings lying in* $\mathrm{AsCpt}(\mathrm{cl}(C))$ *being admissibly homotopic on* $\mathrm{cl}(C)$, *i.e. $F_i = G(., i)$, where $G\colon [0,1] \times \mathrm{cl}(C) \to D$ is a continuous asymptotically compact mapping such that $G(t, x) \neq x$ for all $(t, x) \in [0,1] \times \partial_D C$, then* $\gamma_D(I - F_0, \mathrm{cl}(C)) = \gamma_D(I - F_1, \mathrm{cl}(C))$ *(the homotopy).*

(iii) *If Λ is a topological space, X a quasicomplete LCS, $G\colon \Lambda \times \mathrm{cl}(C) \to D$ a continuous asymptotically compact mapping, $t_0 \in \Lambda$ and $G(t_0, .)$ has no fixed point on $\partial_D C$, then there exists a neighbourhood V of t_0 in Λ such that $G(t, x) \neq x$ and* $\gamma_D(I - G(t, .), \mathrm{cl}(C)) = \gamma_D(I - G(t_0, .), \mathrm{cl}(C))$ *for* $(t, x) \in V \times \partial_D C$.

(iv) *If $C \supseteq \bigcup_{i=1}^{n} C_i$, where C_i, $i = 1, \ldots, n$, are pairwise disjoint (relatively) open subsets of D such that $Fx \neq x$ on* $\mathrm{cl}(C) \setminus \bigcup_{1=i}^{n} C_i$, *then* $\gamma_D(I - F, \mathrm{cl}(C))$ *and* $\gamma_D(I - F, \mathrm{cl}(C_i))$, $i = 1, \ldots, n$, *are defined and*

$$\gamma_D(I - F, \mathrm{cl}(C)) = \sum_{i=1}^{n} \gamma_D(I - F, \mathrm{cl}(C_i)) \qquad \textit{(the additivity)}.$$

(v) $\gamma_D(I - F, \mathrm{cl}(C)) \neq 0$ *implies* $\mathrm{Fix}(F) \neq \varnothing$, *where* $\mathrm{Fix}(F)$ *denotes the set of all fixed point of F in* $\mathrm{cl}(C)$.

(vi) $Q \in \mathrm{CoCl}(X)$, $F^\infty(C) \subseteq Q \subseteq D$ *implies*

$$\gamma_Q(I - F, \mathrm{cl}(Q \cap C)) = \gamma_D(I - F, \mathrm{cl}(C)) .$$

(vii) $\gamma_D(I, \mathrm{cl}(C)) = 1$ *if* $0 \in C$, *and* $\gamma_D(I, \mathrm{cl}(C)) = 0$ *if* $0 \notin \mathrm{cl}(C)$ *(the normalization property).*

(viii) *If, in addition, D and C are symmetric with respect to* 0, $\mathrm{cl}(C)$ *is strictly D-starshaped with respect to 0 (see below), μ is a subadditive mnc in X such that* $\mu(-M) = \mu(M)$, $\pm F(M) \in \mathfrak{M}_\mu$ *and* $\{(1 - t) Fx - tF(-x): 0 \leqq t \leqq 1/ x \in M\} \in \mathfrak{M}_\mu$ *for each* $M \subseteq \mathrm{cl}(C)$, $F \in \mathrm{Dens}(\mathrm{cl}(C), \mu)$ *and*

$$(I - F)(-x) \neq t(I - F)(x) \quad \text{for each} \quad x \in \partial_D C \text{ and } t > 0 , \qquad (*)$$

then $\gamma_D(I - F, \mathrm{cl}(C))$ *is odd. (This is a slight extension of the corresponding result of B. N. Sadowski* [93]*; see his proof.)*

We say that C is strictly *D-starshaped* with respect to 0 iff $tx \in C$ for each $x \in C$ and $t \in [0, 1]$ and $\mathrm{card}\big((R^+x) \cap \partial_D C\big) \leqq 1$ for each $x \in C$, where "card" means "cardinality of".

Let us present some simple lemmas being useful in the following.

Lemma 3. *Let X be a LCS, C a subset of X, $F: C \to X$ and $G: C \to X$ two multivalued mappings and let $H: [0, 1] \times C \to X$ be defined by $H(t, x) = tFx + (1 - t) Gx$. Then*

$$H(t, .)^\infty(C) \subseteq H^\infty([0, 1] \times C) \subseteq (F \cup G)^\infty(C) \quad \text{for all } t \in [0, 1] .$$

Hence, if $F \cup G$ is asymptotically compact, then so are H and each $H(t, .)$. (Let us remember that $(F \cup G)(x) = Fx \cup Gx$.)

Proof. This follows immediately from the definition of the asymptotic range and from the inclusions

$$H(t, M) \subseteq H([0, 1] \times M) \subseteq \mathrm{co}\,\big(F(M) \cup G(M)\big) ,$$

i.e.

$$H(t, .)(M) \subseteq H([0, 1] \times M) \subseteq \mathrm{co}\,(F \cup G)(M) \subseteq \mathrm{cocl}\,(F \cup G)(M) \qquad (1)$$

being true for each $M \in 2^C$. (To prove the lemma set $M = C$.)

Lemma 4. *Let X, C, F, G and H be as in the preceding lemma and μ a monotone* mnc *in X. If $F \cup G \in \mathrm{Dens}(C, \mu)$, then $H \in \mathrm{Dens}([0, 1] \times C, \mu)$ and $H(t, .) \in \mathrm{Dens}(C, \mu)$ for all $t \in [0, 1]$.*

Proof. Use the inclusion (1) of the proof of lemma 1.

The proofs of three following lemmas are also simple and are left to the reader.

Lemma 5. *Let X, Y, be* LCS's, *C a subset of X, $F: C \to Y$ and $G: C \to Y$ two multivalued mappings, μ and ν* mnc's *in X and Y respectively with ν monotone and $k \in R^+$. If $F \cup G \in \mathrm{Bo}(C, k, \mu, \nu)$, then $H \in \mathrm{Bo}([0, 1] \times C, k, \mu, \nu)$ and $H(t, .) \in \mathrm{Bo}(C, k, \mu, \nu)$ for all $t \in [0, 1]$.*

Lemma 6. *Let X, C, F, G, H be as in lemma 4 and μ subadditive* mnc *in X with A_μ ordered. If $F, G \in \mathrm{Dens}(C, \mu)$, then $H \in \mathrm{Dens}([0, 1] \times C, \mu)$ and $H(t, .) \in \mathrm{Dens}(C, \mu)$ for all $t \in [0, 1]$.*

Lemma 7. *Let $X, Y, C, F, G, H, \mu, \nu$ be as in lemma 5 and, in addition, let ν be subadditive. If $F, G \in \mathrm{Bo}\,(C, k, \mu, \nu)$, then $H \in \mathrm{Bo}\,([0, 1] \times C, k, \mu, \nu)$ and $H(t, .) \in \mathrm{Bo}\,(C, k, \mu, \nu)$ for all $t \in [0, 1]$. (Use the implication: $a \leqq b$, $c \leqq d$ in $A_\mu = A_\nu \Rightarrow \sup\,(a, c) \leqq \sup\,(b, d)$.)*

Lemma 8. *Let X, C, F, G, H be as in lemma 3 and μ a monotone* mnc in X *such that $\mu(\{0\} \cup M) = \mu(M)$ for each $M \subseteq F(C)$; let $G = 0$, and $F \in \mathrm{Dens}\,(C, \mu)$. Then $H \in \mathrm{Dens}\,([0, 1] \times C, \mu)$ and $H(t, .) \in \mathrm{Dens}\,(C, \mu)$ for all $t \in [0, 1]$.*

Proof. Use the facts that $\mathrm{co}\,(F \cup G)(M) = \mathrm{co}\,(\{0\} \cup F(M))$ and $\nu(H(t, .)(M)) \leqq \nu(H([0, 1] \times M)) \leqq \nu(\mathrm{co}\,(F \cup G)(M)) = \nu(\mathrm{co}\,(\{0\} \cup F(M))) = \nu(F(M))$ for each $t \in [0, 1]$, $M \in 2^C$ (see also the inclusions (1) in the proof of lemma 3).

Lemma 9. *Let $X, Y, C, F, G, H, k, \mu, \nu$ be as in lemma 5 and let us suppose that $\nu(\{0\} \cup M) = \nu(M)$ for each $M \subseteq F(C)$ and $G = 0$. If $F \in \mathrm{Bo}\,(C, k, \mu, \nu)$, then $H \in \mathrm{Bo}\,([0, 1] \times C, k, \mu, \nu)$ and $H(t, .) \in \mathrm{Bo}\,(C, k, \mu, \nu)$ for all $t \in [0, 1]$.*

Lemma 10. *Let X be a LCS, D a closed convex subset of X, C an open (in the relative topology) subset of D and F, G: $\mathrm{cl}\,(C) \to D$ two continuous mappings such that for each $x \in \mathrm{cl}\,(C)$ there is a convex set $C(x)$ with $C(x) \cap \partial_D C = \emptyset$ (or more generally, $x \notin C(x)$) and $Fx, Gx \in C(x)$ for each $x \in \partial_D C$. If $F \cup G \in \mathrm{AsCpt}\,(\mathrm{cl}\,(C))$, then*

$$\gamma_D(I - F, \mathrm{cl}\,(C)) = \gamma_D(I - G, \mathrm{cl}\,(C)) .$$

Proof. By lemma 3 and the property (ii) of the rotation, it is sufficient to show that $tFx + (1 - t)\,Gx = x$ is not possible for any $t \in [0, 1]$ and $x \in \partial_D C$. But this is clear, because $tFx + (1 - t)\,Gx \in C(x) \in x$.

Remark. lemma 10 remains true with $\mathrm{AsCpt}\,(\mathrm{cl}\,(C))$ replaced by $\mathrm{Dens}\,(\mathrm{cl}\,(C), \mu)$ with μ monotone, or by $\mathrm{Bo}\,(\mathrm{cl}\,(C), k, \mu, \mu)$ with $k < 1$, μ monotone and of type (O, RC).

An application of lemmas 6 and 7 yields

Lemma 11. *Let X, D, C, F, G and $C(x)$ be as in lemma 10. Suppose that either $F, G \in \mathrm{Dens}\,(\mathrm{cl}\,(C), \mu)$ for some subadditive* mnc *μ with A_μ ordered, or $F, G \in \mathrm{Bo}\,(\mathrm{cl}\,(C), k, \mu, \mu)$ for some subadditive* mnc *μ of type (O, RC) and $k < 1$. Then $\gamma_D(I - F, \mathrm{cl}\,(C)) = \gamma_D(I - G, \mathrm{cl}\,(C))$.*

Theorem 3. *Let X be a* LCS, *$D \in \mathrm{CoCl}\,(X)$, C an open (in the relative topology) subset of D containing the origin 0 of X, F: $\mathrm{cl}\,(C) \to D$ a continuous mapping, μ a monotone* mnc *in X such that $\mu(\{0\} \cup M) = \mu(M)$ for each $M \subseteq R(F) = (\mathrm{cl}\,(C))$ and $F \in \mathrm{Dens}\,(\mathrm{cl}\,(C), \mu)$. If $Fx \neq sx$ for each $s > 1$ and each $x \in \partial_D C$ (the Leray-Schauder condition), then F has a fixed point in $\mathrm{cl}\,(C)$. If, in addition $Fx \neq x$ for each $x \in \partial_D C$, then $\gamma_D(I - F, \mathrm{cl}\,(C)) = 1$.*

Proof. Suppose that F has no fixed points on $\partial_D C$. Then the mapping H from lemma 8 is an admissible homotopy between F and O, and hence $\gamma_D(I - F, \mathrm{cl}\,(C)) = \gamma_D(I, \mathrm{cl}\,(C)) = 1$ because $O \in C$ (see the property (vii) of the rotation). Thus, F has a fixed in C (by the property (v) of the rotaton)i. If the assumption made at the beginning of the proof is not true, then F has a fixed point on $\partial_D C$. In each case, F has a fixed point in $\mathrm{cl}\,(C)$.

Now we are in a position to prove an extension of theorem 1 of W. V. Petryshyn [74]. At the same time our proof seems to be simpler (as we can see from the sketch of Petryshyn's proof given in the cited paper).

Theorem 4. *Let X be a* LCS, $D \in \mathrm{CoCl}\,(C)$, *C a bounded open (in the relative topology) subset of D with $0 \in C$, $F\colon \mathrm{cl}\,(C) \to D$ a continuous mapping, μ a monotone* mnc *in X with $\mathfrak{M}_\mu \supseteq \exp\big(\mathrm{cocl}(C \cup F(C))\big)$ and such that $\mu(\{0\} \cup M) = \mu(M)$ for each $M \subseteq R(F) = F\big(\mathrm{cl}\,(C)\big)$. Assume that there exists a sequence $\{k_n\}_{n=1}^{\infty}$ such that $0 < k_n < 1$, $k_n \to 1$ as $n \to \infty$ and $F_n = k_n F \in \mathrm{Dens}\,\big(\mathrm{cl}\,(C), \mu\big)$ for each n. Let us suppose that:*

$$Fx \neq sx \text{ for each } s > 1 \text{ and each } x \in \partial_D C\,. \tag{1}$$

$$0 \in \mathrm{cl_{seq}}\,[(I - F)\,\big(\mathrm{cl}\,(C)\big)] \text{ implies } 0 \in (I - F)\,\big(\mathrm{cl}\,(C)\big)\,. \tag{2}$$

Then F has a fixed point in cl C. *(Here $\mathrm{cl_{seq}}\,(M)$ denotes the sequential closure of M.)*

Proof. The inequalities $0 < k_n < 1$ and $s > 1$ imply $k_n^{-1}s > 1$ and hence $s > 1$ and $x \in \partial_D C$ implies $F_n(x) = k_n\, Fx \neq k_n(k_n^{-1}sx) = sx$. By theorem 3, there exists $x_n \in \mathrm{cl}\,(C)$ such that $F_n(x_n) = x_n$. Then

$$x_n - Fx_n = x_n - k_n^{-1}F_n(x_n) = x_n - k_n^{-1}x_n = (1 - k_n^{-1})\, x_n\,.$$

But $1 - k_n^{-1}$ tends to 0 as $n \to \infty$, $\{x_n\}_{n=1}^{\infty}$ is bounded (since C is so), implying $x_n = Fx_n = (1 - k_n^{-1})\, x_n \to 0$ as $n \to \infty$ and hence $0 \in \mathrm{cl_{seq}}\,[(I - F)\,\big(\mathrm{cl}\,(C)\big)]$. By the condition (2), we conclude that $0 \in (I - F)\,\big(\mathrm{cl}\,(C)\big)$, i.e. F has a fixed point in $\mathrm{cl}\,(C)$.

Remarks. 1. If $\mathrm{cl_{seq}}\,[(I - F)\,\big(\mathrm{cl}\,(C)\big)] = \mathrm{cl}\,[(I - F)\,\big(\mathrm{cl}\,(C)\big)]$, then (2) is equivalent to

$$0 \in \mathrm{cl}\,[(I - F)\,(C)] = \mathrm{cl}\,[(I - F)\,\big(\mathrm{cl}\,(C)\big)] \text{ implies } 0 \in (I - F)\,\big(\mathrm{cl}\,(C)\big). \tag{2'}$$

This is the case if, for example, X is metrizable (or more generally, if the subset $\mathrm{cl}\,(D - D)$ of X is metrizable).

2. Our theorem 4 contains as special case the well-known theorems of Schauder, Rothe, Altman, Darbo [21], Krasnoselski [46, 47], and Sadowski [86]. It also contains some theorems of A. A. Istrăţescu and V. Istrăţescu [36], R. D. Nussbaum [60], W. V. Petryshyn [74] etc. See the results below and W. V. Petryshyn [74].

3. In [74] W. V. Petryshyn consider a seemingly more general condition. But setting $Fx = T(x + x_0) - x_0$ and replacing C by $C - x_0$ his condition reduces to our condition (1) (the Leray-Schauder condition). In the sketch of the proof of his theorem, W. V. Petryshyn [74] states two propositions concerning the topological degree; these propositions can be extended and used to give another proof of our theorem 4.

4. The condition (2) is satisfied if $(I - F)\,\big(\mathrm{cl}\,(C)\big)$ is closed.

Corollary 1. *Let X, D, C, F, μ be as in theorem 3. Assume, for each $x \in \partial_D C$, there are a positively homogeneous function $d_x\colon X \to R^+$ with $d_x(x) > 0$ and a strictly superadditive function $\varphi_x\colon R^+ \to R^+$ (i.e. $a, b > 0$ implies $\varphi_x(x + b) > \varphi_x(a) + \varphi_x(b)$) such that*

$$\varphi_x\big(d_x(Fx)\big) \leqq \varphi_x\big(d_x(x)\big) + \varphi_x\big(d_x(Fx - x)\big) \quad (\text{for } x \in \partial_D C)\,. \tag{*}$$

Then F has a fixed point in $\mathrm{cl}\,(C)$.

Proof. It is sufficient to show that $Fx \neq sx$ whenever $x \in \partial_D C$ and $s > 1$. Assume, by contradiction, that $Fx = sx$ for some $x \in \partial_D C$ and $s > 1$. Then (*) gives:

$$\varphi_x(d_x(sx)) \leqq \varphi_x(d_x(x)) + \varphi_x(d_x((s-1)\,x)) ,$$

i.e.

$$\varphi_x(sd_x(x)) \leqq \varphi_x(d_x(x)) + \varphi_x((s-1)\,d_x(x)) ,$$

a contradiction, since $sd_x(x) = d_x(x) + (s-1)\,d_x(x)$, $d_x(x) > 0$ and $(s-1)$ $d_x(x) > 0$; this completes the proof.

Remarks. 1. If $\varphi: R^+ \to R^+$ is a function such that $\varphi(0) = 0$ and $\varphi(t)\,t^{-1}$ is strictly increasing on $(0, +\infty)$, then is strictly superadditive. See E. Hille and R. S. Phillips [33].

2. Therefore, we can set, for example, $\varphi_x(t) = t^{p(x)}$, $d_x = ||\cdot||$, where $p(x) < 1$ and $||\cdot||$ is a norm on X (the space X need not be normable, i.e. the norm $||\cdot||$ need not induce the topology of X). Setting $p(x) = 2$, we obtain a generalization of Altman's and Rothe's theorems.

Corollary 2. *Let X, D, C, F, μ be as in theorem* 3. *Assume that* cl (C) *is strictly D-starshaped (with respect to* 0*) and $F(\partial_D C) \subseteq$* cl (C). *Then F has a fixed point in* cl (C).

Proof. It is easy to see that the additional hypotheses on C and F imply that F satisfies the Leray-Schauder condition and we can apply theorem 3.

Corollary 3. *Let X be a* LCS, *μ a translation invariant monotone* mnc *in X such that $\mathfrak{M}_\mu$ is translation invariant and $\mu(\{0\} \cup M) = \mu(M)$ for each $M \in \mathfrak{M}_\mu$, $\{C_n\}_{n=1}^\infty$ a sequence of open subsets of X with $0 \in C_n$, $F: X \to X$ a continuous mapping with $F \in$* Dens $($cl $(C_n), \mu)$ *for each n. Let there exist a positive sequence $\{a_n\}_{n=1}^\infty$ tending to $+\infty$ as $n \to +\infty$ such that $d(Fx) \leqq d(x) - a_n$ for each $n \geqq 1$ and each $x \in \partial C_n$, where $d: X \to R^+$ is a subadditive positively homogeneous functional. Then the mapping $T = I - F$ is surjective.*

Proof. It is sufficient to show that the mapping $F_y = F + y$ has a fixed point for each $y \in X$. Let $y \in X$ be given and take n so that $a_n > d(y)$. Since μ is translation invariant and $F \in$ Dens $($cl $(C_n), \mu)$, we have $F_y \in$ Dens $($cl $(C_n), \mu)$. We want to apply theorem 3 to the mapping F_y; we have to prove that F_y satisfies the Leray-Schauder condition on ∂C_n. Let us suppose that $Fx + y = F_y(x) = sx$ for some $x \in \partial C_n$ and $s > 1$. Then:

$$\begin{aligned} 0 \leqq (s-1)\,d(x) &= d(sx) - d(y) + d(y) - d(x) \leqq d(sx - y) + d(y) - d(x) \\ &= d(Fx) + d(y) - d(x) \leqq d(y) - a_n < 0 , \end{aligned}$$

a contradiction. This proves the corollary.

Remark. Let us suppose that X is normable, $||\cdot||$ a norm for X and $d = ||\cdot||$. Then the condition "$||Fx|| \leqq ||x|| - a_n$ for $x \in \partial C_n$, all $n \geqq 1$" is satisfied if, for example, $||Fx|| \leqq k\,||x||$ for some $k < 1$ and $||x||$ sufficiently large (take $C_n = B_X(0, r_n)$ with r_n sufficiently large and $r_n \to +\infty$ as $n \to +\infty$) i.e., if $|F| = \lim\limits_{r \to +\infty} \sup\limits_{||x|| \geqq r} \dfrac{||Fx||}{||x||} < 1$; this the case, if F is asymptotically Fréchet differentiable with $||F'_{as}|| < 1$. Hence, corollary 3 contains the Granas' theorem on quasibounded mappings [28] as a special case even if F is completely continuous; it contains also results of A. Vignoli [106]. The results of A. Vignoli [105] (extensions of intersection theorem of A. Granas [28]) can be easily extended up to our degree of generality (this is left to the reader). All results following theorems 3 and 4 (and many results below) can be somewhat extended using theorem 4 rather than theorem 3 (here it is natural to suppose that $(I - F)(\mathrm{cl}(C))$, respectively $(I - F)(\mathrm{cl}(C_n))$ is closed) an easy exercise. Then, for example, we obtain generalizations of theorems 7 and 7′ of W. V. Petryshyn [73].

Now, let us use the property (viii) of the rotation to derive a surjectivity theorem.

Theorem 5. *Let X be a* NLS, *μ a subadditive translation invariant* mnc *in X such that $\mathfrak{M}_\mu = \mathrm{Bo}(X)$, $\mu(-M) = \mu(M)$ and $\mu(M + K) = \mu(M)$ for each $M \in \mathrm{Bo}(X)$ and each compact segment K beginning at 0. Let $\{C_n\}_{n=1}^{\infty}$ be a sequence of open bounded symmetric and strictly $(X-)$ starshaped (with respect to 0) subsets of X such that $\mathrm{dist}(0, \partial C_n) \to +\infty$ as $n \to +\infty$. Let $F: X \to X$ be a continuous mapping such that $F \in \mathrm{Dens}(\mathrm{cl}(C_n), \mu)$ for all n and $\|Tx\| \to +\infty$ as $\|x\| \to +\infty$, $x \in \bigcup_{n=1}^{\infty} \partial C_n$, where $T = I - F$. Assume that $T(-x) \neq tTx$ for all $x \in \bigcup_{n=1}^{\infty} \partial C_n$ and $t > 0$.*

Then the mapping $T = I - F$ is surjective.

Proof. Let $y \in X$ be arbitrary and set $F_y = F + y$. Since $\|Tx\| \to +\infty$ as $\|x\| \to +\infty$, $x \in \bigcup_{n=1}^{\infty} \partial C_n$, and $\mathrm{dist}(0, \partial C_n) \to +\infty$ as $n \to +\infty$, there exists $n = n(y)$ such that $\|x - Fx\| > \|y\|$ for $x \in \partial C_n$. Set $H(t, x) = Fx + ty = tF_y(x) + (1 - t)Fx$ for $x \in \mathrm{cl}(C_n)$, $t \in [0, 1]$. Then $\|x - H(t, x)\| \geqq \|x - Fx\| - t\|y\| \geqq \|x - Fx\| - \|y\| > 0$ for all $x \in \partial C_n$ and $t \in [0, 1]$. For $M \subseteq \mathrm{cl}(C_n)$, we have:

$$\mu(H([0, 1] \times M)) = \mu(F(M) + [0, 1]y) = \mu(F(M)) .$$

Therefore H is an admissible homotopy between F and F_y, and, by (ii) an (viii), we obtain:

$$\gamma(I - F_y, \mathrm{cl}(C_n)) = \gamma(I - F, \mathrm{cl}(C_n)) = odd \neq 0 .$$

Thus, by the property (v) of the rotation, F_y has a fixed point in C_n, i.e. $Tx = y$ for some $x \in C_n$. Since y was arbitrary in X, we conclude that T is surjective.

Corollary 1. *Let Let X be a* NLS, *μ a* mnc *as in theorem 5, $F: X \to X$ a continuous mapping with $F \in \mathrm{Dens}(C, \mu)$ for $C \in \mathrm{BoCl}(X)$. Assume F is asymptotically Fréchet differentiable and $I - F'_{as}$ is an (topological) isomorphism of X. Then $I - F$ is surjective.*

Proof. Let $k = \|(T'_{as})^{-1}\|^{-1}$; then $\|T'_{as}(x)\| \geqq k\|x\|$ for all $x \in X$. Let $r \in R^+$ be such that $\|Rx\| < \frac{k}{2}\|x\|$ whenever $\|x\| \geqq r$; here $R = T - T'_{as} = F - F'_{as}$. Then

$$\|T'_{as}(x)\| \geqq k\|x\| > \frac{1}{1+t}\frac{\|R(-x)\|}{\|x\|} + \frac{t}{1+t}\frac{\|Rx\|}{\|x\|} \geqq \frac{1}{1+t}\|R(-x) + tRx\|$$

whenever $||x|| \geqq r$ and $t > 0$. This implies $R(-x) \neq tRx + (t+1)\, T'_{as}(x)$ or, in other words, $T(-x) \neq tTx$ if $||x|| \geqq r$ and $t > 0$. We can apply Theorem 5 setting $C_n = B_X(0, r + n - 1)$.

Remark. Let X and Y be LCS's, C a subset of X, $F\colon X \to Y$ a mapping such that $F \in \mathrm{Bo}\,(C, k, \mu, \nu)$ for some $k < 1$ and some mnc's μ and ν in X and Y respectively. Assume that either C is quasicomplete and μ is of type (O, PC) or C is closed and μ is of type (O, RC). Then $F \in \mathrm{Dens}\,(C, \mu, \nu)$, that is, $\mathrm{Bo}\,(C, k, \mu, \nu) \subseteq \mathrm{Dens}\,(C, \mu, \nu)$. This remark will be used below.

The following two lemmas are due to B. N. Sadowski [93].

Lemma 12. *Let X be a Banach space, μ an algebraically subadditive homogeneous* mnc *in X defined on* Bo (X) *and valued in* $[0, +\infty)$ *such that, for any bounded sequence S in X, $\mu(S) = 0$ iff $S \in \mathrm{RelCpt}\,(X)$. Then μ is a pseudonorm on $m(X)$, the Banach space of all bounded sequences of elements of X with the supremum norm, with $(\mu|_{m(X)})^{-1}(0) = k(X)$, the (closed) subspace of $m(X)$, consisting of relatively compact sequences. This pseudonorm induces (canonically) a norm on $X^+ = m(X)/k(X)$.*

Lemma 13. *Let $L\colon X \to X$ be a linear bounded mapping, μ a* mnc *as in the preceding lemma and $R = \lim\limits_{n\to+\infty} (||L^n||^\mu)^{1/n}$, where $||L||^\mu = \sup\{\mu(LM)\colon M \in m(X), \mu(M) = 1\}$ (sup $\varnothing = 0$, by definition). Then*

$$\sigma(L) \cap C_R = \sigma_p(L) \cap C_R$$

where $C_R = \{\lambda \text{ complex}\colon |\lambda| > R\}$, $\sigma(L)$ denotes the spectrum of L and $\sigma_p(L)$ the point spectrum of L. Furthermore, ind $(\lambda I - L) = 0$ *for all $\lambda \in C_R$.*

Lemma 14. *Let X be a Banach space, x a point in X, $r > 0$, $F\colon B_X(x, r) \to X$ a mapping possessing a Fréchet derivative at x and μ a* mnc *in X as in lemma* 12. *Assume, in addition, that μ is monotone translation invariant and $F \in \mathrm{Bo}\,\big(B_X(x, r), k, \mu, \mu\big)$ for some $k \in R^+$. Then*

$$\sigma\big(F'(x)\big) \cap C_k = \sigma_p\big(F'(x)\big) \cap C_k\,.$$

Proof. See the remark above, proposition 6 and lemma 13.

Lemma 15. *Let X, μ be as in lemma* 12, *$F\colon X \to X$ an asymptotically Fréchet differentiable mapping such that $F \in \mathrm{Bo}\,(X, k, \mu, \mu)$ for some $k \in R^+$. Assume, in addition, that μ is monotone and translation invariant. Then*

$$\sigma(F'_{as}) \cap C_k = \sigma_p(F'_{as}) \cap C_k\,.$$

Proof. See lemma 13 and proposition 7.

Now we can state another corollary to theorem 5.

Corollary 2. *Let X be a Banach space, μ a subadditive, algebraically subadditive, homogeneous, translation invariant, regular* mnc *in X defined on* Bo (X) *and valued in R^+, $F\colon X \to X$ an asymptotically Fréchet differentiable continuous mapping such that $F \in \mathrm{Bo}\,(X, k, \mu, \mu)$ for some $k < 1$. If $1 \notin \sigma_p(F'_{as})$, then $I - F$ is surjective.*

Proof. See lemma 15, corollary 1 and the remark following it.

Now, we shall prove some results near to the Fredholm alternative for linear mappings. For some results in this direction see, for example, F. E. Browder

[7], P. Hess [32], R. I. Katschurowski [38—41], M. A. Krasnosjelski [49], S. I. Pohoshajew [76], J. T. Schwartz [94, Theorem 1.22], P. P. Zabreiko and M. A. Krasnosjelski [110]. See also J. Daneš [20], and G. Marinescu [53].

Theorem 6. *Let X be a linear space, Y a Banach space, $X = X_1 \cup X_2$ (a decomposition of X) and $T\colon X \to Y$ a Gateaux differentiable mapping with closed range $R(T) = T(X)$. Suppose that*

$$R(T) \subseteq \Big(\bigcap_{x_1 \in X_1} R(T'(x_1))\Big) \cap \Big(\bigcap_{x_2 \in X_2} \operatorname{cl}\big(R(T'(x_2))\big)\Big) = M .$$

Then $R(T) = R(T'(x_1)) = \operatorname{cl}\big(R(T'(x_2))\big)$ for all $x_1 \in X_1$ and $x_2 \in X_2$.

Proof. Let $Z = \operatorname{cl}\big(\operatorname{sp}(R(T))\big)$. Clearly M contains Z. Since $\dfrac{T(x + t\,h) - T(x)}{t} \in \operatorname{sp}\big(R(T)\big) \subseteq Z$ for all $t \neq 0$, x and h in X, we have $T'(x)\,h \in Z$ for all x and h in X, i.e. $\operatorname{cl}\big(R(T'(x))\big) \subseteq Z$ for all x in X. Since $R(T'(x_1)) \supseteq M \supseteq Z$ for all x_1 in X_1 and $\operatorname{cl}\big(R(T'(x_2))\big) \supseteq M \supseteq Z$ for all x_2 in X_2, it follows that

$$R(T'(x_1)) = Z = \operatorname{cl}\big(R(T'(x_2))\big) \text{ for all } x_1 \text{ in } X_1 \text{ and } x_2 \text{ in } X_2 . \qquad (*)$$

The set $R(T)$ is closed in the closed subspace Z of Y, and we can use the Krasnosjelski-Zabreiko's theorem [110] to the triple $\{X, Z, T\}$ and find out that $R(T) = Z$ which together with (*) gives the assertion of the theorem.

Lemma 16. *Let X be a Banach space and F a continuous mapping from X into X, $T = I - F$. Let us consider the following conditions:*

(a) *$R(T)$ is closed;*

(b) *T is inversely compact, that is, $\{x_n\} \subseteq X$, $\{Tx_n\}$ convergent implies $\{x_n\}$ relatively compact; in other words, K compact in X implies $T^{-1}(K)$ relatively compact;*

(c) *F is demicompact* (see (d)) *and $\|Tx\| \to +\infty$ as $\|x\| \to +\infty$;*

(d) *F is demicompact, that is, $\{x_n\}$ bounded in X with $\{Tx_n\}$ convergent implies $\{x_n\}$ relatively compact:*

(e) *F is μ-densifying on each $C \in$* BoCl (X), *where μ is a monotone* mnc *in X defined on* Bo (X) *satisfying the following condition:*
$S \in$ Bo (X), *K a convergent sequence in X and $F(S) + K \supseteq S$ implies*

$$\mu(F(S)) = \mu(S)$$

(*this last condition if satisfied if, for example, $\mu(M + K) = \mu(M)$ for each $M \in$* Bo (X) *and $K \in$* RelCpt (X));

(f) *F is μ-densifying on each $C \in$* BoCl (X) *where μ is a monotone algebraically subadditive* mnc *in X of type (RC, O).*

Then the following implications hold:

$$\begin{array}{ccccc} (c) & \Rightarrow & (b) & \Rightarrow & (a) \\ & & \Downarrow & & \\ (e) & \Rightarrow & (d) & \Leftarrow & (f) \end{array}$$

Proof is a routine and can be left to the reader.

The following Banach theorem on linear mappings having closed range (see, for example, K. Yosida, Functional Analysis, Springer 1965; Theorem VII.5) is useful in applications of theorem 6.

Theorem 7. *Let X and Y be Banach spaces and T a densely defined linear closed mapping from X into Y. Then the following conditions are equivalent:*
(1) *$R(T)$ is closed (in Y);*
(2) *$R(T^*)$ is closed* (in X^*);
(3) $R(T) = {}^{\perp}N(T^*) \equiv \{y \in Y' : \langle y, y^* \rangle = 0$ *for all* $y^* \in N(T^*)\}$;
(4) $R(T^*) = N(T)^{\perp} \equiv \{x^* \in X^* : \langle x, x^* \rangle = 0$ *for all* $x \in N(T)\}$.
Here $R(T)$ denotes the range of T, $N(T)$ the null space of T and T^* the adjoint mapping to T.

Corollary 1. *Let X and Y be Banach spaces, $T: X \to Y$ a Gateaux differentiable mapping such that*
(1) *$R(T'(x))$ is closed in Y for each x in X;*
(2) *$R(T)$ is closed in Y;*
(3) *the following implication holds for all $x, z \in X$ and $y^* \in Y^*$:*

$$T'(z)^* y^* = 0 \Rightarrow \langle Tx, y^* \rangle = 0.$$

Then the equation $Tx = y$ is solvable for a given y in Y if and only if one (and hence both) of the following two equivalent conditions is satisfied:
(i) *for each x in X the following implication holds:*

$$y^* \in Y^*, \qquad T'(x)^* y^* = 0 \Rightarrow \langle y, y^* \rangle = 0;$$

(ii) *there exists x in X such that the following implication holds:*

$$y^* \in Y^*, \qquad T'(x)^* y^* = 0 \Rightarrow \langle y, y^* \rangle = 0.$$

Therefore, if all $T'(x)$, $x \in X$, are surjective (that is, if all $T'(x)^*$ are injective), then T is surjective.

Proof. See theorems 6 and 7.

Remarks. 1. Let us note that $R(T)$ is closed, where $T: X \to X$ is continuous, if, for example, T is either (1) inversely compact, or (2) it satisfies one of the conditions (d), (e), (f) of lemma 16 and $||Tx|| \to +\infty$ as $||x|| \to +\infty$ (see lemma 16).

2. Corollary 1 extends and gives precision to theorems 2, 3 and 4 of R. I. Katschurowski [41]. (He supposes that T is continuously Fréchet differentiable on X etc. In the notation of theorem 6, corollary says that not only $R(T) = M$ but $R(T) = M = R(T'(x))$ for all x in X). At the same time our proof is simpler and relies heavily on our theorem 6 (being a generalization of Zabreiko-Krasnosjelski's theorem used in its proof).

Theorem 8. *Let X be a Banach space and μ an algebraically subadditive, homogeneous, monotone, translation invariant and regular* mnc *in X defined on* Bo (X) *with values in* $R^+ = [0, +\infty)$. *Let $F: X \to X$ be a Fréchet differentiable mapping lying in the class* Bo (X, k, μ, μ) *for some* $k \in [0, 1)$. *If, for each x in X, 1 is not an eigenvalue of $F'(x)$ and $||Tx|| \to +\infty$ as $||x|| \to +\infty$ (where $T = I - F$), then the mapping $T = I - F$ is surjective.*

Proof. By proposition 6, $F'(x) \in$ Bo (X, k, μ, μ) and hence $||F'(x)||^{\mu} \leqq k < 1$ for all x in X. Hence $1 \notin \sigma_p(F'(x))$ implies $1 \notin \sigma(F'(x))$ for each x in X, i.e. $T'(x)$ is surjective for all x in X. By remark 1 above and the remark before lemma 12, $R(T)$ is closed. Using corollary 1 (or Zabreiko-Krasnosjelski's theorem [110]), we see that T is surjective.

Remarks. 1. It is easy to see that theorem 8 extends theorem 1 of R. I. Katschurowski [41].

2. Theorem 8 is very near to the following theorem of M. A. Krasnosjelski [49]: If X is a Banach space, $F: X \to X$ a *continuously* Fréchet differentiable completely continuous mapping such that $\|x - Fx\| \to +\infty$ as $\|x\| \to +\infty$ and 1 is not an eigenvalue of $F'(x)$, for each x in X, then the equation $x = Fx + y$ has a *unique* solution for each y in X. His proof remains valid if we assume "$F \in \mathrm{Bo}(X, k, \mu, \mu)$ with μ, k as in theorem 8" rather than "F is completely continuous".

3. Further results concerning the "Fredholm alternative" will be given in the second part of this paper.

We finish this section by a consequence of proposition 3.

Proposition 8. *Let X be a* NLS, *$D \in \mathrm{CoCl}(X)$, $0 \in C$ an open (in the relative topology) subset of D with* cl (C) *complete. Let T:* cl $(C) \to D$ *be a φ-contraction, S:* cl $(C) \to D$ *a continuous mapping with $S(C) \in \mathrm{PreCpt}(X)$, and set $F = T + S$. If $Fx \neq sx$ for all $x \in \partial_D C$ and $s > 1$, then F has a fixed point in* cl (C).

(See Corollaries 1 and 2 to theorem 4 for conditions implying the Leray-Schauder condition).

Proof. Use proposition 3, lemma 8 and the properties (ii), (vii) and (v) of the rotation.

§ 3. Applications

This section is devoted to some applications of densifying mappings to nonlinear functional analysis. Many results are generalizations of corresponding results from M. A. Krasnosjelski [47]; some of them are slight proper extensions even in the case of completely continuous mappings.

Proposition 9 and theorem 9 below generalize some existence theorems of C. Avramescu [4], J. K. Hale and M. C. Cruz [31], W. R. Melvin [57]. See also I. A. Rus [84, 85] for a general approach to problems of this kind.

Lemma 17. *If $\varphi: R^+ \to R^+$ is an upper semi-continuous function such that $\varphi(t) < t$ for each $t > 0$, then for each bounded net $(t_\alpha) \subseteq R^+$, the condition $t_\alpha - \varphi(t_\alpha) \to 0$ implies $t_\alpha \to 0$.*

Proof. Let $(t_\alpha) \subseteq R^+$ be a bounded net such that $t_\alpha - \varphi(t_\alpha) \to 0$ and suppose that $t_\alpha \to 0$ is not true. Then there exists a subset (t_β) of (t_α) such that $t_\beta \to \delta > 0$. Then:

$$\varphi(\delta) \geqq \limsup \varphi(t_\beta) = \lim \varphi(t_\beta) = \lim t_\beta = \delta \,,$$

a contradiction to the fact that $\varphi(t) < t$ for each $t < 0$.

Proposition 9. *Let X be a Banach space, Y a compact topological space, $D \in \mathrm{CoCl}(X)$, $0 \in C \subseteq D$ an open (in the relative topology) bounded subset of D, Q:* cl $(C) \to Y$ *a continuous mapping and G:* cl $(C) \times Y \to D$ *a mapping satisfying the following conditions:*

(i) *$G(., y)$ is a φ_y-contraction for each $y \in Y$ with φ_y continuous in addition.*

(ii) *$G(x, y)$ is continuous in $y \in Y$, uniformly with respect to $x \in$* cl (C);

(iii) *$G(x, y) \neq tx$ for each $x \in \partial_D C$, $y \in Y$ and $t > 1$.*

Then there exists a point $x \in$ cl (C) *such that $x = G(x, Qx)$.*

Proof. Since $G(.,y)$ is a φ_y-contraction, it lies in Dens $(\mathrm{cl}(C), \alpha_X)$ and hence it has a unique fixed point, say $f(y)$ in $\mathrm{cl}(C)$ (see proposition 8); that is, $f(y) = G(f(y), y)$. We shall show that the mapping $f\colon Y \to \mathrm{cl}(C)$ is continuous. Indeed, let $y, y' \in Y$ be given. Then:

$$\|fy - fy'\| = \|G(fy, y) - G(fy', y')\| \leqq \|G(fy, y) - G(fy', y)\| + \|G(fy', y) - G(fy', y')\| \leqq \varphi_y(\|fy - fy'\|) + \|G(fy', y) - G(fy', y')\|.$$

If (y_α) is a net in Y converging to y, then the net $(fy_\alpha) \subseteq \mathrm{cl}(C)$ is bounded and

$$0 \leqq \|fy_\alpha - fy\| - \varphi_y(\|fy_\alpha - fy\|) \leqq \|G(fy_\alpha, y_\alpha) - G(fy_\alpha, y)\| \underset{\alpha}{\to} 0\,, \qquad (*)$$

where we have used the condition (ii). Hence

$$\lim_\alpha [\|fy - fy_\alpha\| - \varphi_y(\|fy_\alpha - fy\|)] = 0\,,$$

and therefore, by lemma 17, the bounded net (fy_α) converges to fy, that is, f is continuous. Since Q is continuous and has relatively compact range, $f \circ Q$ is so $\left[\mathrm{cl}\left(f(Q(\mathrm{cl}(C)))\right) = \mathrm{cl}\left(f(\mathrm{cl}(Q(C)))\right) = f(\mathrm{cl}(Q(C)))\right.$, since f is continuous and $\mathrm{cl}\left(Q(C)\right)$ is compact$\left.\right]$. The mapping $f \circ Q$ maps $\mathrm{cl}(C)$ into itself, and, by Schauder fixed point theorem, it has a fixed point, say x_0, in $\mathrm{cl}(C)$. Then:

$$x_0 = f(Q(x_0)) = G(f(Q(x_0))\,, \qquad Q(x_0)) = G(x_0, Q(x_0))\,,$$

that is, the equation $x = G(x, Q(x))$ has a solution in $\mathrm{cl}(C)$.

Theorem 9. *Let (M, d) be a complete bounded metric space (nonempty), X a quasicomplete* LCS, *$D \in \mathrm{CoCl}(X)$, $0 \in C \subseteq D$ an open (in the relative topology) subset of D, $K\colon M \times \mathrm{cl}(C) \to D$ a continuous mapping, $H\colon M \times \mathrm{cl}(C) \to M$ a mapping continuous in the second variable uniformly with respect to the first one. Let us suppose that:*

(i) *$K(M \times C)$ is relatively compact;*
(ii) *$K(x, y) \neq ty$ for all $x \in M$, $y \in \partial_D C$, $t < 1$;*
(iii) *for each $y \in \mathrm{cl}(C)$, $H(.,y)\colon M \to M$ is a φ_y-contraction. Then the system:*

$$\begin{aligned} x &= H(x, y) \\ y &= K(x, y) \end{aligned} \qquad (1)$$

has a solution (x, y) in $M \times \mathrm{cl}(C)$.

Proof. For $y \in \mathrm{cl}(C)$, $H(.,y)\colon M \to M$ is a φ_y-contraction, and hence, by Boyd-Wong's fixed point theorem [5], it has a unique fixed point $f(y)$ in M. Similarly as in the proof of proposition 9 we obtain that for each $y, y' \in \mathrm{cl}(C)$:

$$0 \leqq d(fy', fy) - \varphi_y\left(d(fy', fy)\right) \leqq d\left(H(fy', y'), H(fy', y)\right),$$

and hence, by lemma 17 and the continuity of H in y (uniformly with respect to $x \in M$), this implies the continuity of $f\colon \mathrm{cl}(C) \to M$.

The mapping $h\colon \mathrm{cl}(C) \to \mathrm{cl}(C) \times M$, defined by $h(y) = (fy, y)$ is continuous, and so is the mapping $T = K \circ h\colon \mathrm{cl}(C) \to D$; the range $T(\mathrm{cl}(C)) \subseteq K(M \times \mathrm{cl}(C))$ is relatively compact and $Ty = K(fy, y) \neq ty$ for all $y \in \partial_D C$ and $t > 1$. Thus, by theorem 3, T has a fixed point, say y_0, in $\mathrm{cl}(C)$. Set $x_0 = fy_0$. Then $x_0 = fy_0 = H(fy_0, y_0) = H(x_0, y_0)$, and $y_0 = Ty_0 = K(fy_0, y_0) = K(x_0, y_0)$, i.e. (x_0, y_0) is a solution of the equation (1). (Here we have used the trivial fact that T is α_P-densifying, where P is a defining system of pseudonorms on X.)

Remarks. 1. The condition (iii) of theorem 9 is satisfied in each of the following cases (compare with considerations related to corollaries 1 and 2 to theorem 3):

(a) $K(M \times \mathrm{cl}(C)) \subseteq \mathrm{cl}(C)$ and C strictly D-starshaped with respect to 0;

(b) $K(M \times \partial_D C) \subseteq \mathrm{cl}(C)$ and C strictly D-starshaped with respect to 0;

(c) $\|\cdot\|$ a norm on the linear space Y and there is a system $\{\varphi_y\}_{y\in\partial_D C}$ of strictly superadditive functions from R^+ into R^+ such that:

$$\varphi_y(\|K(x, y)\|) \leqq \varphi_y(\|y\|) + \varphi_y(\|y - K(x, y)\|) \text{ for each } x \in M,\ y \in \partial_D C;$$

(d) $\|\cdot\|$ a norm on the linear space Y, $p\colon \partial_D C \to (1, +\infty)$ a function such that:

$$\|K(x, y)\|^{p(y)} \leqq \|y\|^{p(y)} + \|y - K(x, y)\|^{p(y)} \text{ for each } x \in M,\ y \in \partial_D C;$$

(e) $(.,.)$ is an inner product on the linear space Y and

$$(K(x, y), y) \leqq (y, y) \text{ for all } x \in M,\ y \in \partial_D C\ .$$

2. Theorem 9 remains valid, if $H\colon M \times \mathrm{cl}(C) \to Y$ is a mapping continuous in the second variable uniformly with respect to the first one, where Y is a NLS, and H satisfies one of the following conditions rather than the condition (iii) of theorem 9:

(a) $0 \in \mathrm{Int}\, M$, Y is reflexive, $M \in \mathrm{BoCoCl}(Y)$, and for each $y \in \mathrm{cl}(C)$, $H(x, y) \neq tx$ for all $t > 1$, $x \in \partial M$, and $H(., y)\colon M \to Y$ can be ecxtended to a *uniformly strictly contractive mapping* on M relative to X (let us remember — see W. A. Kirk [43, 44] — that a mapping T of a subset$N$ of a metric space (X, d) into X is uniformly strictly contractive on a subset M of N relative to N, iff for each $x \in N$ there is $\alpha(x) \in [0, 1)$ such that $d(Ty, Tx) \leqq \alpha(x)\, d(y, x)$ for all $y \in M$; T is called a *generalized contraction* iff it is uniformly strictly contractive on N relative to N);

(b) Y is a conjugate Banach space, M a (nonempty) weakly*-compact convex subset of Y, and for each $y \in \mathrm{cl}(C)$, $H(., y)\colon M \to M$ is a generalized contraction;

(c) Y as in (b), M a closed ball in Y, and for each $y \in \mathrm{cl}(C)$ $H(\partial M, y) \subseteq M$ and $H(., y)$ is a generalized contraction;

(d) Y as in (b), N a convex weakly*-closed subset of Y, M (nonempty) convex weakly*-compact subset of N, and for each $y \in \mathrm{cl}(C)$ $H(., y)$ is a generalized contraction with $H(M, y) \subseteq N$ and $H(\partial_N M, y) \subseteq M$;

(e) $\mathcal{T}$ is a Hausdorff topology on Y such that: 1) $(Y, \mathcal{T})$ is a linear topological space, 2) $S \subseteq Y$ convex $\mathcal{T}$-compact implies S (norm) closed, and 3) each ball in Y is $\mathcal{T}$-closed; $M \neq \emptyset$ is a convex $\mathcal{T}$-compact subset of Y, and for each $y \in \mathrm{cl}(C)$ $H(., y)\colon M \to M$ is a generalized contraction;

(f) Y and $\mathcal{T}$ as in (e), M is a $\mathcal{T}$-compact closed ball in Y, and for each $y \in \mathrm{cl}(C)$ $H(., y)$ is a generalized contraction with $H(\partial M, y) \subseteq M$.

Proof. As in the proof of theorem 9 but using W. A. Kirk [43, theorem 1], [44, theorem 1.1 and corollary 1.2], [45, theorem 3.1] and J. Reinermann [83, theorem 1 and corollary 1] respectively, rather than the Boyd-Wong's theorem, we obtain the existence of the mapping $f\colon \mathrm{cl}(C) \to M$ (i.e., $f = H / (f \times id)$). Since $H(., y)$ $(y \in \mathrm{cl}(C))$ is a generalized contraction, we have the inequality:

$$[1 - \alpha_y(fy)]\, ||fy' - fy|| \leqq ||H(fy', y') - H(fy', y)||$$

implying the continuity of f. The rest of the proof follows the corresponding part of the proof of theorem 9.

3. Similar considerations can be made concerning proposition 9. The results of N. A. Assad and W. A. Kirk [3] specialized to single-valued mappings can be used as well but we shall not delay on it. In certain cases, the above assertions can be derived directly from theorem 3 (or proposition 8). Indeed, let $i: M \to M \times \mathrm{cl}\,(C)$ and $j: \mathrm{cl}\,(C) \to M \times \mathrm{cl}\,(C)$ be the natural imbeddings, $F = H \times K: M \times \mathrm{cl}\,(C) \to M \times X$ (respectively $F: M \times \mathrm{cl}\,(C) \to Y \times X$), i.e. $F = i \circ H + j \circ K$. Then $\partial(M \times \mathrm{cl}\,(C)) = (\partial M \times \mathrm{cl}\,(C)) \cup (M \times \partial C)$ and $F \in \mathrm{Dens}\,(M \times \mathrm{cl}\,(C), \mu)$ for a suitable mnc μ in $Y \times X$, etc.

4. Proposition 9 remains valid if the conditions (i) and (ii) are replaced by:

(i') $G(., y)$ is a φ-contraction, for each y in Y, with φ continuous (φ is the same for all $y \in Y$);

(ii') $G(x, y)$ is continuous in $y \in Y$, for each $x \in \mathrm{cl}\,(C)$.

To prove this, it sufficies to transpose y_α and y in (*) of the proof of proposition 9 and repeat the proof.

5. Theorem 9 remains valid if $H: M \times \mathrm{cl}\,(C) \to M$ is supposed to be continuous in the second variable, for each fixed value of the first one rather than uniformly with respect to the first one, and the condition (iii) of theorem 9 is replaced by:

(iii') for each $y \in \mathrm{cl}\,(C)$, $H(., y): M \to M$ is a φ-contraction (φ is the same for all $y \in \mathrm{cl}\,(C)$).

The only change in the proof is of the same character as in the preceding remark.

6. It seems that theorems 1.1A and 1.1B of R. W. Melvin [57] require a small correction. In his notation, $Q(\Gamma)$ should be replaced by $\mathrm{cl}\,(Q(\Gamma))$. Compare with our proposition 9 and remark 4.

Lemma 18. *Let Λ be a compact topological space, X a* LCS, *$C \in \mathrm{Cl}\,(X)$, $F: \Lambda \times C \to X$ a continuous mapping such that $F \in \mathrm{AsCpt}\,(C)$. Let $\mathrm{Fix}\,(F) = \{x \in C: F(\lambda_x, x) = x$ for some $\lambda_x \in \Lambda\}$. Then the set $\mathrm{Fix}\,(F)$ is compact.*

Proof. It is easy to see that $\mathrm{Fix}\,(F) \subseteq F^\infty(\Lambda \times C)$ and therefore $\mathrm{Fix}\,(F)$ is relatively compact. It is sufficient to show that this set is closed. Let (x_α) be a net in $\mathrm{Fix}\,(F)$ converging to a point x. For each α there is $\lambda_\alpha \in \Lambda$ with $F(\lambda_\alpha, x_\alpha) = x_\alpha$. Since Λ is compact, there is a subset (λ_β) of (λ_α) converging to a point λ in Λ. Then the continuity of F implies that $F(\lambda_\beta, x_\beta) \to F(\lambda, x)$. Hence $F(\lambda, x) = x$ and $x \in \mathrm{Fix}\,(F)$.

Lemma 19. *Let D be a convex closed subset of a* LCS *X and C an open (in the relative topology) subset of D. Let $F: \mathrm{cl}\,(C) \to D$ be a continuous mapping such that $F \in \mathrm{AsCpt}\,(\mathrm{cl}\,(C))$ and $Fx \neq x$ for each $x \in \partial_D C$. Let x be an isolated fixed point of F in $\mathrm{cl}\,(C)$ and let C_1 and C_2 be two (relatively) open neighbourhoods of x such that $C_i \subseteq C$ and $\mathrm{cl}\,(C_i) \cap \mathrm{Fix}\,(F) = \{x\}$. Then*

$$\gamma_D(I - F, \mathrm{cl}\,(C_1)) = \gamma_D(I - F, \mathrm{cl}\,(C_2))\,.$$

Proof. Let C_0 be an open (in the relative topology) neighbourhood of the closed set $\mathrm{Fix}(F) \setminus \{x\}$ (see lemma 18), disjoint from both C_1 and C_2. Since $Fx \neq x$ for each $x \in \mathrm{cl}(C) \setminus (C_0 \cup C_i)$ $(i = 1, 2)$, we have by the property (iv) of the rotation:

$$\gamma_D(I - F, \mathrm{cl}(C)) = \gamma_D(I - F, \mathrm{cl}(C_0)) + \gamma_D(I - F, \mathrm{cl}(C_i)) \quad (i = 1, 2) .$$

This equality implies the desired one.

Let X be a LCS, $D \in \mathrm{CoCl}(X)$, and C an (relatively) open subset of D. Let $F\colon \mathrm{cl}(C) \to D$ be a continuous mapping such that $F \in \mathrm{AsCpt}(\mathrm{cl}(C))$ and $Fx \neq x$ for all $x \in \partial_D C$. Let $C_1 \subseteq C$ be an (relatively) open neighbourhood of an isolated fixed point x of F such that $C_1 \cap \mathrm{Fix}(F) = \{x_0\}$. Then we define:

$$\mathrm{ind}_D(x, F) = \gamma_D(I - F, \mathrm{cl}(C_1)) .$$

This definition is correct by lemma 19.

Proposition 10. *Let X be a* LCS, *$D \in \mathrm{CoCl}(X)$, C an (relatively) open subset of D, $F\colon \mathrm{cl}(C) \to D$ a continuous mapping such that $F \in \mathrm{AsCpt}(\mathrm{cl}(C))$ and $Fx \neq x$ for each $x \in \partial_D C$. If the set $\mathrm{Fix}(F)$ is discrete, then it is finite and*

$$\gamma_D(I - F, \mathrm{cl}(C)) = \sum_{x \in \mathrm{Fix}(F)} \mathrm{ind}_D(x, F) .$$

Proof. Since, by lemma 18, $\mathrm{Fix}(F)$ is a discrete compact space, it is finite. Take disjoint (relatively) open neighbourhoods $C_x \subseteq C$ of $x \in \mathrm{Fix}(F)$. Then the property (iv) of the rotation implies:

$$\gamma_D(I-F, \mathrm{cl}(C)) = \sum_{x \in \mathrm{Fix}(F)} \gamma_D(I-F, \mathrm{cl}(C_x)) = \sum_{x \in \mathrm{Fix}(F)} \mathrm{ind}_D(x, F) .$$

Corollary 1. *Let X, D, C, F be as in proposition* 10. *If $x_1, \ldots, x_n$ are isolated fixed points of F, and*

$$\gamma_D(I-F, \mathrm{cl}(C)) \neq \sum_{i=1}^{n} \mathrm{ind}_D(x_i, F) , \qquad (*)$$

then F has a fixed point in C different from x_i, $i = 1, \ldots, n$.

Remark. Clearly, the condition (*) is implied by the following one:

$$\gamma_D(I-F, \mathrm{cl}(C)) \neq \sum_{i=1}^{n} \mathrm{ind}_D(x_i, F) \pmod m ,$$

where m is an integer; especially the case $m = 2$ is used very frequently.

We shall use the following generalization of the multiplicative property of the rotation due to Yu. G. Borisovič ("On the relative rotation of compact vector fields in linear spaces", Trudy seminara po funkc. anal., Vyp. 12, Voronež, 1969, pp. 3—27).

Proposition 11. *Let X_i be* LCS's, *$D_i \in \mathrm{CoCl}(X_i)$, C_i an (relatively) open subset of D_i, $F_i\colon \mathrm{cl}(C_i) \to D_i$ a continuous compact mapping with $F_i(x_i) \neq x_i$ for all $x_i \in \partial_{D_i} C_i$, $i = 1, 2$. Let $X = X_1 \times X_2$, $D = D_1 \times D_2$, $C = C_1 \times C_2$ and $F = F_1 \times F_2$, i.e. $F(x_1, x_2) = (F_1(x_1), F_2(x_2))$. Then $Fx \neq x$ for each $x \in \partial_D C$ and*

$$\gamma_D(I - F, \mathrm{cl}(C)) = \gamma_{D_1}(I - F_1, \mathrm{cl}(C_1)) \cdot \gamma_{D_2}(I - F_2, \mathrm{cl}(C_2)) . \qquad (*)$$

Now, it is easy to extend this proposition to the case of asymptotically compact mappings.

Proposition 12. *Let X_i, D_i, C_i, be as in proposition* 11 *and F_i: $\mathrm{cl}(C_i) \to D_i$ a continuous mapping such that $F_i \in \mathrm{AsCpt}\,(\mathrm{cl}\,(C_i))$ and $F_i(x_i) \neq x_i$ for each $x_i \in \partial_{D_i} C_i$ $(i = 1, 2)$. Let X, D, C, F be defined as proposition* 11. *Then $Fx \neq x$ for each $x \in \partial_D C$ and the equality* (*) *holds.*

Proof. Let $E_i = F_i^\infty(\mathrm{cl}\,(C_i))$, $G_i = (I - F_i)|_{\mathrm{cl}(C_i \cap E_i)}$, $E = E_1 \times E_2$, $G = G_1 \times G_2$. It is easy to see that $F^\infty(\mathrm{cl}\,(C)) = E$ and $(I - F)|_E = G$. We have by the property (i) of the rotation that $\gamma_D(I - F), \mathrm{cl}\,(C)) = \gamma_E(G, \mathrm{cl}\,(E \cap C))$ and $\gamma_{D_i}(I - F_i, \mathrm{cl}\,(C_i)) = \gamma_{E_i}(G_i, \mathrm{cl}\,(E_i \cap C_i))$. This and the equality (*) of proposition 11 applied to $\{I - G_i, E_i, C_i, I - G, E, C\}$ rather than to $\{F_i, D_i, C_i, F, D, C\}$ completes the proof.

If $F_i \in \mathrm{Dens}\,(\mathrm{cl}(C_i), \mu_i)$ for some mnc's μ_i in X_i, then it is possible to define a mnc in X such that $F \in \mathrm{Dens}\,(\mathrm{cl}\,(C), \mu)$. More precisely, the following proposition is true.

Proposition 13. *Let X_i, D_i, C_i be as in proposition* 11 *and F_i: $\mathrm{cl}\,(C_i) \to D_i$ a mapping such that $F_i \in \mathrm{Dens}\,(\mathrm{cl}\,(C_i), \mu_i)$, where μ_i is a monotone* mnc *in X_i. Let us define $\mathfrak{M}_\mu = \exp\,(\mathrm{cocl}\,(C)) \cup \exp\,(\mathrm{cocl}\,(F\,(\mathrm{cl}\,(C))))$ and $A_\mu = A_{\mu_1} \times A_{\mu_2}$, where the order in A_μ is defined as follows: $a \leqq b$ in A_μ iff $a_i \leqq b_i$ $(i = 1, 2)$ (here $a = (a_1, a_2)$, $b = (b_1, b_2)$), and $\mu(M) = (\mu_1\,(\mathrm{proj}_{X_1} M), \mu_2\,(\mathrm{proj}_{X_2} M))$ for each $M \in \mathfrak{M}_\mu$. Then μ is a monotone* mnc *in X and $F \in \mathrm{Dens}\,(\mathrm{cl}\,(C), \mu)$.*

Proof. Firstly, we have to prove that μ is a mnc. Clearly, $\mathfrak{M}_\mu$ is closed under the operation cocl. Let $M \in \mathfrak{M}_\mu$. Then

$$\begin{aligned}\mu(\mathrm{cocl}\,M) &= (\mu_1(\mathrm{proj}_{X_1}\,\mathrm{cocl}\,M), \mu_2(\mathrm{proj}_{X_2}\,\mathrm{cocl}\,M))\\ &= (\mu_1(\mathrm{cocl}\,\mathrm{proj}_{X_1}\,M), \mu_2(\mathrm{cocl}\,\mathrm{proj}_{X_2}\,M))\\ &= (\mu_1(\mathrm{proj}_{X_1}\,M), \mu_2(\mathrm{proj}_{X_2}\,M)) = \mu(M)\,.\end{aligned}$$

Here we have used the equality $\mu_i(\mathrm{proj}_{X_i}\,\mathrm{cocl}\,M) = \mu_i(\mathrm{cocl}\,\mathrm{proj}_{X_i}\,M)$, which follows from the monotonicity of μ_i, the continuity of proj_{X_i} and the following inclusions:

$$\mathrm{proj}_{X_i}\,M \subseteq \mathrm{proj}_{X_i}\,(\mathrm{cocl}\,M) \subseteq \mathrm{cl}\,(\mathrm{proj}_{X_i}\,\mathrm{co}\,M) = \mathrm{cocl}\,(\mathrm{proj}_{X_i}\,M)\,.$$

The proof of the monotonicity of μ can be left to the reader.

Now, we shall prove that $F \in \mathrm{Dens}\,(\mathrm{cl}\,(C), \mu)$. Let $M \subseteq \mathrm{cl}\,(C)$ and $\mu(F(M)) \geqq \mu(M)$. Then, by the definition of μ, we have

$$\mu_i(\mathrm{proj}_{X_i}\,F(M)) \geqq \mu_i(\mathrm{proj}_{X_i}\,M)\,.$$

Since $\mathrm{proj}_{X_i}\,F(M) = F_i(\mathrm{proj}_{X_i}\,M)$, we have

$$\mu_i(F_i(\mathrm{proj}_{X_i}\,M)) \geqq \mu_i(\mathrm{proj}_{X_i}\,M) \quad (i = 1, 2)\,.$$

This implies that $\mathrm{proj}_{X_i}\,M$ $(i = 1, 2)$ are relatively compact. Since

$$M \subseteq \mathrm{cl}\,(\mathrm{proj}_{X_1}\,M) \times \mathrm{cl}\,(\mathrm{proj}_{X_2}\,M)\,,$$

M is relatively compact. Therefore $F \in \mathrm{Dens}\,(\mathrm{cl}\,(C), \mu)$.

Theorem 10. *L et X_i be a metrizable* LCS, *$D_i \in \mathrm{CoCl}\,(X_i)$, C_i an (relatively) open subset of D_i, F_i: $\mathrm{cl}\,(C_i) \to D_i$ a continuous mapping such that $F_i(x_i) \neq x_i$ for each $x_i \in \partial_{D_i} C_i$ and G_i: $\mathrm{cl}\,(C_1) \times \mathrm{cl}\,(C_2) \to D_i$ a continuous mapping $(i = 1, 2)$.*

Let μ_i *be an algebraically subadditive monotone* mnc *in* X_i *such that:*
(i) S_i *a countable subset of* $\partial_{D_i} C_i$, K_i *a countable relatively compact subset of* $(I - F_i)(\partial_{D_i} C_i)$ *implies* K_i, $F_i(S_i) + K_i \in \mathfrak{M}_{\mu_i}$ *and* $\mu_i(K_i) = 0$ $(i = 1, 2)$;
(ii) $M_i \subset \mathrm{cl}\,(C_i)$ *implies* $\exp\,[\mathrm{cocl}\,(F_i(M_i) \cup \varepsilon\, G_i(M_i) \cup \{0\})] \subseteq \mathfrak{M}_{\mu_i}$ *for small* $\varepsilon > 0$ $(i = 1, 2)$.
Assume that $\mu_i(\{0\} \cup G_i(\mathrm{cl}\,(C_1) \times \mathrm{cl}\,(C_2))) = 0$ *and that the sets*

$$G_i(\mathrm{cl}\,(C_1) \times \partial_{D_2} C_2) \quad \text{and} \quad G_i(\partial_{D_1} C_1 \times \mathrm{cl}\,(C_2)) \quad (i = 1, 2)$$

are bounded. If $F_i \in \mathrm{Dens}\,(\mathrm{cl}\,(C_i), \mu_i)$ *and* $\gamma_{D_i}(I - F_i, \mathrm{cl}\,(C_i)) \neq 0$ $(i = 1, 2)$, *then there is a positive number* ε_0 *such that the system*

$$\begin{aligned} x_1 &= F_1(x_1) + \varepsilon_1\, G_1(x_1, x_2) \\ x_2 &= F_2(x_2) + \varepsilon_2\, G_2(x_1, x_2)\,, \end{aligned} \tag{1}$$

has at least one solution in $\mathrm{cl}\,(C_1) \times \mathrm{cl}\,(C_2)$ *for each* $\varepsilon_1, \varepsilon_2 \in [0, \varepsilon_0]$.

Proof. First of all, we prove the following lemma.

Lemma 20. *The set* $(I - F_i)(\partial_{D_i} C_i)$ *is closed in* D_i (*and in* X_i, clearly), $i = 1, 2$.

Proof of the lemma. Let $x_n \in \partial_{D_i} C_i$ be such that $x_n - F_i(x_n) = y_n \to z_i$ as $n \to +\infty$. If we denote $S_i = \{x_n\}_{n=1}^{\infty}$ and $K_i = \{y_n\}_{n=1}^{\infty}$, we have $S_i \subseteq F_i(S_i) + K_i$, and hence

$$\mu_i(S_i) \leqq \mu_i(F_i(S_i) + K_i) \leqq \mu_i(F_i(S_i)) + \mu_i(K_i) = \mu_i(F_i(S_i))\,.$$

Since F_i is μ_i-densifying, the set S_i is relatively compact. Then a subsequence of $\{x_n\}$, which we denote by $\{x_n\}$ again, converges to a point x. From the continuity of F_i and $y_n \to z_i$ it follows that $x - F_i(x) = z_i$, i.e. $z_i \in (I - F_i)(\partial_{D_i} C_i)$, because $x \in \partial_{D_i} C_i$; the set $(I - F_i)(\partial_{D_i} C_i)$ is closed $(i = 1, 2)$.

Proof of the theorem completed. Since the set $(I - F_i)(\partial_{D_i} C_i)$ is closed by lemma 20 and does not contain the origin, three is a convex neighbourhood U_i of 0 in X_i such that $(I - F_i)(\partial_{D_i} C_i) \cap U_i = \emptyset$, i.e.

$$x_i - F_i(x_i) \notin U_i \text{ for each } x_i \in \partial_{D_i} C_i \quad (i = 1, 2)\,.$$

This implies immediately that

$$x - F\,x \notin U = U_1 \times U_2 \quad \text{for each } x \in \partial_D C\,,$$

where $F(x_1, x_2) = (F_1(x_1), F_2(x_2))$. Since the sets

$$G_i(\mathrm{cl}\,(C_1) \times \partial_{D_2} C_2)\,, \qquad G_i(\partial_{D_1} C_1 \times \mathrm{cl}\,(C_2))$$

are bounded, there is a positive number a such that these sets are contained in aU_i $(i = 1, 2)$. Let $\varepsilon_0 = a^{-1}$ and $\varepsilon_i \in [0, \varepsilon_0]$ $(i = 1, 2)$. Let us define G: $\mathrm{cl}\,(C) \to D$ by

$$G(x) = (\varepsilon_1 G_1(x), \varepsilon_2 G_2(x))\,,$$

where we use the notation of proposition 11. From the definition of the number a it follows that

$$tG(x) \in U = U_1 \times U_2 \text{ for each } (t, x) \in [0, 1] \times \partial_D C\,.$$

Hence $H(t, x) = Fx + tGx \neq x$ for each $(t, x) \in [0, 1] \times \partial_D C$.

Now, we shall prove that H is an admissible homotopy between F and $F + G$. It is sufficient to prove that $H \in \operatorname{Dens}\big([0, 1] \times \operatorname{cl}(C), \mu\big)$. Let M be a subset of $\operatorname{cl}(C)$. Then

$$\mu\big(H([0, 1] \times M)\big) \leqq \mu\big(F(M)\big) + \mu\,\big(\operatorname{co}(\{0\} \cup G(M))\big)\;;$$

but we have

$$\begin{aligned} 0 &\leqq \mu_i\big(\operatorname{proj}_{X_i} \operatorname{co}(\{0\} \cup G(M))\big) = \mu_i\big(\operatorname{co} \operatorname{proj}_{X_i}(\{0\} \cup G(M))\big) \\ &= \mu_i\big(\operatorname{proj}_{X_i}(\{0\} \cup G(M))\big) = \mu_i\big(\{0\} \cup \varepsilon_i\, G_i(M)\big) \\ &\leqq \mu_i\big(\{0\} \cup \varepsilon_i G_i(\operatorname{cl}(C))\big) = 0\,. \end{aligned}$$

Hence $\mu\big(H([0, 1] \times M)\big) \geqq \mu(M)$ implies that $\mu\big(F(M)\big) \geqq \mu(M)$; thus M is relatively compact, since $F \in \operatorname{Dens}\big(\operatorname{cl}(C), \mu\big)$ by proposition 13.

We have proved that the homotopy H between F and $F + G$ is admissible, and we can conclude that

$$\gamma_D\big(I - F - G, \operatorname{cl}(C)\big) = \gamma_D\big(I - F, \operatorname{cl}(C)\big)\,.$$

But, by proposition 12,

$$\gamma_D\big(I - F, \operatorname{cl}(C)\big) = \gamma_{D_1}\big(I - F_1, \operatorname{cl}(C_1)\big) \cdot \gamma_{D_2}\big(I - F_2, \operatorname{cl}(C_2)\big) \neq 0\,.$$

Therefore $\gamma_D\big(I - F - G, \operatorname{cl}(C)\big) \neq 0$ and the equation $x = Fx + Gx$ has at least one solution in C, i.e. the system (1) has a solution.

Remark. The hypotheses on μ_i simplify if μ_i is supposed to be regular and G_i completely continuous.

Theorem 11. *Let X be a (real) Banach space, $A\colon X \to X$ a continuous linear mapping and $r_\Phi(A)$ the Fredholm radius of A. Let B be a closed ball in X centered at* 0, *and β a real number such that $|\beta|^{-1} > r_\Phi(A)$ and $\beta^{-1} \notin \sigma_p(A)$. If $\beta A \in \operatorname{AsCpt}(B)$, then*

$$\gamma(I - \beta A, B) = (-1)^{m(A,\beta)} = \operatorname{ind}(0, \beta A)\,,$$

where $m(A, \beta) = \sum_{\lambda \in \sigma} \dim E(\lambda, A)$, $\sigma = \{\lambda \in \sigma_p(A) : |\lambda| > |\beta|^{-1}, \operatorname{sign} \beta = \operatorname{sign} \lambda\}$, and $E(\lambda, A)$ denotes the eigenfunction subspace of A corresponding to the eigenvalue λ.

Proof. Since the set σ is finite (well-known), the subspace $X_1 = \sum_{\lambda \in \sigma} E(\lambda, A)$ is finite dimensional and $\dim X_1 = \sum_{\lambda \in \sigma} \dim E(\lambda, A)$. Let $X = X_1 \oplus X_2$ where X_2 is an appropriate closed linear subspace of X which is invariant with respect to A and denote $B_i = B \cap X_i$, $S_i = \partial B_i$, $B_0 = B_1 \times B_2$ and $A_i = A|_{X_i}$.

It is easy to see that the mapping $H_1\colon [0, 1] \times B_1 \to X_1$ defined by

$$H_1(t, x) = t\beta A_1(x) + 2(1 - t)\, x \quad \text{for} \quad (t, x) \in [0, 1] \times B_1$$

is an admissible homotopy between βA_1 and $2\,I$. Hence we can write

$$\gamma(I - A_1, B_1) = \gamma(I - 2\,I, B_1) = \gamma(-I, B_1) = (-1)^{\dim X_1} = (-1)^{m(A,\beta)}\,.$$

Now, let us define a mapping $H_2\colon [0, 1] \times B_2 \to X_2$ as follows:

$$H_2(t, x) = t\beta A_2(x) \quad \text{for} \quad (t, x) \in [0, 1] \times B_2\,.$$

We shall show that H_2 is an admissible homotopy between O and βA_2. Since A_2 has no eigenvalues λ with $|\lambda| > |\beta|^{-1}$ and $\operatorname{sign}\lambda = \operatorname{sign}\beta$, we have $H_2(t, x) \neq x$ for all $(t, x) \in [0, 1] \times \partial B_2$. It is easy to verify that $H_2 \in \operatorname{AsCpt}([0, 1] \times B_2)$. (Indeed, it is possible to derive that $H_2^\infty([0, 1] \times B_2) \subseteq (\beta A_2)^\infty (B_2) \subseteq (\beta A)^\infty(B)$). Hence $\gamma(I - \beta A_2, B_2) = 1$.

Since $I - \beta A$ has no fixed points in $X \setminus \{0\}$, we have by the property (iv) of the rotation and proposition 12:

$$\gamma(I - \beta A, B) = \gamma(I - \beta A, B_0) = \gamma(I - \beta A_1, B_1) \cdot \gamma(I - \beta A_2, B_2) = (-1)^{m(A,\beta)}.$$

Remark. It is easy to prove the following inclusion:

$$(\beta A)^\infty (B) \subseteq \bigcap_{n=0}^{\infty} (\beta A)^n (B).$$

If $\beta A \in \operatorname{AsCpt}(B)$ for some ball B centered at 0, then this is true for each ball at 0.

Corollary. *Let X be a (real) Banach space, $A: X \to X$ a continuous linear mapping and μ a* mnc *as in lemma* 12. *If $A \in \operatorname{Bo}(X, k, \mu, \mu)$ and $\beta \neq 0$ is a real number such that $|\beta|\, k < 1$ and $\beta^{-1} \notin \sigma_p(A)$, then*

$$\gamma(I - \beta A, B) = (-1)^{m(A,\beta)} \qquad (= \operatorname{ind}(0, \beta A))$$

for each closed ball B in X centered at 0.

Proof. Clearly, $\beta A \in \operatorname{Dens}(B, \mu) = \operatorname{AsCpt}(B)$, and $r_\Phi(A) \leqq k$ by lemma 12. The assertion follows from theorem 11.

Remarks. 1. Let $A: X \to X$ be a continuous linear mapping in a Banach space X. Then $A \in \operatorname{AsCpt}(B)$ (B any ball at 0), if $A^n = K + L$ for some completely continuous linear mapping K, a linear mapping L with $|||L||| < 1$, where $|||\cdot|||$ is a norm in X equivalent with the original one in X, and a positive integer n. Indeed, let χ be the Hausdorff mnc in $(X, |||\cdot|||)$. Then, clearly, $||A^n||^\chi \leqq |||L|||$ and hence $r_\Phi(A) \leqq |||L|||^{1/n} < 1$.

2. Therefore $A \in \operatorname{AsCpt}(B)$ holds if, for example, A^n is completely continuous for some positive integer n.

Theorem 12. *Let X be a Banach space and μ a* mnc *in X as in lemma* 14. *Let C be an open bounded subset of X, $F: C \to X$ a continuous mapping which is Fréchet differentiable at a point x in C. Assume that $A = F'(x)$, $1 \notin \sigma_p(A)$ and $F \in \operatorname{Bo}(C, k, \mu, \mu)$ for some $k < 1$. Then x is an isolated fixed point of F and*

$$\operatorname{ind}(x, F) = (-1)^{m(A,1)} \qquad (= \operatorname{ind}(0, A)).$$

Proof. By lemma 14, we have $1 \in \varrho(A)$, the resolvent set of A. Let $a = ||(I - A)^{-1}||^{-1}$. Then there is a ball $B = B_X(0, r)$ in X such that $x + B \subseteq C$ and

$$y \in x + B \quad \text{implies} \quad ||Fy - Fx - A(y - x)|| \leqq \frac{a}{2} ||y - x||. \tag{1}$$

This inequality easily implies that the following implication holds:

$$y \in x + B \quad \text{implies} \quad ||Fy - y|| \geqq \frac{a}{2} ||y - x||, \tag{2}$$

i.e. x is the only fixed point of F in $x + B$.

Now, we shall show that the mapping $H: [0, 1] \times (x + B) \to X$ defined by $H(t, y) = tFy + (1 - t)\big(x + A(y - x)\big)$ is an admissible homotopy between F and A_1, where A_1 is defined as follows: $A_1(y) = x + A(y - x)$. Suppose that $H(t, y) = y$ for some $(t, y) \in [0, 1] \times (x + \partial B)$. Then $t\big(Fy - x - A(y - x)\big) = A(y - x) - (y - x)$, i.e.

$$t\big(Fy - Fx - A(y - x)\big) = A(y - x) - (y - x)\,, \tag{3}$$

since $x = Fx$. From (3) we obtain

$$t\|Fy - Fx - A(y - x)\| = \|A(y - x) - (y - x)\|\,,$$

which together with (1), (2) and $\|y - x\| = r$ gives

$$\frac{a}{2} \cdot r \geqq t \cdot \frac{a}{2} \cdot r \geqq ar$$

a contradiction. To show that H is admissible it is sufficient to prove that $H \in \mathrm{Bo}\big([0, 1] \times (x + B), k, \mu, \mu\big)$. This follows from lemma 7, since $A_1 \in \mathrm{Bo}\,(X, k, \mu, \mu)$. Hence $\gamma(I - F, x + B) = \gamma(I - A_1, x + B)$. Since $\gamma(I - A_1, x + B) = \gamma(I - A, B)$, the theorem now follows immediately from the corollary above.

Corollary 1. *Let X, C and μ be as in theorem* 12. *Let $F\colon \mathrm{cl}\,(C) \to X$ be a continuous mapping such that $F \in \mathrm{Bo}\,\big(\mathrm{cl}\,(C), k, \mu, \mu\big)$ with $k < 1$. Suppose that F has only a finite number of fixed points in C, say $x_1, \ldots, x_n$, and $Fx \neq x$ for all $x \in \partial C$. Let G be Fréchet differentiable at each point x_i and $A_i = F'(x_i)$. If $1 \notin \sigma_n(A_i)$ for $i = 1, \ldots, n$, then*

$$\gamma\big(I - F, \mathrm{cl}\,(C)\big) = \sum_{i=1}^{n} (-1)^{m(A_i, 1)}\,.$$

Therefore, for example, $\gamma\big(I - F, \mathrm{cl}(C)\big) = n \pmod 2$ (which is usually used in applications).

Proof. See theorem 12 and proposition 10.

Corollary 2. *Let X, C and μ be as in theorem* 12. *Let $F\colon \mathrm{cl}\,(C) \to X$ be a continuous mapping such that $F \in \mathrm{Bo}\,\big(\mathrm{cl}\,(C), k, \mu, \mu\big)$ with $k < 1$. Let $x_1, \ldots, x_n$ be fixed points of F in C and suppose that $Fx \neq x$ for all $x \in \partial C$. If F is Fréchet differentiable at each x_i and $1 \notin \sigma_p\big(F'(x_i)\big)$, $i = 1, \ldots, n$, and*

$$\gamma\big(I - F, \mathrm{cl}\,(C)\big) \neq n \pmod 2\,,$$

then F has at least one fixed point in C different from $x_1, \ldots, x_n$.

Proof. See Corollary 1.

Remark. Corollary 1 can be used in theorem 10. For example, if n is odd, then $\gamma\big(I - F, \mathrm{cl}\,(C)\big) \neq 0$.

Proposition 14. *Let X and μ be as in theorem* 8, *and $F\colon X \to X$ an asymptotically Fréchet differentiable continuous mapping with $F \in \mathrm{Bo}\,(X, k, \mu, \mu)$ for some $k < 1$. If 1 is not an eigenvalue of $A = F'_{as}$, then there is $r_0 > 0$ such that, for each $r \geqq r_0$, $\gamma\big(I - F, B_X(0, r)\big)$ is defined and equals to $\gamma\big(I - A, B_X(0, r)\big) = (-1)^{m(A, 1)}\ \big(= \mathrm{ind}\,(0, A)\big)$.*

Proof. By lemma 15, we have $1 \in \varrho(A)$ (= the resolvent set of A). Let $a = ||(I - A)^{-1}||^{-1}$. Hence: $||x - F'_{as}(x)|| \geqq a||x||$ for all x in X. Let $r_0 > 0$ be so large that $||Fx - F'_{as}(x)|| \leqq \frac{a}{2}||x||$ if $||x|| \geqq r_0$. Then for all $t \in [0, 1]$ and $x \in X$, $||x|| \geqq r_0$, we have $||x - tFx - (1 - t) F'_{as}(x)|| \geqq ||x - F'_{as}(x)|| - ||Fx - F'_{as}(x)|| \geqq a||x|| - \frac{a}{2}||x|| \geqq \frac{a}{2}||x|| \geqq \frac{a}{2} \cdot r_0 > 0$. Let $r \geqq r_0$ and set $C = \operatorname{Int} B_X(0, r)$, $D = X$, $G = A = F'_{as}$ and $C(x) = [Fx, F'_{as}(x)] = \{tFx + (1 - t) F'_{as}(x): t \in [0, 1]\}$ in lemma 11. We conclude

$$\gamma(I - F, B_X(0, r)) = \gamma(I - A, B_X(0, r)) = \operatorname{ind}(0, A) = (-1)^{m(A, 1)}$$

(the last equality follows from theorem 11).

The following proposition can be considered as a generalization of a result of H. Amann [1].

Proposition 15. *Let X be a Banach space and μ a* mnc *in X as in theorem* 8, *$F: X \to X$ an asymptotically Fréchet differentiable continuous mapping with $F \in \operatorname{Bo}(X, k, \mu, \mu)$ for some $k < 1$, and $x_1, \dots, x_n (n \geqq 1)$ different fixed point of F. Assume F is Fréchet differentiable at each x_i. If*

$$1 \notin \sigma_p(F'_{as}) \cup \bigcup_{i=1}^{n} \sigma_p F'(x_i)$$

and n is even, then F has at least one fixed point in X different from $x_1, \dots, x_n$.

Proof. Let r_0 satisfy the conclusion of proposition 14. Then $\gamma(I - F, B_X(0, r_0)) = 1 \pmod 2$, and the proposition follows from corollary 2 to theorem 12.

Theorem 13. *Let X, C and μ be as theorem* 12. *Let $F: [\lambda_0, \lambda_0 + h] \times C \to X$ be a continuous mapping such that $F \in \operatorname{Bo}([\lambda_0, \lambda_0 + h] \times C, k, \mu, \mu)$ where $\lambda_0 \in R$, $h > 0$ and $k < 1$. Suppose that for each $\lambda \in [\lambda_0, \lambda_0 + h]$ $F(\lambda, .)$ is Fréchet differentiable in C and let the Fréchet derivative of $F(\lambda, .)$ at $x \in C$ be denoted by $F'(\lambda, x)$. Suppose that $F': [\lambda_0, \lambda_0 + h] \times C \to B(X)$ is continuous at (λ_0, x_0), where $x_0 \in C$ is fixed and $B(X)$ denotes the Banach space of all continuous linear mappings of X with the usual norm. Let x_0 be a fixed point of $F(\lambda_0, .)$ and $1 \notin \sigma_p(F'(\lambda_0, x_0))$. Furthermore, assume that for each closed ball B centered at x_0 $F(., x)$ is continuous at λ_0 uniformly with respect to $x \in \partial B$.*

Then there exists a closed ball B centered at x_0 and a number $\varepsilon_0 > 0$ such that for each $\lambda \in [\lambda_0, \lambda_0 + \varepsilon_0]$ there is exactly one fixed point $x(\lambda)$ of the mapping $F(\lambda, .)$ in B. Moreover, $x(\lambda) \to x_0$ as $\lambda \to \lambda_0^+$.

Proof. It is easy to see that $1 \in \varrho(F'(\lambda_0, x_0))$. Let $a = ||(I - F'(\lambda_0, x_0))^{-1}||^{-1}$. By theorem 12, x_0 is an isolated fixed point of $F(\lambda_0, .)$ and

$$\operatorname{ind}(x_0, F(\lambda_0, .)) = \operatorname{ind}(0, F'(\lambda_0, x_0)) = j = \pm 1 . \tag{1}$$

Since F' is continuous at (λ_0, x_0), there are $\varepsilon_2 > 0$ and $r_2 > 0$ such that

$$||F'(\lambda, x) - F'(\lambda_0, x_0)|| < a \text{ for each } (\lambda, x) \in [\lambda_0, \lambda_0 + \varepsilon_2] \times B_X(x_0, r_2) . \tag{2}$$

We know that x_0 is an isolated fixed point of $F(\lambda_0, .)$ and hence there is $r_3 > 0$ such that

$$F(\lambda_0, x) \neq x \quad \text{for each } x \in B_X(x_0, r_3) \setminus \{x_0\} . \tag{3}$$

Let $B = B_X(x_0, r)$, where $r = \min\{r_1, r_2, r_3\}$. Since, by (3), $0 \notin (I - F)(\lambda_0, .)(\partial B)$ and the last set is closed by lemma 20, we have

$$d = \inf\{||x - F(\lambda_0, x)||: x \in \partial B\} > 0 . \tag{4}$$

Since $F(\lambda, x)$ is continuous at λ_0 uniformly with respect to $x \in \partial B$, there is $\varepsilon_3 > 0$ such that

$$||F(\lambda, x - F(\lambda_0, x)|| < d \quad \text{for all } (\lambda, x) \in [\lambda_0, \lambda_0 + \varepsilon_3] \times \partial B . \tag{5}$$

Set $\varepsilon_0 = \min\{\varepsilon_1, \varepsilon_2, \varepsilon_3\}$.

Let $\lambda \in [\lambda_0, \lambda_0 + \varepsilon_0]$ be given and let the mapping $H: [0, 1] \times B \to X$ be defined by $H(t, x) = F(\lambda_0 + t(\lambda - \lambda_0), x)$. Suppose that $H(t, x) = x$ for some $(t, x) \in [0, 1] \times \partial B$. Then $F(\lambda_0 + t(\lambda - \lambda_0), x) - F(\lambda_0, x) = x - F(\lambda_0, x)$, and by (4), (5) we obtain:

$$d > ||F(\lambda_0 + t(\lambda - \lambda_0), x) - F(\lambda_0, x)|| = ||x - F(\lambda_0, x)|| \geqq d ,$$

a contradiction. Since $H([0, 1] \times M) \subseteq F([\lambda_0, \lambda_0 + \varepsilon_0] \times M) \subseteq F([\lambda_0, \lambda_0 + h] \times M)$ for each $M \subseteq B$, μ is monotone and $F \in \mathrm{Bo}([\lambda_0, \lambda_0 + h] \times C, k, \mu, \mu)$ we have $H \in \mathrm{Bo}([0, 1] \times B, k, \mu, \mu)$. We have just proved that H is an admissible homotopy on B between $F(\lambda_0, .)$ and $F(\lambda, .)$, and hence, by (1) and Proposition 10:

$$\gamma(I - F(\lambda, .), B) = \gamma(I - F(\lambda_0, .), B) = \mathrm{ind}\,(0, F'(\lambda_0, x_0)) = j \neq 0 . \tag{6}$$

Thus, $F(\lambda, .)$ has fixed points in B. Let x be one of them. Since $||F'(\lambda, x) - F'(\lambda_0, x_0)|| < a$, by (2), we have $1 \in \varrho(F'(\lambda, x))$. By theorem 12, x is an isolated fixed point of $F(\lambda, x)$ and

$$\mathrm{ind}\,(x, F(\lambda, .)) = \mathrm{ind}\,(0, F'(\lambda, x)) . \tag{7}$$

Let us define a mapping $H: [0, 1] \times (B - x_0) \to X$ as follows:

$$H(t, z) = tF'(\lambda, x)(z) + (1 - t)\, F'(\lambda_0, x_0)(z) .$$

Suppose that $H(t, z) = z$ for some $(t, z) \in [0, 1] \times (\partial B - x_0)$. Then

$$t[F'(\lambda, x) - F'(\lambda_0, x_0)](z) = [I - F'(\lambda_0, x_0)](z)$$

and by (2) and the definition of a,

$$a||z|| \leqq ||z - F'(\lambda_0, x_0)(z)|| = t||F'(\lambda, x) - F'(\lambda_0, x_0)||\, ||z|| < a||z||$$

(since $z \neq 0$), a contradiction. We know that $F'(\lambda, x), F'(\lambda_0, x_0) \in \mathrm{Bo}(X, k, \mu, \mu)$, by proposition 6, and $H \in \mathrm{Bo}([0, 1] \times (B - x_0), k, \mu, \mu)$, by lemma 7. This means that H is an admissible homotopy on $B - x_0$ between $F'(\lambda_0, x_0)$ and $F'(\lambda, x)$ and we conclude that

$$\begin{aligned}\mathrm{ind}\,(0, F'(\lambda, x)) &= \gamma(I - F'(\lambda, x), B - x_0) = \gamma(I - F'(\lambda_0, x_0), B - x_0) \\ &= \mathrm{ind}\,(0, F'(\lambda_0, x_0)) = j .\end{aligned}$$

We have proved that each fixed point x of $F(\lambda, .)$ in B is isolated and

$$\mathrm{ind}\,(x, F(\lambda, .)) = j . \tag{8}$$

The set of fixed point of $F(\lambda, .)$ in B is discrete and, by lemma 18, compact. Hence it is finite. Let $x_1, \ldots, x_n$ be all fixed point of $F(\lambda, .)$ in B. By proposition 10, (8) and (6) we have

$$j = \gamma(I - F(\lambda, .), B) = \sum_{i=1}^{n} \operatorname{ind}(x_i, F)(\lambda, .)) = nj ,$$

that is, $n = 1$ (since $j \neq 0$). Thus, $F(\lambda, .)$ has exactly one fixed point, say $x(\lambda)$, in B.

It is easy to see from the considerations made above and the uniqueness of fixed points that $x(\lambda) \to x_0$ as $\lambda \to \lambda_0^+$.

Corollary 1. *If in addition to the hypotheses of theorem 13, F' is continuous on $[\lambda_0, \lambda_0 + h] \times C$ and for each $x \in C$, $\lambda_1 \in [\lambda_0, \lambda_0 + h]$ and each closed ball B centered at x, $F(\lambda, z)$ is continuous at λ_1 uniformly with respect to $z \in \partial B$, then the mapping $x\colon [\lambda_0, \lambda_0 + \varepsilon_0] \to B$ from theorem 13 is continuous.*

Corollary 2. *Let X, C and μ be as in theorem 12. Let $F, G\colon C \to X$ be two continuous mappings such that $F \in \mathrm{Bo}(C, k, \mu, \mu)$ and $G \in \mathrm{Bo}(C, m, \mu, \mu)$ for some $k < 1$ and $m \in R^+$. Let x_0 be a fixed point of F in C. Suppose that F and G are Fréchet differentiable in a neighbourhood of x_0 and both F' and G' are continuous at x_0. If $1 \notin \sigma_p(F'(x_0))$, then there is a closed ball B centered at x_0 and a number $\varepsilon_0 > 0$ such that for each $\lambda \in [-\varepsilon_0, \varepsilon_0]$, the mapping $F + \lambda G$ has only one fixed point $x(\lambda)$ in B. In addition, $x(\lambda) \to x_0$ as $\lambda \to 0$.*

If, in addition to the hypotheses made above, F and G are continuously Fréchet differentiable in a neighbourhood of x_0, then the mapping $x\colon [-\varepsilon_0, \varepsilon_0] \to B$ is continuous.

Proof. Let $\lambda_0 = 0$, $h \in \left(0, \dfrac{1-k}{1+m}\right)$, $k_1 = k + hm$ (< 1) and C_1 a neighbourhood of x_0 contained in C such that G and G' are bounded on C_1. Let us define $F^+, F^-\colon [\lambda_0, \lambda_0 + h] \times C_1 \to X$ as follows:

$$F^+(\lambda, x) = Fx + \lambda Gx , \quad F^-(\lambda, x) = Fx - \lambda Gx .$$

Now, it is easy to see that the quintuplets $\{\lambda_0, h, C_1, k_1, F^+\}$ and $\{\lambda_0, h, C_1, k_1, F^-\}$ satisfy the hypotheses of theorem 13 or corollary 1.

Corollary 3. *Let X, C and μ be as in theorem 12. Let $F\colon C \to X$ be a continuous mapping such that $F \in \mathrm{Bo}(C, k, \mu, \mu)$ for some $k \in R^+$. Let $\lambda_0 \in R$ be a number and $x_0 \in C$ an eigenvector of F corresponding to the characteristic value λ_0: $x_0 = \lambda_0 F x_0$. If $|\lambda_0|\, k < 1$, F is Fréchet differentiable in a neighbourhood of x_0, F' is continuous at x_0 and $\lambda_0^{-1} \notin \sigma_p(F'(x_0))$ (this has a sense also in the case when $\lambda_0 = 0$; then 0^{-1} is not defined and hence the condition is satisfied), then there is a closed ball B centered at x_0 and $\varepsilon_0 > 0$ such that to each $\lambda \in [\lambda_0 - \varepsilon_0, \lambda_0 + \varepsilon_0]$ there is exactly one eigenvector $x(\lambda)$ of F in B corresponding to the characteristic value λ: $x(\lambda) = \lambda F(x(\lambda))$. Moreover, $x(\lambda)$ is continuous at λ_0.*

If, in addition, F is continuously Fréchet differentiable in a neighbourhood of x_0, then the mapping $x\colon [\lambda_0 - \varepsilon_0, \lambda_0 + \varepsilon_0] \to B$ is continuous.

Let us consider a continuous asymptotically compact mapping $F\colon [0, 1] \times \mathrm{cl}(C) \to X$, where X is a LCS, $D \in \mathrm{CoCl}(X)$, and C an (relatively) open subset of D. If $F(t, x) \neq x$ for all $(t, x) \in [0, 1] \times \partial_D C$, then, as we know, $\gamma_D(I - F(0, .), \mathrm{cl}(C)) = \gamma_D(I - F(1, .), \mathrm{cl}(C))$. Hence the following lemma holds:

Lemma 21. *Let* X, D, C *be as above,* $F\colon [0,1] \times \mathrm{cl}(C) \to D$ *a continuous mapping such that* $F \in \mathrm{AsCpt}\big([0,1] \times \mathrm{cl}(C)\big)$ *and* $F(i, x) \neq x$ *for each* $i = 0, 1$ *and each* $x \in \partial_D C$. *If* $\gamma_D\big(I - F(0,.), \mathrm{cl}(C)\big) \neq \gamma_D\big(I - F(1,.), \mathrm{cl}(C)\big)$, *then there are* $t \in (0, 1)$ *and* $x \in \partial_D C$ *such that* $F(t, x) = x$, *that is, there is an eigenvector* x *in* $\partial_D C$ *of* F *corresponding to a characteristic value* $t \in (0, 1)$.

Lemma 22. *Let* X, D, C *be as above,* μ *a monotone* mnc *in* X, $F\colon \mathrm{cl}(C) \to D$ *a continuous mapping such that* $Fx \neq x$ *for each* $x \in \partial_D C$, $\{0\} \cup M \in \mathfrak{M}_\mu$ *and* $\mu(\{0\} \cup M) = \mu(M)$ *for each* $M \in F\big(\mathrm{cl}(C)\big)$, *and* $F \in \mathrm{Dens}\,\big(\mathrm{cl}(C), \mu\big)$. *If one of the following conditions is satisfied:*

(1) $0 \in C$ *and* $\gamma_D\big(I - F, \mathrm{cl}(C)\big) \neq 1$;

(2) $0 \notin \mathrm{cl}(C)$ *and* $\gamma_D\big(I - F, \mathrm{cl}(C)\big) \neq 0$,

then there exists an eigenvector x *of* F *on* $\partial_D C$ *corresponding to a characteristic value* t *greater than* 1.

Proof. Consider the mapping $G\colon [0,1] \times \mathrm{cl}(C) \to X$ defined by $G(t, x) = tFx$. Suppose to the contrary, that $Fx \neq tx$ for each $x \in \partial_D C$ and $t > 1$. As we have shown in the proof of theorem 3, this means that G is an admissible homotopy between O and F on $\mathrm{cl}(C)$. Hence $\gamma_D\big(I - F, \mathrm{cl}(C)\big) = \gamma_D\big(I, \mathrm{cl}(C)\big)$. Now apply the property (vii) of the rotation to obtain a contradiction.

Now, let $F\colon C \to X$ be a mapping where X is a LCS and C an open subset of X. We say that *eigenvectors of* F *constitute a continuous branch* (in B, see below) *at a point* $x_0 \in C$ if there exists an open neighbourhood B of x_0 such that for each neighbourhood $V \subseteq B$ of x_0 there is a point $x \in \partial V$ such that $\lambda Fx = x$ for some $\lambda \in R$.

Using Lemma 22 and the method of the proof of theorem IV.1.4 from the book of M. A. Krasnosjelski [47], the reader can easily derive theorem 14 below which is not stated in full generality here.

Theorem 14. *Let* X *be a Banach space,* $\mu\colon \mathrm{Bo}(X) \to R^+$ *a homogeneous, algebraically subadditive, subadditive and regular* mnc, C *an open subset of* X *and* $F\colon \mathrm{cl}(C) \to X$ *a continuous mapping with* $F \in \mathrm{Bo}\,(\mathrm{cl}(C), k, \mu, \mu)$ *for some* $k \in R^+$. *Let* $x_0 \in C$ *be an isolated fixed point of the mapping* $\lambda_0 F$ *for some* $\lambda_0 \neq 0$ *with* $|\lambda_0|\, k < 1$. *If either* $x_0 \neq 0$ *and* $\mathrm{ind}\,(x_0, \lambda_0 F) \neq 0$ *or* $x_0 = 0$ *and* $\mathrm{ind}\,(x_0, \lambda_0 F) \neq 1$, *then the eigenvectors of* F *constitute a continuous branch in an open neighbourhood* B *of* x_0. *Moreover, the set of characteristic values of* F *corresponding to eigenvectors of* B *contains an open interval centered at* λ_0.

In what follows we are interested in the openness of the range of a mapping of the form $I - F$ where F is densifying.

Proposition 16. *Let* X *be a* LCS, $D \in \mathrm{CoCl}(X)$, C *an (relatively) open subset of* D, μ *a monotone* mnc *in* X, $F\colon \mathrm{cl}(C) \to D$ *a continuous* μ*-densifying mapping and* $y \in X$. *Assume that* $[0, y] \cap (I - F)(\partial_D C) = \emptyset$, $F\big(\mathrm{cl}(C)\big) \subseteq D - y$, $F(M) + [0, y] \in \mathfrak{M}_\mu$ *and* $\mu\big(F(M) + [0, y]\big) = \mu\big(F(M)\big)$ *for each* $M \subseteq \mathrm{cl}(C)$ *(the equality holds if* μ *is algebraically subadditive,* $[0, y] \in \mathfrak{M}_\mu$ *and* $\mu([0, y]) = 0$*). If* $\gamma_D\big(I - F, \mathrm{cl}(C)\big) \neq 0$, *then* $y \in (I - F)(C)$.

Proof. Let $H\colon [0,1] \times \mathrm{cl}(C) \to D$ be defined by $H(t, x) = Fx + ty$. Then $H([0,1] \times M) = F(M) + [0, y]$ for each $M \subseteq \mathrm{cl}(C)$. Hence $\mu\big(H([0,1] \times M)\big) \geqq \mu(M)$ implies $\mu\big(F(M)\big) = \mu\big(F(M) + [0, y]\big) \geqq \mu(M)$, i.e. $M \in \mathrm{RelCpt}\,\big(\mathrm{cl}(C)\big)$, since $F \in \operatorname{Dens}(\operatorname{cl}(C), \mu)$. From $[0, y] \cap (I - F)(\partial_D C) = \emptyset$ it follows that $H(t, x) \neq 0$ for each $(t, x) \in [0, 1] \times \partial_D C$. Thus, H is an admissible homotopy between F and $F + y$, and we conclude that

$$\gamma_D(I - (F + y), \operatorname{cl}(C)) = \gamma_D(I - F, \operatorname{cl}(C)) \neq 0,$$

and $F + y$ must have a fixed point in C, that is $y \in (I - F)(C)$.

Corollary 1. *Suppose, in addition to hypothese of proposition* 16, *that* $F(M) + [0, y'] \in \mathfrak{M}_\mu$ *for each* $M \subseteq \operatorname{cl}(C)$ *and* $y' \in [0, y]$. *If* $\gamma_D(I - F, \operatorname{cl}(C)) \neq 0$, *then* $[0, y] \subseteq (I - F)(C)$.

Proof. It is easy to verify that $F(\operatorname{cl}(C)) \subseteq D - y'$ for all $y' \in [0, y]$ and proposition 16 can be applied.

Corollary 2. *Let* X *be a* LCS, $D \in \operatorname{CoCl}(X)$, C *an (relatively) open subset of* D, V *a convex neighbourhood of* 0 *in* X, $F \colon \operatorname{cl}(C) \to D$ *a continuous* μ*-densifying mapping, where is a monotone* mnc *in* X *such that* $F(M) + [0, y] \in \mathfrak{M}_\mu$ *and* $\mu(F(M) + [0, y]) = \mu(F(M))$ *for all* $M \subseteq \operatorname{cl}(C)$ *and* $y \in V$. *Assume that* $F(\operatorname{cl}(C)) + V \subseteq D$. *Then* $V \subseteq (I - F)(C)$ *provided that* $\gamma_D(I - F, \operatorname{cl}(C)) \neq 0$.

Corollary 3. *Let* X *be a metrizable* LCS, C *an open subset of* X, μ *an algebraically subadditive monotone* mnc *in* X *of type* $(RC, 0)$, $F \colon \operatorname{cl}(C) \to X$ *a continuous* μ*-densifying mapping without fixed points on* ∂C. *Assume that* $M \in \mathfrak{M}_\mu$, $K \in \operatorname{RelCpt}(X)$ *implies* $M + K \in \mathfrak{M}_\mu$. *If* $\gamma(I - F, \operatorname{cl}(C)) \neq 0$, *then* $(I - F)(C)$ *is a neighbourhood of* 0.

Proof. By lemma 20, $(I - F)(\partial C)$ is closed. Since $0 \notin (I - F)(\partial C)$, there is an open convex neighbourhood V of O in X such that $V \cap (I - F)(\partial C) = \emptyset$. Apply corollary 2 with $D = X$.

The same considerations imply the validity of the following

Proposition 17. *Let* X, C, μ, F *be as in corollary* 3. *Let* U *be a connected open subset of* $X \backslash (I - F)(\partial C)$. *Assume that* $M \in \mathfrak{M}_\mu$, $K \in \operatorname{RelCpt}(X)$ *implies* $M + K \in \mathfrak{M}_\mu$. *Then* $\gamma(I - F - y, \operatorname{cl}(C))$ *is constant for* $y \in U$. *(The proof uses the fact that each two points of* U *can be connected by a piece-wise linear curve in* U.*)*

We finish this paper by some extensions of results of R. D. Nussbaum [60]. The proofs are similar and are presented here for sake of completeness.

Lemma 23. *Let* X, Y, Z *be three* LCS's, D *a closed subset of* Z, C *a closed subset of* X *and* T *a compact topological space. Let* $F \colon C \to Y$ *and* $E, G \colon T \times D \to C$ *be mappings such that* F *is continuous and both* E *and* G *are continuous at points of* T *uniformly with respect to points of* D *and continuous at points of* D *for each fixed point of* T. *Let* μ, ν, ζ *be nonnegative* mnc's *in* X, Y, Z *respectively such that:*

(1) $\mathfrak{M}_\zeta \supseteq 2^D$, ζ *is of type* $(0, RC)$;

(2) $\mathfrak{M}_\nu \supseteq \exp F(C)$ *and* $\mathfrak{M}_\nu \supseteq \exp [F(C) + F(C)]$, ν *is algebraically subadditive, subadditive and of type* $(RC, 0)$;

(3) $\mathfrak{M}_\mu \supseteq 2^C$, μ *is continuous, monotone and of type* $(0, RC)$.

Let us suppose that $E(t, .) \in \operatorname{Bo}(D, e_t, \zeta, \mu)$ *and* $G(t, .) \in \operatorname{Bo}(D, g_t, \zeta, \mu)$ *where* $e_t + g_t \leqq 1$ *for all* $t \in T$, *and for each* $M \in 2^C \backslash \operatorname{RelCpt}(C)$ *there is a neighbour-*

hood V of 0 in X such that $\nu\big(F((M+V)\cap C)\big) < \mu(M)$. Let us define a mapping $H\colon T\times D\to Y$ as follows: $H(t,z) = F\big(E(t,z)\big) + F\big(G(t,z)\big)$. Then H is continuous and $H \in \mathrm{Dens}\,(T\times D, \zeta, \nu)$.

Proof. It is easy to see that E, G and hence H are continuous. Let us remark that $F \in \mathrm{Bo}\,(C, 1, \mu, \nu)$, because μ is of type $(0, RC)$, $F \in \mathrm{Dens}\,(C, \mu, \nu)$ (easy), F is continuous, C is closed and ν is of type $(RC, 0)$. Since ζ if of type $(0, RC)$, it is sufficient to show that $M \subseteq D$ and $\zeta(M) > 0$ implies $\nu\big(H(T\times M)\big) < \zeta(M)$. Let $M \subseteq D$ with $\zeta(M) > 0$ be given. Firstly, we claim that for each t in T there is a neighbourhood $B(t)$ of t in T such that

$$\nu\big(H(B(t)\times M)\big) < \zeta(M)\,. \tag{*}$$

Let $t \in T$ be given. Consider two cases:

(i) $\mu\big(E(t,M)\big) = \mu\big(G(t,M)\big) = 0$. Since μ is continuous there is a neighbourhood V of 0 in X such that each of the two following triples of inclusions

$$N\subseteq C,\ N\subseteq E(t,M)+V,\ \ E(t,M)\subseteq N+V\,,$$

$$N\subseteq C,\ N\subseteq G(t,M)+V,\ \ G(t,M)\subseteq N+V$$

implies that $\mu(N) < \frac{1}{2}\zeta(M)$. The continuity of E and G on T uniformly with respect to D implies the existence of a neighbourhood $B(t)$ of t in T such that

$$E\big(B(t)\times M\big)\subseteq E(t,M)+V \quad\text{and}\quad G\big(B(t)\times M\big)\subseteq G(t,M)+V\,.$$

From the inclusion

$$H(B(t)\times M)\subseteq F\big(E(B(t)\times M)\big)+F\big(G(B(t)\times M)\big)$$

we conclude that

$$\nu\big(H(B(t)\times M)\big)\leqq \nu\Big(F\big(E(B(t)\times M)\big)\Big)+\nu\Big(F\big(G(B(t)\times M)\big)\Big)$$
$$\leqq \mu\big(E(B(t)\times M)\big)+\mu\big(G(B(t)\times M)\big)\leqq \mu\big((E(t,M)+V)\cap C\big)$$
$$+\,\mu\big((G(t,M)+V)\cap C\big) < 2\cdot\tfrac{1}{2}\zeta(M) = \zeta(M)\,.$$

(ii) At least one of the numbers $\mu\big(E(t,M)\big)$ and $\mu\big(G(t,M)\big)$ is positive. Suppose that $\mu\big(E(t,M)\big) = e > 0$, for example. By the hypothesis there is a neighbourhood V of 0 in X such that

$$v = \nu\Big(F\big((E(t,M)+V)\cap C\big)\Big) < \mu\big(E(t,M)\big) = e\,.$$

Since μ is continuous, we can suppose that V is so small that $N\subseteq C$, $N \subseteq G(t,M)+V$ and $G(t,M)\subseteq N+V$ implies $|\mu(N)-\mu\big(G(t,M)\big)| < e - v$. Owing to the type of continuity of E and G there is a neighbourhood $B(t)$ of t in T such that $E\big(B(t)\times M\big)\subseteq E(t,M)+V$ and $G\big(B(t)\times M\big)\subseteq G(t,M)+V$. Then

$$\nu\big(H(B(t)\times M)\big)\leqq \nu\Big(F\big(E(B(t)\times M)\big)\Big)+\nu\Big(F\big(G(B(t)\times M)\big)\Big)$$
$$\leqq \nu\Big(F\big(C\cap(E(t,M)+V)\big)\Big)+\nu\Big(F\big(C\cap(G(t,M)+V)\big)\Big)\leqq v$$
$$+\,\mu\big(C\cap(G(t,M)+V)\big) < v + [\mu(G(t,M)) + e - v] = \mu\big(E(t,M)\big)$$
$$+\,\mu\big(G(t,M)\big)\leqq (e_t+g_t)\,\zeta(M)\leqq \zeta(M)\,.$$

Thus, in each case (*) holds. The open covering $\{B(t)\}_{t\in T}$ of the compact space T contains a finite subcovering, say $\{B(t_i)\}_{i=1}^n$. Then:

$$\nu(H(T \times M)) = \nu\left(\bigcup_{i=1}^{n} H(B(t_i) \times M)\right) = \max\{\nu(H(B(t_i) \times M)) : i = 1, \dots, n\}$$

$< \zeta(M)$. The lemma is proved.

Remark. If H is defined by $H(t, z) = F(E(t, z)) - F(G(t, z))$, then the conclusion of lemma 23 holds provided ν is, in addition, even on $F(C)$: $\nu(M) = \nu(-M)$ for each $M \subseteq F(C)$ (and $\exp(-F(C))$ and $\exp(F(C) - F(C))$ are contained in $\mathfrak{M}_\mu$).

Now we can prove an invariance of domain theorem.

Theorem 15. *Let X be a* NLS, *C a bounded open symmetric subset of X with* cl (C) *strictly (X—) starshaped (with respect to 0), μ a nonnegative regular algebraically subadditive homogeneous subadditive continuous* mnc *in X and F:* cl $(C) \to X$ *a continuous mapping such that $F(0) = 0$ and for each $M \subseteq$* cl (C), *$M \notin$* RelCpt (X), *there is a neighbourhood V of in 0 in X satisfying $\mu(F((M + V) \cap \text{cl}(C))) < \mu(M)$. Assume that $\mathfrak{M}_\mu$ contains* exp cl (C), *sets $-M$, $M + K$ for all $M \in \mathfrak{M}_\mu$ and $K \in$* RelCpt (X), *and sets $F(M) - F(N)$ for all $M, N \subseteq$* cl (C). *If the mapping $T = I - F$ is one-to-one on each set $\left[-\frac{x}{2}, 0\right] \cup \left[\frac{x}{2}, x\right]$, $x \in \partial C$, then the set $(I - F)(C) = T(C)$ is a neighbourhood of 0.*

Proof. The proof is a standard one. Let us remark that $T = I - F$ has no zeroes on ∂C, since $T(0) = 0$, T is injective on each set $\left[-\frac{x}{2}, 0\right] \cup \left[\frac{x}{2}, x\right]$, $x \in \partial C$, and $0 \in C$. Since $(I - F)(\partial C)$ is closed, by lemma 20, and does not contain 0, there is a convex neighbourhood W of 0 in X such that $(I - F)(\partial C) \cap W = \emptyset$. We shall show that $(I - F)(C) \supseteq W$ which will prove the theorem.

Let $y \in W$ be given and set $L = F + y$. Since $F, L \in$ Dens $(\text{cl}(C), \mu)$ the mapping H: $[0, 1] \times \text{cl}(C) \to X$, defined by $H(x, t) = tLx + (1 - t)Fx$, lies in Dens $([0, 1] \times \text{cl}(C), \mu)$, by lemma 6. If $H(t, x) = x$ for some $(t, x) \in [0, 1] \times \partial C$, then $x - Fx = ty$, which yields a contradiction: $W \not\ni x - Fx = ty \in W$. Hence H, being continuous, is an admissible homotopy between F and L and we conclude that

$$\gamma(I - L, \text{cl}(C)) = \gamma(I - F, \text{cl}(C)) . \tag{1}$$

Now, let us define another mappings E, G, H: $[0, 1] \times \text{cl}(C) \to X$: $E(t, x) = \left(1 - \frac{t}{2}\right)x$, $G(t, x) = -\frac{t}{2}x$, $H(t, x) = F(E(t, x)) - F(G(z, x)) = F\left(\left(1 - \frac{t}{2}\right)x\right) - F\left(-\frac{t}{2}x\right)$. Since μ is homogeneous, we have $E(t, .) \in$ Bo $(\text{cl}(C), e_t, \mu, \mu)$ and $G(t, .) \in$ Bo $(\text{cl}(C), g_t, \mu, \mu)$ where $e_t = 1 - \frac{t}{2}$, $g_t = \frac{t}{2}$, and hence $e_t + g_t = 1$ for each $t \in [0, 1]$. By the Remark following Lemma 23: $H \in$ Dens $([0, 1] \times \text{cl}(C), \mu)$. E and G are continuous on $[0, 1]$ uniformly with respect to cl (C) since C is bounded. If $H(t, x) = x$ for some $(t, x) \in [0, 1] \times \partial C$, then we obtain easily that $T(E(t,x)) = T(G(t,x))$, yielding a contradiction, since $E(t;x) \in \left[\frac{x}{2}, x\right]$, $G(t,x) \in \left[-\frac{x}{2}, 0\right] \ni 0$ and $x \neq 0$. Hence H is an admissible homotopy between $H(0,.) = F$ and $H(1,.) = J$. Since J is odd and satisfies the conditions of the property (viii) of the rotation, we have

$$\gamma(I - F, \mathrm{cl}\,(C)) = \gamma(I - J, \mathrm{cl}\,(C)) \equiv 1 \pmod 2\,. \tag{2}$$

(1) and (2) yield $\gamma(I - L, \mathrm{cl}\,(C)) \neq 0$. Therefore, L has a fixed point in C, that is, there exists a point $x \in C$ such that $x = Lx = Fx + y$, i.e. $y \in T(C)$. Since $y \in W$ was arbitrary, we have $W \subseteq T(C)$.

Corollary. *Let X be a Banach space, C an open subset of X, $F: C \to X$ a continuous mapping with $T = I - F$ injective. Let $\mu: \mathrm{Bo}\,(X) \to R^+ = [0, +\infty)$ a regular algebraically subadditive homogeneous subadditive continuous* mnc. *Assume that for each $x \in C$ there exists a neighbourhood U of x, $U \subseteq C$, such that for any $M \subseteq U$ with $M \notin \mathrm{RelCpt}\,(X)$ there is a neighbourhood V of 0 in X satisfying the inequality $\mu(F((M + V) \cap U)) < \mu(M)$. Then the set $T(C) = (I - F)\,(C)$ is open.*

Proof follows easily from theorem 15 and it can be left to the reader.

Remark. It is easy to verify that the condition concerning the inequality is satisfied, if each point $x \in C$ possesses a neighbourhood W with $F \in \mathrm{Bo}\,(W, k, \mu)$ for some $k = k(x) < 1$. It is not hard to see that the condition concerning the inequality is equivalent to the following one provided F is locally uniformly continuous on C:
each point $x \in C$ possesses a neighbourhood W such that $F \in \mathrm{Dens}\,(W, \mu)$.

References

[1] Amann, H., Existence of multiple solutions for nonlinear elliptic boundary value problems. Indiana Univ. Math. J. **21** (1972) 925–935.

[2] —, On the number of solutions of nonlinear equations in ordered Banach spaces. J. Funct. Anal. **11** (1972) 346–384.

[3] Assad, N. A., and Kirk, W. A., Fixed point theorems for set-valued mappings of contractive type.

[4] Avramescu, C., Asupra unei teoreme de punct fix. Studii Cerc. Mat. **22** (1970) 215–221.

[5] Boyd, D. W., and Wong, J. S. W., On nonlinear contractions. Proc. Amer. Math. Soc. **20** (1969) 458–464.

[6] Browder, F. E., On the fixed point index for continuous mappings of locally connected spaces. Summa Brasil. Math. **4** (1960) 253–293.

[7] —, Normal solvability and the Fredholm alternative for mappings into infinite dimensional manifolds. J. Funct. Anal. **8** (1971) 250–274.

[8] Browder, F. E., and Nussbaum, R. D., The topological degree for noncompact nonlinear mappings in Banach spaces. Bull. Amer. Math. Soc. **74** (1968) 671–677.

[9] Cellina, A., On the existence of solutions of ordinary differential equations in Banach spaces. Funkc. Ekvacioj **14** (1971) 129–136.

[10] —, On the local existence of solutions of ordinary differential equations. Bull. Acad. Polon. Sci. **20** (1972) 293–296.

[11] —, A theorem on nonempty intersection. Bull. Acad. Polon. Sci. **20** (1972) 471–472.

[12] Daneš, J., Some fixed point theorems. Comment. Math. Univ. Carolinae **9** (1968) 223–235.

[13] —, Generalized concentrative mappings. Summer School on Fixed Points, Krkonoše, Czechoslovakia, Aug. 31–Sept. 6, 1969.

[14] —, Generalized concentrative mappings and their fixed points. Comment. Mat. Univ. Carolinae **11** (1070) 115–136.

[15] —, Fixed point theorems, Nemyckii and Uryson operators, and continuity of nonlinear mappings. Comment. Math. Univ. Carolinae **11** (1970) 481–500.

[16] —, Some fixed point theorems in metric and Banach spaces. Comment. Math. Univ. Carolinae **12** (1971) 37–52.

[17] —, Surjectivity and fixed point theorems (Preliminary communication). Comment. Math. Univ. Carolinae **13** (1972) 181–184.

[18] —, Some existence theorems for nonlinear equations of Hammerstein type, Comment. Math. Univ. Carolinae **13** (1972) 397–401.

[19] —, Some fixed point and existence theorems. EQUADIFF III = Czechoslovak Conference on Differential Equations and Their Applications. Brno, Aug. 28–Sept. 1, 1972.

[20] —, A geometric theorem useful in nonlinear functional analysis. Boll. Un. Mat. Italiana **6** (1972) 369–375.

[21] Darbo, G., Punti uniti in transformazioni a condominio non compatto. Rend. Sem. Mat. Univ. Padova **24** (1955) 84–92.

[22] Fenske, C., Analytische Theorie des Abbildungsgrades für Abbildungen in Banachräumen. Math. Nachr. **48** (1971) 279–290.

[23] —, Learay-Schauder Theorie für eine Klasse differenzierbarer Abbildungen in Banachräumen. Gesellschaft f. Math. u. Datenverarb. Bonn, Nr. 48 (1971).

[24] Fitzpatrick, P. M., A generalized degree for uniform limits of A-proper mappings. J. Math. Anal. Appl. **35** (1971) 536–552.

[25] Furi, M., and Martelli, M., Some fixed-points for multi-valued mappings in topological vector spaces. Ann. Mat. Pura Appl. **92** (1972) 169–175.

[26] Goldenstein, L. S., Gochberg, I. Z., und Markus, A. S., Исследование некоторых свойств линейных ограниченных операторов в связи с их q-нормой. Уч. зап. Кишиневск. ун-та **29** (1957) 29—36.

[27] Goldenstein, L. S., und Markus, A. S., О мере некомпактности ограниченных множеств и линейных операторов. Исслед. по алгебре и мат. анал. Кишинев, (1965) 45—54.

[28] Granas, A., The theory of compact vector fields and some of its applications to topology of functional spaces (I). Rozprawy Mat. XXX, Warszawa (1962).

[29] Gurari, W. I., О дифференциальных свойствах модулей выпуклости банаховых пространств, Мат. исслед. Кишинев **2** (1967) 141—148.

[30] Hadžić, O., Theorems on continuous dependence of the fixed point on parameters and applications to differential equations in locally convex spaces. Review Research Sci. Univ. Novi Sad **1** (1971) 3–14.

[31] Hale, J. K., and Cruz, M. A., Existence, uniqueness and continuous dependence for hereditary systems. Ann. Mat. Pura Appl. **85** (1970) 63–82.

[32] Hess, P., On the Fredholm alternative for nonlinear functional equations in Banach spaces. Proc. Amer. Math. Soc. **33** (1972) 55–61.

[33] Hille, E., and Phillips, R. S., Functional Analysis and Semigroups. Amer. Math. Soc. Colloq. Publ. XXXI, New York (1957).

[34] Himmelberg, C. J., and van Vleck, F. S., Fixed points of semicondensing multifunctions. Boll. Un. Mat. Ital. **5** (1972) 187–194.

[35] Istrăţescu, A., and Istrăţescu, V., On the theory of fixed points for some classes of mappings. II. Rev. Roum. Math. Pures Appl. **16** (1971) 1073—1076.

[36] —, —, On the theory of fixed points for some classes of mappings. III. Rend. Accad. Naz. Lincei **49** (1970) 43—46.

[37] —, —, On the theory of fixed points for some classes of mappings. V. Rend. Accad. Naz. Lincei **51** (1971) 162—167.

[38] Katschurowski, R. I., О разрешимости некоторых классов нелинейных уравнений. ДАН СССР **189**, 1 (1969) 35—37.

[39] —, К теории Фредгольма для нелинейных операторных уравнений. ДАН СССР **192**, 5 (1970) 969—972.

[40] —, О нелинейных операторах, областями значений которых являются подпространства. ДАН СССР **196** (1971) 508—511.

[41] —, Обобщение теорем Фредгольма и теорем о линейных операторах с замкнутой областью значений на некоторые классы нелинейных операторов. ДАН СССР **197** (1971) 520—523.

[42] Kamenski, T. A., u. Myschkis, A. D., Краевые задачи для нелинейного дифференциального уравнения с отклоняющимся аргументом нейтрального типа. Дифф. урав. **8** (1972) 2171—2179.

[43] Kirk, W. A., On nonlinear mappings of strongly semicontractive type. J. Math. Anal. Appl. **27** (1969) 409—412.

[44] —, Mappings of generalized contractive type. J. Math. Anal. Appl. **32** (1970) 567—572.

[45] —, Fixed point theorems for nonlinear nonexpansive and generalized contraction mappings. Pacific J. Math. **38** (1971) 89—94.

[46] Krasnosjelski, M. A., Два замечания о методе последовательных приближений, УМН **10**, 1 (1955) 123—127.

[47] —, Топологические методы в теории интегральных уравнений. Москва (1956). (Engl. Übersetzung: Topological methods in the theory of nonlinear integral equations. Pergamon Press, Oxford (1964)).

[48] —, Положительные решения операторных уравнений. Москва (1962). (Engl. Übersetzung: Positive solutions of operator equations, P. Noordhoff Ldt. 1964).

[49] —, On some fixed point principles.

[50] Kuratowski, C., Sur les espaces complets. Fund. Math. **15** (1930) 301—309.

[51] —, Topology. Vol. I, Academic Press, New York and PWN, Warszawa (1966).

[52] Lebow, A., and Schechter, M., Semigroups of operators and measures of noncompactness. J. Funct. Anal. **7** (1971) 1—26.

[53] Marinescu, G., Tratat de Analiză Funcţională, Vol. II. Edit. Acad. Rep. Romania 1972.

[54] Massa, S., and Roux, D., Applicazioni densificanti e teoremi di punto unito. Boll. Un. Mat. Ital. **4** (1971) 835—840.

[55] Melamed, W. B., О вычислении вращения вполне непрерывного векторного поля в критическом случае. Сибирск. матем. журн. **2** (1961) 414—427.

[56] Melamed, W. B., u. Perow, A. I., Обобщение теоремы М. А. Красносельского о полной непрерывности производной Фреше вполне непрерывного оператора. Сибирск. матем. журн. **4** (1963) 702—704.

[57] Melvin, W. R., Some extensions of the Krasnojelski fixed point theorems. J. Diff. Equations **11** (1972) 335—348.

[58] Montagnana, M., and Vignoli, A., On quasi-convex mappings and fixed point theorems. Boll. Un. Mat. Ital. **4** (1971) 870—878.

[59] Moore, R. H., and Nashed, M. Z., Local and asymptotic approximations of nonlinear operators by $(k_1 ..., k_N)$-homogeneous operators (to appear). (MRC Techn.-Rep. 1179, Univ. of Wisconsin, 1972.)

[60] Nussbaum, R. D., The fixed point index and fixed point theorems for k-set-contractions. Ph. D. Dissertation. Univ. of Chicago (1969).

[61] —, The fixed point index and asymptotic fixed point theorems for k-set-contractions. Bull. Amer. Math. Soc. **75** (1969) 490–495.

[62] —, Some fixed point theorems. Bull. Amer. Math. Soc. **77** (1971) 360–365.

[63] —, Estimates for the number of solutions of operator equations. Applicable Analysis **1** (1971) 183–200.

[64] —, Existence and uniqueness theorems for some functional differential equations of neutral type. J. Diff. Equations **11** (1972) 607–623.

[65] —, Asymptotic fixed point theorems for local condensing mappings. Math. Ann. **191** (1971) 181–195.

[66] —, An asymptotic fixed point theorem.

[67] —, The radius of the essential spectrum. Duke Math. J. **38** (1970) 473–478.

[68] —, The ball intersection properties for Banach spaces (to appear).

[69] —, Some asymptotic fixed point theorems.

[70] —, A generalization of the Ascoli theorem and an application to functional differential equations. J. Math. Appl. (to appear).

[71] —, Periodic solutions of some nonlinear autonomous functional differential equations.

[72] Obuchowski, V. V., On some fixed point principles for multivalued densifying operators. Trudy mat. fak. VGU, vyp. 4, Voronež (1971) 70–79 (in Russian).

[73] Petryshyn, W. V., Structure of the fixed point sets of k-set-contractions. Archive Rat. Mech. Anal. **40** (1971) 312–328.

[74] —, A new fixed point theorem and its applications. Bull. Amer. Math. Soc. **78** (1972) 225–229.

[75] —, Note on the structure of fixed point sets of 1-set-contractions. Proc. Amer. Math. Soc. **31** (1972) 189–194.

[76] Pochoshajew, S. I., О слабо нелинейных гиперболических системах. Труды моск. матем. о-ва **23** (1970) 61–76.

[77] Reich, S., Fixed points in locally convex spaces. Math. Z. **125** (1972) 17–31.

[78] —, A fixed point theorem in locally convex spaces. Bull. Calcutta Math. Soc. **63** (1971) 199–200.

[79] —, Characteristic vectors of nonlinear operators. Rend. Accad. Naz. Lincei **50** (1971) 682–685.

[80] —, A fixed point theorem. Rend. Accad. Naz. Lincei **51** (1971) 26–28.

[81] —, Fixed points of multi-valued functions. Rend. Accad. Naz. Lincei **51** (1971) 32–35.

[82] —, Fixed points in complete metric spaces. Rend. Accad. Naz. Lincei **51** (1971) 270–273.

[83] Reinermann, J., Fixed point theorems for generalized contractions. Comment. Math. Univ. Carolinae **13** (1972) 617–627.

[84] Rus, I. A., Asupra punctelor fixe ale aplicatiilor definite pe un produs cartezian, I.: Structuri algebrice. Studii Cerc. Mat. **24** (1972) 891–896.

[85] —, Asupra punctelor fixe ale aplicatiilor definite pe un produs cartezian, II.: Spatii metrice. Studii Cerc. Mat. **24** (1972) 897–904.

[86] Sadowski, B. N., Об одном принципе неподвижной точки. Функц. анализ и его прилож. **1** (1967) 74–76.

[87] —, Локальные теоремы существования для обыкновенных дифференциальных уравнений в банаховых пространствах. Пробл. мат. анал. сложн. систем, вып. **1**, Воронеж, (1967) 70—74.

[88] —, О мерах некомпактности и уплотняющих операторах. Пробл. мат. анал. сложн. систем, вып. **2**, Воронеж, (1968) 89—119.

[89] —, On local solvability of ordinary differential equations in a Banach space. Prob. mat. anal. složn. sistem, vyp. **3**, Voronež (1968) 232—243 (in Russian).

[90] —, Some remarks on densifying mappings and measures of noncompactness. Trudy mat. fak. VGU, Vyp. **1**, Voronež (1970) 112—124 (in Russian).

[91] —, On differential equations with uniformly continuous right-hand side. Trudy NII mat. VGU, Vyp. **1**, Voronež (1970), 128—136 (in Russian).

[92] —, Применение топологических методов в теории периодических решений нелинейных дифференциально-операторных уравнений нейтрального типа. ДАН СССР **200** (1971) 1037—1040.

[93] —, Предельно компактные и уплотняющие операторы. УМН **27** (1972) 1, 81—146.

[94] Schwartz, J. T., Nonlinear Functional Analysis. Gordon and Breach Sci. Publishers, New York (1969).

[95] Steinlein, H., Ein Satz über den Leray-Schauderschen Abbildungsgrad. Math. Z. **126** (1972) 176—208.

[96] —, Über die verallgemeinerten Fixpunktindizes von iterierten verdichtenden Abbildungen.

[97] Thomas, J. W., Upper and lower bounds for the number of solutions of functional equations involving k-set-contractions.

[98] —, A bifurcation theorem for k-set-contraction.

[99] —, The multiplicity of an operator is a special case of the topological degree for k-set-contractions.

[100] Thomas, J. W., and O'Neil, T., On the equivalence of multiplicity and the generalized topological degree. Trans. Amer. Math. Soc. **167** (1972) 333—345.

[101] Vidossich, G., On the structure of the set of solutions of nonlinear equations. J. Math. Anal. Appl. **34** (1971) 602—617.

[102] —, Applications of topology to analysis: On the topological properties of the set of fixed points of nonlinear operators. Conference Semin. Mat. Univ. Bari **126**, (1971) 1—62.

[103] —, Existence, uniqueness and approximation of fixed points as a generic property.

[104] Vignoli, A., On α-contractions and surjectivity. Boll. Un. Mat. Ital. **4** (1971) 1—10.

[105] —, An intersection theorem in Banach spaces. Rend. Accad. Naz. Lincei **49** (1970) 180—183.

[106] —, On quasibounded mappings and nonlinear functional equations. Rend. Accad. Naz. Lincei **50** (1971) 114—117.

[107] Webb, J. R. L., On a characterization of k-set-contractions. Rend. Accad. Naz. Lincei **50** (1971) 686—689.

[108] —, A fixed point theorem and applications to functional equations in Banach spaces. Boll. Un. Mat. Ital. **4** (1971) 775—788.

[109] Zabreiko, P. P., u. Krasnosjelski, M. A., Вычисление индекса неподвижной точки векторного поля. Сибирск. мат. журн. **5** (1964) 509—531.

[110] —, О разрешимости нелинейных операторных уравнений. Функц. анализ и его приложен. **5** (1971) 42—44.

[111] —, Итерации операторов и неподвижные точки. ДАН СССР **196** (1971) 1006—1009.

[112] Petryshyn, W. V., Remarks on condensing and k-set-contractive mappings. J. Math. Anal. Appl. **39** (1972) 717–741.
[113] —, Surjectivity theorems for odd mappings of A-proper type. Math. Ann. **192** (1971) 155–172.
[114] —, On nonlinear equations involving pseudo-A-proper mappings and their uniform limits with applications. J. Math. Anal. Appl. **38** (1972) 672–720.

Anschrift des Autors:

Mathematical Institute
of Charles University
18600 Prague 8-Karlín
Sokolovská 83
Czechoslovakia

Topics in nonlinear spectral theory

SVATOPLUK FUČÍK, MILAN KUČERA, JIŘÍ SOUČEK, VLADIMÍR SOUČEK, Prag

What do we mean under nonlinear spectral theory? We can define this subject as the investigation of problem about operators, which belongs to linear spectral theory in linear case. Now, what are the main branches of linear spectral theory? They may be listed under three general headings:
A. Solvability of the equation $\lambda u - Au = f$.
B. The investigation of the set Λ of all eigenvalues.
C. Spectral decomposition.

Section 1 deals with the nonlinear analog of the solvability of the equation $\lambda u - Au = f$ in the dependence of real parameter λ. It is so-called Fredholm alternative for nonlinear operators. The set Λ of all eigenvalues for nonlinear operators is investigated in Section 2 and 4. Section 4 is based on the infinite-dimensional analog of Morse-Sard theorem (see Section 3). As a consequence of the investigation of the structure of the set Λ we obtain a result about the number of the bifurcation points for nonlinear operators (Section 5). Abstract results we apply to the concrete problems in Section 6, 7, 8 and 9. Some open problems are mentioned in Section 10. Note here that the result for nonlinear operators analogous to C is not obtained up to this time. There exists only one paper (by our meaning), namely the paper by H. Flashka [8], which deals with such a problem in very special case.

1. Fredholm alternative

This Section deals with the solving of nonlinear operator's equations $\lambda T(x) - S(x) = f$ in the dependence of the real parameter λ, where T and S are nonlinear operators defined on a real Banach space X with values in the dual space X^*. Theorems of the following type are obtained: The equation $\lambda T(x) - S(x) = f$ is solvable for each $f \in X^*$ provided the equation $\lambda T(x) - S(x) = 0$ has the trivial solution only. It is easy but interesting that in the nonlinear alternative the asymptotes of T and S play the essential role. Let us define first: an operator $F_0: X \to X^*$ is said to be a-homogeneous, $a > 0$, if $F_0(tu) = t^a F_0(u)$ for each $u \in X$ and all $t \geq 0$. An operator $F: X \to X^*$ has a-asymptote $F_0: X \to X^*$ in the infinity if

$$\lim_{||x|| \to \infty} \frac{||F(x) - F_0(x)||}{||x||^a} = 0$$

and F_0 is a-homogeneous.

Theorem 1. ([39, 40, 42]). *Let X be a real, reflexive and separable Banach space. Let $T, S: X \to X^*$ have a-asymptotes T_0, S_0, respectively. Let T, T_0 be bounded (they map bounded sets in X into bounded sets in X^*), demicontinuous: $x_n \to x$ (strong convergence) $\Rightarrow Tx_n \rightharpoonup Tx$, $T_0x_n \rightharpoonup T_0x$ (weak convergence).*

Let S, S_0 be completely continuous (they are continuous and map bounded sets into compact sets), T_0, S_0 be odd, and let $tT + (1-t)T_0$ satisfy for $0 \leq t \leq 1$ the condition (S): if $x_n \rightharpoonup x$ and $(tTx_n + (1-t)T_0x_n - tTx - (1-t)T_0x, x_n - x) \to 0$, then $x_n \to x$.

Then $(\lambda T - S)X = X^*$ and

$$||u|| \leq c\,(1 + ||\lambda Tu - Su||)^{1/a} \tag{1.1}$$

provided $\lambda \neq 0$ is not an eigenvalue for the couple (T_0, S_0) (i.e., $(\lambda T_0 - S_0)u = 0 \Rightarrow u = 0$). If (1.1) is satisfied, then λ is not an eigenvalue.

The proof is based on the properties of Leray-Schauder degree, namely, on the Borsuk-Ulam theorem about the degree of an odd mapping.

Remark that it is not true the following assertion: let T_0 and S_0 be a-homogeneous operators satisfying all assumptions of Theorem 1. If $\lambda T_0 - S_0$ is onto, then λ is not an eigenvalue of the couple (T_0, S_0) (see Example in the paper [15]).

First result about Fredholm alternative for nonlinear operators were independently obtained by S. I. Pochoshajew [44] and J. Nečas [38]. S. I. Pochoshajew supposed that T and S are the odd and a-homogeneous operators and the Banach space X has a Schauder basis (he supposed slightly weaker assumption than the existence of the basis). J. Nečas introduced the notion of the asymptotes (quasihomogeneous and strongly quasihomogeneous operators) and he proved the Fredholm alternative for such operators. For his method it is substantial that the Banach space X is a complex space. M. Kučera [27] proved a result similar to [38] for real space, his condition "being near to homogeneity" being stronger than those in [38]. Preceding papers discourse only on the operators the domain of which is a Banach space X, the range being in its dual space X^*. The Fredholm alternative in the case $Y \neq X^*$ is discussed in the papers [43, 9, 10, 11].

Other generalizations and various definitions of a-asymptotes were given by P. Hess [20, 21, 22] and R. L. J. Webb [49]. By anolther method and for the operators near to linear on the Fredholm alternative was proved by R. I. Katschurowski [23].

2. Ljusternik-Schirelmann theory

Let f and g be two nonlinear functionals defined on a real Banach space X. Consider the eigenvalue problem

$$\lambda f'(u) = g'(u)\,, \qquad u \in M_r(f) = \{x \in X: f(x) = r\} \tag{2.1}$$

($r > 0$ is a prescribed number, f' and g' denote Fréchet derivatives of f and g, respectively).

The solution $u \in M_r(f)$ of the problem (2.1) is the critical point of the functional g with respect to the manifold $M_r(f)$. The value of the functional g at the critical point is called the critical level. The real number λ for which the eigenvalue problem (2.1) has a solution is said to be an eigenvalue.

The Ljusternik-Schnirelmann theory, one of the most interesting part of the nonlinear spectral analysis, deals with the investigation of the set $\Gamma(r)$ of all critical levels. This theory is a nonlinear analog of the Courant-Weinstein principle whose importance is well known.

The basic question is, under which conditions the set $\Gamma(r)$ and the set $\Lambda(r)$ of all eigenvalues have the same properties as in the linear case, i.e., the set $\Gamma(r)$ (resp. $\Lambda(r)$) is a sequence of positive numbers which converge to zero.

L. A. Ljusternik and L. Schnirelmann proved that the set $\Gamma(r)$ (resp. $\Lambda(r)$) is at least countable (see [31, 32, 33]). For Hilbert spaces this theorem is concluded in the book of M.A. Krasnosjelskii [24]. For Banach spaces such a theorem was proved by E. S. Citlanadze [5], F. E. Browder [4] and S. Fučík-J. Nečas [14]. E. S. Citlanadze and F. E. Browder used the notion "the category of the sets" in the sense of L. Schnirelmann.

E. S. Citlanadze used for the proof the theory of ordinary differential equations in Banach spaces. He supposed that the operators f' and g' are lipschitzian. The proof in [14] is based on the notion "the order of the sets" which is similar to "the genus" introduced in [24]. In this proof "a rather complicated ε, δ-gymnastics" is necessary. On the other hand, we need only that f', g', in some sense, are uniformly continuous what gives the theoretical background for application to the differential and integral equations in the spaces of the type L_p with $p > 1$ (and not only $p \geqq 2$ if Lipschitz condition for f', g' is supposed).

Various considerations in this field are e.g. in the papers by M. S. Berger [3], A. Ambrosetti [1] and J. Naumann [35, 36, 37].

First we recall the following definition: The closed subset K of a Banach space X is said to be a set of the first order (we denote $\operatorname{ord} K = 1$) if $x \in K \Rightarrow -x \notin K$. The set $K \subset X$ is said to be n-th order ($\operatorname{ord} K = n$) if there exist closed sets $K_1, \ldots, K_n$ such that $\operatorname{ord} K_i = 1$ $(i = 1, \ldots, n)$, $\bigcup_{i=1}^{n} K_i = K$ and n is a minimal integer with the previous property. Put $\operatorname{ord} \emptyset = 0$ ($\emptyset$ denotes the empty set) and $\operatorname{ord} K = \infty$ if K is not a set of the n-th order for arbitrary n (for instance, if $0 \in K$, then $\operatorname{ord} K = \infty$).

For arbitrary integer k let $V_k(r)$ be a system of all compact sets $K \subset M_r(f)$ such that $\operatorname{ord} K \geqq k$. Set

$$\hat{g}(K) = \inf_{x \in K} g(x) \tag{*}$$

for each $K \in V_k(r)$ and

$$\gamma_k(r) = \sup_{K \in V_k(r)} \hat{g}(K)\,. \tag{**}$$

The main goal of this section is that $\gamma_k(r) \in \Gamma(r)$, precisely the following

Theorem 2. *Let f and g be two even functionals defined on a reflexive infinite-dimensional Banach space X with Fréchet derivatives f' and g' on X. Let the*

following assumptions be valid:

$$f(x) = 0 \Leftrightarrow x = 0\,, \tag{2.2}$$

$$\lim_{\|x\|\to\infty} f(x) = \infty\,, \tag{2.3}$$

$$\inf_{x\in M_r(f)} \big(f'(x), x\big) \geqq c(r) > 0 \quad \textit{for each} \quad r > 0, \tag{2.4}$$

where $c(r)$ *is monotone function,*

$$f'\colon X \to X^* \textit{ is a bounded operator,} \tag{2.5}$$

$$x_n \rightharpoonup x,\ f'(x_n) \to w \Rightarrow x_n \to x\,, \tag{2.6}$$

$$g(x) \geqq 0 \textit{ for each } x \in X\,, \tag{2.7}$$

$$g(x) = 0 \Leftrightarrow x = 0\,, \tag{2.8}$$

$$g'(x) = 0 \Leftrightarrow x = 0\,, \tag{2.9}$$

$$g'\colon X \to X^* \textit{ is a completely continuous mapping,} \tag{2.10}$$

$$x_n \rightharpoonup x_n,\ g'(x_n) \to 0,\ 0 < t_n < 1 \Rightarrow g'(t_n x_n) \to 0\,, \tag{2.11}$$

for each $\mu > 0$ *and* $r > 0$ *there exists* $\delta > 0$ *such that the inequalities*

$$\|f'(x+h) - f'(x)\| \leqq \mu\,, \qquad \|g'(x+h) - g'(x)\| \leqq \mu \tag{2.12}$$

hold for each $x \in M_r(f)$ *and all* $h \in X$ *with* $\|h\| \leqq \delta$.

Then for each positive integer k *and all* $r > 0$ *there exists* $x_k \in M_r(f)$ *such that*

$$\frac{(g'(x_k), x_k)}{(f'(x_k), x_k)} f'(x_k) = g'(x_k) \tag{2.13}$$

and

$$g(x_k) = \gamma_k(r)\,. \tag{2.14}$$

Moreover, if X *is a Banach space with a Schauder basis, then*

$$\lim_{k\to\infty} \gamma_k(r) \downarrow 0\,, \qquad x_k \rightharpoonup 0\,. \tag{2.15}$$

(For the proof see [14, theorem 1 and 2] and [18, remark 2].)

The forms (*) and (**) are sometimes called minimax process. The critical level which we obtain by minimax process we shall called as LS-critical level and if $x_k \in M_r(f)$ is the corresponding critical point for which the relation (2.14) holds, then the number

$$\lambda_k = \frac{(g'(x_k), x_k)}{(f'(x_k), x_k)}$$

is said to be LS-eigenvalue.

If X is a Hilbert space and the operator f' is the identity, g' is linear and the assumptions of Theorem 2 are valid, then each critical level is also LS-critical level. This is the Courant-Weinstein principle (see e.g. J. Naumann [35]). In the nonlinear case this is not generally true (see Example in [15]).

3. Morse-Sard theorem

The well-known theorem about real-valued functions, so called Morse or Morse-Sard theorem, says that if Ω is an open subset of Euclidean N-space R_N and $f \in C^N(\Omega)$ is a real function, then the Lebesgue measure of the set $f(B)$ is zero, where $B = \{x \in \Omega: \operatorname{grad} f(x) = 0\}$. For this theorem see e.g. A. P. Morse [34] and A. J. Dubowicki [7]. It is proved in [28] that if $f \in C^{k,\alpha}(\Omega)$ (k is a positive integer, $\alpha \in \langle 0, 1\rangle$), then the set $f(B)$ has $n/(k+\alpha)$-dimensional Hausdorff measure zero. If $f \in C^\infty(\Omega)$, then the set $f(B)$ has s-dimensional Hausdorff measure zero for each $s > 0$, but this set need not be countable. It is proved in [46], that in the case of real-analytic function f (i.e., each point $w \in \Omega$ has an open neighborhood U such that the function f has a power series expansion in U), the set $f(B)$ is countable.

In the sequel we wish to give an analogous assertion in infinite-dimensional Banach spaces. As the countraexample of I. Kupka [29] shows, in the whole generality such assertion is not true. I. Kupka constructed the functional $f \in C^\infty$ on the separable Hilbert space such that the set $f(B)$ has nonzero Lebesgue measure. S. I. Pochoshajew [45] introduced the notion of "Fredholm functional" and he proved under some additional assumptions the set $f(B)$ has a zero measure. One from his assumptions is that

$$\sup_{x \in \Omega} \dim \operatorname{Ker} f''(x) < \infty$$

and this assumption is very difficult to verify.

In this section we give the Morse-Sard theorems for real-analytic and $C^{k,\alpha}$ Fredholm functionals in infinite-dimensional Banach spaces. Analoguous assertion can be proved for functionals which derivative has a finite-dimensional range. The proofs are given in [17]. The case $f \in C^{k,\alpha}$ is solved in the paper [13].

First we recall the linear operator A defined on the Banach space X with values in Banach space Y is said to be Fredholm operator if the following conditions are fulfilled:

(i) $R = A(X)$ is a closed subspace of Y,
(ii) Y/R has a finite dimension,
(iii) $\dim A^{-1}(0) = N(A) < \infty$.

Let f be a functional defined on an open subset Ω of a Banach space X with Fréchet derivative f'. Denote

$$B = \{x \in \Omega: f'(x) = 0\}.$$

Theorem 3. *Let $f: \Omega \to R_1$ be a real-analytic functional (resp. $f \in C^{k+1,\alpha}(\Omega)$) on an open set Ω in a Hilbert space H. Suppose that $f''(x_0)$ is a Fredholm operator at the point $x_0 \in B$.*

Then there exists a neighborhood $V(x_0) \subset \Omega$ such that $f(B \cap V(x_0))$ contains only one point (resp. the set $f(B \cap V(x_0))$ has $N(f''(x_0))/(k+\alpha)$-dimensional Hausdorff measure zero).

Moreover, in the real-analytic case the set $f(B)$ is countable provided H is separable and $f''(x)$ is Fredholm for each $x \in B$. In the $C^{k+1,\alpha}$-case the set $f(B_n)$ has $n/(k+\alpha)$-dimensional Hausdorff measure provided H is separable, where

$$B_n = \{y \in B: f''(y) \text{ is Fredholm and } N(f''(y)) \leq n\}.$$

Theorem 4. *Let f be a real-analytic functional (resp. $f \in C^1(\Omega)$) defined on an open subset Ω of a Banach space X and let Y be another Banach space. Suppose that there exists a bilinear form $\langle .,. \rangle$ on $X \times Y$ such that for fixed $x \in X$, $\langle x, . \rangle$ is continuous on Y and $\langle x, y_0 \rangle = 0$ for all $x \in X \Rightarrow y_0 = 0$. For each $x \in \Omega$ let there exist $F(x) \in Y$ (under our assumptions there exists only one) such that*

$$f'(x)\, h = \langle h, F(x) \rangle \tag{3.1}$$

for each $h \in X$.

Let the operator $F: \Omega \to Y$ be real-analytic (resp. $F \in C^{k,\alpha}(\Omega)$). Let $x_0 \in B$. Suppose that

$$A = F'(x_0) \quad \text{is Fredholm operator.} \tag{3.2}$$

Then there exists a neighborhood $U(x_0)$ in X of the point x_0 such that $f(B \cap U(x_0))$ is a one-point set (resp. the set $f(B \cap U(x_0))$ has $N(A)/(k+\alpha)$-dimensional Hausdorff measure zero).

Moreover, $f(B)$ is countable (resp. $f(B_n)$ has $n/(k+\alpha)$-dimensional Hausdorff measure zero, where

$$B_n = \{y \in B: F'(y) \text{ is Fredholm and } N(F'(y)) \leqq n\})$$

provided X is separable and the condition (3.2) holds for each $x_0 \in B$.

4. Converse of Ljusternik-Schnirelmann theory

As it was remarked in section 2 by the minimax process we do not obtain all critical levels in the nonlinear case. However, the situation is not so bad. Under certain conditions it can be proved again that $\Gamma(r)$ is a sequence of positive numbers converging to zero. This is the main goal of the investigation in this section. While the determination of the lower bound for the number of the points of the set $\Gamma(r)$ is based on topological methods, the upper bound is found on the basis of the Morse-Sard theorem.

Theorem 5 ([15]). *Let us suppose:*

$$f \text{ is a real-analytic functional on a Hilbert space } H, \tag{4.1}$$

$$f(0) = 0,\ f(u) > 0 \text{ for all } u \neq 0\,, \tag{4.2}$$

there exists a continuous and nondecreasing function $c_1(t) > 0$ for $t > 0$ such that for all $u, h \in H$

$$d^2 f(u, h, h) \geqq c_1(f(u))\, \|h\|^2\,, \tag{4.2}$$

$$\text{the set } M_r(f) \text{ is bounded,} \tag{4.4}$$

$$f' \text{ is a bounded mapping,} \tag{4.5}$$

$$\inf_{x \in M_r(f)} (f'(x), x) \geqq c_2 > 0\,, \tag{4.6}$$

$$g \text{ is a real-analytic functional on } H, \tag{4.7}$$

$$g' \text{ is strongly continuous (it maps weakly convergent sequences onto strongly convergent ones),} \tag{4.8}$$

$$g(u) \neq 0 \Rightarrow g'(u) \neq 0\,. \tag{4.9}$$

Then the set $\Gamma(r)$ of all critical levels is at most countable and its only cummulation point can be zero.

Theorem 6 ([13]). *Let f, g be two functionals defined on a real Hilbert space H. Suppose $f, g \in C^{k+1,\alpha}(H)$. Let x_0 be a critical point of the functional g with respect to the manifold $M_r(f)$ and let λ_0 be a corresponding eigenvalue.*

Then under assumptions $f'(x_0) \neq 0$ and $\lambda_0 f''(x_0) - g''(x_0)$ is Fredholm there exists a neighborhood $V(x_0)$ of the point x_0 such that the set $g(S \cap V(x_0))$ has $[N(\lambda_0 f''(x_0) - g''(x_0)) + 1]/(k+\alpha)$-dimensional Hausdorff measure zero, where S is the set of all critical points of the functional g with respect to the manifold $M_r(f)$.

Moreover, if $(f'(x), x) \neq 0$ for $x \in S$ and S_n is the set of all $y \in S$ such that the operator

$$x \mapsto \frac{(g'(y), y)}{(f'(y), y)} f''(x) - g''(x)$$

is Fredholm at the point y and

$$N\left(\frac{(g'(y), y)}{(f'(y), y)} f'' - g''\right) \leqq n ,$$

then the set $g(S_n)$ has $(n+1)/(k+\alpha)$-dimensional Hausdorff measure zero.

Theorem 7 ([16, 13]). *Let X_1, X_2, X be three real Banach spaces such that $X_1 \subset X$ and let the identity mapping from X_1 into X be continuous. Let X be a reflexive Banach space. Suppose that $\langle .,. \rangle$ is a bilinear form on $X_1 \times X_2$ continuous on X_2 for fixed $h \in X_1$ and such that the following implication holds:*

$$\langle h, x\rangle = 0 \quad \text{for each } h \in X_1 \Rightarrow x = 0 .$$

Let f, g be Fréchet differentiable functionals on X. Denote by S the set of all critical points of the functional g with respect to the manifold $M_r(f)$. Let

$$f, g \text{ be real-analytic on } X_1 \ (\text{resp. } f, g \in C^{k+1,\alpha}(X_1)) . \tag{4.10}$$

Suppose that for each $x \in X_1$ there exists a couple $F(x), G(x) \in X_2$ (if there exists at least one couple, then it is unique) such that

$$df(x, h) = \langle h, F(x)\rangle , \tag{4.11}$$

$$dg(x, h) = \langle h, G(x)\rangle \tag{4.12}$$

for each $h \in X_1$ and let

$$F, G\colon X_1 \to X_2 \text{ be real-analytic on } X_1 \ (\text{resp. } F, G \in C^{k,\alpha}(X_1)) \tag{4.13}$$

and

$$G\colon X_1 \to X_2 \text{ be completely continuous on } X_1 . \tag{4.14}$$

Denote $S_1 = \{x \in M_r(f) \cap X_1$: there exists $\lambda \in R_1$ such that

$$\lambda df(x, h) = dg(x, h) \text{ for each } h \in X_1\} .$$

Let $x_0 \in S_1$ and let $\lambda_0 \in R_1$ be the corresponding eigenvalue from the definition of the set S_1. Suppose $\lambda_0 \neq 0$ and

$$F'(x_0)\colon X_1 \to X_2 \text{ is Fredholm} . \tag{4.15}$$

Then there exists a neighborhood $U(x_0)$ in X_1 of the point x_0 such that the set $g(S_1 \cap U(x_0))$ is a one-point set (resp. has $(N(\lambda_0 F'(x_0) - G'(x_0)) + 1))/(k + \alpha)$-dimensional Hausdorff measure zero.

Moreover: a. *Real-analytic case. If for each $\delta > 0$ the set $S(\delta)$ is a compact subset of X_1, where*

$$S(\delta) = \{x \in M_r(f) : \text{there exists } \lambda,\ |\lambda| \geqq \delta,\ \text{such that } \lambda f'(x) = g(x)\}\,, \tag{4.16}$$

$$g' : X \to X^* \text{ is a strongly continuous}\,, \tag{4.17}$$

$$g(u) \neq 0 \Rightarrow g'(u) \neq 0\,, \tag{4.18}$$

$$f' : X \to X^* \text{ is continuous and bounded,} \tag{4.19}$$

$$f'(x) \neq 0 \quad \text{for} \quad x \in M_r(f)\,, \tag{4.20}$$

$$\lim_{\|x\| \to \infty} f(x) = \infty\,. \tag{4.21}$$

Then the set $g(S) - \{0\}$ is isolated and thus the set $\Gamma(r)$ is at most countable and the only possible accumulation point can be zero.

b. *$C^{k,\alpha}$-case. Let X_1 be separable. Denote by S_n the set of all $y \in S \cap X_1$ such that the mapping*

$$x \mapsto \frac{(g'(y), y)}{(f'(y), y)} F'(x) - G'(x)$$

is Fredholm at the point y and

$$N\left(\frac{(g'(y), y)}{(f'(y), y)} F' - G'\right) \quad \text{at } y \text{ is } \leqq n\,.$$

Then the set $g(S_n)$ has $(n + 1)/(k + \alpha)$-dimensional Hausdorff measure zero.

Real-analytic case of Theorem 7 with slightly modified assumptions is proved in [18]. Summarizing, if the assumptions of Theorem 2 and real-analytic Theorem 7 are fulfilled together, we obtain the set $\Gamma(r)$ is a sequence of positive numbers converging to zero. Let the functional f be a-homogeneous and g be b-homogeneous. Then

$$\Gamma(r) = r^{b/a}(a/b)\,\Lambda(1)\,, \tag{4.22}$$

for $(f'(u), u) = a\,f(u)$ and $(g'(u), u) = b\,g(u)$.

Thus from the relation (4.22) it follows that under homogenity of functionals f and g and the assumptions of Theorem 2 and 7 the set of all eigenvalues is a sequence $\{\lambda_n\}$, $\lambda_n \downarrow 0$.

5. Remark to M. A. Krasnoselskii's main bifurcation theorem

Suppose that X and Y are real Banach spaces and T, S are the mappings of X into Y such that $T(0) = S(0) = 0$. Define $\lambda_0 \in R_1$ to be bifurcation point of the couple (T, S) if, for every $\varepsilon, \delta > 0$ there exist $x \in X$, $\lambda \in R_1$ with $0 < \|x\| \leqq \varepsilon$ and $|\lambda - \lambda_0| \leqq \delta$ such that $\lambda T(x) = S(x)$.

In many papers it is considered the case $X = Y$, T is the identity operator $(T = I)$ and $S = B + C$, where B is the Fréchet derivative of the operator S

at the point 0 and

$$\lim_{||x|| \to 0} \frac{||C(x)||}{||x||} = 0 .$$

It is known (see M. A. Krasnosjelski [24, p. 195]) that if B is completely continuous linear operator, then the number $\lambda_0 \in R_1$ which is the bifurcation point of the couple (I, S) is the eigenvalue of the operator B. If the operator S is completely continuous, then the operator B is completely continuous, too and, each eigenvalue of the odd multiplicity of the operator B is the bifurcation point of the couple (I, S) (see [24, p. 199]). It is possible that the eigenvalue of even multiplicity of the operator B is not the bifurcation point of the couple (I, S) (see example in [24, p. 196]). But if the operator S is a gradient of some functional g, then each eigenvalue of the operator B is the bifurcation point of the couple (I, S); precisely:

Suppose that $S\big(S(0) = 0\big)$ is a completely continuous nonlinear operator on the real Hilbert space X which is the Fréchet derivative of the weakly continuous functional $g\big(g(0) = 0\big)$. Let the functional g be uniformly differentiable on some neighborhood of the point 0. Denote by B the Fréchet derivative of the operator S at the point 0.

Then each eigenvalue of the operator B is a bifurcation point of the couple (I, S). (It is so-called M. A. Krasnosjelski's Main Bifurcation Theorem the proof of which is given in [24, p. 332] and in [25].) In the following theorem we consider the equations of the type $\lambda T(x) = S_1(x) + S_2(x)$, where T, S_1, S_2 are operators on a Banach space X with values in the dual space X^*. We suppose that T and S_1 are a-homogeneous operators and S_2 is a nonlinear "higher order" operator.

Theorem 8 ([19]). *Let f, g_1, g_2 be three even Fréchet differentiable functionals on a reflexive infinitedimensional Banach space X with a Schauder basis. Let f and $g = g_1 + g_2$ satisfy all assumptions of Theorem 2 for all sufficiently small $r > 0$, let f, g_1 be a -homogeneous and let the functional g_1 satisfy* (2.7)—(2.10) *and* (2.12). *Suppose, moreover:*

$$\lim_{r \to 0+} r^{-1} \sup_{x \in M_r(f)} |g_2(x)| = 0 , \tag{5.1}$$

$$\lim_{||x|| \to 0} ||x||^{1-a} g_2'(x) = 0 . \tag{5.2}$$

Denote by B the set of all bifurcation points of the couple (f', g'), $\Lambda_0(1)$ (resp. $\Lambda(1)$) the set of all LS-eigenvalues (resp. eigenvalues) [for definition see Section 2] of the couple (f', g_1').

Then $\Lambda_0(1) \subset B \subset \Lambda(1)$.

If X is a Hilbert space and the operator f' is the identity, g_1' is linear and assumptions of theorem 8 are fulfilled, then we obtain the assertion of M. A. Krasnosjelski's main bifurcation theorem, for $\Lambda_0(1) = \Lambda(1)$ according to Courant-Weinstein principle. Generally, $\Lambda_0(1) \neq \Lambda(1)$ (see Section 2). To obtain the upper (resp. lower) bound for the number of the points in the set B it is sufficient to give the conditions for the number of points in the set $\Lambda(1)$ (resp. $\Lambda_0(1)$). With such problems Sections 2 and 4 deal.

6. Nonlinear integral equations

This section deals with the solvability and spectrum of nonlinear integral equations which are sometimes called the equations of Lichtenstein type (or generalized Hammerstein type). The study of such equations is based on Sections 1–5.

Theorem 9 ([18]). *Let n be an odd integer, Ω bounded domain in R_N with boundary $\partial\Omega$,* means $\partial\Omega = 0$. *Suppose that $K(s, t_1, \dots, t_n)$ is continuous function on $\bar{\Omega} \times \bar{\Omega}_n$, where $\Omega_n = \Omega \times \cdots \times \Omega$ (n-times). Let K be symmetric in all variables and let $\varepsilon > 0$.*

Then there exists at most countable set $\Lambda \subset R_1$ such that the equation

$$\lambda \left(u^n(s) + \varepsilon \left(\int_\Omega u^2(x)\, dx \right)^{\frac{n-1}{2}} u(s) \right) - \int_{\Omega_n} K(s, t_1, \dots, t_n)\, u(t_1) \cdots u(t_n)\, dt_1 \cdots dt_n = 0 \tag{6.1}$$

has a nontrivial solution $u \in L_{n+1}(\Omega)$ iff $\lambda \in \Lambda$.

Morevoer, only possible limit point of the set Λ is zero.

For the proof of theorem 9 set in theorem 7:

$$X_3 = L_{n+1}(\Omega)\,, \qquad X_1 = X_2 = C(\bar{\Omega})\,, \qquad r = 1\,,$$

$$f(u) = 1/(n+1) \left[\int_\Omega (u(x))^{n+1}\, dx + \varepsilon \left(\int_\Omega u^2(x)\, dx \right)^{\frac{n+1}{2}} \right],$$

$$g(u) = 1/(n+1) \int_\Omega \int_{\Omega_m} K(s, t_1, \dots, t_n)\, u(s)\, u(t_1) \cdots u(t_n)\, ds\, dt_1 \cdots dt_n\,,$$

$\langle .\,, . \rangle$ the inner product in $L_2(\Omega)$.

Theorem 10 ([18, 39]). *Let K be the kernel on $\bar{\Omega} \times \bar{\Omega}_n$ introduced in Theorem 9. Let $\varepsilon \geqq 0$ and $h(t_1, \dots, t_n, u_1, \dots, u_n)$ be function defined on $\Omega_n \times R_n$, satisfying Caratheodory condition and the growth condition*

$$|h(t_1, \dots, t_n, u_1, \dots, u_n)| \leqq c \sum_{k=1}^{n} |u_k|^n + d(t_1, \dots, t_n)\,, \tag{6.2}$$

where $c > 0$ and $d \in L_{1+(1/n)}(\Omega_n)$.

Moreover, suppose the asymptotic condition

$$|\tau^{-n} h(t_1, \dots, t_n, \tau u_1, \dots, \tau u_n) - a(t_1, \dots, t_n)\, u_1 \cdots u_n| \leqq c(\tau) \sum_{k=1}^{n} |u_k| + d'(\tau, t_1, \dots, t_n) \tag{6.3}$$

with $\lim_{\tau\to\infty} c(\tau) = 0$,

$$\lim_{\tau\to\infty} \int_{\Omega_n} |d'(\tau, t_1, \dots, t_n)|^{1+(1/n)}\, dt_1 \cdots dt_n = 0$$

and $a \in L_\infty(\Omega_n)$.

Under assumptions (6.2) *and* (6.3) *the equation*

$$\lambda \left(u^n(s) + \varepsilon \left(\int_\Omega u^2(x)\, dx \right)^{\frac{n-1}{2}} u(s) \right) - \int_{\Omega_n} K(s, t_1, \dots, t_n)\, h(t_1, \dots, t_n, u(t_1), \dots, u(t_n))\, dt_1, \dots, dt_n = z(s) \tag{6.4}$$

with $\lambda \neq 0$ *is solvable in* $L_{n+1}(\Omega)$ *for each* $z \in L_{1+(1/n)}(\Omega)$ *provided the equation*

$$\lambda \left(u^n(s) + \varepsilon \left(\int_\Omega u^2(x)\, dx \right)^{\frac{n-1}{2}} u(s) \right)$$

$$- \int_{\Omega_n} K(s, t_1, \dots, t_n)\, a(t_1, \dots, t_n)\, u(t_1) \cdots u(t_n)\, dt_1 \cdots dt_n = 0 \qquad (6.5)$$

has in $L_{n+1}(\Omega)$ *the trivial solution only.*

This theorem follows from theorem 1 setting

$$(T_0 u)(s) = u^n(s) + \varepsilon \left(\int_\Omega u^2(x)\, dx \right)^{\frac{n-1}{2}} u(s)\,,$$

$$(Su)(s) = \int_{\Omega_n} K(s, t_1, \dots, t_n)\, h\big(t_1, \dots, t_n, u(t_1), \dots, u(t_n)\big)\, dt_1 \cdots dt_n\,,$$

$$(S_0 u)(s) = \int_{\Omega_n} K(s, t_1, \dots, t_n)\, a(t_1, \dots, t_n)\, u(t_1) \cdots u(t_n)\, dt_1 \cdots dt_n\,.$$

Theorem 11 ([18]). *Let all assumptions of theorem 9 be fulfilled and let*

$$\int_{\Omega_{n+1}} K(s, t_1, \dots, t_n)\, u(s)\, u(t_1) \cdots u(t_n)\, ds\, dt_1 \cdots dt_n > 0$$

for $u \in L_{n+1}(\Omega)$, $u \neq 0$.

Then the set Λ *is precisely countable and the only one limit point of the set* Λ *is zero.*

If $\varepsilon = 0$, then we can apply theorem 2 and theorem 7 is not applicable. The results of section 5 may be apply to the bifurcation points of the equation

$$\lambda \left(u^n(s) + \varepsilon \left(\int_\Omega u^2(x)\, dx \right)^{\frac{n-1}{2}} u(s) \right) = \int_{\Omega_n} K_n(s, t_1, \dots, t_n)\, u(t_1) \cdots u(t_n)\, dt_1 \cdots dt_n$$

$$+ \sum_{\substack{j=n+2 \\ j=1 \,(\mathrm{mod}\, 2)}}^{m} \int_{\Omega_j} K_j(s, t_1, \dots, t_j)\, u(t_1) \cdots u(t_j)\, dt_1 \cdots dt_j\,.$$

7. Ordinary differential equations

Since this paper does not deal with the problems in the most general case we give in following sections only examples of equations satisfying assumption of previous sections. The general case is solved in the papers referred above.

Example 1. Let h be a real function, $h \in C^{k,\alpha}(R_1)$. We seek a weak solution $u \in \mathring{W}_2^m(\langle 0, 1 \rangle)$ of the boundary value problem

$$\left. \begin{aligned} &\lambda(-1)^m u^{(2m)} + h(u) = 0\,, \\ &u^{(i)}(0) = u^{(i)}(1) = 0\,, \qquad i = 0, 1, \dots, m-1\,. \end{aligned} \right\} \qquad (7.1)$$

According to Theorem 6 the set $\Gamma(r)$ has $(m+1)/(k+\alpha)$-dimensional Hausdorff measure zero for each $r > 0$.

Example 2. The same assertion as in example 1 is for boundary value problem

$$\left.\begin{aligned} &\lambda[(-1)^m u^{(2m)}(t) + h_1(t, u(t))] + h_2(t, u(t)) = 0 \,, \\ &u^{(i)}(0) = u^{(i)}(1) = 0 \,, \qquad i = 0, 1, \ldots, m-1 \,, \end{aligned}\right\} \tag{7.2}$$

under some additional assumptions on the functions h_1, h_2.

Example 3. Consider the boundary value problem

$$\left.\begin{aligned} &\lambda(-1)^m u^{(2m)} + |u|^{q-2} u = 0 \,, \\ &u^{(i)}(0) = u^{(i)}(1) = 0 \,, \qquad i = 0, 1, \ldots, m-1 \end{aligned}\right\} \tag{7.3}$$

with $q > 2$. Set

$$f(u) = 1/2 \int_0^1 |u^{(m)}(t)|^2 \, dt \,,$$

$$g(u) = 1/q \int_0^1 |u(t)|^2 \, dt \,.$$

The functional f is 2-homogeneous and g is q-homogeneous. Moreover, $f, g \in C^{k,\alpha}(\mathring{W}_2^m(\langle 0, 1\rangle))$, where $k = [q]$, $\alpha = q - [q]$ ($[q]$ denotes entire part of q). From theorem 6 we obtain: the set of all critical levels $\Gamma(r)$ and the set of all eigenvalues $\Lambda(r)$ have $(m+1)/(q-1)$-dimensional Hausdorff measure zero. If q is an odd integer, then theorem 5 together with theorem 2 are applicable and thus the set $\Lambda(r)$ is a sequence of positive numbers converging to zero.

Example 4. By a quite different method it was proved ([41, 42]) that the set of all eigenvalues for the nonlinear Sturm-Liouville equation

$$-\lambda[(a|u'|^{p-2} u')' + b|u|^{p-2} u] - c|u|^{p-2} u = 0$$

with suitable boundary value conditions is a sequence converging to zero. The same result for a fourth order ordinary differential equation is included in the work [26].

Note that such equations are not included in the theory given in Section 4, for the conditions about Fredholmian second derivative of considered functionals are not fulfilled.

8. Partial differential equations

Example 5. The boundary value problem

$$\left.\begin{aligned} &-\lambda \sum_{i=1}^N \frac{\partial}{\partial x_i}\left(\frac{\partial u}{\partial x_i}\right)^3 - |u|^{m-1} u = f \,, \\ &u = 0 \quad \text{on} \quad \partial\Omega \end{aligned}\right\} \tag{8.1}$$

has for $1 \leq m < 3$ the weak solution $u \in \mathring{W}_4^1(\Omega)$ for each $f \in (\mathring{W}_4^1(\Omega))^*$ and all $\lambda \neq 0$. For $m = 3$ the same problem has a weak solution for arbitrary $f \in (\mathring{W}_4^1(\Omega))^*$ provided

$$\left.\begin{aligned} &-\lambda \sum_{i=1}^N \frac{\partial}{\partial x_i}\left(\frac{\partial u}{\partial x_i}\right)^3 - |u|^2 u = 0 \,, \\ &u = 0 \quad \text{on} \quad \partial\Omega \end{aligned}\right\} \tag{8.2}$$

has a trivial solution only. On the problem (8.2) the Ljusternik-Schnirelmann theory is applicable, but the converse not.

Example 6. The same result as in example 5 takes place for the problem

$$\left.\begin{aligned} &-\lambda \sum_{i=1}^{N} \frac{\partial}{\partial x_i}\left(\frac{\partial u}{\partial x_i}\right)^3 - (1 + |u|^{m-1})\, u = f\,, \\ &u = 0 \quad \text{on} \quad \partial\Omega\,. \end{aligned}\right\} \tag{8.3}$$

Example 7. Consider the weak solution of the Dirichlet boundary value problem

$$\left.\begin{aligned} &\lambda(-1)^{m+1}\, \Delta^m u + g(u) = 0\,, \\ &D^\alpha u = 0 \quad \text{on} \quad \partial\Omega\,, \qquad |\alpha| \leqq m-1\,. \end{aligned}\right\} \tag{8.4}$$

If $2\,m < N$ we suppose that g is a polynomial function of the degree $k < \dfrac{N+2\,m}{N-2\,m}$. Then the assumptions of theorem 2 and theorem 7 (real-analytic) are satisfied. The same problem can be solved on the base of our abstract theorems in the case $2\,m \geqq N$, too.

Example 8. Consider the boundary value problem

$$\left.\begin{aligned} &-\lambda \sum_{i=1}^{N} \frac{\partial}{\partial x_i}\left[\left(1 + \sum_{j=1}^{N} \left|\frac{\partial u}{\partial x_j}\right|^2\right)^{(m/2)-1} \frac{\partial u}{\partial x_i}\right] - (1 + u^2)^{(m/2)-1}\, u = 0\,, \\ &u = 0 \quad \text{on} \quad \partial\Omega\,, \end{aligned}\right\} \tag{8.5}$$

$1 < m < \infty$. The problem (8.5) satisfies all assumptions of real-analytic theorem 7 and theorem 2 (the existence of Schauder bases in $\mathring{W}^k_m(\Omega)$ is proved in [12, 48]).

9. Integrodifferential equations

Unfortunately, we do not know an example of boundary value problem for nonlinear differential (ordinary or partial) equation to be satisfied together all assumptions of Fredholm alternative for homogeneous operators, Ljusternik-Schnirelmann theory and its converse. Such examples can be given for integral (section 6) and for boundary value problem for integrodifferential equations (see [47]).

Example 9. Consider two functionals

$$\left.\begin{aligned} &f(u) = \sum_{i=1}^{N} \int_\Omega \left(\frac{\partial u}{\partial x_i}\right)^4 dx + \left(\int_\Omega u^2\, dx\right)^2, \\ &g(u) = \int_\Omega u^4\,. \end{aligned}\right\} \tag{9.1}$$

Then all assumptions of theorem 1, theorem 2 and real-analytic theorem 7 are fulfilled. Thus we have: $(\lambda f' - g')\,(\mathring{W}^1_4) = (\mathring{W}^1_4)^*$ provided $0 \neq \lambda \notin \Lambda(1)$. Moreover, $\Lambda(1)$ is a sequence of positive numbers converging to zero. The big role in the proof the regularity of the weak solution plays. Analogous example can be given for higher order integrodifferential equations solved on the domains in two dimensional space.

10. Open problems

Problem 1. Up to this time it is not solved problem about the range of nonlinear equations $\lambda f' - g'$ in the case that λ is an eigenvalue of the "asymptotic equation" $\lambda f'_0 - g'_0$. Some examples are given in the paper [2], where the range is not whole X^* and it is closed in X^*, and in the papers [30, 50], where the range of certain differential operators is open and not whole X^*. Nobody know to describe the set of all functions $f \in W_2^{-1}(0, 1)$ for which there exists a weak solution of the boundary value problem

$$\left.\begin{aligned} -\lambda u'' - u + \frac{u}{1 + |u|^2} &= f\,, \\ u(0) = u(1) &= 0 \end{aligned}\right\} \tag{10.1}$$

provided λ is an eigenvalue of the equation

$$\left.\begin{aligned} \lambda u'' + u &= 0\,, \\ u(0) = u(1) &= 0\,. \end{aligned}\right\}$$

Problem 2. It is unknown under which assumptions if there exists an eigenvalue $\lambda = 0$ of $\lambda f' - g'$ (f and g are a-homogeneous) to be not the range of $\lambda f' - g'$ whole of X^*. In the finite-dimensional spaces this is also unknown.

Problem 3. Let f be a-homogeneous, g be b-homogeneous. Under certain assumptions (to be valid the conditions in theorem 1) the range of $\lambda f' - g'$ is whole of X^* provided $\lambda \neq 0$ and $a > b$. For $a = b$ we have the Fredholm alternative. What is in the case $a < b$? Such typical example is the following boundary value problem

$$\left.\begin{aligned} u'' + u^3 &= f\,, \\ u(0) = u(1) &= 0\,. \end{aligned}\right\} \tag{10.2}$$

It can be proved that there exists $\varrho > 0$ such that for $\|f\|_C < \varrho$ the problem (10.1) is solvable. Moreover, for $f = \text{const.}$, too. If $f(0) = f(1) = 0$ and $f \in C(\langle 0, 1 \rangle)$ the problem (10.1) is solvable (see [6]). Is it solvable for each $f \in C$? (Note that each Galerkin's approximate exists. Does a sequence of Galerkin's approximants converge?)

Problem 4. Does exist an example of boundary value problem for nonlinear differential equation to be satisfied all assumptions of theorem 1, 2 and 7?

References

[1] Ambrosetti, A., Teoria di Lusternik-Schnirelman su variet à con bordo negli spazi di Hilbert. Rend. Sem. Mat. Univ. Padova XLV (1971) 336–353.

[2] Ambrosetti, A., and Prodi, G., On the inversion of some differentiable mappings with singularities between Banach spaces. Proc. of Czech. summer school, Babylon (1971). Theory of Nonlinear Operators, pp. 9–28.

[3] Berger, M. S., An eigenvalue problem for nonlinear elliptic partial differential equations. Trans. Amer. Math. Soc. **120** (1965) 145–184.

[4] Browder, F. E., Infinite-dimensional manifolds and nonlinear eigenvalue problems. Ann. of Math. **82** (1965) 459–477.

[5] Citlanadse, Je. S., Теоремы существования точек минимакса в пространствах Банаха и их приложения. Труды матем. о-ва **2** (1953) 235—274.

[6] Conti, R., and Sansone, G., Non-linear differential equations. Pergamon Press, New York (1964).

[7] Dubowizki, A. Ja., О множествах точек вырождения. Известия АН СССР **31** (1967) 27—36.

[8] Flashka, H., A nonlinear spectral theorem for abstract Nemitsky Operators. Indiana Univ. Math. Journ. **21** (1971) 265–276.

[9] Fučík, S., Fredholm alternative for nonlinear operators in Banach spaces and its applications to the differential and integral equations, Comment. Math. Univ. Carolinae **11** (1970) 271–284.

[10] —, Fredholm alternative for nonlinear operators in Banach spaces and its applications to differential and integral equations. Čas. Pěst. Mat. **96** (1971) 371–390.

[11] —, Note on the Fredholm alternative for nonlinear operators. Comment. Math. Univ. Carolinae **12** (1971) 213–226.

[12] Fučík, S., John, O., and Nečas, J., On the existence of Schauder bases in Sobolev spaces. Comment. Math. Univ. Carolinae **13** (1972) 163–175.

[13] Fučík, S., Kučera, M., Nečas, J., Souček, J., and Souček, V., Morse-Sard theorem in infinitedimensional spaces and the investigation of critical levels. Čas. Pěst. Mat. **99** (1974).

[14] Fučík, S., and Nečas, J., Ljusternik-Schnirelmann theorem and nonlinear eigenvalue problems. Math. Nachr. **53** (1972) 277–289.

[15] Fučík, S., Nečas, J., Souček, J., and Souček, V., Upper bound for the number of eigenvalues for nonlinear operators. Ann. Scuola Norm. Sup. Pisa **27** (1973), 53–71.

[16] —, Upper bound for the number of critical levels for nonlinear operators in Banach spaces of the type of second order nonlinear partial differential operators. Journ. Funct. Anal. **11** (1972), 314–333.

[17] —, New infinite dimensional versions of Morse-Sard theorem. Boll. U. Mat. Ital. **6** (1972), 317–322.

[18] —, Note to nonlinear spectral theory: Application to the nonlinear integral equations of the Lichtenstein type. Math. Nachr. **58** (1973), 257–267.

[19] —, Remark to M. A. Krasnoselskii's main bifurcation theorem. Arch. Rat. Mech. Anal. (to appear).

[20] Hess, P., Nonlinear functional equations in Banach spaces and homotopy arguments, Bull. Amer. Math. Soc. **77** (1971), 211–215.

[21] —, On nonlinear mappings of monotone type homotopic to odd operators. Journ. Funct. Anal. **11** (1972), 138–167.

[22] —, On a method of singular perturbation type for proving the solvability of nonlinear equations in Banach spaces. Math. Z., (to appear).

[23] Katschurowski, R. I., Регулярные точки, спектр и собственные функции нелинейных операторов. ДАН СССР **188**, 2 (1969) 274—277.

[24] Krasnosjelski, M. A., Топологические методы в теории нелинейных интегральных уравнений. Москва (1956).

[25] —, Применение вариационных методов в задаче о точках бифуркации. Матем. сб. **33**, 1 (1953) 199—214.

[26] Kratochwil, A. and Nečas, I., О дискретности спектра нелинейного уравнения Штурма-Лиувилля четвертого порядка. Comment. Math. Univ. Carolinae **12** (1971) 639–653.

[27] Kučera, M., Fredholm alternative for nonlinear operators, Comment. Math. Univ. Carolinae **11** (1970) 337–363.

[28] —, Hausdorff measures of the set of critical values of functions of the class $C^{k,\lambda}$. Comment. Math. Univ. Carolinae **13** (1972) 333–350.

[29] Kupka, I., Counterexample to the Morse-Sard theorem in the case of infinite-dimensional manifolds. Proc. Amer. Math. Soc. **16** (1965) 954–957.

[30] Landesman, E. M., and Lazer, A. C., Nonlinear perturbations of linear elliptic boundary value problems at resonance. J. Math. Mech. **19** (1970) 609–623.

[31] Ljusternik, L. A., Об одном классе нелинейных операторов в гильбертовом пространстве. Известия АН СССР, сер. матем. 5, 1939, 357—364.

[32] Ljusternik, L. A., and Schnirelman, I. T., Применение топологии к экстремальным задачам. Труды 2-го Всесоюзного матем. с'езда, т. **1** (1934) 224—237.

[33] —, Топологические методы в вариационных задачах и ихириложения к дифференциальной геометрии поверхностей. УМН **2**, 1 (1947) 166—217.

[34] Morse, A. P., The behaviour of a function on its critical set. Ann. of Math. **40** (1939) 62–70.

[35] Naumann, J., A characterization of the eigenvalues of a completely continuous selfadjoint operators, Comment. Math. Univ. Carolinae **13** (1972) 63–78.

[36] —, Remarks on nonlinear eigenvalue problems, Proc. Czech. summer school, Babylon (1971). Theory of Nonlinear Operators, pp. 61–84.

[37] —, Lusternik-Schnirelman-Theorie und nichtlineare Eigenwertprobleme. Math. Nachr. **53** (1972) 303–336.

[38] Nečas, J., Sur l'alternative de Fredholm pour les opérateurs non-linéaires avec applicationa aux problémes aux limites. Ann. Scuola Norm. Sup. Pisa, XXIII (1969) 331–345.

[39] —, Remark on the Fredholm alternative for nonlinear operators with application to nonlinear equations of generalized Hammerstein type, Comment. Math. Univ. Carolinae **13** (1972) 109–120.

[40] —, Fredholm alternative for nonlinear operators and application to partial differential equations and integral equations. Čas. Pěst. Mat. **97** (1972) 65–71.

[41] —, О дискретности спектра нелинейного уравнения Штурма-Лиувилля. ДАН СССР **201** (1971) 1045—1048.

[42] —, Spectral analysis of nonlinear operators, Proc. Czech. summer school, Babylon (1971). Theory of Nonlineare Operators, pp. 85–119.

[43] Petryshyn, W. V., Nonlinear equations involving noncompact operators. Proc. Symp. Pure Math. Vol. 18, Part 1, Nonlinear functional analysis, Rhode Island (1970).

[44] Pochoshajew, S. I., О разрешимости нелинейных уравнений с нечетными операторами. Функцион. анализ. и его прилож. **1** (1967) 66—73.

[45] —, О множестве критических значений функционалов. Мат. сб. **75**, 117 (1968) 106—111.

[46] Souček, J., and Souček, V., The Morse-Sard theorem for real-analytic functions. Comment. Math. Univ. Carolinae **13** (1972) 45–51.

[47] Stará, J., Fučík, S., Nečas, J., Souček, J., and Souček, V., Note to nonlinear spectral theory: Application to boundary value problems for integrodifferential equations, (in progress).

[48] Triebel, H., Über die Existenz von Schauderbasen in Sobolew-Besov-Räumen. Isomorphiebeziehungen. Studia Math. XLIV (1973), 83–100.

[49] Webb, J. R. L., A fixed point theorem and application to functional equations in Banach spaces, Boll. U. Mat. Ital. **4** (1971) 755–788.

[50] Williams, S. A., A sharp sufficient condition for solution of a nonlinear boundary value problem. J. Diff. Eq. **8** (1970) 580–586.

Anschrift der Autoren:

Svatopluk Fučík, Vladimir Soucek, Department of Math. Analysis, Faculty of Math. and Physics, Charles University, 83 Sokolovská, Prague 8-Karlín, Czechoslovakia.

Milan Kučera, Jiří Souček, Mathematical Instutite, Czechoslovak Academy of Sciences, 25 Žitná, Prague 1, Czechoslovakia.

Zur Fehlerabschätzung bei Gleichungen mit nichtlinearen Potentialoperatoren

HERBERT GAJEWSKI, Berlin

1. Einleitung

Sei im folgenden stets X ein reller, reflexiver Banachraum und X^* der zu X duale Raum. Wir bezeichnen durch $||\cdot||$ bzw. $||\cdot||_*$ die Norm in X bzw. X^* und durch (f, x) den Wert des linearen Funktionals $f \in X^*$ im Element $x \in X$. Wir betrachten die Gleichung

$$Ax = f\,, \tag{1}$$

dabei sei $A \in (X \to X^*)$ ein (möglicherweise) nichtlinearer Operator und $f \in X^*$ ein vorgegebenes Element. Wir setzen zunächst einmal voraus:

1. A ist hemistetig, d. h., die Form $\big(A(x + ty), h\big)$ ist für beliebige feste $x, y, h \in X$ stetig bezüglich $t \in [0, 1]$;
2. A ist stark monoton, d. h. $(Ax - Ay, x - y) \geqq m||x - y||^2$, $\forall x, y \in X$, $0 < m = \text{const.}$;

dann hat Gleichung (1) genau eine Lösung, die wir im folgenden durch x_0 bezeichnen wollen.

Definition: A heißt *Potentialoperator*, wenn ein Funktional $F \in (X \to R^1)$ derart existiert, daß gilt

$$\lim_{t\to 0} \frac{F(x + th) - F(x)}{t} = (Ax, h),\ \forall x, h \in X\,.$$

Sei A im folgenden Potentialoperator, dann gilt wegen der Hemistetigkeit von A offenbar $F(x) = \int_0^1 (Atx, x)\,dt$. Für das Funktional $G(x) = F(x) - (f, x)$ gilt mit beliebigem $x \in X$

$$\begin{aligned} G(x) - G(x_0) &= \int_0^1 \big(A(x_0 + t(x - x_0)), x - x_0\big)\,dt - (f, x - x_0) \\ &= \int_0^1 \big(A(x_0 + t(x - x_0)) - Ax_0, t(x - x_0)\big)\frac{dt}{t} \\ &\geqq m||x - x_0||^2 \int_0^1 t\,dt = \frac{m}{2}\,||x - x_0||^2\,. \end{aligned}$$

Daraus finden wir einerseits die bekannte Aussage

$$G(x_0) = \min_{x\in X} G(x)\,, \tag{2}$$

zum anderen für eine beliebige Näherung $\bar{x}$ von x_0 die Fehlerabschätzung

$$||\bar{x} - x|| \leqq \left[\frac{2}{m}\left(G(\bar{x}) - G(x_0)\right)\right]^{1/2}. \tag{3}$$

Die praktische Verwendung von (3) setzt die Kenntnis geeigneter Methoden voraus, den Minimalwert $d = G(x_0)$ nach unten abzuschätzen. Von Bittner [1, 2] stammt die Idee, dazu ein zu (2) „reziprokes" Variationsproblem heranzuziehen und letzteres mittels Aussagen der von Fenchel begründeten Dualitätstheorie konvexer Funktionale herzuleiten.

Wir werden in unserem Vortrag einige der von Bittner [12] veröffentlichten Ergebnisse vereinfachen und ergänzen. Schlüssel dafür ist eine gegenüber [1, 2] neue — wie es scheint die natürliche — Charakterisierung des zum Potential F eines Potentialoperators A im Fenchelschen Sinn konjugierten Funktionals und zwar als Potential des zu A inversen Operators. Grundlegend für diese Charakterisierung ist die zuerst von Langenbach [5] bewiesene Feststellung, daß — unter geeigneten Voraussetzungen an A bzw. A^{-1} mit A auch A^{-1} (überhaupt) Potentialoperator ist. Diese Aussage werden wir in einer für unsere Zwecke geeigneten Form neu beweisen. Bevor wir zur detaillierten Darstellung der angekündigten Ergebnisse kommen, gehen wir kurz und kritisch auf die möglichen Alternativen zu den hier in Rede stehenden, auf Dualitätssätzen beruhenden Fehlerabschätzungen bei nichtlinearen Gleichungen der Form (1) ein.

a) Defektabschätzungen: Aus der starken Monotonie von A für beliebiges $x \in X$ sofort

$$m||x - x_0||^2 \leqq (Ax - Ax_0, x - x_0) = (Ax - f, x - x_0) \leqq ||Ax - f||_* \, ||x - x_0|| \,.$$

Für eine beliebige Näherung $\bar{x}$ von x_0 erhalten wir daraus die Fehlerabschätzung

$$||\bar{x} - x_0|| \leqq \frac{1}{m} ||A\bar{x} - f||_* \,. \tag{4}$$

Diese Fehlerabschätzung hat den wesentlichen Mangel, daß die praktische Berechnung des Ausdrucks $||A\bar{x} - f||_*$ die Lösung der Gleichung $Jx = A\bar{x} - f$ (J = Dualitätsabbildung von X) nötig macht, es gilt dann $||A\bar{x} - f||_* = ||x||$. Bei der in den Anwendungen häufig gegebenen Raumkonstellation $X \subset H \subset X^*$ mit stetiger, dichter Einbettung von X in H gilt zwar im oft realisierten Spezialfall $A\bar{x} - f \in H$, $||A\bar{x} - f||_* \leqq ||A\bar{x} - f||_H$, jedoch folgt im allgemeinen nicht einmal für stetige $A \in (X \to X^*)$, $||A\bar{x} - f||_H \to 0$, wenn $||\bar{x} - x_0|| \to 0$. Die Methode der kleinsten Fehlerquadrate, deren Näherungen $\bar{x}$ nach Definition den Defekt $||A\bar{x} - f||_H$ minimieren, hat bei nichtlinearen Gleichungen der Form (1) insbesondere gegenüber dem Galerkinverfahren eine Reihe theoretischer und numerischer Nachteile.

b) Fehlerabschätzung auf der Grundlage der Regularitäts- und Approximationstheorie: Sei $M \subset X$ ein endlich dimensionaler Teilraum von X. Sei $\bar{x} \in M$ die Lösung der Funktionalgleichung

$$(Ax, h) = (f, h)\,, \qquad h \in M\,. \tag{5}$$

(Die Existenz und Einzigkeit der als Galerkinnäherung bezüglich des Unterraums M anzusehenden Lösung $\bar{x}$ von (5) ist bei unseren Voraussetzungen über A offenbar gesichert.)

Für beliebiges $v \in M$ gilt

$$0 = (A\bar{x} - Ax_0, \bar{x} - v) = (A\bar{x} - Ax_0, \bar{x} - x_0) + (A\bar{x} - Ax_0, x_0 - v)$$
$$\geqq m\|\bar{x} - x_0\|^2 - \|A\bar{x} - Ax_0\|_* \|x_0 - v\| .$$

Ist zusätzlich A Lipschitz-stetig, dann folgt daraus

$$\|\bar{x} - x_0\| \leqq \frac{L}{m} \|x_0 - v\| . \tag{6}$$

Diese Beziehung kann man wie folgt zur Fehlerabschätzung benutzen, sofern man über zusätzliche, d. h. über die Inklusion $x_0 \in X$ hinausgehende Informationen bezüglich der Regularität von x_0 verfügt. Sei $X_1 \subset X$ ein weiterer Banachraum mit der Norm $\|\cdot\|_1$ und sei bekannt, daß $x_0 \in X_1$ und $\|x_0\|_1 \leqq C$. Dann finden wir aus (6) sofort

$$\|\bar{x} - x_0\| \leqq C \cdot d(M, X_1) , \qquad d(M, X_1) = \sup_{\|x\|_1 \leqq 1} \inf_{v \in M} \|x - v\| . \tag{7}$$

Die Abschätzung von $d(M, X_1)$ ist ein Problem der Approximationstheorie. Insbesondere für aus sogenannten Spline-Funktionen erzeugte Familien von Approximationsräumen $\{M_h\}$ (h ist hier ein reeller Parameter, der bei Randwertproblemen die Unterteilung des zugrundeliegenden Gebiets charakterisiert) hat man die entsprechenden Größen $d(M_h, X_1)$ in den letzten Jahren intensiv untersucht und asymptotische Abschätzungen für $h \to 0$ angegeben. Der Benutzung von (4) zur effektiven Fehlerabschätzung steht jedoch in der Regel die Tatsache entgegen, daß auch bei Vorliegen geeigneter Regularitätsaussagen und Apriori-Abschätzungen bezüglich x_0, die in den genannten asymptotischen Abschätzungen auftretenden Konstanten nicht explizit bekannt sind.

2. Potentialoperatoren und ihre Inversen

Zunächst stellen wir einige Hilfsaussagen zusammen:

Lemma 1. *Sei $u \in ([0, 1] \to X)$ eine Lipschitz-stetige Abbildung. Dann existiert für fast alle $t \in [0, 1]$ die starke Ableitung $u'(t)$*, d. h. $\lim_{s \to t} \left\| \frac{u(s) - u(t)}{s - t} - u'(t) \right\| = 0$ *für fast alle $t \in [0, 1]$.*

Einen Beweis dieses Lemmas findet man z. B. in [4].

Lemma 2. *Seien $u \in ([0, 1] \to X)$ und $w \in ([0, 1] \to X^*)$ Lipschitz-stetige Abbildungen. Dann gilt*

$$\big(w(1), u(1)\big) - \big(w(0), u(0)\big) = \int_0^1 \big(w'(t), u(t)\big)\, dt + \int_0^1 \big(w(t), u'(t)\big)\, dt .$$

Beweis: Die Funktion $\varphi(\cdot) = \big(w(\cdot), u(\cdot)\big)$ ist wegen

$$|\varphi(t) - \varphi(s)| = |\big(w(t), u(t)\big) - \big(w(s), u(s)\big)|$$
$$\leqq |\big(w(t) - w(s), u(t)\big)| + |\big(w(s), u(t) - u(s)\big)|$$
$$\leqq |t - s| \,(L_w \max_{0 \leqq t \leqq 1} \|u(t)\| + L_u \max_{0 \leqq t \leqq 1} \|w(t)\|_*$$

Lipschitz-stetig also absolut stetig. Deswegen existiert für fast alle $t \in [0, 1]$ die Ableitung $\varphi'(t)$, und es gilt

$$\varphi(1) - \varphi(0) = \int_0^1 \varphi'(t)\, dt .$$

Andererseits gilt auf Grund von Lemma 1 für fast alle $t \in [0, 1]$

$$\begin{aligned}\varphi'(t) &= \lim_{s\to t} \frac{1}{(s-t)}\big((w(s), u(s)) - (w(t), u(t))\big)\\ &= \lim_{s\to t}\left\{\left(\frac{w(s)-w(t)}{(s-t)}, u(s)\right) - \left(w(t), \frac{u(s)-u(t)}{s-t}\right)\right\}\\ &= \big(w'(t), u(t)\big) + \big(w(t), u'(t)\big) .\end{aligned}$$

Damit ist Lemma 2 bewiesen.

Lemma 3. *Ein demistetiger Operator $A \in (X \to X^*)$ ist dann und nur dann Potentialoperator mit dem Potential $F(x) = \int_0^1 (Atx, x)\, dt$, wenn die Bedingung*

$$\int_0^1 \big(Au(t), u'(t)\big)\, dt = \int_0^1 (Atx, x)\, dt - \int_0^1 (Aty, y)\, dt$$

für jede Lipschitz-stetige Abbildung $u \in ([0, 1] \to X)$ mit $u(0) = y$, $u(1) = x$ erfüllt ist.

Beweis: Notwendigkeit: Wegen der Demistetigkeit von A gilt

$$\|A\big(u(s) + \tau(u(t) - u(s))\big)\|_* \leqq M , \qquad \forall \tau, t, s \in [0, 1] .$$

Wir betrachten die reelle Funktion $\varphi(t) = F\big(u(t)\big)$. Wegen

$$\begin{aligned}|\varphi(t) - \varphi(s)| = |F\big(u(t)\big) - F\big(u(s)\big)| &= \left|\int_0^1 \big(A(u(s) + \tau(u(t) - u(s)), u(t) - u(s)\big)\, d\tau\right|\\ &\leqq M\|u(t) - u(s)\| \leqq M \cdot L_u|t - s|\end{aligned}$$

ist φ Lipschitz-stetig also absolut stetig. Folglich gilt

$$\int_0^1 (Atx, x)\, dt - \int_0^1 (Aty, y) = \varphi(1) - \varphi(0) = \int_0^1 \varphi'(t)\, dt .$$

Andererseits finden wir mit geeignetem $\tau_0 \in [0, 1]$ auf Grund des Mittelwertsatzes der Integralgleichung

$$\begin{aligned}\varphi'(t) &= \lim_{s\to t} \frac{\varphi(s) - \varphi(t)}{s - t} = \lim_{t\to s}\left(A\big(u(t) + \tau_0(u(s) - u(t))\big), \frac{u(s) - u(t)}{s - t}\right)\\ &= \big(Au(t), u'(t)\big)\end{aligned}$$

und damit $\int_0^1 \big(Au(t), u'(t)\big) = \int_0^1 (Atx, x)\, dt - \int_0^1 (Aty, y)\, dt.$

Hinlänglichkeit: Wir setzen $u(s) = x + sth$ und finden mit geeignetem $s_0 \in [0, 1]$

$$\lim_{t\to 0} \frac{F(x+th) - F(x)}{t} = \lim_{t\to 0} \frac{1}{t} \left[\int_0^1 (As(x+th), x+th)\, ds - \int_0^1 (Asx, x)\, ds \right]$$

$$= \lim_{t\to 0} \frac{1}{t} \int_0^1 (Au(s), u'(s))\, ds = \lim_{t\to 0} \frac{1}{t} \int_0^1 (Au(s), th)\, ds$$

$$= \lim_{t\to 0} (Ax + s_0 th, h) = (Ax, h) .$$

Damit ist Lemma 3 vollständig bewiesen.

Bemerkung 1. Offenbar kann man Lemma 3 auch wie folgt formulieren: Ein demistetiger Operator ist genau dann Potentialoperator mit dem Potential $F(x) = \int_0^1 (Atx, x)\, dt$, wenn für jede Lipschitz-stetige Funktion $u \in ([0, 1] \to X)$ mit $u(0) = u(1)$ gilt $\int_0^1 (Au(t), u'(t))\, dt = 0$.

Satz 1 (Langenbach [5]). *Sei $A \in (X \to X^*)$ ein demistetiger Operator mit Lipschitz-stetigem Inversen $A^{-1} \in (X^* \to X)$. Dann ist die Potentialität von A hinreichend für die Potentialität von A^{-1}.*

Beweis. Sei $w \in ([0, 1] \to X^*)$ eine beliebige Lipschitz-stetige Abbildung mit $w(0) = w(1)$. Auf Grund von Bemerkung 1 genügt es zu zeigen

$$\int_0^1 (w'(t), A^{-1}w(t))\, dt = 0 .$$

Wegen der Lipschitz-Stetigkeit von A^{-1} ist die Abbildung $u(t) = A^{-1} w(t)$ Lipschitz-stetig. Außerdem gilt offenbar $u(0) = u(1)$. Wegen Lemma 2 finden wir unter nochmaliger Benutzung von Bemerkung 1

$$0 = (w(1), u(1)) - (w(0), u(0)) = \int_0^1 (w'(t), u(t))\, dt + \int_0^1 (w(t), u'(t))\, dt$$

$$= \int_0^1 (w'(t), A^{-1}w(t))\, dt + \int_0^1 (Au(t), u'(t))\, dt = \int_0^1 (w'(t), A^{-1}w(t))\, dt .$$

Damit ist Satz 1 bewiesen.

Wie der folgende Satz zeigt, lassen sich die Potentiale

$$F(x) = \int_0^1 (Atx, x)\, dt, \; x \in X \quad \text{und} \quad F^*(x^*) = \int_0^1 (x^*, A^{-1}tx^*)\, dt , \quad x^* \in X^*$$

zueinander inverser Potentialoperatoren $A \in (X \to X^*)$ und $A^{-1} \in (X^* \to X)$ als im Sinne der Fenchelschen Dualitätstheorie zueinander konjugierte Funktionale charakterisieren.

Satz 2. *Sei $A \in (X \to X^*)$ ein hemistetiger, monotoner Potentialoperator mit Lipschitz-stetigem Inversen $A^{-1} \in (X^* \to X)$ und $A0 = 0$. Dann gilt*

$$F^*(x^*) = \max_{x \in X} \{(x^*, x) - F(x)\} \tag{8}$$

$$F(x) = \max_{x^* \in X^*} \{(x^*, x) - F^*(x^*)\} . \tag{9}$$

Beweis: Auf Grund von Satz 1 ist A^{-1} Potentialoperator. Wir beweisen (8). Sei $u(t) = A^{-1}tx^*$ und $x_0 = A^{-1}x^*$. Wegen der Monotonie von A realisiert x_0 das Minimum des Funktionals $F(x) - (x^*, x)$ auf X. Unter Benutzung von Lemma 2 und Lemma 3 finden wir daher

$$\begin{aligned} F^*(x^*) &= \int_0^1 (x^*, A^{-1}tx^*)\,dt = \int_0^1 \big((Au(t))', u(t)\big)\,dt \\ &= (Ax_0, x_0) - (A0, 0) - \int_0^1 (Au(t), u'(t))\,dt \\ &= (x^*, x_0) - F(x_0) + F(0) = (x^*, x_0) - F(x_0) = \max_{x \in X} \{(x^*, x) - F(x)\}. \end{aligned}$$

Wir beweisen (9): A^{-1} ist mit A monoton. Deswegen realisiert $x_0^* = Ax$ das Minimum des Funktionals $F^*(x^*) - (x^*, x)$ auf X^*. Sei jetzt $u(t) = A^{-1}tx_0^*$, dann finden wir wiederum unter Benutzung von Lemma 2 und 3:

$$\begin{aligned} \max_{x^* \in X^*} \{(x^*, x) - F^*(x^*)\} &= (x_0^*, x) - \int_0^1 (x_0^*, A^{-1}tx_0^*)\,dt \\ &= (x_0^*, x) - \int_0^1 \big((Au(t))', u(t)\big)\,dt \\ &= (x_0^*, x) - (Ax, x) + (A0, 0) + \int_0^1 \big(Au(t), u'(t)\big)\,dt \\ &= \int_0^1 \big(Au(t), u'(t)\big)\,dt = F(x) - F(0) = F(x). \end{aligned}$$

Damit ist Satz 2 vollständig bewiesen.

3. Dualität und Fehlerabschätzung

Wir kommen jetzt zu der einleitend angekündigten Fehlerabschätzung mittels Dualitätsaussagen. Dazu sei Y ein weiterer reeller reflexiver Banachraum mit der Norm $|\cdot|$ und Y^* der zu Y duale Raum mit der Norm $|\cdot|_*$. Durch $\langle y^*, y \rangle$ bezeichnen wir den Wert des linearen Funktionals $y^* \in Y^*$ im Element $y \in Y$. Der Operator A aus Gleichung (1) besitze jetzt die Darstellung

$$A = L^* A_0 L. \tag{10}$$

Dabei sei: $L \in (X \to Y)$ ein linearer stetiger Operator mit $|Lx| = ||x|| \; \forall x \in X$; $A_0 \in (Y \to Y^*)$ ein (möglicherweise) nichtlinearer Operator; und schließlich $L^* \in (Y^* \to X^*)$ der zu L adjungierte Operator. O. B. d. A. setzen wir voraus $A_0 0 = 0$. (Widrigenfalls betrachten wir an Stelle von (1) die Gleichung $\bar{A}u = f - L^*A_0 0$ mit $\bar{A} = L^*\bar{A}_0 L$, $\bar{A}_0 y = A_0 y - A_0 0$.) Sei $g \in Y^*$ ein beliebiges Element mit $L^*g = f$. (Die Existenz eines solchen Elements folgt aus der Tatsache, daß wegen $|Lx| = ||x||$ die Gleichung $Lx = 0$ nur die Nullösung besitzt. Die Konstruktion von g ist in konkreten Fällen in der Regel hinreichend einfach).

Satz 3. *Sei A_0 hemistetig, stark monoton, Potentialoperator und $A_0 0 = 0$. Sei weiter x_0 die Lösung von Gleichung* (1). *Dann gilt mit den Bezeichnungen*

$$K = \{y | y = Lx, x \in X\}, \quad K^* = \{y^* | L^* y^* = 0\}; \quad y_0 = Lx_0, \quad y_0^* = A_0 y_0 - g;$$

$$\Phi(y) = \int_0^1 \langle A_0 ty, y\rangle \, dt - \langle g, y\rangle, \quad \Psi(y^*) = -\int_0^1 \langle g + y^*, A_0^{-1} t (g + y^*)\rangle \, dt$$

die Dualitätsaussage

$$\Phi(y_0) = \min_{x \in K} \Phi(y) = \max_{y^* \in K^*} \Psi(y^*) = \Psi(y_0^*) . \tag{11}$$

Beweis: Seien $F_0(y) = \int_0^1 \langle A_0 ty, y\rangle \, dt$ bzw. $F_0^*(y^*) = \int_0^1 \langle y^*, A^{-1} ty^*\rangle \, dt$ die Potentiale von A_0 bzw. A_0^{-1}. (A_0^{-1} ist offenbar Lipschitz-stetig bezüglich $L_{A_0^{-1}} = 1/m$). Satz 2 und (2) liefern

$$\begin{aligned} \Psi(y_0^*) &= -\int_0^1 \langle g + y_0^*, A_0^{-1} t (g + y_0^*)\rangle \, dt = -\max_{y \in Y} (\langle g + y_0^*, y\rangle - F_0(y)) \\ &= -(\langle g + y_0^*, y_0\rangle - F_0(y_0)) = F_0(y_0) - \langle g, y_0\rangle - (L^* y_0^*, x_0) \\ &= F(y_0) - \langle g, y_0\rangle = \Phi(y_0) . \end{aligned}$$

Andererseits gilt für beliebige $y = Lx$, $y^* \in K^*$ auf Grund von Satz 2

$$\begin{aligned} \Phi(y) - \Psi(y^*) &= F_0(y) - \langle g, y\rangle + F_0^*(g + y^*) \\ &= F_0(y) - \langle g, y\rangle + \max_{y \in Y} (\langle g + y^*, y\rangle - F_0(y)) \\ &\geqq \langle y^*, y\rangle = \langle y^*, Lx\rangle = \langle L^* y^*, x\rangle = 0 . \end{aligned}$$

Beide Beziehungen zusammen ergeben (11). Damit ist Satz 3 bewiesen.

Satz 4. *Es seien die Voraussetzungen von Satz 3 erfüllt. Dann gilt mit den in Satz 3 eingeführten Bezeichnungen für beliebige $y = Lx$, $x \in X$, $y^* \in K^*$ die Fehlerabschätzung*

$$\|x - x_0\| = |y - y_0| \leqq \left[\frac{2}{m} (\Phi(y) - \Psi(y^*))\right]^{1/2}, \tag{12}$$

$$m = \textit{Monotoniekonstante von } A_0 .$$

Ist zusätzlich A_0 Lipschitz-stetig, dann gilt auch die duale Fehlerabschätzung

$$|y^* - y_0^*| \leqq L_0 \left[\frac{2}{m} (\Phi(y) - \Psi(y^*))\right]^{1/2}, \tag{13}$$

$$L_0 = \textit{Lipschitz-Konstante von } A_0 .$$

Beweis: Es gilt wegen Lemma 3

$$\Phi(y) - \Psi(y_0) = \int_0^1 (A_0(y_0 + t(y - y_0)), y - y_0)\, dt - (f, x - x_0)$$

$$= \int_0^1 \langle A_0(y_0 + t(y - y_0)), y - y_0\rangle\, dt - \langle A_0 y_0, y - y_0\rangle$$

$$= \int_0^1 \langle A_0(y_0 + t(y - y_0)) - A_0 y_0, t(y - y_0)\rangle \frac{dt}{t} \geqq m|y - y_0|^2 \int_0^1 t\, dt$$

$$= \frac{m}{2} |y - y_0|^2 .$$

Daraus folgt unter Benutzung von (11) die Abschätzung (12).

Zum Beweis von (13) bemerken wir zunächst, daß A_0^{-1} wegen der Lipschitz-Stetigkeit und starken Monotonie von A_0 ebenfalls stark monoton und zwar bezüglich der Konstanten m/L_0^2 ist. Wir finden daher

$$\Psi(y_0^*) - \Psi(y^*) = F_0^*(g + y^*) - F_0^*(g + y_0^*)$$

$$= \int_0^1 \langle y^* - y_0^*, A_0^{-1}(g + y_0^* + t(y^* - y_0^*))\rangle\, dt - \langle y^* - y_0^*, A_0^{-1}(g + y_0^*) - y_0\rangle$$

$$= \int_0^1 \langle t(y^* - y_0^*), A_0^{-1}(g + y_0^* + t(y^* - y_0^*)) - A_0^{-1}(g + y_0^*)\rangle \frac{dt}{t} - \langle y^* - y_0^*, Lx_0\rangle$$

$$\geqq \frac{m}{L_0^2} |y^* - y_0^*|_*^2 \int_0^1 t\, dt - (L^*(y^* - y_0^*), x_0)$$

$$= \frac{m}{2L_0^2} |y^* - y_0^*|^2 .$$

Unter Benutzung von (11) finden wir daraus (13).

Bemerkung 2. Die Darstellung (10) ist typisch für elliptische Differentialgleichungen in divergenter Form. (vgl. z. B. [1—3])

Bemerkung 3. Bei vielen konkreten Randwertproblemen besitzt nicht nur die Lösung x_0 von (1) bzw. $y_0 = Lx_0$, sondern auch das Element $y_0^* = A_0 y_0 - g$ selbständige praktische Bedeutung. (Übrigens können häufig auch die Funktionale Φ und Ψ sowie die Vergleichsmengen K und K^* praktisch interpretiert werden.) Deswegen ist es sinnvoll, simultan Näherungen $\bar{x}$ für x_0 und $\bar{y}^*$ für y_0^* zu berechnen und diese mittels (12) und (13) abzuschätzen. Zur numerischen Konstruktion derartiger Näherungen eignen sich die üblichen Verfahren zur Konstruktion von Minimalfolgen konvexer Funktionale. (Ritzsche Methode, Methode des steilsten Abstiegs usw.)

Bemerkung 4. Sei speziell Y Hilbertraum und $Y^* = Y$ sowie $A_0^{-1} = A_0 = I =$ identischer Operator in Y. In diesem Fall kann y_0 offenbar als Lösung des Minimumproblems

$$|g - y|^2 = \min_{y \in K}$$

und y_0^* als Lösung des Minimumproblems

$$|g + y_0^*|_*^2 = \min_{y^* \in K^*}$$

charakterisiert werden. Praktisch heißt das, y_0 ist die Projektion von g auf K, $-y_0^*$ ist die Projektion von g auf K^*. Bei Einsetzung Ritzscher Näherungen y_n bzw. y_n^* von y_0 bzw. y_0^* anstelle von y bzw. y^* reduziert sich (12) auf die im Zusammenhang mit der Methode der orthogonalen Projektion bekannte Fehlerabschätzung.

Literatur

[1] Bittner, L., Abschätzungen bei Variationsmethoden mit Hilfe von Dualitätssätzen I. Num. Math. **11** (1968) 129–143.

[2] Bittner, L., Abschätzungen bei Variationsmethoden mit Hilfe von Dualitätssätzen II. Num. Math. **16** (1971) 285–303.

[3] Gajewski, H., Über einige Fehlerabschätzungen bei Gleichungen mit monotonen Potentialoperatoren in Banach-Räumen. Mber. DAW **12** (1970) 571–580.

[4] Komura, Y., Nonlinear semi-groups in Hilbert space. J. Math. Soc. Japan **19** (1967) 493–507.

[5] Langenbach, A., Über Potentialoperatoren und ihre Inversen. Mber. DAW **12** (1970).

Anschrift des Autors:

ZIMM der AdW der DDR
108 Berlin, Mohrenstr. 39

Zum Galerkinverfahren für Evolutionsgleichungen

KONRAD GRÖGER, Berlin

0. Einführung

In der vorliegenden Arbeit befassen wir uns mit Sätzen über Anfangswertaufgaben zu Evolutionsgleichungen der Form

$$u' + Au = f$$

für Volterraoperatoren A, die einen Funktionenraum $X = L^p(S; V) \cap L^{p_0}(S; H)$, $1 < p \leqq p_0 < \infty$, in dessen dualen X^* abbilden (zur Erläuterung der Begriffe und Bezeichnungen siehe Abschnitt 1). Unser besonderes Interesse gilt dabei Aussagen über die Konvergenz des Galerkinverfahrens. Die in dieser Arbeit erhaltenen Sätze verallgemeinern Ergebnisse von Lions-Strauss [11], Browder [1], Gajewski [3, 4] und Gröger [5, 6, 7].

Die Arbeit besteht aus drei Abschnitten. Im ersten Abschnitt führen wir die benötigten Begriffe und Bezeichnungen ein. Im zweiten Abschnitt beweisen wir unter relativ schwachen Monotonie- und Stetigkeitsvoraussetzungen bezüglich A Aussagen über Existenz und Einzigkeit von Lösungen der untersuchten Aufgaben sowie über die Konvergenz des Galerkinverfahrens. Im dritten Abschnitt zeigen wir, daß die Ergebnisse des zweiten Abschnitts durch Regularitätsaussagen und schärfere Konvergenzaussagen für das Galerkinverfahren ergänzt werden können, wenn man an den Operator A entsprechend stärkere Voraussetzungen stellt.

1. Bezeichnungen und Begriffe

Es sei $S = [0, T]$ ein endliches reelles Intervall und E ein Banachraum mit dem dualen Raum E^*.

Definition 1. Mit $L^p(S; E)$, $1 \leqq p \leqq \infty$, bezeichnen wir den linearen Raum der auf S definierten Bochner-meßbaren Funktionen u mit Werten in E, für die

$$||u||^p_{L^p(S;E)} = \int_S ||u(t)||^p_E \, dt < \infty$$

ist (für $p = \infty$ ist die übliche Modifikation vorzunehmen). Mit der Norm $||u||_{L^p(S;E)}$ ist $L^p(S; E)$ ein Banachraum.

Definition 2. Mit $C^l(S; E)$ bezeichnen wir den Banachraum der auf S definierten, im Fréchetschen Sinne l-mal stetig differenzierbaren Funktionen mit Werten in E mit der üblichen Norm. Für $C^0(S; E)$ schreiben wir $C(S; E)$.

Definition 3. Mit $C^l_w(S; E)$ bezeichnen wir den linearen Raum der auf S definierten Funktionen u mit Werten in E, für die gilt

$$(f, u(\cdot)) \in C^l(S) \quad \text{für jedes } f \in E^* .$$

Mit $u^{(i)}$, $i = 1, \ldots, l$, bezeichnen wir die durch

$$(f, u^{(i)}(\cdot)) = \frac{d^i}{dt^i}(f, u(\cdot)) \quad \text{für jedes } f \in E^*$$

definierten Funktionen. Mit Hilfe der Halbnormen

$$p(u; i, f) = \sup_{t \in S} |(f, u^{(i)}(t))| , \qquad i = 0, \ldots, l , \qquad f \in E^* ,$$

wird $C^l_w(S; E)$ zu einem lokalkonvexen topologischen Vektorraum.

Für $C^0_w(S; E)$ schreiben wir $C_w(S; E)$.

Definition 4. Es sei $D(S)$ der Raum der Grundfunktionen im Sinne von L. Schwartz. Wir bezeichnen den Raum der stetigen linearen Abbildungen von $D(S)$ in den Banachraum E mit $D^*(S; E)$ und nennen die Elemente dieses Raums *Distributionen über S mit Werten in E*. Wir fassen $D^*(S; E)$ als lokalkonvexen Raum auf, und zwar definieren wir die Topologie durch die Halbnormen

$$p(u; f, \varphi) = |(f, u(\varphi))| , \qquad u \in D^*(S; E), f \in E^*, \varphi \in D(S) .$$

Für $u \in D^*(S; E)$ bezeichnen wir die durch

$$u'(\varphi) = - u(\varphi') , \qquad \varphi \in D(S) ,$$

definierte stetige Abbildung u' von $D(S)$ in E als *Ableitung* von u.

Bemerkung 1. Einer Funktion $u \in L^p(S; E)$ wird durch die Vorschrift

$$v_u(\varphi) = \int_S \varphi(t)\, u(t)\, dt , \qquad \varphi \in D(S) ,$$

eine Distribution v_u über S mit Werten in E zugeordnet. Die Zuordnung $u \to v_u$ ist eineindeutig. Diese Feststellung gestattet die Identifizierung von u und v_u.

Im folgenden sei H ein reeller Hilbertraum mit dem Skalarprodukt $(.,.)$ und der Norm $|\cdot|$. V sei ein in H stetig eingebetteter und in H dicht liegender separabler[1]) und reflexiver Banachraum mit der Norm $||\cdot||$. Wir identifizieren H mit seinem dualen Raum H^* und kommen so zur Inklusion

$$V \subset H \subset V^* ;$$

dabei ist V^* der zu V duale Raum. Die Norm in V^* bezeichnen wir mit $||\cdot||_*$. Das Skalarprodukt zwischen Elementen aus V^* und V bezeichnen wir wie das Skalarprodukt in H mit $(.,.)$. Diese Bezeichnung wird durch die Identifizierung von H und H^* gerechtfertigt.

Es seien zwei reelle Zahlen p und p_0 mit der Eigenschaft $1 < p \leqq p_0 < \infty$ gegeben. Wir setzen $X = L^p(S; V) \cap L^{p_0}(S; H)$. X ist mit der Norm

$$||u||_X = ||u||_{L^p(S; V)} + ||u||_{L^{p_0}(S; H)} , \qquad u \in X ,$$

[1]) Die Voraussetzung der Separabilität ist nicht wesentlich, sie ist jedoch in den praktisch interessierenden Fällen gegeben und erleichtert die Beweisführung.

ein Banachraum. Der zu X duale Raum X^* ist dann $L^q(S; V^*) + L^{q_0}(S; H)$, $\frac{1}{p} + \frac{1}{q} = \frac{1}{p_0} + \frac{1}{q_0} = 1$, mit der Norm

$$\|f\|_{X^*} = \inf_{\substack{f_1+f_2=f \\ f_1 \in L^q(S;\, V^*) \\ f_2 \in L^{q_0}(S;\, H)}} \max\left(\|f_1\|_{L^q(S;\, V^*)}, \|f_2\|_{L^{q_0}(S;\, H)}\right), \qquad f \in X^* .$$

Das folgt daraus, daß für einen reflexiven Banachraum E der zu $L^p(S; E)$ duale Raum mit $L^q(S; E^*)$ identifiziert werden kann (siehe Edwards [2]). Das Skalarprodukt $\langle f, u\rangle$ von $f \in X^*$ und $u \in X$ wird durch

$$\langle f, u\rangle = \int_S \big(f(t), u(t)\big)\, dt$$

gegeben.

Mit X_t bezeichnen wir den Raum, der sich ergibt, wenn man in der Definition von X das Intervall S durch $[0, t]$ ersetzt, d. h., es ist $X_t = L^p([0, t]; V) \cap L^{p_0}([0, t]; H)$ mit der entsprechenden Norm.

Weiter setzen wir $W = \{u | u \in X, u' \in X^*\}$; dabei bezeichnet u' die Ableitung von u im Sinne des Distributionenraums $D^*(S; V^*)$. Mit der Norm

$$\|u\|_W = \|u\|_X + \|u'\|_{X^*}, \qquad u \in W,$$

wird W zu einem Banachraum.

Bemerkung 2. Ist $p_0 = p$, so ist $X = L^p(S; V)$ mit einer zur üblichen Norm in $L^p(S; V)$ äquivalenten Norm. Entsprechend ist $X^* = L^q(S; V^*)$.

Bemerkung 3. Der Raum W ist in den Raum $C(S; H)$ stetig eingebettet, und für $u, v \in W$ gilt die folgende Formel der partiellen Integration:

$$\big(u(t), v(t)\big) - \big(u(0), v(0)\big) = \int_0^t \left\{\big(v'(s), u(s)\big) + \big(u'(s), v(s)\big)\right\} ds, \qquad t \in S. \quad (1)$$

Für den Spezialfall $p = p_0$ ist diese Tatsache beispielsweise bei Raviart [12] vermerkt. Im allgemeinen Fall kann sie dadurch bewiesen werden, daß man zeigt, daß $C^1(S; V)$ in W dicht liegt und daß die Formel (1) für $u, v \in C^1(S; V)$ richtig ist.

Definiton 5. Eine Abbildung $A \in (X \to X^*)$[1]) nennen wir *koerzitiv*, wenn sie der Bedingung

$$\lim_{\|u\|_X \to \infty} \frac{1}{\|u\|_X} \langle Au, u\rangle = +\infty$$

genügt.

Definition 6. Ist X_1 ein Raum von Funktionen über S mit Werten in E_1 und X_2 ein Raum von Funktionen über S mit Werten in E_2, so bezeichnen wir

[1]) Für beliebige Mengen M und N bezeichnen wir mit $(M \to N)$ die Gesamtheit der auf M definierten Abbildungen mit Werten in N.

eine Abbildung $A \in (X_1 \to X_2)$ als *Volterraoperator*, wenn für jedes $t \in S$ aus $u, v \in X_1$ und

$$u(s) = v(s) \qquad \text{für fast alle} \quad s \in [0, t]$$

folgt

$$(Au)(s) = (Av)(s) \qquad \text{für fast alle} \quad s \in [0, t] .$$

2. Existenz- und Einzigkeitssätze; Galerkinverfahren

Satz 1. *Es sei $A \in (X \to X^*)$ ein hemistetiger, monotoner*[1]) *und koerzitiver Volterraoperator. Dann besitzt die Aufgabe*

$$u' + Au = f , \qquad u(0) = a , \qquad u \in X , \tag{2}$$

für beliebige $f \in X^*$ und $a \in H$ genau eine Lösung u. Es ist $u \in W \subset C(S; H)$, und die Zuordnung $a \to u$ ist als Abbildung von H in $C(S; H)$ stetig.

Bemerkung 4. Da u' als Ableitung von u in $D^*(S; V^*)$ aufgefaßt wird, hat die Gleichung $u' + Au = f$ in $D^*(S; V^*)$ einen Sinn. Genügt $u \in X$ dieser Gleichung, so ist $u' = -Au + f \in X^*$ und damit $u \in W \subset C(S; H)$. Das rechtfertigt die gestellte Anfangsbedingung $u(0) = a \in H$.

Für den Beweis von Satz 1 ist das folgende Lemma nützlich.

Lemma 1. *Ist $A \in (X \to X^*)$ ein monotoner und koerzitiver Volterraoperator, so gilt für jedes $t \in S$*

$$\lim_{\|u\|_{X_t} \to \infty} \frac{1}{\|u\|_{X_t}} \int_0^t ((Au)(s), u(s))\, ds = +\infty , \tag{3}$$

$$\int_0^t ((Au)(s) - (Av)(s), u(s) - v(s))\, ds \geqq 0 \qquad \text{für} \quad u, v \in X . \tag{4}$$

Beweis. Es sei

$$u_t(s) = \begin{cases} u(s) & \text{für} \quad 0 \leqq s \leqq t , \\ 0 & \text{für} \quad t < s \leqq T . \end{cases}$$

Da A Volterraoperator ist, gilt $(Au)(s) = (Au_t)(s)$ für fast alle $s \in [0, t]$. Außerdem ist $\|u_t\|_X = \|u\|_{X_t}$. Daher folgt die Behauptung (3) direkt aus der Koerzitivität von A. Analog folgt die Behauptung (4) aus der Monotonie von A. Damit ist Lemma 1 bewiesen.

Den Beweis von Satz 1 führen wir mit Hilfe des Galerkinverfahrens. Bevor wir zum eigentlichen Beweis kommen, definieren wir das Galerkinverfahren für die Aufgabe (2) und beweisen einige Lemmata über die Lösungen der Galerkingleichungen.

Es sei $\{h_1, h_2, \ldots\}$ ein in V und damit auch in H vollständiges System linear unabhängiger Elemente. H_n sei die lineare Hülle von $\{h_1, \ldots, h_n\}$, versehen mit dem von H induzierten Skalarprodukt. Wir identifizieren H_n mit dem dazu

---

[1]) Zur Definition von Hemistetigkeit, Demistetigkeit und Monotonie siehe beispielsweise Kato [9].

dualen Raum H_n^*. Weiter sei $X_n = L^{p_0}(S; H_n)$ und $X_n^* = L^{q_0}(S; H_n)$. Das Skalarprodukt von $f \in X_n^*$ und $u \in X_n$ wird durch $\int_S (f(t), u(t))\, dt$ gegeben. Es kann daher ebenfalls mit $\langle f, u\rangle$ bezeichnet werden.

Jedem Operator $A \in (X \to X^*)$ ordnen wir durch die Vorschrift

$$\langle A_n u, v\rangle = \langle Au, v\rangle \qquad \text{für jedes} \quad v \in X_n \tag{5}$$

einen Operator $A_n \in (X_n \to X_n^*)$ zu.

Es sei $\{a_n\}$, $n = 1, 2, \ldots$, eine in H gegen a konvergierende Folge mit der Eigenschaft $a_n \in H_n$. Schließlich sei $f_n \in X_n^*$ erklärt durch die Beziehung

$$\langle f_n, v\rangle = \langle f, v\rangle \qquad \text{für jedes} \quad v \in X_n\,. \tag{6}$$

Zur Aufgabe (2) definieren wir für $n = 1, 2, \ldots$ die entsprechenden Galerkingleichungen

$$u_n' + A_n u_n = f_n\,, \qquad u_n(0) = a_n\,, \qquad u_n \in X_n\,. \tag{7}$$

Lemma 2. *Unter den Voraussetzungen von Satz* 1 *besitzt die Aufgabe* (7) *für jedes n genau eine Lösung $u_n \in W$. Die Folge $\{u_n\}$ ist in $C(S; H)$ und in X beschränkt und die Folge $\{Au_n\}$ in X^*.*

Bemerkung 5. Wird $A \in (X \to X^*)$ durch eine Familie von Operatoren $A(t) \in (V \to V^*)$ gegeben, d. h. gilt $(Au)(t) = A(t)\, u(t)$, so handelt es sich bei der Aufgabe (7) um ein Anfangswertproblem für ein System gewöhnlicher Differentialgleichungen. Die Existenz einer Lösung folgt dann aus dem bekannten Existenzsatz von Carathéodory.

Für den Existenzbeweis im allgemeinen Fall benötigen wir das

Lemma 3. *Es sei $b \in H_n$, $S_0 = [t_0, T]$ und C_b die in $C(S_0; H_n)$ abgeschlossene Teilmenge derjenigen $v \in C(S_0; H_n)$, für die $v(t_0) = b$ ist. B sei ein demistetiger Volterraoperator aus $(C_b \to L^{q_0}(S_0; H_n))$, der der Voraussetzung*

$$\|Bv\|_{L^{q_0}(S_0;\, H_n)} \leqq M\,, \qquad M = \text{const}, \qquad \textit{für} \quad v \in C_b \tag{8}$$

genügt. Dann besitzt die Aufgabe

$$v(t) = b - \int_{t_0}^{t} (Bv)(s)\, ds \qquad \textit{für jedes} \quad t \in S_0$$

eine Lösung $v \in C_b$.

Beweis. Wir setzen für jede natürliche Zahl m

$$\delta_m = \frac{T - t_0}{m} \qquad \text{und} \qquad t_{i,m} = t_0 + i\,\delta_m\,, \qquad i = 0, \ldots, m\,.$$

Wir definieren $v_m \in C_b$ nacheinander in den Intervallen $[t_{i,m}, t_{i+1,m}]$ durch folgende rekursiv zu verstehende Vorschrift:

$$v_m(t) = b \qquad \text{für} \quad t_0 \leqq t \leqq t_{1,m}\,,$$

$$v_m(t) = v_m(t_{i,m}) - \int_{t_{i-1,m}}^{t-\delta_m} (Bv_{i,m})(s)\, ds \qquad \text{für} \quad t_{i,m} < t \leqq t_{i+1,m}\,, \quad i = 1, \ldots, m-1;$$

dabei sei

$$v_{i,m}(s) = \begin{cases} v_m(s) & \text{für} \quad t_0 \leqq s \leqq t_{i,m}\,, \\ v_m(t_{i,m}) & \text{für} \quad t_{i,m} < s \leqq T\,. \end{cases}$$

Da B Volterraoperator ist, folgt aus dieser Definition

$$v_m(t) = \begin{cases} b & \text{für} \quad t_0 \leqq t < t_{1,m}\,, \\ b - \int\limits_{t_0}^{t-\delta_m} (Bv_m)(s)\,ds & \text{für} \quad t_{i,m} < t \leqq T\,. \end{cases}$$

Wir zeigen nun, daß die Folge $\{v_m\}$ in C_b kompakt ist. Aus (9) und (8) folgt für $t_2 \geqq t_1$:

$$|v_m(t_2) - v_m(t_1)| \leqq \left(\int\limits_{t_1}^{t_2} dt\right)^{1/p_0} \cdot \left(\int\limits_{S} |(Bv_m)(t)|^{q_0}\,dt\right)^{1/q_0} \leqq (t_2 - t_1)^{1/p_0} M\,.$$

Für alle m ist $v_m(t_0) = b$. Daher folgt aus der letzten Abschätzung sowohl die gleichgradige Stetigkeit als auch die gleichmäßige Beschränktheit der Funktionen v_m und damit die Kompaktheit der Folge $\{v_m\}$ in C_b.

Es sei $\{v_{m_j}\}$, $j = 1, 2, \ldots$, eine in C_b konvergente Teilfolge von $\{v_m\}$ und v der Grenzwert dieser Folge. Nach (9) ist für $t > t_{1,m_j}$

$$v_{m_j}(t) = b - \int\limits_{t_0}^{t} (Bv_{m_j})(s)\,ds + \int\limits_{t-\delta_{m_j}}^{t} (Bv_{m_j})(s)\,ds\,. \tag{10}$$

Wegen (8) gilt

$$\left|\int\limits_{t-\delta_{m_j}}^{t} (Bv_m)(s)\,ds\right| \leqq \delta_{m_j}^{1/p_0} M\,,$$

d. h., das zweite Integral in (10) konvergiert für $j \to \infty$ gegen 0. Für $t \in S_0$ und $x \in H_n$ ist

$$l(w) = \left(\int\limits_{t_0}^{t} w(s)\,ds, x\right), \qquad w \in L^{q_0}(S_0; H_n)\,,$$

ein stetiges lineares Funktional auf $L^{q_0}(L_0; H_n)$. Auf Grund der Demistetigkeit von B ist also

$$\lim_{j\to\infty} \left(\int\limits_{t_0}^{t} (Bv_{m_j})(s)\,ds, x\right) = \left(\int\limits_{t_0}^{t} (Bv)(s)\,ds, x\right) \qquad \text{für jedes} \quad x \in H_n\,,$$

d. h.

$$\lim_{j\to\infty} \int\limits_{t_0}^{t} (Bv_{m_j})(s)\,ds = \int\limits_{t_0}^{t} (Bv)(s)\,ds\,.$$

Aus (10) ergibt sich folglich durch den Grenzübergang $j \to \infty$:

$$v(t) = b - \int\limits_{t_0}^{t} (Bv)(s)\,ds\,.$$

Damit ist Lemma 3 bewiesen.

Bemerkung 6. Der Beweis von Lemma 3 ist eine leichte Modifikation des bekannten Beweises von Tonelli für den Satz von Carathéodory.

Beweis von Lemma 2. Wir nehmen zunächst einmal an, zu $t_1 \in S$ und jedem n sei eine Funktion $u_n \in L^{p_0}([0, t_1]; H_n)$ gegeben, für die gilt:

$$u_n(t) = a_n - \int_0^t \{(A_n u_n)(s) - f_n(s)\}\, ds \qquad \text{für } t \in [0, t_1]\,. \tag{11}$$

Aus (11) folgt für $t \in [0, t_1]$:

$$\frac{1}{2}(|u_n(t)|^2 - |a_n|^2) = \int_0^t (u_n'(s), u_n(s))\, ds = \int_0^t (-(Au_n)(s) + f(s), u_n(s))\, ds\,.$$

Folglich ist

$$\frac{1}{2}|u_n(t)|^2 + \int_0^t ((Au_n)(s) - (A0)(s), u_n(s))\, ds \leqq \frac{1}{2}|a_n|^2 + \|f - A0\|_{X^*} \|u_n\|_{X_t}\,. \tag{12}$$

Daraus ergibt sich wegen der Koerzitivität von A (vgl. (3))

$$\|u_n\|_{X_{t_1}} \leqq K_1 \tag{13}$$

und

$$\|u_n\|_{C([0,t_1];H)} \leqq K_2\,; \tag{14}$$

dabei sind K_1 und K_2 von n und t_1 unabhängige Konstanten. Aus (12), (13) und (14) folgt unter Benutzung eines Ergebnisses von Browder und Hess (siehe Hess [8]) die Abschätzung

$$\|Au_n\|_{X_{t_1}^*} \leqq K_3\,, \tag{15}$$

wobei auch K_3 nicht von n und t_1 abhängt.

Wir beweisen nun, daß die Lösung $u_n \in L^{p_0}([0, t_1]; H_n)$ von (11) eindeutig bestimmt ist. Genügt auch $v_n \in L^{p_0}([0, t_1]; H_n)$ der Gleichung (11), so gilt für $t \in [0, t_1]$:

$$\frac{1}{2}|u_n(t) - v_n(t)|^2 = \int_0^t (u_n'(s) - v_n'(s), u_n(s) - v_n(s))\, ds$$

$$= -\int_0^t ((Au_n(s) - (Av_n)(s), u_n(s) - v_n(s))\, ds \leqq 0\,.$$

Daraus folgt $u_n = v_n$.

Wir kommen zum Beweis der Existenz einer Lösung der Aufgabe (7) für gegebenes n. Es sei S_1 die Menge der $t_1 \in S$, für die (11) eine zu $L^{p_0}([0, t_1]; H_n)$ gehörige Lösung besitzt. Wir schließen dabei zunächst nicht aus, daß S_1 allein aus dem Punkt $t = 0$ besteht. S_1 ist offenbar von der Form $[0, t_0)$ oder $[0, t_0]$. Wir beweisen, daß der erste Fall nicht eintritt, d. h., daß t_0 zu S_1 gehört. Da (11) für jedes $t_1 \in [0, t_0)$ genau eine Lösung $u_n \in L^{p_0}([0, t_1]; H_n)$ besitzt, können wir davon ausgehen, daß uns eine im Intervall $[0, t_0]$ definierte Funktion u_n gegeben ist, die (11) mit t_0 anstelle von t_1 genügt und die für jedes $t_1 \in [0, t_0)$ zu $L^{p_0}([0, t_1]; H_n)$ gehört. Aus (13) folgt $\|u_n\|_{X_{t_0}} \leqq K_1$ und damit $u_n \in L^{p_0}([0, t_0]; H_n)$. Daher ist, wie behauptet, $t_0 \in S_1$.

Wir führen nun die Annahme $t_0 < T$ zum Widerspruch. Es sei $b = u_n(t_0)$ und

$$w(t) = \begin{cases} u_n(t) & \text{für } 0 \leqq t \leqq t_0\,, \\ b & \text{für } t_0 < t < T\,. \end{cases}$$

7 Nonlinear Operator

$A \in (X \to X^*)$ ist als hemistetige und monotone Abbildung nach Kato [9] auch demistetig. Folglich ist auch A_n, aufgefaßt als Abbildung von $C(S; H_n)$ in $X_n^* = L^{q_0}(S; H_n)$, demistetig. Daher existieren Zahlen $r > 0$ und M derart, daß gilt

$$||A_n u - f_n||_{X_n^*} \leqq M \qquad \text{für} \quad ||u - w||_{C(S;H_n)} \leqq r \tag{16}$$

(demistetige Operatoren sind lokal beschränkt). Für beliebiges $v \in C_b$ (vgl. Lemma 3) definieren wir $u \in C(S; H_n)$ durch

$$u(t) = \begin{cases} u_n(t) & \text{für} \quad 0 \leqq t \leqq t_0\,, \\ v(t) & \text{für} \quad t_0 < t \leqq T \quad \text{und} \quad |v(t) - b| \leqq r\,, \\ b + r\dfrac{v(t) - b}{|v(t) - b|} & \text{für} \quad t_0 < t \leqq T \quad \text{und} \quad |v(t) - b| > r \end{cases} \tag{17}$$

und $B \in (C_b \to L^{q_0}(S_0; H_n))$, $S_0 = [t_0, T]$, durch

$$(Bv)(t) = (A_n u - f_n)(t)\,, \qquad t_0 \leqq t \leqq T\,. \tag{18}$$

Da die Zuordnung $v \to u$ als Abbildung von C_b in $C(S; H_n)$ stetig ist, ist B ein demistetiger Volterraoperator.

Aus (18), (17) und (16) folgt außerdem

$$||Bv||_{L^{q_0}(S_0;H_n)} \leqq ||A_n u - f_n||_{L^{q_0}(S;H_n)} \leqq M \qquad \text{für} \quad v \in C_b\,.$$

Der Operator B genügt also den Voraussetzungen von Lemma 3. Folglich besitzt die Aufgabe

$$v(t) = b - \int_{t_0}^{t} (Bv)(s)\, ds$$

eine Lösung von $v \in C_b$. Wegen $v(t_0) = b$ gilt für hinreichend kleines $\delta > 0$:

$$|v(t) - b| \leqq r \qquad \text{für} \quad t_0 \leqq t \leqq t_0 + \delta\,.$$

Wir setzen die in $[0, t_0]$ definierte Funktion u_n folgendermaßen fort:

$$u_n(t) = v(t) \qquad \text{für} \qquad t_0 < t \leqq t_0 + \delta\,.$$

Dann ist für $t \in [t_0, t_0 + \delta]$

$$(Bv)(t) = (A_n u_n - f_n)(t)$$

und

$$u_n(t) = b - \int_{t_0}^{t} (Bv)(s)\, ds = u_n(t_0) - \int_{t_0}^{t} (A_n u_n - f_n)(s)\, ds$$
$$= a_n - \int_{0}^{t} (A_n u_n - f_n)(s)\, ds\,.$$

Es existiert also eine zu $L^{p_0}([0, t_0 + \delta]; H_n)$ gehörige Lösung der Aufgabe (11) für $t_1 = t_0 + \delta$. Das widerspricht der Definition von t_0. Folglich war die Annahme $t_0 < T$ falsch, d. h., die Aufgabe (11) besitzt für $t_1 = T$ eine Lösung $u_n \in L^{p_0}(S; H_n)$. Diese Funktion u_n ist Lösung der Aufgabe (7).

Die Beschränktheit der Folge $\{u_n\}$ in $C(S; H)$ und X folgt aus (14) und (13) und die Beschränktheit der Folge $\{Au_n\}$ in X^* aus (15).

Damit ist Lemma 2 bewiesen.

Lemma 4. *Die nach Lemma 2 existierende Folge* $\{u_n\}$ *von Galerkinlösungen besitzt eine Teilfolge* $\{u_{n_j}\}$, $j = 1, 2, \ldots$, *mit folgenden Eigenschaften:*

(a) $\{u_{n_j}\}$ *konvergiert in* X *schwach gegen ein* $u \in X$.

(b) $\{u_{n_j}(T)\}$ *konvergiert in* H *schwach gegen ein* $z \in H$.

(c) $\{Au_{n_j}\}$ *konvergiert in* X^* *schwach gegen ein* $v \in X^*$.

(d) *Für die Grenzelemente gilt* $u \in W$, $u(0) = a$, $u(T) = z$ *und*

$$u' + v = f\,. \tag{19}$$

Beweis. Die Existenz einer Teilfolge $\{u_{n_j}\}$ mit den Eigenschaften (a), (b) und (c) folgt aus den Beschränktheitsaussagen von Lemma 2. Auf Grund von (7) gilt für $\varphi \in D(S)$ und $x \in H_n$

$$\left(\int_S \varphi(t)\left(u'_{n_j}(t) + (Au_{n_j})(t)\right) dt, x\right) = \langle u'_{n_j} + Au_{n_j}, \varphi x\rangle = \langle f, \varphi x\rangle = \left(\int_S \varphi(t) f(t)\, dt, x\right), \quad \text{falls} \quad n_j \geqq n\,. \tag{20}$$

Wegen (a) konvergiert die Folge $\{u_{n_j}\}$ auch in $D^*(S; V^*)$ gegen u. Daher konvergiert die Folge $\{u'_{n_j}\}$ in $D^*(S; V^*)$ gegen u'. Aus (20) ergibt sich also für $j \to \infty$ (wenn wir mit $u'(\varphi)$ den Wert der Distribution u' für $\varphi \in D(S)$ bezeichnen):

$$(u'(\varphi), x) = \left(\int_S \varphi(t)\left(-v(t) + f(t)\right) dt, x\right) \quad \text{für} \quad x \in \bigcup_n H_n\,. \tag{21}$$

Da $\bigcup_n H_n$ in V dicht liegt, folgt aus (21)

$$u'(\varphi) = \int_S \varphi(t)\left(-v(t) + f(t)\right) dt\,.$$

Das bedeutet, daß u' als Element von X^* aufgefaßt werden kann und daß die Beziehung (19) gilt. Folglich ist $u \in W$. Für beliebiges $x \in \bigcup_n H_n$ gilt wegen (19) und (7)

$$\begin{aligned}
&\int_S \left(u'(t), (T-t)\,x\right) dt = \int_S \left(f(t) - v(t), (T-t)\,x\right) dt \\
&= \lim_{j\to\infty} \int_S \left(f(t) - (Au_{n_j})(t), (T-t)\,x\right) dt = \lim_{j\to\infty} \int_S \left(u'_{n_j}(t), (T-t)\,x\right) dt \\
&= \lim_{j\to\infty} \left\{\int_S (u_{n_j}(t), x)\, dt - (a_{n_j}, Tx)\right\} = \int_S (u(t), x)\, dt - (a, Tx) \\
&= \int_S \left(u'(t), (T-t)\,x\right) dt + \left(u(0) - a, Tx\right).
\end{aligned}$$

Da $\bigcup_n H_n$ in H dicht liegt, folgt daraus $u(0) = a$. Entsprechend kann $u(T) = z$ gezeigt werden.

Damit ist Lemma 4 bewiesen.

Beweis von Satz 1. Aus (7) und (19) folgt

$$\begin{aligned}
&\overline{\lim_{j\to\infty}}\, \langle Au_{n_j}, u_{n_j}\rangle = \overline{\lim_{j\to\infty}}\, \langle -u'_{n_j} + f, u_{n_j}\rangle \\
&= \overline{\lim_{j\to\infty}} \left\{\frac{1}{2}|a_{n_j}|^2 - \frac{1}{2}|u_{n_j}(T)|^2 + \langle f, u_{n_j}\rangle\right\} \\
&\leqq \frac{1}{2}|a|^2 - \frac{1}{2}|u(T)|^2 + \langle f, u\rangle = \langle -u', u\rangle + \langle f, u\rangle = \langle v, u\rangle\,.
\end{aligned}$$

Aus dieser Abschätzung sowie aus den Aussagen (a) und (c) von Lemma 4 ergibt sich $Au = v$ (siehe beispielsweise Lions [10, S. 180]). Setzt man dieses Ergebnis in (19) ein, so erhält man

$$u' + Au = f .$$

Da außerdem $u \in W$ und $u(0) = a$ ist (siehe Lemma 4), ist damit bewiesen, daß u Lösung der Aufgabe (2) ist.

Sind u_1, u_2 zwei Lösungen der Gleichung $u' + Au = f$, so gilt:

$$\frac{1}{2}\left(|u_1(t) - u_2(t)|^2 - |u_1(0) - u_2(0)|^2\right)$$

$$= \int_0^t (u_1'(s) - u_2'(s), u_1(s) - u_2(s))\, ds$$

$$= - \int_0^t ((A u_1)(s) - (Au_2)(s), u_1(s) - u_2(s))\, ds \leqq 0 .$$

Daraus folgt die Einzigkeit der Lösung der Aufgabe (2) und die behauptete stetige Abhängigkeit der Lösung vom Anfangswert. Damit ist Satz 1 bewiesen.

Satz 2. *Es seien die Voraussetzungen von Satz* 1 *erfüllt. Bezeichnen wir mit* u *die Lösung der Aufgabe* (2) *und mit* u_n *die Lösungen der Galerkingleichungen* (7), *so gilt:*

(a) $u_n \to u$ *in* $C(S; H)$;

(b) $u_n \to u$ *schwach in* X;

(c) $Au_n \to Au$ *schwach in* X^*;

(d) $\int_0^t ((Au_n)(s) - (Au)(s), u_n(s) - u(s))\, ds \to 0 , \quad t \in S .$

Bemerkung 7. Es gibt Operatoren $A \in (X \to X^*)$, für die aus den Eigenschaften (a)—(d) die starke Konvergenz der Folge $\{u_n\}$ gegen u in X folgt. Ist X uniform konvex, so gehören dazu beispielsweise die Operatoren $A \in (X \to X^*)$, für die gilt:

$$\langle Au - Av, u - v\rangle \geqq (\alpha(\|u\|_X) - \alpha(\|v\|_X))(\|u\|_X - \|v\|_X) , \quad u, v \in X;$$

dabei soll α eine monoton wachsende reelle Funktion sein.

Beweis von Satz 2. Die Folge $\{u_n\}$ ist nach Lemma 2 in X beschränkt. Aus dem Beweis von Satz 1 und Lemma 4 geht hervor, daß als Grenzwert der in X schwach konvergenten Teilfolgen von $\{u_n\}$ nur die eindeutig bestimmte Lösung der Aufgabe (2) in Frage kommt. Daher konvergiert die Folge $\{u_n\}$ in X schwach gegen u. Analog beweist man, daß die gesamte Folge $\{Au_n\}$ in X^* schwach gegen Au konvergiert.

Zum Beweis der Aussage (a) wählen wir eine in W gegen die Lösung u konvergierende Folge $\{v_n\}$ mit $v_n \in C^1(S; H_n)$. Daß eine solche Folge existiert, kann man unter Benutzung der Dichtheit von $C^1(S; V)$ in W und des Weierstraßschen Approximationssatzes beweisen. Dann gilt für $t \in S$:

$$\frac{1}{2}\left(|u_n(t) - v_n(t)|^2 - |u_n(0) - v_n(0)|^2\right) = \int_0^t \left(u_n'(s) - v_n'(s), u_n(s) - v_n(s)\right) ds$$

$$= \int_0^t \left(-(Au_n)(s) + (Au)(s) + u'(s) - v_n'(s), u_n(s) - v_n(s)\right) ds$$

$$= \int_0^t \{\left(-(Au_n)(s) + (Au)(s), u_n(s) - u(s)\right)$$
$$+ \left(-(Au_n)(s) + (Au)(s), u(s) - v_n(s)\right) + \left(u'(s) - v_n'(s), u_n(s) - v_n(s)\right)\} ds$$

$$\leqq -\int_0^t \left((Au_n)(s) - (Au)(s), u_n(s) - u(s)\right) ds$$
$$+ \|Au_n - Au\|_{X^*} \|u - v_n\|_X + \|u' - v_n'\|_{X^*} \|u_n - v_n\|_X$$

$$\leqq -\int_0^t \left((Au_n)(s) - (Au)(s), u_n(s) - u(s)\right) ds + K\|u - v_n\|_W, \quad K = \text{const}.$$

Dabei haben wir erneut von der Beschränktheit der Folgen $\{u_n\}$ und $\{Au_n\}$ in X bzw. X^* Gebrauch gemacht. Wegen der Stetigkeit der Einbettung von W in $C(S; H)$ gilt

$$\lim_{n\to\infty} |u_n(0) - v_n(0)| = 0.$$

Daher folgt aus der obigen Abschätzung auf Grund der Wahl der Folge $\{v_n\}$:

$$\lim_{n\to\infty} \|u - u_n\|_{C(S;H)} \leqq \lim_{n\to\infty} \left(\|u - v_n\|_{C(S;H)} + \|v_n - u_n\|_{C(S;H)}\right) = 0.$$

Außerdem ergibt sich

$$\lim_{n\to\infty} \int_0^t \left((Au_n)(s) - (Au)(s), u_n(s) - u(s)\right) ds = 0.$$

Damit ist Satz 2 bewiesen.

Bemerkung 8. Die Sätze 1 und 2 bleiben auch dann richtig, wenn man die Forderung der Koerzitivität von $A \in (X \to X^*)$ durch die etwas schwächere Forderung

$$\lim_{\substack{\|u\|_{L^p(S;V)}\to\infty \\ u \in X}} \frac{1}{\|u\|_{L^p(S;V)}} \langle Au, u\rangle = +\infty$$

ersetzt. Das kann man durch eine leichte Modifikation des ersten Teils des Beweises von Lemma 2 zeigen.

Bemerkung 9. Ist $p_0 \geqq 2$, so ist $X = L^p(S; V) \cap L^{p_0}(S; H) \subset L^2(S; H) \subset L^q(S; V^*) + L^{q_0}(S; H) = X^*$. Dabei ist X in $L^2(S; H)$ und $L^2(S; H)$ in X^* stetig eingebettet. Wir bezeichnen in diesem Falle mit I die (stetige) Einbettung von X in X^*.

Satz 3. *Es sei $p_0 \geqq 2$ und $A \in (X \to X^*)$ ein hemistetiger Operator. Für ein $\lambda \geqq 0$ und beliebige $u, v \in X$ gelte:*

$$\int_S e^{-2\lambda t}\left((Au)(t) - (Av)(t) + \lambda(u(t) - v(t)), u(t) - v(t)\right) dt \geqq 0 \qquad (22)$$

und

$$\lim_{\|u\|_X \to \infty} \frac{1}{\|u\|_X} \int_S e^{-2\lambda t} \big((Au)(t) + \lambda u(t), u(t)\big)\, dt = +\infty .$$

Dann besitzt die Aufgabe

$$u' + Au = f\,, \qquad u(0) = a\,, \qquad u \in X\,, \tag{24}$$

für beliebige $f \in X^*$ und $a \in H$ genau eine Lösung u. Es ist $u \in W$, und die Zuordnung $a \to u$ ist als Abbildung von H in $C(S; H)$ stetig.

Konvergiert die Folge $\{a_n\}$, $a_n \in H_n$, in H gegen a, so besitzt die Aufgabe (7) für jedes n genau eine Lösung u_n, und für die Folge $\{u_n\}$ der Galerkinnäherungen gilt:

(a) $u_n \to u$ *in* $C(S; H)$;

(b) $u_n \to u$ *schwach in* X;

(c) $Au_n \to Au$ *schwach in* X^*;

(d) $\int_0^t \big((Au_n)(s)\big) - \big(Au(s), u_n(s) - u(s)\big)\, ds \to 0\,, \quad t \in S\,.$

Folgerung. *Ist $p_0 \geqq 2$ und $A + \lambda I$, $\lambda \geqq 0$, ein hemistetiger, monotoner und koerzitiver Volterraoperator aus $(X \to X^*)$, so gelten sämtliche Aussagen von Satz 3.*

Beweis der Folgerung. Es genügt zu zeigen, daß unter den gegebenen Voraussetzungen die Beziehungen (22) und (23) gelten. Diese Beziehungen ergeben sich aus der Monotonie und der Koerzitivität von $A + \lambda I$, wenn man das folgende Lemma auf

$$g(s) = \big((Au)(s) - (Av)(s) + \lambda(u(s) - v(s)), u(s) - v(s)\big)$$

bzw. auf

$$g(s) = \big((Au)(s) - (A0)(s) + \lambda u(s), u(s)\big)$$

anwendet.

Lemma 5. *Aus $g \in L^1(s)$ und*

$$\int_0^t g(s)\, ds \geqq 0 \qquad \textit{für jedes} \quad t \in S$$

folgt für $\lambda \geqq 0$:

$$\int_0^t e^{-2\lambda s} g(s)\, ds \geqq e^{-2\lambda T} \int_0^t g(s)\, ds \geqq 0\,.$$

Beweis. Durch partielle Integration ergibt sich

$$\int_0^t e^{-2\lambda s} g(s)\, ds = \int_0^t 2\lambda e^{-2\lambda s} \int_0^s g(\sigma)\, d\sigma\, ds + e^{-2\lambda t} \int_0^t g(s)\, ds \geqq e^{-2\lambda T} \int_0^t g(s)\, ds \geqq 0\,.$$

Beweis von Satz 3. Wir führen den vorliegenden Fall auf den in den Sätzen 1 und 2 behandelten durch eine Transformation zurück. Wir setzen

$$u_\lambda(t) = e^{-\lambda t} u(t), \; f_\lambda(t) = e^{-\lambda t} f(t)$$

und

$$(A_\lambda u_\lambda)(t) = e^{-\lambda t} (Au)(t) + \lambda u_\lambda(t)\,. \tag{25}$$

Der so definierte Volterraoperator A_λ gehört offensichtlich zu $(X \to X^*)$. Außerdem sieht man, daß u genau dann Lösung der Aufgabe (24) ist, wenn für u_λ gilt:

$$u_\lambda' + A_\lambda u_\lambda = f_\lambda\,, \quad u_\lambda(0) = a\,, \quad u_\lambda \in X\,. \tag{26}$$

Unter Benutzung von (22) und (23) ergibt sich in elementarer Weise, daß $A_\lambda \in (X \to X^*)$ hemistetig, monoton und koerzitiv ist. Daher folgt der erste Teil von Satz 3 aus Satz 1.

Die Galerkingleichungen zu (26) lauten

$$u_{\lambda n}' + A_{\lambda n} u_{\lambda n} = f_{\lambda n}\,, \quad u_{\lambda n}(0) = a_n\,, \quad u_{\lambda n} \in X_n; \tag{27}$$

dabei sind $A_{\lambda n} \in (X_n \to X_n^*)$ und $f_{\lambda n} \in X_n^*$ in Analogie zu (5) und (6) durch

$$\left.\begin{aligned} \langle A_{\lambda n} u, v\rangle &= \langle A_\lambda u, v\rangle\,, \\ \langle f_{\lambda n}, v\rangle &= \langle f_\lambda, v\rangle \end{aligned}\right\} \quad \text{für jedes} \quad v \in X_n$$

definiert. Man kann direkt nachrechnen, daß $u_n \in X_n$ genau dann Lösung der Aufgabe (7) ist, wenn $u_{\lambda n}(t) = e^{-\lambda t} u_n(t)$ Lösung der Aufgabe (27) ist.

Aus Lemma 2 und Satz 2 folgt daher, daß die Galerkingleichungen (7) eindeutig lösbar sind und daß für die Funktionen $u_{\lambda n}$ folgendes gilt:

$$u_{\lambda n} \to u_\lambda \text{ in } C(S; H);$$

$$u_{\lambda n} \to u_\lambda \text{ schwach in } X;$$

$$A_\lambda u_{\lambda n} \to A_\lambda u_\lambda \text{ schwach in } X^*;$$

$$\int_0^t \big((A_\lambda u_{\lambda n})(s) - (A_\lambda u_\lambda)(s), u_{\lambda n}(s) - u_\lambda(s)\big)\, ds \to 0\,, \quad t \in S;$$

dabei bezeichnet u_λ die Lösung der Aufgabe (26). Aus diesen Eigenschaften der Folge $\{u_{\lambda n}\}$ folgen in einfacher Weise die Behauptungen (a)—(d) von Satz 3.

Damit ist Satz 3 vollständig bewiesen.

Bemerkung 10. Eine einfache Realisierung der Voraussetzungen von Satz 3 ist bereits in der Folgerung zu diesem Satz angegeben worden. Außerdem lassen sich bestimmte Operatordifferentialgleichungen zweiter Ordnung auf Evolutionsgleichungen zurückführen, die den Voraussetzungen von Satz 3 genügen. Darauf werden wir an anderer Stelle eingehen.

3. Regularitätsaussagen und Galerkinverfahren

Wir befassen uns in diesem Abschnitt mit Evolutionsgleichungen der Form

$$u' + Au = 0\,,$$

für die der Operator $A \in (X \to X^*)$ folgende Darstellung besitzt:

$$(Au)(t) = A(t)\, u(t) \quad \text{für} \quad u \in X,\ t \in S; \tag{28}$$

dabei bezeichnet $\{A(t)\}$, $t \in S$, eine Familie von Operatoren aus $(V \to V^*)$.

Satz 4. *Der Operator $A \in (X \to X^*)$ besitze die Darstellung* (28). *Es sei entweder A hemistetig, monoton und koerzitiv oder $p_0 \geqq 2$ und $A + \lambda I \in (X \to X^*)$ hemistetig, monoton und koerzitiv für ein $\lambda \geqq 0$. Außerdem gelte für eine auf $[0, \infty)$ definierte stetige nichtfallende Funktion ϱ:*

$$\left.\begin{array}{l} |(A(t)\,x - A(s)\,x, y)| \leqq |t - s|\,\varrho(|x|)\,\big(|y| + |(A(s)\,x, y)|\big) \\ \text{für beliebige } s, t \in S \text{ und } x, y \in V\,. \end{array}\right\} \tag{29}$$

Dann besitzt die Aufgabe

$$\left.\begin{array}{l} u'(t) + A(t)\,u(t) = 0 \quad \text{für jedes } t \in S\,, \\ u(0) = a\,, \quad u \in C^1_w(S; H) \cap C_w(S; V) \end{array}\right\} \tag{30}$$

zu jedem $a \in V$ mit $A(0)\,a \in H$ genau eine Lösung. Gelten für die Folge $\{a_n\}$, $a_n \in H_n$, die Beziehungen

$$\lim_{n\to\infty} |a_n - a| = 0 \tag{31}$$

und

$$\sup_n |A(0)\,a_n| < \infty\,, \tag{32}$$

so besitzen die Lösungen u_n der Galerkingleichungen

$$\left\{\begin{array}{l} \big(u_n'(t) + A(t)\,u_n(t), h_i\big) = 0\,, \qquad i = 1, \dots, n\,, \\ u_n(0) = a_n \end{array}\right. \tag{33}$$

folgende Eigenschaften:

(a) $\quad u_n \to u \quad$ *in* $\quad C_w(S; V)$;

(b) $\quad u_n'(t) \to u'(t) \quad$ *schwach in H für jedes $t \in S$.*

Bemerkung 11. Ist $A(0)\,a \in H$, so kann die Bedingung (32) etwa dadurch erfüllt werden, daß man $h_1 = a$ wählt (falls $a \neq 0$ ist) und $a_n = a$ setzt für alle n. Umgekehrt folgt aus dem Beweis von Satz 4, daß $A(0)\,a \in H$ ist, wenn eine Folge $\{a_n\}$, $a_n \in H_n$, mit den Eigenschaften (31) und (32) existiert.

Bemerkung 12. Die Lösbarkeit der Aufgabe (30) ist unter geringfügig anderen Voraussetzungen bereits in einer früheren Arbeit [6] (auf anderem Wege) bewiesen worden. Neu sind die Aussagen über die Konvergenz des Galerkinverfahrens, die bisher nur für den Fall, daß V ein Hilbertraum ist, bekannt waren (vgl. Gajewski [4]).

Dem Beweis von Satz 4 stellen wir einige Lemmata voran. Wir befassen uns zunächst mit dem Zusammenhang zwischen den Eigenschaften des Operators $A \in (X \to X^*)$ und den Eigenschaften der entsprechenden Operatoren $A(t) \in (V \to V^*)$.

Lemma 6. *Ist $A \in (X \to X^*)$ ein monotoner Operator, der den Voraussetzungen* (28) *und* (29) *genügt, so sind die Operatoren $A(t) \in (V \to V^*)$, $t \in S$, monoton.*

Beweis. Es sei $t \in S$, $t + h \in S$ und $h > 0$. Wir setzen

$$u(s) = \begin{cases} x & \text{für } \; t \leqq s \leqq t + h\,, \\ 0 & \text{sonst} \end{cases}$$

und

$$v(s) = \begin{cases} y & \text{für} \quad t \leqq s \leqq t + h\,, \\ 0 & \text{sonst}; \end{cases}$$

dabei bezeichnen x und y gegebene Elemente aus V. Dann gilt

$$\begin{aligned} 0 &\leqq \langle Au - Av, u - v\rangle = \int_t^{t+h} \big(A(s)\,x - A(s)\,y, x - y\big)\,ds \\ &= \int_t^{t+h} \big(A(s)\,x - A(t)\,x - A(s)\,y + A(t)\,y, x - y\big)\,ds + h\big(A(t)\,x - A(t)\,y, x - y\big) \\ &\leqq h^2 \left\{\varrho(|x|)\big(|x - y| + |(A(t)\,x, x - y)|\big) + \varrho(|y|)\big(|x - y| + |(A(t)\,y, x - y)|\big)\right\} \\ &\quad + h\big(A(t)\,x - A(t)\,y, x - y\big)\,. \end{aligned}$$

Aus dieser Ungleichung erhält man die Beziehung

$$0 \leqq \big(A(t)\,x - A(t)\,y, x - y\big)\,,$$

indem man durch h dividiert und den Grenzübergang $h \to 0$ ausführt.

Damit ist Lemma 6 bewiesen.

Lemma 7. *Ist $A \in (X \to X^*)$ ein demistetiger Operator, der den Voraussetzungen* (28) *und* (29) *genügt, so sind die Operatoren $A(t) \in (V \to V^*)$, $t \in S$, demistetig.*

Beweis. Es sei $x \in V$ und $\{x_i\}$ eine in V gegen x konvergierende Folge. Wir setzen

$$v(s) = x\,, \quad v_i(s) = x_i \quad \text{für} \quad s \in S\,.$$

Dann ist

$$\lim_{i\to\infty} \|v_i - v\|_X = 0\,.$$

Weiter setzen wir für gegebene $y \in V$ und $t, t + h \in S$, $h > 0$:

$$w_h(s) = \begin{cases} y & \text{für} \quad t \leqq s \leqq t + h\,, \\ 0 & \text{sonst}\,. \end{cases}$$

Dann gilt

$$\begin{aligned} &|(A(t)\,x_i - A(t)\,x, y)| \\ &= \left|\frac{1}{h}\int_t^{t+h} \big(A(t)\,x_i - A(s)\,x_i + A(s)\,x_i - A(s)\,x + A(s)\,x - A(t)\,x, y\big)\,ds\right| \\ &\leqq \int_S \{\varrho(|x_i|)\big(|w_h(s)| + |(A(s)\,v_i(s), w_h(s))|\big) \\ &\quad + \varrho(|x|)\big(|w_h(s)| + |(A(s)\,v(s), w_h(s))|\big)\}\,ds + \frac{1}{h}|\langle Av_i - Av, w_h\rangle| \\ &\leqq K\,(1 + \|Av_i\|_{X^*} + \|Av\|_{X^*})\,\|w_h\|_X + \frac{1}{h}|\langle Av_i - Av, w_h\rangle| \end{aligned}$$

für eine von i und h unabhängige Konstante K. Da A demistetig ist und die Beziehung

$$\|w_h\|_X = \|y\|\,h^{1/p} + |y|\,h^{1/p_0}$$

gilt, wird die rechte Seite der letzten Ungleichung bei geeigneter Wahl von h für hinreichend große i beliebig klein.

Damit ist Lemma 7 bewiesen.

Lemma 8. *Unter den Voraussetzungen von Satz 4 gehört die Lösung u_n der Galerkingleichungen* (33) *zu $C^1(S; H_n)$, und die Folge $\{u_n\}$ ist in $C^1(S; H_n)$ und in $C(S; V)$ beschränkt.*

Beweis. Wir wissen bereits aus dem vorangegangenen Abschnitt, daß die Galerkingleichungen (33) genau eine Lösung $u_n \in W \subset C(S; H)$ besitzen.

Der Operator $A + \lambda I$ ist monoton und hemistetig und daher nach einem Ergebnis von Kato [9] demistetig. Falls A monoton ist, ist hier und im folgenden $\lambda = 0$ zu setzen. Nach Lemma 7 sind auch die Operatoren $A(t) \in (V \to V^*)$ demistetig. Daraus und aus (29) folgt, wie man mühelos einsieht, daß $(A(t)\, x, y)$ bei festem $y \in V$ in $\{t, x\} \in S \times V$ stetig ist. Das Differentialgleichungssystem (33) erfüllt also die Voraussetzungen des Existenzsatzes von Peano. Folglich gehört u_n zu $C^1(S; H_n)$. Die Beschränktheit der Folge $\{u_n\}$ in $C(S; H)$ ist bereits bekannt (vgl. die Sätze 2 und 3).

Zum Beweis der Beschränktheit der Folge $\{u_n'\}$ in $C(S; H)$ betrachten wir die durch

$$w_n(t) = u_n(t + h) - u_n(t), \; h > 0\,,$$

definierte Funktion w_n. Für diese gilt unter den gegebenen Voraussetzungen

$$\frac{1}{2}\left(|w_n(t)|^2 - |w_n(0)|^2\right) = \int_0^t \left(w_n'(s), w_n(s)\right) ds$$

$$= - \int_0^t \left(A(s + h)\, u_n(s + h) - A(s)\, u_n(s), w(s)\right) ds$$

$$\leqq - \int_0^t \left(A(s + h)\, u_n(s) - A(s)\, u_n(s) - \lambda w_n(s), w_n(s)\right) ds$$

$$\leqq \int_0^t \left\{h\varrho(|u_n(s)|)\left(|w_n(s)| + |(A(s)\, u_n(s), w_n(s))|\right) + \lambda |w_n(s)|^2\right\} ds$$

$$\leqq K \int_0^t h\left(1 + |u_n'(s)| + \frac{1}{h}\, |w_n(s)|\right) |w_n(s)|\, ds;$$

dabei ist K eine von n und t unabhängige Konstante. Wir haben hierbei von Lemma 6 für $A + \lambda I \in (X \to X^*)$ Gebrauch gemacht. Nach Division durch h^2 und Ausführung des Grenzübergangs $h \to 0$ ergibt sich

$$\frac{1}{2}\left(|u_n'(t)|^2 - |u_n'(0)|^2\right) \leqq K \int_0^t \left(1 + 2\, |u_n'(s)|\right) |u_n'(s)|\, ds \leqq \int_0^t \left(1 + 3\, |u_n'(s)|^2\right) ds\,.$$

Durch Anwendung des Gronwallschen Lemmas finden wir

$$|u_n'(t)| \leqq K_0(|u_n'(0)| + 1)\,, \quad K_0 = \text{const}\,. \tag{34}$$

Für $x \in H_n$ ist

$$|(u_n'(0), x)| = |(A(0)\, a_n, x)| \leqq |A(0)\, a_n| \cdot |x| .$$

Daher folgt aus (34) und (32) die Beschränktheit der Folge $\{u_n'\}$ in $C(S; H)$.

Wir wählen nun $t \in S$ und eine natürliche Zahl n und setzen

$$v(s) = u_n(t) \quad \text{für jedes} \quad s \in S .$$

Dann gilt auf Grund des bereits Bewiesenen

$$\begin{aligned}\langle Av + \lambda v, v\rangle &= \int_S (A(s)\, u_n(t) - A(t)\, u_n(t) - u_n'(t) + \lambda u_n(t), u_n(t))\, ds \\ &\leqq \int_S \{T\varrho(|u_n(t)|)\,(|u_n(t)| + |(u_n'(t), u_n(t))|) - (u_n'(t), u_n(t)) \\ &\quad + \lambda |u_n(t)|^2\}\, ds \leqq K_1\end{aligned}$$

für eine von n und t unabhängige Konstante K_1. Daraus folgt wegen der Koerzitivität von $A + \lambda I$

$$\|v\|_X = \|u_n(t)\|\, T^{1/p} + |u_n(t)|\, T^{1/p_0} \leqq K_2 , \qquad K_2 = \text{const} .$$

Also ist die Folge $\{u_n\}$ in $C(S; V)$ beschränkt.

Damit ist Lemma 8 bewiesen.

Beweis von Satz 4. 1. Es sei $z \in V^*$ und $\{z_i\}$ eine in V^* gegen z konvergierende Folge von Elementen aus H. Dann gilt wegen der Beschränktheit der Folge $\{u_n\}$ in $C(S; V)$:

$$\begin{aligned}|(z, u_n(t) - u_m(t))| &\leqq |(z - z_i, u_n(t) - u_m(t))| + |(z_i, u_n(t) - u_m(t))| \\ &\leqq K\, \|z - z_i\|_* + |(z_i, u_n(t) - u_m(t))| .\end{aligned}$$

Aus der Konvergenz der Folge $\{u_n\}$ in $C(S; H)$ folgt, daß die rechte Seite dieser Ungleichung bei geeigneter Wahl von i für hinreichend große n und m unabhängig von t beliebig klein wird. Damit ist die Konvergenz der Folge $\{u_n\}$ in $C_w(S; V)$ bewiesen. Ihr Grenzwert kann nur die nach Satz 1 bzw. Satz 3 existierende Lösung u der Aufgabe

$$u' + Au = 0 , \; u(0) = a , \; u \in X ,$$

sein. Folglich ist $u \in C_w(S; V)$.

2. Für gegebenes $t \in S$ können wir aus der Folge $\{u_n'(t)\}$ auf Grund ihrer Beschränktheit in H eine Teilfolge $\{u_{n_j}'(t)\}$, $j = 1, 2, \ldots$, auswählen, die in H schwach gegen ein $y \in H$ konvergiert. Für diese Folge und beliebiges $x \in \bigcup_n H_n$ finden wir unter Benutzung der Monotonie von $A(t) + \lambda I \in (V \to V^*)$ (vgl. Lemma 6):

$$\begin{aligned}0 &\leqq \lim_{j\to\infty} (A(t)\, x - A(t)\, u_{n_j}(t) + \lambda(x - u_{n_j}(t)), x - u_{n_j}(t)) \\ &= \lim_{j\to\infty} (A(t)\, x + u_{n_j}'(t) + \lambda(x - u_{n_j}(t)), x - u_{n_j}(t)) \\ &= (A(t)\, x + y + \lambda(x - u(t)), x - u(t)) .\end{aligned}$$

Daraus folgt in der üblichen Weise

$$y = - A(t)\, u(t) .$$

Alle in H schwach konvergenten Teilfolgen von $\{u_n'(t)\}$ haben also das Grenzelement $-A(t)\,u(t)$. Daraus geht hervor, daß die gesamte Folge $\{u_n'(t)\}$ in H schwach gegen $-A(t)\,u(t)$ konvergiert.

3. Aus dem soeben Bewiesenen und der Beschränktheit der Folge $\{u_n'(t)\}$ in $C(S; H)$ ergibt sich

$$\sup_{t \in S} |A(t)\,u(t)| < \infty \,. \tag{35}$$

Es sei nun $t \in S$ beliebig und $\{t_i\}$ eine gegen t konvergierende Folge von Punkten aus S. Nach (29) ist für beliebiges $x \in V$:

$$|(A(t)\,u(t_i) - A(t_i)\,u(t_i), x)| \leqq |t - t_i|\,\varrho(|u(t_i)|)\,\big(|x| + |(A(t_i)\,u(t_i), x)|\big)\,.$$

Wegen (35) folgt daraus $A(t)\,u(t_i) \in H$ und

$$|A(t)\,u(t_i)| \leqq K_1 \tag{36}$$

für eine von i und t unabhängige Konstante K_1 sowie

$$\lim_{i \to \infty} |A(t)\,u(t_i) - A(t_i)\,u(t_i)| = 0\,. \tag{37}$$

Wegen (36) gibt es eine Teilfolge $\{A(t)\,u(t_{i_j})\}$, $j = 1, 2, \ldots$, der Folge $\{A(t)\,u(t_i)\}$, die in H schwach gegen ein $y \in H$ konvergiert. Für diese Teilfolge und beliebiges $x \in V$ gilt

$$\begin{aligned} 0 &\leqq \lim_{j \to \infty} \big(A(t)\,x - A(t)\,u(t_{i_j}) + \lambda(x - u(t_{i_j})), x - u(t_{i_j})\big) \\ &= \big(A(t)\,x - y + \lambda(x - u(t)), x - u(t)\big)\,. \end{aligned}$$

Daraus folgt ähnlich wie oben $y = A(t)\,u(t)$ und

$$A(t)\,u(t_i) \to A(t)\,u(t) \text{ schwach in } H\,. \tag{38}$$

Aus (37) und (38) ergibt sich schließlich

$$Au \in C_w(S; H)\,. \tag{39}$$

4. Für beliebiges $x \in H$ gilt auf Grund der vorangegangenen Ergebnisse

$$\big(u(t) - a, x\big) = \lim_{n \to \infty} \big(u_n(t) - a_n, x\big) = \lim_{n \to \infty} \int_0^t \big(u_n'(s), x\big)\,ds = -\int_0^t \big(A(s)\,u(s), x\big)\,ds\,.$$

Aus dieser Beziehung folgt wegen der Stetigkeit des letzten Integranden (vgl. (39)) die Zugehörigkeit von u zu $C_w^1(S; H)$ und die Gültigkeit von

$$u'(t) + A(t)\,u(t) = 0 \quad \text{für jedes} \quad t \in S\,.$$

Damit ist Satz 4 vollständig bewiesen.

Satz 5. *Es seien die Voraussetzungen von Satz 4 erfüllt. Außerdem gelte für eine monoton wachsende Funktion α mit $\alpha(0) = 0$:*

$$\left.\begin{array}{l} \big(A(t)\,x - A(t)\,y + \lambda(x - y), x - y\big) \geqq \alpha(\|x - y\|) \\ \textit{für beliebige } x, y \in V \textit{ und } t \in S\,. \end{array}\right\} \tag{40}$$

Dann gehört für jedes $a \in V$ mit $A(0)\,a \in H$ die nach Satz 4 existierende Lösung der Aufgabe (30) *zu* $C(S;V)$. *Für die Lösungen u_n der Galerkingleichungen* (33) *gilt, wenn die Bedingungen* (31) *und* (32) *erfüllt sind:*

(a) $\quad u_n \to u$ *in* $C(S;V)$;

(b) $\quad u_n \to u$ *in* $C^1_w(S;H)$.

Beweis. Es ist $|A(t)\,u(s)| \leqq K_1$ für eine von s und t unabhängige Konstante K_1 (vgl. (36)). Daraus folgt unter Benutzung von (40)

$$\begin{aligned}2K_1|u(t)-u(s)| &\geqq \big(A(t)\,u(t)-A(t)\,u(s),\,u(t)-u(s)\big)\\ &\geqq \alpha\big(||u(t)-u(s)||\big)-\lambda|u(t)-u(s)|^2.\end{aligned}$$

Hieraus ergibt sich wegen $u \in C(S;H)$ und der Eigenschaften von α die Zugehörigkeit von u zu $C(S;V)$.

$A(0)$ ist demistetig (vgl. Lemma 7) und daher lokal beschränkt. Folglich existiert ein $r > 0$ mit der Eigenschaft

$$\sup_{||x|| \leqq r} ||A(0)\,x||_* < \infty .$$

Wegen (29) ist dann auch

$$\begin{aligned}\sup_{||x|| \leqq r} ||A(t)\,x||_* &\leqq \sup_{||x|| \leqq r} \big(||A(t)\,x - A(0)\,x||_* + ||A(0)\,x||_*\big)\\ &\leqq \sup_{||x|| \leqq r} \big(K\varrho(|x|)\,(1+||A(0)\,x||_*) + ||A(0)\,x||_*\big) < \infty, \quad K = \text{const}.\end{aligned}$$

Daher gilt

$$\begin{aligned}||A(t)\,u_n(t)||_* &= \frac{1}{r}\sup_{||x|| \leqq r}\big(A(t)\,u_n(t),\,x\big)\\ &= \frac{1}{r}\sup_{||x|| \leqq r}\big\{\big(A(t)\,u_n(t)-A(t)\,x+\lambda(u_n(t)-x),\,x-u_n(t)\big)\\ &\qquad -\big(u_n'(t),\,u_n(t)\big)+\big(A(t)\,x-\lambda(u_n(t)-x),\,x-u_n(t)\big)\big\}\\ &\leqq \frac{1}{r}\Big(K_1+\sup_{||x|| \leqq r}||A(t)\,x||_*\,(r+||u_n(t)||)+\lambda\sup_{||x|| \leqq r}|u_n(t)-x|^2\Big) \leqq K_2;\end{aligned}$$

dabei sind K_1 und K_2 von n und t unabhängige Konstanten. Von dieser Abschätzung werden wir unten Gebrauch machen.

Zur Lösung u der Aufgabe (30) wählen wir eine Folge $\{v_n\}$, $v_n \in C(S;H_n)$, die in $C(S;V)$ gegen u konvergiert. Daß eine solche Folge existiert, läßt sich leicht mit Hilfe des Weierstraßschen Approximationssatzes zeigen. In Anbetracht des bereits Bewiesenen gilt

$$\begin{aligned}0 &= \big(u_n'(t)-u'(t)+A(t)\,u_n(t)-A(t)\,u(t),\,u_n(t)-v_n(t)\big)\\ &= \big(u_n'(t)-u'(t),\,u_n(t)-v_n(t)\big)+\big(A(t)\,u_n(t)-A(t)\,u(t),\,u_n(t)-u(t)\big)\\ &\quad +\big(A(t)\,u_n(t)-A(t)\,u(t),\,u(t)-v_n(t)\big)\\ &\geqq -K||u_n-v_n||_{C(S;H)}+\alpha\big(||u_n(t)-u(t)||\big)-\lambda||u_n-u||^2_{C(S;H)}\\ &\quad -K||u-v_n||_{C(S;V)};\end{aligned}$$

dabei ist K eine von n und t unabhängige Konstante. Auf Grund der Wahl der Folge $\{v_n\}$ und der Konvergenz der Folge $\{u_n\}$ gegen u in $C(S; H)$ folgt daraus

$$u_n \to u \quad \text{in} \quad C(S; V) . \tag{41}$$

Aus (41) und (29) sowie der Demistetigkeit der $A(t)$ folgt, wie wir in [6] gezeigt haben,

$$Au_n \to Au \quad \text{in} \quad C_w(S; V^*) . \tag{42}$$

Es sei jetzt $x \in H$ beliebig und $\{x_i\}$ eine in H gegen x konvergierende Folge mit $x_i \in H_i$. Dann gilt für $n \geqq i$

$$\begin{aligned} |(u_n'(t) + A(t)\,u(t), x)| &\leqq |(u_n'(t) + A(t)\,u(t), x - x_i)| + |(u_n'(t) + A(t)\,u(t), x_i)| \\ &\leqq K|x - x_i| + (-A(t)\,u_n(t) + A(t)\,u(t), x_i)| . \end{aligned}$$

Wegen (42) wird die rechte Seite dieser Ungleichung bei geeigneter Wahl von i für hinreichend große n unabhängig von t beliebig klein. Folglich gilt

$$u_n' \to -Au = u' \quad \text{in} \quad C_w(S; H) .$$

Zusammen mit der Konvergenz $u_n \to u$ in $C(S; H)$ liefert diese Feststellung die Konvergenzaussage

$$u_n \to u \quad \text{in} \quad C_w^1(S; H) .$$

Damit ist Satz 5 bewiesen.

Literatur

[1] Browder, F. E., Nonlinear functional analysis and nonlinear partial differential equations. Differ. equat. and their applic., J. Acta Fac. rerum natur. Univ. Comen. (1967) 45–64.

[2] Edwards, R. E., Functional analysis, New York 1965.

[3] Gajewski, H., Zur Approximation nichtlinearer Evolutionsgleichungen durch abstrakte Differentialgleichungen mit lipschitzstetigen Operatoren. Math. Nachr. **48** (1970) 377–385.

[4] —, Zur Konvergenz des Galerkinverfahrens bei nichtlinearen Evolutionsgleichungen. Math. Nachr. **49** (1971) 261–266.

[5] Gröger, K., Zur approximativen Lösung nichtlinearer Evolutionsgleichungen. Math. Nachrichten **49** (1971) 375–388.

[6] —, Zur Regularität und Approximation von Lösungen nichtlinearer Evolutionsgleichungen. Dissertation (Akad. d. Wiss. d. DDR).

[7] —, Zur Theorie nichtlinearer Differentialgleichungen erster und zweiter Ordnung mit Gedächtnis in Banach- bzw. Hilberträumen. Math. Nachr.

[8] Hess, P., Nonlinear functional equations in Banach spaces and homotopy arguments. Bull. Amer. Math. Soc. **77** (1971) 211–215.

[9] Kato, T., Demicontinuity, hemicontinuity and monotonicity II. Bull. Amer. Math. Soc. **73** (1967) 886–889.

[10] Lions, J. L., Quelques méthodes de résolution des problèmes aux limites non linéaires. Paris 1969.

[11] Lions, J. L., and Strauss, W. A., Some non-linear evolution equations. Bull. Soc. Math. France **93** (1965) 43–96.

[12] Raviart, P. A., Sur la résolution et l'approximation de certaines équations paraboliques non linéaires dégénérées. Arch. Rat. Mech. Anal. **25** (1967) 64–80.

Anschrift des Autors:

ZIMM der AdW der DDR
108 Berlin, Mohrenstr. 39

Positive Fixpunkte parameterabhängiger Operatoren

ROLF HÜNLICH

1. Einleitung

In der vorliegenden Arbeit untersuchen wir positive Lösungen von Operatorgleichungen in einem Banachraum, wobei die Operatoren nichtlinear von einem Parameter abhängen. Im Hinblick auf spätere Anwendungen (z. B. zur Untersuchung der plastischen Knickung von Stäben) benutzen wir recht spezielle Voraussetzungen.

Im zweiten Abschnitt behandeln wir lineare Operatoren, im dritten Abschnitt nichtlineare Operatoren. Als Beispiel geben wir im vierten Abschnitt eine Klasse nichtlinearer Integralgleichungen an, auf die die Ergebnisse angewendet werden können.

E sei ein reeller Banachraum, $||\cdot||$ bezeichne die Norm. $K \subset E$ sei ein Kegel. Wir führen in E eine Halbordnung ein: $x \geqq y$ genau dann, wenn $x - y \in K$. Gilt $x \geqq y$ und $x \neq y$, so schreiben wir $x > y$, gilt nicht $x \geqq y$, so schreiben wir $x \not\geqq y$. Die lineare Hülle von K stimme mit E überein. Die Norm sei monoton, d. h. aus $x \geqq y \geqq \Theta$ folgt $||x|| \geqq ||y||$.

Folgendes Lemma werden wir häufig benutzen [2].

Lemma 1.1. *Es sei $x, y \in K$, $y \neq \Theta$. Dann existiert eine Zahl $t_0 \geqq 0$ mit der Eigenschaft, daß $x \geqq ty$ für alle $t \leqq t_0$ und $x \not\geqq ty$ für $t > t_0$ ist. Gibt es eine Zahl $a > 0$ mit $x \geqq ay$, so wird $t_0 \geqq a$. Gilt $x \not\geqq y$, so wird $t_0 \in [0, 1)$.*

$u_0 > \Theta$ sei ein fixierter Element. B sei ein Operator von E in sich, $B\Theta = \Theta$.

Definition. B heißt u_0-beschränkt, wenn für jedes $x \in K$, $x \neq \Theta$, eine positive Zahl $b(x)$ mit der Eigenschaft existiert, daß $Bx \leqq b(x)\, u_0$ wird.

B heißt u_0-monoton, wenn für alle $x, y \in K$, $x - y > \Theta$, eine positive Zahl $a(x, y)$ mit der Eigenschaft existiert, daß $Bx - By \geqq a(x, y)\, u_0$ wird. B heißt konkav, wenn für jedes $x \in K$, $x \neq \Theta$, und für jedes $t \in (0, 1)$ $B(tx) \geqq tBx$ wird.

2. Lineare parameterabhängige Operatoren

Als erstes beschäftigen wir uns mit linearen Operatoren, die von einem Parameter $p \in J$ abhängen, $J \subset R^1$ sei ein Intervall. Für jedes $p \in J$ sei $u_0(p) \in K$, $u_0(p) \neq \Theta$. Folgende Voraussetzungen seien erfüllt.

(A1) Für jedes $p \in J$ sei $A(p)$ ein linearer vollstetiger Operator von E in sich.

(A2) Für jedes $p \in J$ sei $A(p)$ $u_0(p)$-beschränkt und $u_0(p)$-monoton.

Aus (A2) folgt speziell

$$b_0(p)\, u_0(p) \geqq A(p)\, u_0(p) \geqq a_0(p)\, u_0(p)\,, \qquad a_0(p) > 0, \qquad p \in J\,. \tag{2.1}$$

Der Operator $A(p)$ bildet K in sich ab. Wir fragen nach Werten des Parameters p, für die der Operator $A(p)$ einen in K liegenden nichttrivialen Fixpunkt besitzt. Wir suchen also Lösungen des Problems

$$A(p)\, h = h\,, \qquad h \in K\,, \qquad ||h|| = 1\,. \tag{2.2}$$

Zunächst fixieren wir ein $p \in J$ und betrachten als Hilfsaufgabe das Eigenwertproblem

$$A(p)\, h = \lambda h\,, \qquad h \in K\,, \qquad ||h|| = 1\,, \qquad \lambda > 0\,. \tag{2.3}$$

Dieses Eigenwertproblem ist unter allgemeineren Voraussetzungen z. B. in [2] untersucht worden, siehe auch [3]. Wir fassen einige Ergebnisse zusammen.

Lemma 2.1. *Für jedes fixierte $p \in J$ besitzt das Eigenwertproblem* (2.3) *genau eine Lösung $h_0(p)$, $\lambda_0(p)$ mit*

$$a_0(p) \leqq \lambda_0(p) \leqq b_0(p)\,, \tag{2.4}$$

$$a_1(p)\, u_0(p) \leqq h_0(p) \leqq b_1(p)\, u_0(p)\,, \qquad a_1(p) > 0\,. \tag{2.5}$$

Das Spektrum des Operators $A(p)$ liegt im Kreis $|\lambda| \leqq \lambda_0(p)$.

Für jedes $x \in K$, $x \neq \Theta$, sind die Ungleichungen $A(p)\, x > \lambda(p)\, x$, $A(p)\, x < \lambda(p)\, x$ unmöglich.

Wir untersuchen nun die Eigenschaften der Funktion $\lambda_0(p)$. Dazu setzen wir folgendes voraus.

(A3) Für jedes $p_1 \in J$ und für jedes $\varepsilon > 0$ existiere ein $\delta(\varepsilon, p_1) > 0$ mit der Eigenschaft, daß $||A(p_1) - A(p_2)|| < \varepsilon$ für alle $p_2 \in J$ mit $|p_1 - p_2| < \delta(\varepsilon, p_1)$ wird.

(A4) Für jedes $x \in K$, $x \neq \Theta$, und für alle $p_1, p_2 \in J$ mit $p_1 > p_2$ sei $A(p_1)\, x > A(p_2)\, x$.

Lemma 2.2. *Die Funktion $\lambda_0(p)$ ist auf J streng monoton wachsend.*

Beweis. Es sei $p_1, p_2 \in J$, $p_1 > p_2$. Mit h_1, λ_1 bzw. h_2, λ_2 bezeichnen wir die entsprechenden Lösungen des Eigenwertproblems (2.3). Zunächst folgt aus (A4), (A2) und (2.5)

$$\lambda_1 h_1 = A(p_1)\, h_1 \geqq A(p_2)\, h_1 \geqq a h_2\,, \qquad a > 0\,.$$

Deshalb gibt es ein maximales $t_0 > 0$, so daß $h_1 \geqq t_0 h_2$ wird. Weiter erhalten wir

$$\lambda_1 h_1 = A(p_1)\, h_1 - A(p_2)\, h_1 + A(p_2)\, h_1 \geqq A(p_1)\, h_1 - A(p_2)\, h_1 + t_0 \lambda_2 h_2$$

und

$$A(p_1)\, \lambda_1 h_1 = \lambda_1^2 h_1 \geqq A(p_2)\, (A(p_1)\, h_1 - A(p_2)\, h_1) + t_0 \lambda_2^2 h_2 \geqq (a + t_0 \lambda_2^2)\, h_2\,, \quad a > 0\,.$$

Folglich ist

$$h_1 \geqq (t_0 \lambda_2^2/\lambda_1^2 + a/\lambda_1^2)\, h_2\,.$$

Nehmen wir an, es sei $\lambda_1 \leqq \lambda_2$; dann wird $t_0 \lambda_2^2/\lambda_1^2 + a/\lambda_1^2 > t_0$, im Widerspruch zur Maximalität von t_0. Also gilt $\lambda_1 > \lambda_2$, q.e.d.

Lemma 2.3. *Die Funktion $\lambda_0(p)$ ist auf J stetig.*

Beweis. Wir fixieren ein beliebiges $p_0 \in J$, es sei

$$A(p_0)\, h_0 = \lambda_0 h_0\,, \quad h_0 \in K\,, \quad \|h_0\| = 1\,, \quad \lambda_0 > 0\,.$$

Ferner wählen wir eine Folge $\{p_n\}$ mit $p_n \in J$, $\lim\limits_{n\to\infty} p_n = p_0$, es sei

$$A(p_n)\, h_n = \lambda_n h_n\,, \quad h_n \in K\,, \quad \|h_n\| = 1\,, \quad \lambda_n > 0\,.$$

Zunächst nehmen wir an, daß die Folge $\{p_n\}$ monoton wächst. Dann ist auch die Folge $\{\lambda_n\}$ monoton wachsend, ferner gilt $0 < \lambda_1 \leqq \lambda_n \leqq \lambda_0 < +\infty$. Die Folge $\{\lambda_n\}$ ist also konvergent, $\lim\limits_{n\to\infty} \lambda_n = \lambda_* > 0$. Aus $\|h_n\| = 1$ folgt, daß die Folge $\{A(p_0)\, h_n\}$ kompakt ist. Es existiert also ein $u_* \in K$ mit $\lim\limits_{i\to\infty} A(p_0)\, h_{n_i} = u_*$. Aus der Abschätzung

$$\|A(p_{n_i})\, h_{n_i} - u_*\| \leqq \|A(p_{n_i}) - A(p_0)\| + \|A(p_0)\, h_{n_i} - u_*\|$$

und aus (A3) folgt $\lim\limits_{i\to\infty} A(p_{n_i})\, h_{n_i} = u_*$. Damit erhalten wir

$$\lim_{i\to\infty} h_{n_i} = \lim_{i\to\infty} A(p_{n_i})\, h_{n_i}/\lambda_{n_i} = u_*/\lambda_* = h_*$$

mit $h_* \in K$, $\|h_*\| = 1$. Schließlich ist

$$\lim_{i\to\infty} \|A(p_{n_i})\, h_{n_i} - A(p_0)\, h_*\| = 0$$

und folglich

$$A(p_0)\, h_* = \lambda_* h_*\,, \quad h_* \in K\,, \quad \|h_*\| = 1\,, \quad \lambda_* > 0\,.$$

Da dieses Eigenwertproblem eindeutig lösbar ist, muß $\lambda_* = \lambda_0$ sein, so daß wir $\lim\limits_{n\to\infty} \lambda_n = \lambda_0$ erhalten.

Nun betrachten wir den Fall, daß die Folge $\{p_n\}$ monoton fällt. Dann ist die Folge $\{\lambda_n\}$ monoton fallend, und es gilt $0 < \lambda_0 \leqq \lambda_n \leqq \lambda_1$. Deshalb existiert wieder ein $\lambda_* > 0$ mit $\lim\limits_{n\to\infty} \lambda_n = \lambda_*$. Wie eben läßt sich zeigen, daß $\lambda_* = \lambda_0$ sein muß. Qed.

Lemma 2.4. *Es existiere ein $p_1 \in J$ mit $b_0(p_1) \leqq 1$ und ein $p_2 \in J$ mit $a_0(p_2) \geqq 1$. Dann besitzt die Gleichung $\lambda_0(p) = 1$ auf J genau eine Lösung $p_0 \in [p_1, p_2]$.*

Der Beweis folgt unmittelbar aus der Ungleichung (2.4) und aus den angegebenen Eigenschaften der Funktion $\lambda_0(p)$ auf J.

Wir fassen die bisherigen Ergebnisse in einem Satz zusammen.

Satz 2.1. *Der Operator $A(p)$ genüge den Voraussetzungen* (A1) *bis* (A4). *Ferner existiere ein $p_1 \in J$ mit $b_0(p_1) \leqq 1$ und ein $p_2 \in J$ mit $a_0(p_2) \geqq 1$. Dann gibt es ein $p_0 \in [p_1, p_2]$, für das das Problem* (2.2) *genau eine Lösung h_0 besitzt.*

Für $p \in J$, $p \neq p_0$, ist das Problem (2.2) *unlösbar.*

Für $p \in J$, $p < p_0$, liegt das Spektrum des Operators $A(p)$ im Kreis $|\lambda| < 1$.

Für $x \in K$, $x \neq \Theta$, sind die Ungleichungen $A(p_0)\, x > x$ und $A(p_0)\, x < x$ unmöglich.

Wir betrachten nun zwei Operatoren $A_0(p)$ und $A_\infty(p)$, die beide den Voraussetzungen (A1) bis (A4) genügen. Ferner sei $p_* \in J$ fixiert und

(A5) $A_0(p)\,x = A_\infty(p)\,x$ für $x \in K$, $p \in J$, $p \leqq p_*$;

$A_0(p)\,x > A_\infty(p)\,x$ für $x \in K$, $x \neq \Theta$, $p \in J$, $p > p_*$.

Mit $h_0(p)$, $\lambda_0(p)$ bzw. $h_\infty(p)$, $\lambda_\infty(p)$ bezeichnen wir die Lösungen der entsprechenden Eigenwertprobleme (2.3).

Lemma 2.5. *Für $p \in J$, $p \leqq p_*$, gilt $\lambda_0(p) = \lambda_\infty(p)$; für $p \in J$, $p > p_*$, gilt $\lambda_0(p) > \lambda_\infty(p)$.*

Beweis. Für $p \in J$, $p \leqq p_*$, ist die Behauptung offensichtlich. Nun sei $p \in J$, $p > p_*$. Zunächst gilt

$$\lambda_0(p)\,h_0(p) = A_0(p)\,h_0(p) \geqq A_\infty(p)\,h_0(p) \geqq a h_\infty(p)\,, \qquad a > 0\,,$$

so daß ein maximales $t_0 > 0$ mit $h_0(p) \geqq t_0 h_\infty(p)$ existiert. Weiter erhalten wir

$$\lambda_0(p)\,h_0(p) \geqq A_0(p)\,h_0(p) - A_\infty(p)\,h_0(p) + t_0\lambda_\infty(p)\,h_\infty(p)$$

und schließlich

$$h_0(p) \geqq \left(t_0\lambda_\infty^2(p)/\lambda_0^2(p) + a/\lambda_0^2(p)\right) h_\infty(p)\,.$$

Da t_0 maximal ist, muß $\lambda_0(p) > \lambda_\infty(p)$ sein, qed.

Lemma 2.6. *Es existiere ein $p_1 \in J$ mit $b_0(p_1) \leqq 1$ und ein $p_2 \in J$ mit $a_\infty(p_2) \geqq 1$. Dann besitzen die Gleichungen $\lambda_0(p) = 1$ bzw. $\lambda_\infty(p) = 1$ auf J jeweils genau eine Lösung $p_0 \in [p_1, p_2]$ bzw. $p_\infty \in [p_1, p_2]$. Dabei ist entweder $p_0 = p_\infty \leqq p_*$ oder $p_* < p_0 < p_\infty$.*

Der Beweis folgt aus Lemma 2.4 und Lemma 2.5.

3. Nichtlineare parameterabhängige Operatoren

Wir wenden uns nun nichtlinearen parameterabhängigen Operatoren zu. Folgende Voraussetzungen seien erfüllt.

(B1) Für jedes $p \in J$ sei $B(p, .)$ ein vollstetiger Operator von E in sich.

(B2) Für jedes $p \in J$ sei $B(p, .)$ $u_0(p)$-beschränkt, $u_0(p)$-monoton und konkav.

(B3) Für jedes $p_1 \in J$ und für jedes $\varepsilon > 0$ existiere ein $\delta(\varepsilon, p_1) > 0$ mit der Eigenschaft, daß für alle $x \in K$, $x \neq \Theta$, $\|B(p_1, x) - B(p_2, x)\| < \varepsilon\|x\|$ für alle $p_2 \in J$ mit $|p_1 - p_2| < \delta(\varepsilon, p_1)$ wird.

(B4) Für alle $p_1, p_2 \in J$ mit $p_1 > p_2$ und für jedes $x \in K$, $x \neq \Theta$, sei $B(p_1, x) > B(p_2, x)$.

(B5) Für jedes $p \in J$ besitze $B(p, .)$ die starke asymptotische Ableitung nach dem Kegel $A_\infty(p)$. $A_\infty(p)$ sei $u_0(p)$-monoton. Für alle $p_1, p_2 \in J$ mit $p_1 > p_2$ und für jedes $x \in K$, $x \neq \Theta$, sei $A_\infty(p_1)\,x > A_\infty(p_2)\,x$.

(B6) Zusätzlich gelte

$$B(p, x) = A_0(p)\,x \text{ für } p \in J,\ p \leqq p_*;\ x \in K; \tag{3.1}$$

$$B(p, x) = A_0(p)\,x \text{ für } p \in J,\ p > p_*;\ x \in K,\ \|x\| \leqq r_0(p),\ r_0(p) > 0; \tag{3.2}$$

$$B(p, tx) \neq tB(p, x) \text{ für } p \in J,\ p > p_*;\ x \in K,\ \|x\| > r_0(p);\ t \in (0, 1)\,. \tag{3.3}$$

Dabei sei $p_* \in J$ fixiert und $A_0(p)$ ein linearer Operator von E in sich.

Aus diesen Voraussetzungen ergeben sich zunächst einige weitere Eigenschaften der Operatoren $A_0(p)$ und $A_\infty(p)$.

Lemma 3.1. *Die Operatoren* $A_0(p)$ *und* $A_\infty(p)$ *genügen den Voraussetzungen* (A1) *bis* (A5). *Ferner gilt die Ungleichung*

$$A_0(p)\,x \geqq B(p, x) \geqq A_\infty(p)\,x\,, \quad p \in J\,, \quad x \in K\,. \tag{3.4}$$

Beweis. (1) Für jedes $x \in K$ gilt $A_0(p)\,x \geqq B(p, x)$.

Wegen (3.1), (3.2) bleibt nur der Fall $p > p_*$, $||x|| > r_0(p)$ zu untersuchen. Wir setzen $t = r_0(p)/||x||$, offensichtlich ist $t \in (0, 1)$. Aus (3.2) und (B2) folgt dann

$$A_0(p)\,x - B(p, x) = \frac{1}{t}\left(tA_0(p)\,x - B(p, tx)\right) + \frac{1}{t}\left(B(p, tx) - tB(p, x)\right) \geqq \Theta\,.$$

(2) Für jedes $x \in K$ gilt $B(p, x) \geqq A_\infty(p)\,x$.

Für $x = \Theta$ ist die Behauptung offensichtlich. Nun sei $x \neq \Theta$ und $t > 1$. Dann gilt wegen (B2)

$$B(p, x) - A_\infty(p)\,x = \frac{1}{t}\left(B(p, tx) - tA_\infty(p)\,x\right) + \left(B(p, x) - \frac{1}{t}\,B(p, tx)\right)$$
$$\geqq \frac{1}{t}\left(B(p, tx) - tA_\infty(p)\,x\right).$$

Für $t \to \infty$ strebt die rechte Seite dieser Ungleichung nach Θ (siehe (B5)), woraus die Behauptung folgt. Damit ist die Ungleichung (3.4) bewiesen.

(3) $A_0(p)$ genügt den Voraussetzungen (A1) bis (A4).

$A_0(p)$ ist auf E vollstetig (siehe [2]). Es sei $x > \Theta$. Wir setzen $t = r_0(p)/||x||$ und erhalten mit (3.1), (3.2), (B2)

$$A_0(p)\,x = \frac{1}{t}\,B(p, tx) \leqq bu_0(p)\,, \qquad b > 0\,.$$

Ferner sei $x, y \in K$, $x - y > \Theta$. Dann gilt wegen (3.4) und (B2)

$$A_0(p)\,x - A_0(p)\,y \geqq B(p, x - y) \geqq au_0(p)\,, \qquad a > 0\,.$$

$A_0(p)$ ist also $u_0(p)$-beschränkt und $u_0(p)$-monoton. Schließlich sei $x \in K$, $x \neq \Theta$, $p_1, p_2 \in J$. Dann gilt mit $t = \min\,(r_0(p_1), r_0(p_2))/||x||$

$$A_0(p_1)\,x - A_0(p_2)\,x = \frac{1}{t}\left(B(p_1, tx) - B(p_2, tx)\right).$$

Mit (B3) und (B4) bestätigen sich die Aussagen (A3) und (A4).

(4) $A_\infty(p)$ genügt den Voraussetzungen (A1) bis (A4).

$A_\infty(p)$ ist ebenfalls auf E vollstetig (siehe [2]). Die $u_0(p)$-Beschränktheit folgt aus (3.4), die $u_0(p)$-Monotonie wurde in (B5) vorausgesetzt. Nun sei $x \in K$,

$x \neq \Theta$, $p_1, p_2 \in J$ und $t > 0$. Dann gilt

$$A_\infty(p_1)\, x - A_\infty(p_2)\, x = \frac{1}{t} A_\infty(p_1)\,(tx) - \frac{1}{t} B(p_1, tx) + \frac{1}{t} B(p_2, tx)$$

$$- \frac{1}{t} A_\infty(p_2)\,(tx) + \frac{1}{t} B(p_1, tx) - \frac{1}{t} B(p_2, tx)\,.$$

Zu einem vorgegebenen $\varepsilon > 0$ existiert nach (B3) ein $\delta > 0$, so daß

$$\|A_\infty(p_1)\, x - A_\infty(p_2)\, x\| \leqq \frac{1}{t} \|A_\infty(p_1)\,(tx) - B(p_1, tx)\|$$

$$+ \frac{1}{t} \|A_\infty(p_2)\,(tx) - B(p_2, tx)\| + \varepsilon\|x\|$$

wird, sofern $|p_1 - p_2| < \delta$ ist. Lassen wir in dieser Ungleichung t nach Unendlich streben, so erhalten wir (A3). (A4) schließlich wurde in (B5) vorausgesetzt.

(5) (A5) ist erfüllt.

Für $p \leqq p_*$ folgt die Behauptung aus (B5) und (3.1). Nun sei $p > p_*$. Aus (3.4) folgt $A_0(p)\, x \geqq A_\infty(p)\, x$ für alle $x \in K$. Es gebe ein $x > \Theta$ mit $A_0(p)\, x = A_\infty(p)\, x$. Wegen (3.4) ist dann

$$tA_0(p)\, x = B(p, tx) \quad \text{für alle } t > 0\,.$$

Wir setzen $t = t_1 t_2$, $t_1 \in (0, 1)$, $t_2 > r_0(p)/\|x\|$; (3.3) führt dann auf den Widerspruch

$$tA_0(p)\, x = B(t_1 t_2 x) \neq t_1 B(p, t_2 x) = tA_0(p)\, x\,.$$

Folglich ist $A_0(p)\, x > A_\infty(p)\, x$ für alle $x > \Theta$, $p > p_*$. Damit ist das Lemma vollständig bewiesen, qed.

Mit $\lambda_0(p)$ bzw. $\lambda_\infty(p)$ bezeichnen wir die positiven Eigenwerte des für die Operatoren $A_0(p)$ bzw. $A_\infty(p)$ formulierten Eigenwertproblems (2.3). Es gebe ein $p_1 \in J$ mit $b_0(p_1) \leqq 1$ und ein $p_2 \in J$ mit $a_\infty(p_2) \geqq 1$. Dann sind die entsprechenden Probleme (2.2) eindeutig lösbar:

$$A_0(p_0)\, h_0 = h_0\,, \qquad h_0 \in K\,, \qquad \|h_0\| = 1\,, \tag{3.5}$$

$$A_\infty(p_\infty)\, h_\infty = h_\infty\,, \qquad h_\infty \in K\,, \qquad \|h_\infty\| = 1\,, \tag{3.6}$$

$$p_0, p_\infty \in [p_1, p_2]\,, \qquad p_0 \leqq p_\infty\,. \tag{3.7}$$

Der Operator $B(p, .)$ bildet K in sich ab. Wir fragen nach Werten des Parameters p, für die der Operator $B(p, .)$ einen in K liegenden Fixpunkt besitzt:

$$B(p, x) = x\,, \qquad x \in K\,. \tag{3.8}$$

Wegen $B(p, \Theta) = \Theta$ besitzt dieses Problem immer die triviale Lösung $x = \Theta$. Im folgenden interessieren uns gerade nichttriviale Lösungen des Problems.

Lemma 3.2. $x = x(p)$, $p \in J$, *sei eine nichttriviale Lösung des Problems* (3.8). *Dann gilt*

$$a_2(p)\, u_0(p) \leqq x(p) \leqq b_2(p)\, u_0(p) \quad \text{mit} \quad a_2(p) > 0\,, \qquad p_0 \leqq p \leqq p_\infty\,. \qquad (3.9)$$

Beweis. Die erste Behauptung folgt aus der Tatsache, daß der Operator $B(p, .)$ $u_0(p)$-beschränkt und $u_0(p)$-monoton ist. Zum Beweis der zweiten Behauptung gehen wir vom Gegenteil aus. Es sei zunächst $p < p_0$. Dann gilt

$$x = B(p, x) \leqq A_0(p)\, x \leqq A_0(p_0)\, x \leqq bh_0\,, \qquad b > 0\,.$$

Deshalb gibt es ein maximales $t_0 > 0$ mit $h_0 \geqq t_0 x$. Weiter wird

$$h_0 \geqq t_0 x + A_0(p_0)\, h_0 - A_0(p)\, h_0\,.$$

Wir wenden auf diese Ungleichung des Operators $A_0(p_0)$ an und erhalten $h_0 \geqq t_0(1 + a)\, x$ mit $a > 0$. Das widerspricht aber der Maximalität von t_0. Nun sei $p > p_\infty$. Aus

$$x = B(p, x) \geqq A_\infty(p)\, x \geqq A_\infty(p_\infty)\, x \geqq ah_\infty\,, \qquad a > 0\,,$$

folgt die Existenz eines maximalen $t_0 > 0$ mit $x \geqq t_0 h_\infty$. Damit wird

$$x \geqq A_\infty(p)\, x \geqq t_0(h_\infty + A_\infty(p)\, h_\infty - A_\infty(p_\infty)\, h_\infty)\,.$$

Wenden wir auf diese Ungleichung den Operator $A_\infty(p_\infty)$ an, so erhalten wir $x \geqq t_0(1 + a)\, h_\infty$ mit $a > 0$, im Widerspruch zur Maximalität von t_0. Qed.

Nach Lemma 2.6 haben wir zwei Fälle zu unterscheiden: $p_\infty = p_0 \leqq p_*$ und $p_\infty > p_0 > p_*$. Im ersten Fall kommen wir zu folgendem Ergebnis.

Satz 3.1. *Es sei* $p_\infty = p_0 \leqq p_*$. *Für* $p = p_0$ *besitzt das Problem* (3.8) *nur Lösungen der Form* $x = ch_0$, $c \in [0, \infty)$. *Für* $p \in J$, $p \neq p_0$, *besitzt das Problem* (3.8) *nur die triviale Lösung.*

Im weiteren befassen wir uns nur noch mit dem zweiten Fall.

Lemma 3.3. *Für* $p = p_0$ *besitzt das Problem* (3.8) *nur Lösungen der Form* $x(p_0) = cr_0(p_0)\, h_0$ *mit* $c \in [0, 1]$.
Für $p = p_\infty$ *besitzt das Problem* (3.8) *nur die triviale Lösung.*

Beweis. Die Existenz der trivialen Lösung ist offensichtlich. Nun sei $B(p_0, x) = x$, $x > \Theta$. Die Annahme $||x|| > r_0(p_0)$ führt auf einen Widerspruch. Denn es sei $t \in (0, 1)$; dann wird wegen (3.4), (B2) und (3.3)

$$A_0(p_0)\,(tx) \geqq B(p_0, tx) > tB(p_0, x) = tx\,,$$

also $A_0(p_0)\, x > x$, $x > \Theta$, was wegen der letzten Aussage von Satz 2.1 unmöglich ist. x löst also das Problem

$$A_0(p_0)\, x = x\,, \qquad x > \Theta\,, \qquad ||x|| \leqq r_0(p_0)\,.$$

Aus Satz 2.1 folgt $x = cr_0(p_0)\, h_0$ mit $c \in (0, 1]$, womit die erste Behauptung bewiesen ist. Jetzt sei $B(p_\infty, x) = x$, $x > \Theta$. Die Annahme $||x|| \leqq r_0(p_\infty)$ führt auf einen Widerspruch zu Satz 2.1; denn dann wäre

$$A_0(p_\infty)\, h = h\,, \qquad h = x/||x|| \in K\,, \qquad ||h|| = 1\,, \qquad p_\infty \neq p_0\,.$$

Also ist $||x|| > r_0(p_\infty)$. Wir wählen ein $t > 1$ und erhalten

$$A_\infty(p_\infty)(tx) \leqq B(p_\infty, tx) < tB(p_\infty, x) = tx ,$$

also $A_\infty(p_\infty)\, x < x$, $x > \Theta$, was wegen Satz 2.1 ebenfalls unmöglich ist. Qed.

Lemma 3.4. *Für jedes $p \subset (p_0, p_\infty)$ besitzt das Problem* (3.8) *genau eine nichttriviale Lösung $x(p)$. Es ist $||x(p)|| > r_0(p)$.*

Beweis. $p \in (p_0, p_\infty)$ sei beliebig fixiert. Wir definieren den Operator

$$Cx = \begin{cases} B(p, x) & ||x|| \geqq r_0(p), \\ A_0(p)\, x + \big(r_0(p) - ||x||\big)\, h_0 & ||x|| \leqq r_0(p). \end{cases}$$

Der Operator C ist auf E vollstetig und bildet K in sich ab; C besitzt die starke asymptotische Ableitung nach dem Kegel $A_\infty(p)$, und das Spektrum des Operators $A_\infty(p)$ liegt im Kreis $|\lambda| < 1$. Deshalb besitzt C auf K mindestens einen Fixpunkt x (siehe [2]).

Nehmen wir nun an, es sei $||x|| \leqq r_0(p)$, also

$$A_0(p)\, x + sh_0 = x , \qquad s = r_0(p) - ||x|| \geqq 0 . \tag{3.10}$$

$s = 0$ ist unmöglich; denn dann wäre

$$A_0(p)\, h = h , \quad h = x/r_0(p) \in K , \quad ||h|| = 1 , \quad p \neq p_0 ,$$

im Widerspruch zu Satz 2.1. Aus (3.10) folgt, daß es ein maximales $t_0 > 0$ mit $x \geqq t_0 h_0$ gibt. Ebenfalls aus (3.10) folgt

$$x = sh_0 + A_0(p)\, x \geqq sh_0 + t_0 A_0(p_0)\, h_0 = (t_0 + s)\, h_0$$

im Widerspruch zur Maximalität von t_0. Deshalb ist $||x|| > r_0(p)$ und x Fixpunkt von $B(p, .)$.

Wir beweisen nun die Einzigkeit. Es sei

$$x_i = B(p, x_i) , \quad x_i \in K , \quad ||x_i|| > r_0(p) , \quad i = 1, 2 , \quad x_1 \neq x_2 .$$

O. B. d. A. gilt dann $x_1 \geqq x_2$, und es gibt ein maximales $t_0 \in [0, 1)$ mit $x_1 \geqq t_0 x_2$. Aus (3.9) folgt $t_0 > 0$. $x_1 = t_0 x_2$ ist unmöglich, denn dann wäre

$$x_1 = B(p, x_1) = B(p, t_0 x_2) > t_0 B(p, x_2) = x_1 .$$

Da $B(p, .)$ $u_0(p)$-monoton und konkav ist, gilt deshalb

$$x_1 = B(p, x_1) \geqq B(p, t_0 x_2) + a u_0(p) \geqq (t_0 + a/b_2)\, x , \quad a, b_2 > 0 ,$$

was der Maximalität von t_0 widerspricht. Q.e.d.

Wir fassen die bisherigen Ergebnisse in einem Satz zusammen.

Satz 3.2. *Es sei $p_\infty > p_0 > p_*$. Für $p \in J$ und $p < p_0$ oder $p \geqq p_\infty$ besitzt das Problem* (3.8) *nur die triviale Lösung.*

Für $p = p_0$ besitzt das Problem (3.8) *nichttriviale Lösungen der Form $x(p_0) = c r_0(p_0)\, h_0$ mit $c \in (0, 1]$.*

Für jedes $p \in (p_0, p_\infty)$ *besitzt das Problem* (3.8) *genau eine nichttriviale Lösung* $x(p)$ *mit* $||x(p)|| > r_0(p)$.

Satz 3.3. *Es sei* $p_\infty > p_0 > p_*$. *Die nichttriviale Lösung* $x(p)$, $p \in (p_0, p_\infty)$, *des Problems* (3.8) *hat folgende Eigenschaften:*

(1) $x(p_2) \geqq x(p_1) \geqq r_0(p_0)\, h_0$, $||x(p_2)|| > ||x(p_1)|| > r_0(p_0)$ *für* $p_\infty > p_2 > p_1 > p_0$;

(2) $x(p) \in C((p_0, p_\infty), E)$;

(3) $\lim\limits_{p \to p_0} ||x(p)|| = r_0(p_0)$;

(4) $\lim\limits_{p \to p_\infty} ||x(p)|| = +\infty$.

Beweis. Wir definieren die Funktion

$$z(p) = \begin{cases} x(p) & \text{für} \quad p \in (p_0, p_\infty)\,, \\ r_0(p_0)\, h_0 & \text{für} \quad p = p_0\,. \end{cases}$$

Wegen Satz 3.2 ist $z(p)$, $p \in [p_0, p_\infty)$ die eindeutig bestimmte Lösung des Problems

$$B(p, z) = z\,, \qquad z \in K\,, \qquad ||z|| \geqq r_0(p)\,. \tag{3.11}$$

(1) Wir zeigen $z(p_2) \geqq z(p_1)$, $||z(p_2)|| > ||z(p_1)||$ für $p_\infty > p_2 > p_1 \geqq p_0$. Daraus folgt dann die erste Behauptung. Sei also

$$B(p_i, z_i) = z_i\,, \qquad z_i \in K\,, \qquad ||z_i|| \geqq r_0(p_i)\,, \qquad i = 1, 2\,, \qquad p_\infty > p_2 > p_1 \geqq p_0,$$

und $z_2 \not\geqq z_1$. Dann gibt es ein maximales $t_0 \in (0, 1)$ mit $z_2 \geqq t_0 z_1$. $z_2 = t_0 z_1$ ist unmöglich, denn dann wäre

$$z_2 = B(p_2, z_2) > B(p_1, t_0 z_1) \geqq t_0 z_1 = z_2\,.$$

Deshalb erhalten wir

$$z_2 \geqq B(p_1, z_2) \geqq B(p_1, t_0 z_1) + a u_0(p_1) \geqq (t_0 + a/b_2)\, z_1\,, \qquad a, b_2 > 0\,.$$

im Widerspruch zur Maximalität von t_0. Wegen $z_2 \geqq z_1$, $p_2 > p_1$ ergibt sich weiter

$$z_2 \geqq B(p_2, z_1) = z_1 + B(p_2, z_1) - B(p_1, z_1) > z_1\,,$$
$$z_2 \geqq B(p_1, z_2) \geqq B(p_1, z_1) + a u_0(p_1) \geqq (1 + a/b_2)\, z_1\,, \qquad a, b_2 > 0\,.$$

Hieraus folgt schließlich $||z_2|| > ||z_1||$.

(2) Wir zeigen $\lim\limits_{p \to p^0 + 0} z(p) = z(p^0)$ für $p^0 \in [p_0, p_\infty)$.

Sei also $p^0 \in [p_0, p_\infty)$ fixiert und

$$B(p^0, z^0) = z^0\,, \qquad z^0 \in K\,, \qquad ||z^0|| \geqq r_0(p^0)\,.$$

Wir wählen eine Folge $\{p_n\}$ mit $p_n \in (p^0, p_\infty)$, $p_{n+1} \leqq p_n$, $\lim\limits_{n \to \infty} p_n = p^0$, es sei

$$B(p_n, z_n) = z_n\,, \qquad z_n \in K\,, \qquad ||z_n|| \geqq r_0(p_n)\,.$$

Wegen $p^0 \leqq p_{n+1} \leqq p_n \leqq p_1$ ist $z^0 \leqq z_{n+1} \leqq z_n \leqq z_1$ und $||z^0|| \leqq ||z_n|| \leqq ||z_1||$. Deshalb ist die Folge $\{B(p^0, z_n)\}$ monoton fallend und kompakt, also konvergent, $\lim_{n\to\infty} B(p^0, z_n) = y^0$. Aus der Abschätzung

$$||z_n - y^0|| \leqq ||B(p_n, z_n) - B(p^0, z_n)|| + ||B(p^0, z_n) - y^0||$$

und aus Voraussetzung (B3) folgt $\lim_{n\to\infty} z_n = y^0$, dabei ist $y^0 \in K$, $||y^0|| \geqq r_0(p^0)$. Schließlich wird

$$||B(p^0, y^0) - y^0|| \leqq ||B(p^0, y^0) - B(p^0, z_n)|| + ||B(p^0, z_n) - y^0|| ,$$

so daß wir durch Grenzübergang

$$B(p^0, y^0) = y^0 , \quad y^0 \in K , \quad ||y^0|| \geqq r_0(p^0)$$

erhalten. Da das Problem (3.11) eindeutig lösbar ist, folgt $y^0 = z^0$, also $\lim_{n\to\infty} z_n = z^0$.

(3) Wie bei (2) kann gezeigt werden, daß $\lim_{p\to p^0-0} z(p) = z(p^0)$ für jedes $p^0 \in (p_0, p_\infty)$ ist. Insgesamt folgt dann $z(p) \in C([p_0, p_\infty), E)$, woraus sich unter Beachtung der Definition von $z(p)$ die zweite und dritte Behauptung ergeben.

(4) Wir beweisen die letzte Behauptung. Es sei $p_0 < p_n < p_\infty$, $p_{n+1} \geqq p_n$, $\lim_{n\to\infty} p_n = p_\infty$ und

$$B(p_n, x_n) = x_n , \quad x_n \in K , \quad x_n \neq \Theta .$$

Die Folge $\{||x_n||\}$ ist monoton wachsend, $||x_n|| \geqq ||x_1|| > 0$. Nehmen wir an, es sei $||x_n|| \leqq c < \infty$. Dann finden wir wieder ein y^0 mit $\lim_{n\to\infty} x_n = y^0$ und $B(p_\infty, y^0) = y^0$, $y^0 \in K$, $y^0 \neq \Theta$.

Das widerspricht aber der ersten Aussage von Satz 3.2. Folglich ist $\lim_{p\to p_\infty} ||x(p)|| = \infty$. Qed.

Bemerkung 1. Ein Vergleich mit den üblichen Definitionen (siehe z. B. [1]) zeigt, daß p_0 Bifurkationspunkt und p_∞ asymptotischer Bifurkationspunkt des Operators $B(p, .)$ ist.

Bemerkung 2. Für jedes fixierte $p \in (p_0, p_\infty)$ läßt sich die nichttriviale Lösung $x(p)$ des Problems (3.8) durch die in [2] beschriebenen Iterationsverfahren konstruieren.

4. Eine Klasse nichtlinearer Integralgleichungen

$\Omega \subset R^n$ sei ein beschränktes abgeschlossenes Gebiet, $J = (0, \infty) \subset R^1$. Vorgegeben seien zwei Funktionen $K(p, x, y)$, $P(p, r)$ mit folgenden Eigenschaften:

(K1) Für jedes $p \in J$ sei $K(p, x, y) \in C(\Omega \times \Omega)$, $K(p, x, y) \not\equiv 0$ und $K(p,x,y) \geqq 0$.

(K2) Für jedes $p \in J$ und für jedes Teilgebiet $\Omega_1 \subset \Omega$ mit mes $\Omega_1 > 0$ existiere ein $e_0(p, \Omega_1) > 0$ mit der Eigenschaft

$$\int_{\Omega_1} K(p, x, y)\, dy \geqq e_0(p, \Omega_1)\, u_0(p, x) \qquad \text{für alle} \quad x \in \Omega ,$$

$$u_0(p, x) = \int_{\Omega} K(p, x, y)\, dy . \tag{4.1}$$

(K3) Zu jedem $p_1 \in J$ und zu jedem $\varepsilon > 0$ existiere ein $\delta(\varepsilon, p_1) > 0$, so daß $\max_{\Omega\times\Omega} |K(p_1, x, y) - K(p_2, x, y)| < \varepsilon$ wird für alle $p_2 \in J$ mit $|p_1 - p_2| < \delta$.

(K4) Für alle $p_1, p_2 \in J$ mit $p_1 > p_2$ sei $K(p_1, x, y) \not\equiv K(p_2, x, y)$ und $K(p_1, x, y) \geqq K(p_2, x, y)$.

(P1) Für jedes $p \in J$ sei $P(p, r) \in C(-\infty, \infty)$, $P(p, -r) = -P(p, r)$.

(P2) Für jedes $p \in J$ sei $P(p, r_1) > P(p, r_2)$ für $r_1 > r_2 \geqq 0$ und $P(p, tr) \geqq tP(p, r)$ für $r \geqq 0$, $t \in (0, 1)$.

(P3) Zu jedem $p_1 \in J$ und zu jedem $\varepsilon > 0$ existiere ein $\delta(\varepsilon, p_1) > 0$, so daß $|P(p_1, r) - P(p_2, r)| \leqq \varepsilon r$ wird für alle $p_2 \in J$ mit $|p_1 - p_2| < \delta$, $r \geqq 0$.

(P4) Für alle $p_1, p_2 \in J$ mit $p_1 \geqq p_2$ sei $P(p_1, r) \geqq P(p_2, r)$, $r \geqq 0$.

(P5) Für jedes $p \in J$ sei $\lim_{r\to\infty} P(p, r)/r = P_\infty(p) > 0$.

(P6) Zusätzlich sei $p_* > 0$ und

$$\begin{array}{llll} P(p, r) = P_0(p)\, r & p \in (0, p_*]\,, & r \geqq 0\,, & \\ P(p, r) = P_0(p)\, r & p > p_*\,, & 0 \leqq r \leqq r_0(p)\,, & r_0(p) > 0\,, \\ P(p, tr) \neq tP(p, r) & p > p_*\,, & r > r_0(p)\,, & t \in (0, 1)\,. \end{array}$$

Wir erwähnen noch einige weitere Eigenschaften der Funktion $P(p, r)$:

$$P_0(p)\, r \geqq P(p, r) \geqq P_\infty(p)\, r\,, \quad p \in (0, \infty)\,, \quad r \geqq 0; \tag{4.2}$$

$$P_\infty(p_1) \geqq P_\infty(p_2)\,, \quad p \in (0, \infty)\,, \quad p_1 \geqq p_2\,. \tag{4.3}$$

Es sei $E = C(\Omega)$; die Norm ist durch $||u|| = \max_{\Omega} |u(x)|$ gegeben. Die Menge der nichtnegativen stetigen Funktionen $C_+(\Omega)$ bildet in E einen Kegel K. $u \geqq \Theta$ bedeutet $u(x) \geqq 0$ für alle $x \in \Omega$, $u > \Theta$ bedeutet $u(x) \geqq 0$ für alle $x \in \Omega$ und $u(x) \neq 0$ für mindestens ein $x \in \Omega$. Die lineare Hülle von K stimmt mit E überein, die Norm ist monoton. $u_0(p)$ sei durch (4.1) gegeben, offensichtlich gilt $u_0(p) > \Theta$. Wir betrachten den Operator

$$B(p, u) = \int_\Omega K(p, x, y)\, P(p, u(y))\, dy\,, \quad u \in E\,, \quad p \in J\,, \tag{4.4}$$

für den wir die Voraussetzungen (B1) bis (B6) nachweisen wollen.

(B1) Die Behauptung folgt aus (K1) und (P1).

(B2) $p \in J$ sei fixiert; ferner sei $u > \Theta$. Dann gilt

$$B(p, u) \leqq \max_\Omega \left(P(p, u(y))\right) u_0(p) = bu_0(p)\,, \quad b > 0\,.$$

Nun sei $u, v \in K$, $u - v > \Theta$, also $u(y) \geqq v(y) \geqq 0$, $u(y) \not\equiv v(y)$. Dann gibt es ein Teilgebiet $\Omega_1 \subset \Omega$, mes $\Omega_1 > 0$, so daß $P(p, u(y)) - P(p, v(y)) \geqq c > 0$ für alle $y \in \Omega_1$ wird. Mit (K2) erhalten wir

$$B(p, u) - B(p, v) \geqq c \int_{\Omega_1} K(p, x, y)\, dy \geqq au_0(p)\,, \quad a > 0\,.$$

Ebenso folgt aus (P2), daß $B(p, .)$ konkav ist.

(B3) Für alle $p_1, p_2 \in J$ und für jedes $u \in K$ gilt

$$B(p_1, u) - B(p_2, u) = \int_\Omega \big(K(p_1, x, y) - K(p_2, x, y)\big) P\big(p_1, u(y)\big)\, dy$$
$$+ \int_\Omega K(p_2, x, y) \big(P(p_1, u(y)) - P(p_2, u(y))\big)\, dy\,.$$

Es sei $\varepsilon > 0$ beliebig fixiert. Wir setzen

$$c(p_1) = \max\Big(P_0(p_1), \max_{\Omega\times\Omega} K(p_1, x, y)\Big) > 0\,,$$

$$\varepsilon_1 = c(p_1)\big(\sqrt{1 + \varepsilon/(c^2(p_1) \operatorname{mes} \Omega)} - 1\big) > 0\,.$$

Wegen (K3) und (P3) finden wir ein $\delta_1(\varepsilon_1, p_1) = \delta(\varepsilon, p_1)$ mit der Eigenschaft, daß

$$\max_{\Omega\times\Omega} |K(p_1, x, y) - K(p_2, x, y)| < \varepsilon_1\,,$$

$$|P(p_1, u(y)) - P(p_2, u(y))| \leqq \varepsilon_1 |u(y)|$$

wird, sofern $|p_1 - p_2| < \delta$ ist. Unter Berücksichtigung von (4.2) schätzen wir die Integranden in (4.5) ab und erhalten

$$||B(p_1, u) - B(p_2, u)|| < \varepsilon_1 c(p_1) \operatorname{mes} \Omega ||u|| + \big(\varepsilon_1 + c(p_1)\big)\, \varepsilon_1 \operatorname{mes} \Omega ||u|| = \varepsilon ||u||$$

für $|p_1 - p_2| < \delta$, $u > \Theta$.

(B4) Nun sei $p_1 > p_2$, $u > \Theta$. Dann sind die Integranden in (4.5) wegen (K4) und (P4) nichtnegativ. Außerdem gibt es mindestens ein $x \in \Omega$ und ein Teilgebiet $\Omega_1 \subset \Omega$ mit mes $\Omega_1 > 0$, so daß für $y \in \Omega_1$ der erste Integrand in (4.5) positiv ist. Daraus folgt die Behauptung.

(B5) Wegen (P5) besitzt $B(p, .)$ die starke asymptotische Ableitung

$$A_\infty(p)\, u = P_\infty(p) \int_\Omega K(p, x, y)\, u(y)\, dy\,, \quad u \in E\,, \quad p \in J\,. \tag{4.6}$$

$A_\infty(p)$ ist $u_0(p)$-monoton, da $P_\infty(p) > 0$ vorausgesetzt wurde. Die letzte Eigenschaft von (B5) folgt aus (4.3).

(B6) Diese Eigenschaften folgen aus (P6). Dabei ist

$$A_0(p)\, u = P_0(p) \int_\Omega K(p, x, y)\, u(y)\, dy\,, \quad u \in E\,, \quad p \in J\,. \tag{4.7}$$

Zur Untersuchung der nichttrivialen Lösungen der Gleichung

$$B(p, u) = u\,, \quad u \in K\,, \quad p \in J$$

lassen sich also die erzielten Ergebnisse heranziehen.

In gleicher Weise kann der Fall $K = C_-(\Omega)$ betrachtet werden, wobei $C_-(\Omega)$ die Menge der auf Ω stetigen nichtpositiven Funktionen bezeichnet.

Literatur

[1] Krasnosjelski, M. A., Топологические методы в теории нелинейных интегральных уравнений. Москва (1956).
[2] —, Положительные решения операторных уравнений. Москва (1962).
[3] Sabreiko, P. P., u. a., Интегральные уравнения. Москва (1968).

Anschrift des Autors:

ZIMM der AdW der DDR
108 Berlin, Mohrenstr. 39

Regularity results for non linear integro-differential operators in two dimensions

O. JOHN und J. STARÁ, Prag

§ 1. Introduction

The purpose of this paper is to prove regularity of the weak solutions of the Dirichlet problem for a non linear elliptic integro-differential equation in two dimensions. This question arises naturally in the non linear spectral theory (see [2]).

The functionals f, g used in this paper represent a slight generalization of the simple example:

$$B = \mathring{W}_p^k; \qquad g(u) = \|u\|_{L_p}^p; \qquad f_1(u) = \|u\|_{W_p^k}^p; \qquad f_2(u) = \|u\|_{W_2^k}^2 .$$

We shall consider a weak solution u of the eigen-value problem

$$\lambda(f_1(u) + f_2^{p/2}(u))' = g'(u) \tag{1.1}$$

and prove that the set

$$B_\delta = \{u \in \mathring{W}_p^k; \qquad \|u\|_{W_2^k} = 1; \quad u \text{ solves (1.1) with } \lambda \in \langle -\delta, \delta\rangle\}$$

is a compact set in $C^{2k,\mu}$.

Lemmas 2.1; 2.2 concluded in § 2 give the apriori estimates of solution u. § 3 contains the main theorem on the regularity that is proved by usual homotopic method and its corollary.

§ 2. Apriori estimates

Let us put the problem considered here in the following way:

Ω is a bounded domain in E_2 with infinitely smooth boundary ($\bar{\Omega}$ is the closure of Ω). It is our object to prove estimates for the solution $u \in \mathring{W}_p^k \cap W_q^{2k+1}$ of the equation

$$\sum_{|\alpha| \leq k} (-1)^{|\alpha|} D^\alpha c_\alpha(x, \delta u(x)) = \sum_{|\alpha| \leq k} (-1)^{|\alpha|} D^\alpha f_\alpha , \tag{2.1}$$

where $\delta u = \{D^\gamma u\}_{|\gamma| \leq k}$.

Let us denote

$$\varkappa = \operatorname{card} \{\alpha; |\alpha| \leq k\} ,$$

$$\Theta(\xi) = \Big(\sum_{|\alpha| \leq k} (\xi_\alpha)^2\Big)^{1/2}, \qquad \forall \xi = \{\xi_\alpha\}_{|\alpha| \leq k} \in E_\varkappa ,$$

$$\Theta(u) = \Big(\sum_{|\alpha| \leq k} (D^\alpha u)^2\Big)^{1/2}, \qquad \forall\, u \in W_p^k .$$

A. Growth conditions. For every α; $|\alpha| \leqq k$ the functions c_α belong to $C^1(\bar{\Omega} \times E_\varkappa)$ and are non linear with polynomial growth (of the order $p - 1$) in the second variable, i.e. let us suppose that there exists $C > 0$ such that for every $x \in \bar{\Omega}$, $\xi \in E_\varkappa$

$$|c_\alpha(x, \xi)| \leqq C\big(1 + \theta^{p-1}(\xi)\big) ,$$

$$\left|\frac{\partial c_\alpha}{\partial x_i}(x, \xi)\right| \leqq C\big(1 + \theta^{p-1}(\xi)\big); \qquad i = 1, 2 ,$$

$$\left|\frac{\partial c_\alpha}{\partial \xi_\beta}(x, \xi)\right| \leqq C\big(1 + \theta^{p-2}(\xi)\big); \qquad \forall \beta;\ |\beta| \leqq k$$

Moreover, $\forall \alpha, \beta$; $|\alpha| = |\beta| = k$ $\quad \dfrac{\partial c_\alpha}{\partial \xi_\beta} = \dfrac{\partial c_\beta}{\partial \xi_\alpha}$.

B. Ellipticity. For the considered solution u there exists a constant γ (depending on u) such that the inequality

$$\sum_{|\alpha| = |\beta| = k} \frac{\partial c_\alpha}{\partial \xi_\beta}\big(x, \delta u(x)\big)\, \eta_\alpha \eta_\beta \geqq \gamma(u)\, \big(1 + \theta^{p-2}(u)\big)\, |\eta|^2$$

holds for every $x \in \bar{\Omega}$, $\eta \in E_\varkappa$.

C. Coerciveness. There exist two positive constants C_1, C_2 so that

$$\forall x \in \bar{\Omega}, \xi \in E_\varkappa , \qquad \sum_{|\alpha| \leqq k} \xi_\alpha c_\alpha(x, \xi) \geqq C_1 \theta^p(\xi) - C_2 .$$

D. Suppose that conditions A, B, C are satisfied uniformly with regard to an orthonormal transformation of coordinate system.

Using — with slight modifications — the method of J. Nečas (see [3]), we can obtain the following

Lemma 2.1. *Let the conditions A, B, C, D be satisfied with $p \geqq 2$, let $q > 2$. Then for every $m, n \in N$ there exists a positive constant $C(m, n)$ such that for every solution u of* (2.1) *such that*

(i) $$u \in \mathring{W}_p^k \cap W_q^{2k+1} ,$$

(ii) $$\gamma(u) > \frac{1}{m},$$

(iii) $$\|f\|_{1,q} < n$$

the inequality $\|u\|_{k+1,q} \leqq C(m, n)$ *holds.* (We denote $\|u\|_{k,l} = \|u\|_{W_l^k}$ and $\|f\|_{j,l} = \sum_{|\beta| \leqq k} \|f_\beta\|_{j,l}$.)

The coefficients c_α satisfy all the conditions required in [3], [4], however, the condition of ellipticity is not valid. But for fixed u the whole proceeding can be employed step by step with $\gamma(u)$ (or $1/m$) instead of the constant of ellipticity.

Lemma 2.2. *Let the assumptions of lemma 2.1 be satisfied, let* $\forall\alpha;\ |\alpha| \leqq k$, $c_\alpha \in C^{k+1}(\bar{\Omega} \times E_\varkappa)$, *let* $q \geqq 4$, $k > 1$. *Then for every* $m, n \in N$ *there exists a positive constant* $C(m, n)$ *such that for every solution* u *of the equation* (2.1) *such that*

(i) $$u \in \mathring{W}_p^k \cap W_q^{2k+1},$$

(ii) $$\gamma(u) > \frac{1}{m},$$

(iii) $$\|f\|_{k+1,q} < n$$

the inequality $\|u\|_{2k+1,q} \leqq C(m, n)$ *holds.*

Remark. The assumptions on q, k and c_α are required in such a strong form for the reason of commodity only.

Proof. Let us write every $\beta;\ |\beta| = k$ as $\beta = \beta' + \beta''$, where $|\beta'| = k - 1$; $|\beta''| = 1$.

Integrating by parts in (2.1) we obtain

$$\int_\Omega \sum_{|\alpha|=|\beta|=k} D^{\beta'} v \cdot \frac{\partial c_\beta}{\partial \xi_\alpha} D^{\alpha+\beta''} u = \int_\Omega \sum_{|\beta|=k} D^{\beta'} v\, g^1_{\beta'} + \sum_{|\beta|<k} D^\beta v\, g^2_\beta\,, \tag{2.2}$$

where

$$g^2_\beta = f_\beta - c_\beta; \qquad g^1_{\beta'} = D^{\beta''} f_\beta - \left(D^{\beta''} c_\beta + \sum_{|\alpha|<k} \frac{\partial c_\beta}{\partial \xi_\alpha} D^{\alpha+\beta''} u\right).$$

According to lemma 2.1 $g^1_{\beta'}, g^2_\beta \in W^1_{q/2}$.

Repeating the same procedure with all indices $\beta;\ |\beta| = k - 1$ we get a linear equation in the form

$$\sum_{\substack{|\alpha|=k+2\\ |\beta|=k-2}} D^\beta\big((\tilde{c}_{\alpha\beta})\, D^\alpha u\big) = \sum_{|\beta|\leqq k-2} D^\beta h_\beta\,, \tag{2.3}$$

with Hölder-continuous coefficients $\tilde{c}_{\alpha\beta}$. The right parts h_β conclude the derivatives of $g^1_{\beta'}, g^2_\beta$ which belong to $L_{q/2}$ according to lemma 2.1 and

$$\frac{\partial^2 c_\beta}{\partial \xi_\alpha\, \partial \xi_\beta} \cdot D^\alpha \frac{\partial u}{\partial x_i} \cdot D^\gamma \frac{\partial u}{\partial x_j}; \qquad |\alpha| \leqq k; \quad |\gamma| \leqq k$$

and

$$\frac{\partial^2 c_\beta}{\partial \xi_\alpha\, \partial x_i} \cdot D^\alpha \frac{\partial u}{\partial x_j}; \qquad |\alpha| \leqq k$$

which are also in $L_{q/2}$. We can use the results of Agmon, Douglis, Nirenberg (see [1]) and obtain

$$\|u\|_{k+2,q/2} \leqq C$$

where C depends on estimates of the foregoing lemma, m, n and C^2-norms of coefficients. Using the Sobolev embedding theorem we can improve the summability of the right part and we estimate

$$\|u\|_{k+2,q} \leqq C\,.$$

The coefficients $\tilde{c}_{\alpha\beta}$ are continuously differentiable and we repeat all the proceeding. Finally, we get the estimate of u in W_q^{2k+1} by a constant which depends on the estimate of Lemma 2.1; m; n and C^{k+1}-norms of coefficients c_α.

§ 3. Main theorem

Let us now consider two functionals f_1, f_2 defined on $\mathring{W}_p^k(\Omega)$: $\forall u \in \mathring{W}_p^k$

$$f_1(u) = \int_\Omega F_1(x, \delta u(x))\, dx\,, \qquad f_2(u) = \int_\Omega F_2(x, \delta u(x))\, dx\,.$$

A'. Let us suppose that $F_i \in C^{k+2}(\bar{\Omega} \times E_\varkappa)$, let us denote $\dfrac{\partial F_i}{\partial \xi_\alpha} = a_\alpha^i$; $\dfrac{\partial^2 F_i}{\partial \xi_\alpha\, \partial \xi_\beta} = a_{\alpha\beta}^i$. Let a_α^i $(i = 1, 2;\ |\alpha| \leqq k)$ satisfy the growth conditions with the exponents $p_2 = 2$, $p_1 = p \geqq 2$, i.e. let us suppose that there exists $C > 0$ such that for every $x \in \bar{\Omega}$, $\xi \in E_\varkappa$, $\forall \alpha, \beta$; $|\alpha| \leqq k$, $|\beta| \leqq k$; $i, j = 1, 2$

$$|a_\alpha^i(x, \xi)| \leqq C \cdot \theta^{p_i - 1}(\xi)\,,$$

$$\left|\frac{\partial a_\alpha^i}{\partial x_j}(x, \xi)\right| \leqq C\big(1 + \theta^{p_i - 1}(\xi)\big)\,,$$

$$\left|\frac{\partial a_\alpha^i}{\partial \xi_\beta}(x, \xi)\right| \leqq C\big(1 + \theta^{p_i - 2}(\xi)\big)\,.$$

B'. Let us suppose that there exists a positive constant γ such that $\forall x \in \bar{\Omega}$, $\xi \in E_\varkappa$; $i = 1, 2$

$$\sum_{\substack{|\alpha| \leqq k \\ |\beta| \leqq k}} a_{\alpha\beta}^i(x, \xi)\, \eta_\alpha \eta_\beta \geqq \gamma |\eta|^2\, \theta^{p_i - 2}(\xi)\,.$$

D.' Let us suppose that the conditions A', B' are satisfied uniformly with regard to an orthonormal transformation of coordinate system in E_2.

Let us consider the equation

$$\forall v \in \mathring{W}_p^k \qquad (f_1 + f_2^{p/2})'\,(u)\,(v) = \int_\Omega g \cdot v\,. \tag{3.1}$$

Theorem 3.1. *Let conditions A', B', D' be satisfied, $k > 1$. Then for every $m, n \in N$ there exists a constant $C(m, n)$ such that for every solution $u \in \mathring{W}_p^k$ of the equation* (3.1) *satisfying*

(i) $$A(u) = \frac{p}{2} \cdot f_2^{\frac{p}{2} - 1}(u) > \frac{1}{m}$$

(ii) $$\|g\|_{1,q} < nu\,,$$

the inequality $\|u\|_{2k+1,q} \leqq C(m, n)$ holds.

Proof. For every $v \in \mathring{W}_p^k$ we can write (3.1) in the form

$$\left(f_1'(u) + A(u) f_2'(u)\right)(v) = \int_\Omega g \cdot v$$

or, if we denote $a_\alpha^1(u) + A(u)\, a_\alpha^2(u) = b_\alpha(u)$ in the form

$$\int_\Omega \sum_{|\alpha| \leq k} D^\alpha v\, b_\alpha(u) = \int_\Omega g \cdot v\,. \tag{3.2}$$

Let us define for every $t \in \langle 0, 2\rangle$ the operator Φ_t: $\mathring{W}_{p_t}^k \to (\mathring{W}_{p_t}^k)'$ so that $\forall u, v \in \mathring{W}_{p_t}^k$

$$\langle \Phi_t(u), v\rangle = \int_\Omega \sum_{|\alpha| \leq k} D^\alpha v(x)\, c_\alpha^t\left(x, \delta u(x)\right) dx\,, \tag{3.3}$$

where

$$t \in \langle 0, 1\rangle \Rightarrow p_t = 2 + t(p-2)$$

$$C_\alpha^t\left(x, \delta u(x)\right) = D^\alpha u(x) \cdot [\theta^{p-2}(u)\,(x) + A(u)]^t;$$

$$t \in \langle 1, 2\rangle \Rightarrow p_t = p$$

$$C_\alpha^t\left(x, \delta u(x)\right) = (t-1)\, b_\alpha\left(x, \delta u(x)\right) + (2-t)\, C_\alpha^1\left(x, \delta u(x)\right).$$

Now, the assumptions of lemma 2.2 are satisfied uniformly with regard to $t \in \langle 0, 2\rangle$, (i) implies $\gamma(u) > C/m$. Moreover, there exist the functions $\{f_\beta\}_{|\beta|=k}$ such that $\forall v \in D(\Omega)$

$$\int_\Omega g \cdot v = \int_\Omega \sum_{|\beta|=k} D^\beta v \cdot f_\beta$$

and $||f_\beta||_{k+1,q} \leq C||g||_{1,q} < C \cdot n$ (see [3]).

Let us denote $\mathcal{P} = \{t \in \langle 0, 2\rangle$; the assertion of theorem 3.1 is valid for Φ_t, i.e. $\forall m, \tilde{n} \in N$; $u \in \mathring{W}_{p_t}^k$ such that $A(u) > 1/m$ $\Phi_t(u) = g$, where $||g||_{1,q} < \tilde{n}$, $||u||_{2k+1,q} \leq C(m, \tilde{n})\}$. $\left(C(m, \tilde{n})\right.$ is independent of $t \in \langle 0, 2\rangle$.$\left.\right)$

Now, $\mathcal{P} \neq \emptyset$, as $0 \in \mathcal{P}$. $\left(\Phi_0(u) = \sum_{|\alpha| \leq k} (-1)^{|\alpha|} D^{2\alpha} u\right.$ and the assertion is included in [1].$\left.\right)$

(1) *$\mathcal{P}$ is closed.*

Let $t_n \in \mathcal{P}$ converge to t_0, let $u \in \mathring{W}_{p_{t_0}}^k$, $Au > 1/m$, $\Phi_{t_0}(u) = g$, $||g||_{1,q} < \tilde{n}$. All Φ_t are strongly monotone on $\mathring{W}_{p_t}^k$, therefore there exist $u_n \in \mathring{W}_{p_{t_n}}^k$ such that

$$\Phi_{t_n}(u_n) = g\,.$$

Let us prove that there exists a subsequence u_{n_k} and $m_1 \in N$ such that $\forall k \in N$ $A(u_{n_k}) > 1/m_1$. On the contrary, if $A(u_n) \to 0$ then (using the monotonicity of f_2) $u_n \to 0$ in W_2^k and (using (3.3) and the monotonicity of f_1)

$$||u_n||_{k, p_{t_n}} \to 0$$

$u \neq 0$ implies that $g \neq 0$, $g \in W_q^1$ and so we can find a function $\varphi \in D(\Omega)$ so that

$$0 < \varepsilon \leq \int_\Omega g \cdot \varphi = \langle \Phi_{t_n}(u_n), \varphi\rangle \leq ||\varphi||_{k, p_{t_n}} \cdot ||u_n||_{k, p_{t_n}}^{p_{t_n}},$$

but the last term tends to zero. $\{t_{n_k}\}_{k=1}^{\infty} \in \mathcal{P}$ and it implies $u_{n_k} \in W_q^{2k+1}$ and $||u_{n_k}||_{2k+1,q} \leqq C(m_1, \tilde{n})$. Thus $\{u_{n_k}\}_{k=1}^{\infty}$ forms a relatively compact set in W_q^{2k} and there exists a subsequence (let us denote it u_{n_k}, too) which tends to u in W_q^{2k} and it implies that $\exists k_0 \; \forall k > k_0$

$$A(u_{n_k}) > \frac{1}{m} \Rightarrow ||u||_{2k+1,q} < C(m, \tilde{n}) .$$

(2) *$\mathcal{P}$ is open.*

Let $t_0 \in \mathcal{P}$. Let us denote $u_1 \in \mathring{W}^k_{p_{t_0}}$ the solution of the equation $\Phi_{t_0}(u) = g$ and $m_1 \in N$ such that $A(u_1) > 1/m_1$. Let $\mathcal{V} = \{u \in \mathring{W}^k_p \cap W_q^{2k+1};\; ||u||_{2k+1,q} \leqq C(m_1, \tilde{n})\}$. Let us define $\mathcal{A} = \Phi_{t_0}^{-1}(\Phi_{t_0} - \Phi_t + g)$. For $t - t_0$ sufficiently small

$$||\Phi_t(u) - \Phi_{t_0}(u)||_{1,q} < u - ||g||_{1,q} .$$

$\Phi_{t_0}^{-1}$ is a continuous mapping from W_q^1 in W_2^k so that for sufficiently small $t - t_0$ also $A(u) > 1/m_1$ for all $u \in \mathcal{V}$.

The weakly continuous operator $\mathcal{A}$ transforms $\mathcal{V}$ into $\mathcal{V}$ and thus it has a fixed point u_2.

$$||u_2||_{2k+1,q} < C(m_1, \tilde{n}) , \qquad \Phi_t(u_2) = q$$

and it implies that $u \neq 0$, so $\forall\, m \in N$ such that $A(u_2) > 1/m$ (and according to Lemma 2.2)

$$||u_2||_{2k+1,q} < C(m, \tilde{n}) .$$

But this implies that $\mathcal{P} = \langle 0, 2\rangle$, and the theorem 3.1 is valid for $t = 2$, which completes the proof.

Theorem 3.2. *Let all the assumptions of theorem 3.1 be satisfied; let $h \in C^2(\bar{\Omega} \times E_1)$, satisfying the growth conditions A. Let us denote*

$$g(u) = \int_\Omega h(x, u(x))\, dx .$$

Then the set $B_\delta = \{u \in \mathring{W}^k_p;\; f_2(u) = 1.\; u$ is a solution of the equation

$$\lambda(f_1 + f_2^{p/2})'\,(u) = g'(u) \tag{3.3}$$

with $\lambda \notin \langle -\delta, \delta\rangle\}$ is a relatively compact set in $C^{2k,\mu}$.

Proof. $u \in B_\delta$ solves the equation (3.2) with the right part $G(x) = \frac{1}{\lambda} \cdot \frac{\partial h}{\partial \xi}\,(x, u(x))$. From growth conditions

$$||G||_{1,q} \leqq C(\delta) \cdot ||u^{p-1}||_{1,q} .$$

Using the Sobolev theorem, the existence theorem and strong monotonicity of f_2, we obtain that the assumptions (i), (ii) of theorem 3.1 are satisfied with m, n depending on δ, k, p and on the estimates of f_1, f_2, g. Theorem 3.1 affirms that

$$u \in B_\delta \Rightarrow ||u||_{2k+1,q} < C_q \;\forall q \geqq 1 .$$

The compactness of embedding from W_q^{2k+1} into $C^{2k,\mu}$ completes the proof.

References

[1] Agmon, S., Douglis, A., and Nirenberg, L., Estimates near boundary for solutions of elliptic partial differential equations satisfying general boundary conditions 1. Comm. Pure Appl. Math. **12** (1959) 623–727.
[2] Fučík, S., and Souček, V., this volume.
[3] Nečas, J., Sur la régularité des solutions variationnelles des équations elliptiques non-linéaires d'ordre $2k$ en deux dimensions. Annali Sc. Norm. Sup. Pisa, fasc. III (1967) 427–457.
[4] Stará, J., Regularity results for non linear elliptic systems in two dimensions. Annali Sc. Norm. Sup. Pisa Vol. XXV, fasc. I (1971) 163–190.

Adresse der Autoren:
Department of Mathematics
Charles University, Prague
Prague 8-Karlin

The Rothe method and nonlinear parabolic equations of arbitrary order

J. KAČUR, Bratislava

In this paper we present a review of the results contained in the author's lecture on Summer school in GDR in 1972. The detailed proofs will be given in the paper [1], to appear.

Introduction. We deal with the first initial-boundary value problem for nonlinear parabolic equations of the form

$$\frac{\partial u}{\partial t} + \sum_{i \in M} (-1)^{|i|} D^i a_i(t, x, Du) = f(x, t) \tag{1}$$

in a domain $Q \equiv \Omega \times \langle 0, T \rangle$, where $\Omega \subset E^N$ (N-dimensional Euclidean space) is a bounded domain with the Lipschitzian boundary $\partial\Omega$. M is a subset of multiindices $i \equiv (i_1, \ldots, i_N)$ ($i_1, \ldots, i_N$ are nonnegative integers and $|i| = \sum_{s=1}^{N} i_s$) with $|i| \leq k$.

We denote $D^i \equiv \dfrac{\partial^{|i|}}{\partial x_1^{i_1} \cdots \partial x_N^{i_N}}$.

Let us consider the following initial-boundary conditions

$$u(x, 0) = u_0(x, 0) \quad \text{and} \quad \frac{\partial^s u(x, t)}{\partial \nu^s} = \frac{\partial^s u_0(x, t)}{\partial \nu^s} \tag{2}$$

for $s = 0, 1, \ldots, k - 1$ and $(x, t) \in \partial\Omega \times \langle 0, T \rangle$, where ν is an exterior normal with respect to $\partial\Omega$.

The case $u_0(x, t) \equiv u_0(x)$ and $a_i(t, x, \xi)$ $a_i(x, \xi)$, $i \in M$ is studied in the part I of [1], and the general case is studied in II of [1] under somewhat stronger assumptions than in I.

To solve the problem (1), (2), we apply the idea of Rothe in the following way:

We solve the following system of elliptic equations

$$\frac{u_j - u_{j-1}}{\Delta t_j} + \sum_{i \in M} (-1)^{|i|} D^i a_i(t_j, x, Du_j) = f(x, t_j)\,, \tag{1'}$$

$$\frac{\partial^s u_j(x)}{\partial \nu^s} = \frac{\partial^s u_0(x, t_j)}{\partial \nu^s} \quad \text{for} \quad x \in \partial\Omega,\ s = 0, 1, \ldots, k - 1\,, \tag{2'}$$

where $\{t_j\}_{j=1}^{m}$ is a partition of the interval $\langle 0, T \rangle$, $\sum_{s=1}^{m} \Delta t_s = T$ and $u_0(x) \equiv u_0(x, 0)$. By means of the weak solutions u_j, $j = 1, \ldots, m$ of the problem (1'), (2') we construct the Rothe's function $u^{(m)}(x, t)$ in Q as follows:

$$u^{(m)}(x, t) = u_{j-1}(x) + \frac{t - t_{j-1}}{\Delta t_j}\left(u_j(x) - u_{j-1}(x)\right)$$

for $t_{j-1} \leqq t \leqq t_j$ and $j = 1, 2, \dots, m$.

For the sequence $u^{(m)}$ we obtain the a priori estimates independent on $\{t_j\}_{j=1}^m$, which will guarantee that this sequence is bounded in the norm of a certain reflexive space of the Orlicz-Sobolev type (see [2]). In order to prove that a weak limit $u(x, t)$ of the sequence $\{u^{(m)}\}$ is a weak solution of (1), (2) we use the „trick of Minty-Browder" for monotone operators.

In the equation (1) or (1') we impose only a small restrictions on the growth of $a_i(t, x, \xi)$ in the vector ξ and the existence of $\{u_j\}_{j=1}^m$ is proved in the author's papers [2] and [3].

The advantage of the method used is that we obtain a weak solution which has better regularity properties in the variable t then the weak solution of F. E. Browder [4] and J. A. Dubinski [5]. Farther we prove that $u^{(m)}$ converges in the norm of the space $C(\langle 0, T\rangle, L_2(\Omega))$ to the solution $u(x, t)$ of (1), (2) and, in addition to it, the estimate

$$\|u^{(m)} - u\|_{C(\langle 0,T\rangle, L_2(\Omega))} \leqq C \cdot \max_{i=1,2,\dots,m} \sqrt{\Delta t_i}$$

takes place, which is interesting also from the numerical point of view. The used method enables us to use the direct variational methods in the parabolic boundary value problems. The attention to this fact has been called, e.g., by P. P. Mosolow [6], K. Rektorys [7] etc.

The Rothe's method (method of lines) has already been used in many papers, e.g., in [8], [9], [10] and [11]. The authors in the mentioned papers considered the linear parabolic equations of the second order and used the maximum principle. In [8] and [10] even a class of nonlinear equations has been considered.

Definitions and assumptions. $a_i(t, x, \xi)$ are real continuous functions defined for $x \in \bar{\Omega}$ and $|\xi| < \infty$, where ξ is a real vector $\xi \in E^d$ with $d = \text{card } M$. We suppose that $M \supset \{i, |i| = k\}$, which can be yet weakened. The growth of $a_i(t, x, \xi)$ in ξ is described by means of functions of certain class $\mathfrak{M}_3$, which is essentially larger than the class of polynomials $|u|^p$ — see [3]. $\mathfrak{M}_3$ is the set of all real continuous functions $g(u)$ such that:

I. $ug(u)$ is convex and even for $|u| \geqq u_1$ and $\lim\limits_{u\to\infty} \left(ug(u)\right)' = \infty$.

II. There exists a constant $c(2)$ such that

$$g(2u) \leqq c(2) \cdot g(u) \quad \text{for each} \quad u \geqq u_1 .$$

III. There exists $l > 1$ such that

$$g(u) \leqq \frac{1}{2} g(lu) \quad \text{for each } u \geqq u_1 ,$$

where u_1 is a sufficiently large positive number.

Let $g_i(u) \in \mathfrak{M}_3$, $i \in M$ be such that $g_i(u) \leqq g_j(u)$ (or $g_i(u) \geqq g_j(u)$) for $u \geqq u_1$ and $i, j \in M$. Then the growth conditions are of the form

$$|a_i(t, x, \xi)| \leqq C\Big(1 + \sum_{j\in M} \min\big(|g_i(\xi_j)|, |g_j(\xi_j)|\big)\Big), \qquad i \in M. \tag{3}$$

The condition (3) can be yet weakened — see [2] (relation 2.1). For the ellipticity (monotonicity) we shall suppose

$$\sum_{i\in M} (\xi_i - \zeta_i)\,[a_i(t, x, \xi) - a_i(t, x, \zeta)] > 0 \quad \text{for} \quad \xi \neq \zeta. \tag{4}$$

The boundary value problem (1′), (2′) is considered in the following space $W(\Omega)$:

$$W(\Omega) \equiv \{u \in L_2(\Omega);\ D^i u \in L^*_{G_i}(\Omega), \quad \text{for each } i \in M\},$$

where $D^i u$ are distribution derivatives and $D^0 u \equiv u$. $L^*_{G_i}(\Omega)$ is the Orlicz's space generated by the function $G_i(u) = g_i(u) \cdot u$, where $g_i(u)$ is from (3), $i \in M$. ($L_p(\Omega)$ is Orlicz's space generated by $|u|^p$.) The norm in $W(\Omega)$ is defined by the relation

$$||u||_{W(\Omega)} = \sum_{i\in M} ||D^i u||_{G_i} + ||u||_{L_2},$$

where $||\cdot||_{G_i}$ is the corresponding Orlicz's norm in the space $L^*_{G_i}(\Omega)$.

For coerciveness we shall suppose

$$\sum_{i\in M} \xi_i a_i(t, x, \xi) \geqq c_1 \sum_{i\in M} \xi_i g_i(\xi_i) - c_2. \tag{5}$$

The condition (5) can be weakened, since the problem (1′), (2′) is a Dirichlet problem — see [3] (§ 5).

Let us denote by $C^\infty(\bar{Q})$ the set of all functions defined on Q having derivatives of all orders in x, t, extendable continuously on $\bar{Q}$. Let $C_0^\infty(Q)$ be a subset of all functions in $C^\infty(\bar{Q})$ which have support in Q ($\sup v \subset Q$, for $v \in C^\infty(\bar{Q})$). Let us consider the space $\mathring{W}(\Omega) \equiv \overline{C_0^\infty(Q)}$, where the closure is taken in the norm of the space $W(\Omega)$.

Analogously as $W(\Omega)$ we define the space

$$W(Q) \equiv \{u \in L_2(Q);\ D^i u \in L^*_{G_i}(Q) \text{ for } i \in M\};$$

where $D^i u$ are distributive derivatives in the variable x. The norm in $W(Q)$ is defined by the same way as for $W(\Omega)$. Farther, we shall consider the space

$$F \equiv F(Q) \equiv \left\{u \in L_2(Q);\ \frac{\partial u}{\partial t} \in L_2(Q)\right\}$$

with the norm

$$||u||_F = ||u||_{L_2} + \left\|\frac{\partial u}{\partial t}\right\|_{L_2}$$

where $\partial u/\partial t$ is the distributive derivative. Let us denote by $B(Q)$ the space

$$B \equiv B(Q) \equiv W(Q) \cap F(Q)$$

with the norm

$$||u||_B = ||u||_{W(Q)} + ||u||_F.$$

Let S be the set $\partial\Omega \times \langle 0, T\rangle \cup \partial\Omega \times \{0\}$. Consider the set $\mathring{U} \equiv \{v \in C^\infty(\bar{Q});$ $\sup v \cap S = \emptyset\}$. We define the space $\mathring{B} = \bar{\mathring{U}}$, where the closure is taken in the norm of the space B. At last we denote $\mathring{V} = \bar{\mathring{B}}$, where the closure is taken in the norm of the space $W(Q)$. For simplicity we denote $(u, v)_Q = \int\limits_Q uv\,dx\,dt$ for $u, v \in L_2(Q)$ and $a_Q(u, v) = \int\limits_Q \sum\limits_{i\in M} D^i v a_i(t, x, Du)\,dx\,dt$ for $u, v \in W(Q)$.

We define a weak solution of the problem (1), (2) as follows:

Definition. Let be $u_0(x, t) \in B$ and $f(x, t) \in L_2(Q)$. Then u is a weak solution of (1), (2), iff $u - u_0 \in \mathring{V}$ and

$$\left(\frac{\partial u}{\partial t}, v\right)_Q + a_Q(u, v) = (f, v)_Q \tag{6}$$

holds for all $v \in \mathring{V}$.

The used method requires more regularity properties of $u_0(x, t)$ and $f(x, t)$ than the previous definition. In this paper we shall assume, that $u_0(x, t) \in B$ and $f(x, t) \in F$. (For $f \in F$ the trace $f(x, t_j)$ is well defined.) But, in general (the case II) we shall suppose

$$\frac{\partial^2}{\partial t^2} D^\alpha u_0(x, t) \in C(\bar{Q}) \qquad \text{for all } |\alpha| \leqq k \tag{7}$$

$$\left\|\frac{\partial f}{\partial t}\right\|_{L_2(Q)} \leqq C \qquad \text{a.e. in } \langle 0, T\rangle \text{ and} \tag{8}$$

$$\frac{\partial a_i(t, x, \xi)}{\partial \xi_j} = \frac{\partial a_j(t, x, \xi)}{\partial \xi_i} \qquad \text{for all } i, j \in M\,. \tag{9}$$

The condition (7) can be weakened. The positive constants we denote by c with or without indices and dependence of c on the parameter ε we denote by $C(\varepsilon)$.

Results

Lemma 1. *The spaces $W(\Omega)$, $W(Q)$, B and F are reflexive Banach spaces. $C^\infty(\Omega)$ is dense set in $W(\Omega)$ and $C^\infty(Q)$ is dense set in $W(Q)$, B and F.*

From the definition of the class $\mathfrak{M}_3$ it follows $W(\Omega) \subset W_{k,p}(\Omega)$ (Sobolev space) for suitable $p > 1$. Let u_j, $j = 1, 2, \ldots, m$ be the weak solutions (1′), (2′).

Lemma 2. *If the conditions* (3), (4) *and* (5) *are satisfied, then*

$$\|u_j\|_{W(Q)} \leqq C(u_0, f) \quad \textit{for all } j = 1, 2, \ldots, m\,,$$

where $c(u_0, f)$ is independent on $\{t_j\}_{j=1}^m$.

The a priori estimate of $\partial u^{(m)}/\partial t$ in the norm of the space $L_2(Q)$, where $u^{(m)}$ is Rothe's function is the crucial point of the work. We shall assume

$$D^i a_i\big(0, x, Du_0(x, 0)\big) \in L_2(Q)\,, \quad \text{for all } i \in M\,. \tag{10}$$

Lemma 3. *Let* (3,) (4), (5) *and* (10) *be satisfied. If* $u_0(x, t) \equiv u_0(x)$ *and* $a_i(t, x, \xi) \equiv a_i(x, \xi)$, $i \in M$, *then*

$$\left\| \frac{u_j - u_{j-1}}{\Delta t_j} \right\|_{L_2(\Omega)} \leqq C(u_0, f)\,, \quad \text{for all } j = 1, 2, \dots, m\,,$$

where $c(u_0, f)$ *is independent on* $\{t_j\}_{j=1}^m$.

In general case, if u_0 and a_i depend on t then we prove lemma 3 by using the idea of P. P. Mosolow [6] under some additional assumptions.

Lemma 4. *Suppose that* $\dfrac{\partial^s a_i(t, x, \xi)}{\partial t^p\, \partial \xi_\alpha^q\, \partial \xi_\beta^r}$, *where* $s = 0, 1, 2$ *and* $p + q + r = s$ *are continuous in all variables,* $i, \alpha, \beta \in M$ *and satisfy the growth condition* (3). *If* (3), (4), (5), (7), (8), (9) *and* (10) *are satisfied, then the estimate*

$$\left\| \frac{u_j - u_{j-1}}{\Delta t_j} \right\|_{L_2(\Omega)} \leqq c(u_0, f)\,, \quad \text{for all } j = 1, 2, \dots, m\,,$$

takes place, where $c(u_0, f)$ *is independent on* $\{t_j\}_{j=1}^m$.

The proof of Lemma 4 is technically rather complicated.

Theorem 1. *If the assumptions of lemma* 3 *or lemma* 4 *are satisfied, then there exists a unique weak solution of* (1), (2).

We sketch the idea of the proof. lemma 2 and lemma 3 (or lemma 4) imply that the sequence of Rothe's functions $\{u^{(m)}\}$ is bounded in the norm of the space B. Due to lemma 1 a subsequence of $\{u^{(m)}\}$ weakly converges (in the space B) to $u(x, t) = u_0(x, t) + z(x, t)$, where $z(x, t) \in \mathring{B}$. Using the "trick of Minty-Browder" we realize a limiting process in (1') in order to prove, that $u(x, t)$ is a weak solution of (1), (2).

The a priori estimates from lemma 2 and lemma 3 (or lemma 4) imply that the weak solution of (1), (2) is more regular then the theorem 1 states.

Theorem 2. *Suppose that the assumptions of lemma 3 (or lemma 4) are satisfied. Then*

$$u(x, t) \in L_\infty(\langle 0, T\rangle, W(\Omega))$$

and

$$\frac{\partial u(x, t)}{\partial t} \in L_\infty(\langle 0, T\rangle, L_2(\Omega))\,,$$

where u *is the weak solution of* (1), (2).

The following theorem deals with the problem, in what sense the weak solution of (1), (2) satisfies the initial-boundary condition (2).

Theorem 3. *If* $z(x, t) \in \mathring{B}$, *then* $z(x, \delta) \to z(x, 0) = 0$ *for* $\delta \to 0$ *in the space* $L_2(\Omega)$ *and*

$$\left.\frac{\partial^s z(x, t)}{\partial \nu^s}\right|_{\partial\Omega \times \langle 0, T\rangle} = 0\,, \qquad s = 0, 1, \dots, k-1\,, \ \text{a.e. in } \langle 0, T\rangle\,,$$

where (11) *holds in the sense of the traces of the space*

$$W(\Omega) \subset W_{k,p}(\Omega) \quad (\text{for a suitable } p > 1)\,.$$

From the numerical point of view is interesting:

Theorem 4. *The sequence of Rothe's functions* $\{u^{(m)}\}$ *converges in the norm of the space* $C(\langle 0, T\rangle, L_2(\Omega))$ *to the weak solution of the problem* (1), (2) *and the following estimate*

$$\|u^{(m)} - u\|_{C(\langle 0,T\rangle, L_2(\Omega))} \leqq C \max_{i=1,2,\ldots,m} \sqrt{\Delta t_i} \tag{12}$$

takes place.

The constant in (12) can be determined from the estimations in Lemma 2 and Lemma 3 (or Lemma 4).

At last we present some examples in which the previous results can be applied.

Examples

1. $$\frac{\partial u}{\partial t} + \sum_{i \in M} (-1)^{|i|} D^i[l_i(x, t)\, g_i(D^i u)] = f(x, t)\,,$$

$0 < C \leqq l_i(x, t) \in C^2(\bar{Q})$, $g_i(u) \in C^2(E^1)$, $g_i(u) \in \mathcal{M}_3$ and $g_i'(u) > 0$ for $u \in E^1$, where $i \in M$.

2. $$\frac{\partial u}{\partial t} - \sum_{i=1}^{N} \frac{\partial}{\partial x_i}\left(l_i(x, t) \left|\frac{\partial u}{\partial x_i}\right|^{p-2} \frac{\partial u}{\partial x_i}\right) = f(x, t)\,,$$

where $p \geqq 2$, $0 < C < l_i(x, t) \in C^1(\bar{Q})$ and $\partial^2 l_i/\partial t^2 \in L_\infty(Q)$. The condition (2) is given by $u_0(x) \in W(\Omega)$.

3. $$\frac{\partial u}{\partial t} - \Delta u + a(t, x, u) = f(x, t)\,,$$

$$|a(t, x, s)| \leqq C(1 + g(s)), \qquad s a(t, x, s) \leqq C_1\, s g(s) - C_2, \tag{i}$$

where $g(s) \in \mathcal{M}_3$.

$$\frac{\partial a(t, x, s)}{\partial s} > 0 \quad \text{for} \quad s \in E^1\,.$$

$a(t, x, s)$ and its partial derivatives up to the second order are continuous and satisfy (i).

References

[1] Kačur, J., The Rothe method and nonlinear parabolic boundary value problems of arbitrary order. I, II. Czech. Math. J.

[2] —, On existence of the weak solution for non-linear partial differential equations of elliptic type. I. Comment. Math. Univ. Carolinae **11**, 1 (1970).

[3] —, On existence of the weak solution for non-linear partial differential equations of elliptic type. II. Comment. Math. Univ. Carolinae **13**, 2 (1972).

[4] Browder, F. E., Strongly nonlinear parabolic boundary value problems. Amer. Journ. of Math. **86**, 2 (1964).

[5] Dubinski, J. A., Квазилинейные еллиптические и параболические уравнения любого порядка. УМН, 25, вып. **1**, 139 (1968) 45—90.

[6] Mosolow, P. P., Вариационные методы в нестационарных задачах. (Параболический случай). Изв. АН СССР, **34** (1970) 425—457.
[7] Rektorys, K., On application of direct variational methods to the solution of parabolic boundary value problems of arbitrary order in the space variables. Czech. Math. J. **21** 96 (1971) 318—339.
[8] Ladishenskaja, O. A., Решение в целом первой краевой задачи для квазилинейных параболических уравнений. ДАН СССР **102** (1956) 636—639.
[9] Wentzel, T. D., Первая краевая задача для квазилинейново уравнения со многими пространственными переменными. Матем. сб. **41**, 83 (1957) 499—520.
[10] Tschshou-Ji-Lin, Краевые задачи для нелинейных параболических уравнений. Мат. сб. **47**, 89—4, (1959) 431—484.
[11] Ильин, А. М., Калашников, А. С., Олейник, О. А., Линейные уравнения второго порядка параболического типа. УМН **17**, вып. 3, (1962) 3—146.

Anschrift des Autors:
J. Kačur
Vilová 15
Bratislava
Czechoslovakia

Näherungsverfahren für einige nichtlineare Probleme

REINHARD KLUGE, Berlin

Wir geben hier für einige nichtlineare Probleme, wie das Anfangswertproblem für nichtlineare gewöhnliche Differentialgleichungen erster Ordnung im Banach-Raum und vor allem für nichtlineare Variationsungleichungen mit monotonen Operatoren und konvexen Funktionalen, Aussagen über die Existenz von Lösungen und deren näherungsweisen Bestimmung mittels Verfahren in der Form von Folgen einfacherer Probleme oder Iterationsverfahren von Folgen dieser einfacheren Probleme.

Punkt 1 enthält Ergebnisse über Iterationsverfahren bei Folgen von Abbildungen in metrischen Räumen, Punkt 2 die Anwendung eines dieser Ergebnisse auf gewöhnliche Differentialgleichungen im Banachraum, Punkt 3 beinhaltet vor allem die näherungsweise Lösung von Variationsungleichungen, die weite Klassen nichtlinearer analytischer Probleme erfassen und in Punkt 4 wird eine Erweiterung der Iterationsverfahren aus Punkt 3, die auf den allgemeinen Verfahren von Punkt 1 basieren, für Minimumprobleme ohne und mit Nebenbedingungen und für Operatorgleichungen und -ungleichungen vom Hilbertraum auf allgemeinere Banachräume vorgenommen.

Hinweise auf die Literatur erfolgen in den einzelnen Punkten.

1. Iterationsverfahren bei Folgen von Abbildungen

Sei M ein metrischer Raum mit der Metrik $r(.,.)$ und (A_i) eine Folge von Abbildungen $A_i \in (M \to M)$. Wir betrachten Iterationsverfahren bei Folgen von Abbildungen

$$x_i = A_i x_{i-1}, \quad x_0 \in M. \tag{1}$$

Dazu setzen wir

$$T_n = A_n A_{n-1} \cdots A_2 A_1, \quad n = 1, 2, \dots .$$

Satz 1. *T_n sei für jedes n Lipschitz-stetig mit der Konstanten $L(T_n) \to 0$ für $n \to \infty$. Weiter existiere ein $u_0 \in M$ mit $r(u_0, T_n u_0) \to 0$ für $n \to \infty$. Dann konvergiert das Iterationsverfahren* (1) *in M gegen u_0. Dabei gilt*

$$r(u_0, x_n) \leqq r(u_0, T_n u_0) + L(T_n)\, r(u_0, x_0), \quad n = 1, 2, \dots . \tag{2}$$

Beweis: Es gilt $x_i = T_i x_0$ und damit

$$r(u_0, x_n) \leqq r(u_0, T_n u_0) + r(T_n u_0, T_n x_0) \leqq r(u_0, T_n u_0) + L(T_n)\, r(u_0, x_0) \to 0$$

für $n \to \infty$, q. e. d.

Bemerkungen. 1. Unter den Voraussetzungen der Existenz von Fixpunkten u_n für die T_n und $r(u_n, u_0) \to 0$ ist Satz 1 ein Spezialfall von Satz 2 in [21].

2. Bezüglich Verfahren (1) für gewisse „lokal"-Lipschitz-stetige Abbildungen vgl. ebenfalls [21].

Korollar 1. *Sei M vollständig und (A_i) eine Folge von (strikt) kontraktiven Abbildungen $A_i \in (M \to M)$ mit den Eigenschaften* (i) *und* (ii):

(i) *Die Fixpunkte y_i der A_i konvergieren in M gegen ein $u_0 \in M$.*
(ii) *Es ist $L(A_i) \leqq L < 1$ für $i = 1, 2, \ldots$.*

Dann konvergiert das Iterationsverfahren (1) *gegen u_0. Dabei gilt für jedes n*

$$r(u_0, x_n) \leqq r(u_0, y_n) + \sum_{i=1}^{n-1} L^{n-i} r(y_{i+1}, y_i) + L^n[r(y_1, u_0) + r(u_0, x_0)] . \quad (3)$$

Beweis: T_n ist hier strikt kontraktiv mit $L(T_n) = L^n \to 0$ für $n \to \infty$. Weiter gilt wegen $r(y_{i+1}, y_i) \to 0$ für $i \to \infty$ und $\sum_{i=1}^{n-1} L^{n-i} c_i \to 0$ für $i \to \infty$ unter der Voraussetzung $c_i \geqq 0$, $c_i \to 0$ für $i \to \infty$ und $L < 1$

$$\begin{aligned} r(u_0, T_n u_0) &\leqq r(u_0, y_n) + r(y_n, T_n u_0) = r(u_0, y_n) + r(A_n y_n, A_n T_{n-1} u_0) \\ &\leqq r(u_0, y_n) + L r(y_n, T_{n-1} u_0) \leqq r(u_0, y_n) + L\{ r(y_n, y_{n-1}) \\ &\quad + r(y_{n-1}, T_{n-1} u_0)\} \leqq \cdots \leqq r(u_0, y_n) + \sum_{i=1}^{n-1} L^{n-i} r(y_{i+1}, y_i) \\ &\quad + L^n r(y_1, y_0) \to 0 \quad \text{für} \quad n \to \infty . \end{aligned}$$

Satz 1 liefert die Aussagen, q. e. d.

Bemerkungen. 3. In den Voraussetzungen von Korollar 1 sei mit einer festen Zahl $L_1 < 1$

$$r(u_0, y_i) \leqq c_0 L_1^i , \qquad i = 1, 2, \ldots . \quad (4)$$

Dann gilt

$$r(u_0, x_i) \leqq c L_0^i$$

mit $L_0 = \max\{L, L_1\}$ für $L \neq L_1$ $(i = 1, 2, \ldots)$, und $L_0 = L + \varepsilon L$ mit $0 < \varepsilon < (1 - L)/L$ für $L = L_1$ (und i derart, daß $i \leqq (1 + \varepsilon)^i$). Die Aussagen über die Konvergenzgeschwindigkeit (4) der Folge (y_n) gegen u_0 übertragen sich folglich auf das Iterationsverfahren (1) [19].

4. Die Aussage von Korollar 1 über die Konvergenz von (1) wurde für Banachräume in [6] bewiesen. In der hier angegebenen erweiterten Form findet sie sich in [26] und für Banachräume mit (3) und Stabilitätsaussagen in [12].

Aus Korollar 1 resultiert (vgl. [12, 26]): Sei M vollständig und (A_i) eine Folge von kontraktiven Abbildungen $A_i \in (M \to M)$, für die die entsprechende Folge der Fixpunkte y_i $(y_i = A_i y_i,\ i = 1, 2, \ldots)$ in M gegen $u_0 \in M$ konvergiert. Dann konvergiert das Iterationsverfahren

$$x_i = (A_i)^{n(i)} x_{i-1} , \qquad x_0 \in M , \quad i = 1, 2, \ldots \quad (5)$$

mit $n(i) = [e_0/(1 - L(A_i))]$ und $e_0 > 0$ in M ebenfalls gegen u_0. Zusätzlich gilt für (5) die Abschätzung (3) mit $L = (1/e)^{e_0}$ (e bezeichnet hier die Eulersche Zahl und $[a]$ für $a > 0$ die kleinste ganze Zahl, die größer oder gleich a ist).

5. Analoga zu Korollar 1 und zu den in Bemerkung 4 formulierten Aussagen für gewisse „lokal-“kontraktive Abbildungen finden sich in [26, 27].

Korollar 2. *Sei M vollständig, $T \in (M \to M)$, T^n für $n = 1, 2, \ldots$ lipschitzstetig mit $L(T^n) \to 0$ für $n \to \infty$. Dann konvergiert das Iterationsverfahren*

$$x_i = Tx_{i-1}\,, \qquad x_0 \in M\,, \tag{6}$$

in M gegen den einzigen Fixpunkt u_0 von T. Dabei gilt

$$r(u_0, x_n) \leqq L(T^n)\, r(u_0, x_0) \qquad n = 1, 2, \ldots\,. \tag{7}$$

Beweis: Die Existenz genau eines Fixpunktes u_0 von T folgt aus $L(T^{n_0}) < 1$ für eine gewisse natürliche Zahl n_0 (vgl. [5]). Wir setzen $T_n = T^n$, $n = 1, 2, \ldots$. Dann gilt

$$r(u_0, T_n u_0) = r(u_0, T^n u_0) = r(u_0, u_0) = 0\,.$$

Die Aussagen folgen damit aus Satz 1, q. e. d.

Bemerkungen. 6. Korollar 2 ist eine spezielle Variante von Satz 1 in [21]. Andererseits resultiert es aus folgender Aussage: Sei M vollständig und T^p für eine natürliche Zahl p strikt kontraktiv. Dann konvergiert das Verfahren (6) in M gegen den einzigen Fixpunkt von T (zum Beweis vgl. [5]). Für Anwendungen ist Korollar 2 jedoch ausreichend.

7. In den Voraussetzungen von Korollar 2 gilt unter der zusätzlichen Voraussetzung

$$\sum_{n=1}^{+\infty} L(T^n) < +\infty \tag{8}$$

die Abschätzung

$$r(u_0, x_n) \leqq r(Tx_0, x_0) \sum_{i=n}^{+\infty} L(T^i)\,, \; n = 1, 2, \ldots\,, \tag{9}$$

(zum Beweis vgl. [21]).

8. Sei M vollständig und $T \in (M \to M)$ strikt kontraktiv. Dann konvergiert (6) gegen den einzigen Fixpunkt u_0 von T, und es gilt (9) in der Form

$$r(u_0, x_n) \leqq L(T)^n\, r(Tx_0, x_0)/(1 - L(T)) \qquad \text{(vgl. z. B. [35])}. \tag{10}$$

Diese Aussagen folgen offensichtlich aus Korollar 2 und Bemerkung 7.

2. Gewöhnliche Differentialgleichungen im Banachraum

Wir wenden nun Satz 1 auf abstrakte Differentialgleichungen im Banachraum B an. Dazu betrachten wir das folgende Problem:

Gesucht wird eine abstrakte Funktion $x(t)$: $[0, T] \to B$ ($T > 0$, fixiert) mit den Eigenschaften: 1. $x \in C(0, T; B)$, 2. f. f. a. (bezüglich des Lebesgueschen Maßes) $t \in [0, T]$ existiert die Ableitung $x'(t)$, und es gilt

$$x'(t) = f(t, x(t)) \qquad \text{f. f. a. } t \text{ und } x(0) = x_0 \in B \tag{11}$$

mit $f(t, .) \in (B \to B)$ f. f. a. t.

x heißt Lösung von (11). (11) erfaßt u. a. gemischte Probleme für pseudoparabolische partielle Differentialgleichungen. Unter den Voraussetzungen

$(V_1)_p$: Für $x \in C(0, T; B)$ gilt $f(t, x(t)) \in L_p(0, T; B)$, $p \in [1, +\infty)$ (die Integration wird hier im Sinne von Bochner benutzt; vgl. z. B. [8]).

und

(V_2): Mit einer von t unabhängigen Konstanten L gilt

$$\|f(t, x) - f(t, y)\| \leqq L\|x - y\| \quad \text{f. f. a. } t \text{ und für } x, y \in B$$

besitzt (11) genau eine Lösung x mit $x' \in L_p(0, T; B)$ (vgl. [20]).

Ist

(V_3): Für $x \in C(0, T; B)$ gilt $f(t, x(t)) \in C(0, T; B)$

anstelle von $(V_1)_p$ erfüllt, so existiert ebenfalls eine Lösung x mit $x' \in C(0, T; B)$ ([35]).

Zur Lösung von (11) wurden Iterationsverfahren in [20, 21], Projektionsverfahren für den Fall des Hilbertraumes in [20] und Projektions-Iterationsverfahren für den Fall des Hilbertraumes in [20, 21] begründet. Zur Lösung allgemeinerer Differentialgleichungen und -ungleichungen im Hilbertraum vom Standpunkt der Monotonie-Theorie über Korollar 1 und die Aussage von Bemerkung 4 vgl. [28] und [29] (in diesen Proceedings).

Wir zeigen nun hier auf der Basis von Satz 1, daß die Einschränkung der Projektions-Iterationsverfahren für (11) in [20, 21] auf den Hilbertraum nicht notwendig ist. Dazu betrachten wir die Abbildung

$$(Ax)(t) \equiv x_0 + \int_0^t f(s, x(s))\, ds \quad \text{für} \quad x \in C(0, T; B).$$

Wir formulieren nun eine Voraussetzung an den Raum B.

(K): Sei (B_n) eine Folge von Unterräumen B_n von B. Für jedes n existiere ein Projektionsoperator P_n von B auf B_n mit $\|P_n\| \leqq k$, $n = 1, 2, \ldots$, und $\|(I - P_n)h\| \to 0$ für $n \to \infty$ und $h \in B$.

(K) ist z. B. erfüllt, wenn B eine Schauderbasis besitzt (vgl. [35]). Unter den Voraussetzungen (K), $(V_1)_p$ (oder (V_3)) und (V_2) setzen wir $T_n = A_n \cdots A_1$ mit $A_i = P_i A$. Es gilt für $y \in C(0, T; B)$

$$(A_i y)(t) = P_i x_0 + \int_0^t P_i f(s, x(s))\, ds .$$

Weiter ist für $z \in C(0, T; B)$

$$\|T_1 y(t) - T_1 z(t)\| \leqq k \int_0^t \|f(s, y(s)) - f(s, z(s))\|\, ds \leqq kL \int_0^t \|y(s) - z(s)\|\, ds$$
$$\leqq kLt\, \|y - z\|_c$$

und durch vollständige Induktion

$$\|T_n y - T_n z\|_c \leqq (kLT)^n\, \|y - z\|_c / n!\,, \qquad n = 1, 2, \ldots . \tag{12}$$

Hier bezeichnet $\|\cdot\|_c$ die Norm in $C(0, T; B)$ und im Weiteren $\|\cdot\|_p$ die Norm von $L_p(0, T; B)$.

Nach (12) ist T_n für jedes n Lipschitz-stetig als Abbildung von $C(0, T; B)$ in $C(0, T; B)$ mit $L(T_n) = (kLT)^n/n! \to 0$ für $n \to \infty$. Wir prüfen nun, ob $\|T_n x - x\|_c \to 0$ für $n \to \infty$ und die Lösung x von (11). Es gilt

$$\|T_2 x(t) - x(t)\| = \|T_2 x(t) - T^2 x(t)\| = \|P_2 T P_1 x(t) - T T x(t)\|$$

$$\leqq k \|T P_1 T x - T^2 x\| + \|P_2 T^2 x - T^2 x\| \leqq \|P_2 T^2 x - T^2 x\|_c + kLt \|Tx - P_1 T x\|_c$$

und weiter durch vollständige Induktion

$$\|T_n x(t) - x(t)\| \leqq \sum_{i=1}^{n} (kLt)^{n-i} \|T^i x - P_i T^i x\|_c/(n-i)!\,,$$

damit auch

$$\|T_n x - x\|_c \leqq \sum_{i=1}^{n} (kLT)^{n-i} \|x - P_i x\|_c/(n-i)!\,. \tag{13}$$

Satz 2. *Unter den Voraussetzungen* (K), $(V_1)_p$ *(oder* (V_3)*) und* (V_2) *konvergiert das Projektions-Iterationsverfahren*

$$y_n = (P_n A)\, y_{n-1}\,, \qquad y_0 \in C(0, T; B)\,,$$

in $C(0, T; B)$ *gegen die einzige Lösung von* (11). *Außerdem konvergiert die Folge der Ableitungen* y_n' *in* $L_p(0, T; B)$ *(oder* $C(0, T; B)$*) gegen* x'.

Beweis. Für jedes t gilt $\|x(t) - P_i x(t)\| \to 0$, folglich nach dem Satz von Dini $\|x - P_i x\|_c \to 0$ für $i \to \infty$. Aus (13) resultiert dann $\|T_n x - x\|_c \to 0$ für $n \to \infty$. Zusammen mit (12) folgt nun aus Satz 1 die Aussage über die Konvergenz des Projektions-Iterationsverfahrens. Wir gehen nun zur Konvergenz der Ableitungen über. Es gilt $y_n(t) = P_n x_0 + \int_0^t P_n f(s, y_{n-1}(s))\, ds$, folglich $y_n'(t) = P_n f(t, y_{n-1}(t))$ und damit

$$\|x'(t) - y_n'(t)\| \leqq \|f(t, x(t)) - P_n f(t, y_{n-1}(t))\| \leqq \|f(t, x(t)) - P_n f(t, x(t))\| + kL \|x - y_{n-1}\|_c\,,$$

auf grund der Minkowski-Ungleichung, $\|x - y_{n-1}\|_c \to 0$ und des Lebesgueschen Satzes im Fall $(V_1)_p$

$$\|x' - y_n'\|_p \leqq \left(\int_0^T \|(I - P_n)\, f(t, x(t))\|^p\, dt\right)^{1/p} + kL \|x - y_{n-1}\|_c\, T^{1/p} \to 0$$

$$\text{für } n \to \infty\,.$$

Im Fall (V_3) gilt auf grund des Satzes von Dini

$$\|x' - y_n'\|_c \leqq \|f(t, x(t)) - P_n f(t, x(t))\|_c + kL \|x - y_{n-1}\|_c \to 0$$

$$\text{für } n \to \infty\,, \text{ q. e. d.}$$

Bemerkungen. 9. Die Aussage von Satz 2 läßt sich analog zu [21] auf Differentialgleichungen beliebiger Ordnung erweitern.

10. Die numerische Realisierung des Projektions-Iterationsverfahrens aus Satz 2 reduziert sich für endlichdimensionale B_n (ohne Berücksichtigung der Berechnung von „Fourier"-Koeffizienten) auf die Bestimmung von Lebesgue-Integralen mit variabler oberer Grenze.

3. Nichtlineare Variationsungleichungen

3.1. Sei B ein reeller Banachraum, B^* der zu B adjungierte Raum, (w, v) der Wert des Funktionals $w \in B^*$ auf dem Element $v \in B$, $C \subset B$ nicht leer, konvex und abgeschlossen, $T \in (B \to B^*)$, $g \in B^*$ und $f \in (B \to R_1)$ mit $f(x) \not\equiv +\infty$ für $x \in C$.

Wir betrachten das folgende Problem: Gesucht ist ein $u \in C$ mit

$$(g - Tu, u - v) \geqq f(u) - f(v)\,, \qquad v \in C\,, \tag{14}$$

(14) heißt nichtlineare Variationsungleichung und erfaßt speziell Operatorungleichungen

$$(g - Tu, u - v) \geqq 0\,, \qquad v \in C\,, \tag{15}$$

Operatorgleichungen

$$Tu = g\,, \tag{16}$$

Fixpunktprobleme

$$Au = u \tag{17}$$

und Minimumprobleme für differenzierbare (T ist der Gradient eines Funktionals und $f = 0$) wie nicht differenzierbare Funktionale (ohne und mit Nebenbedingungen)

$$f(u) = \min_{v \in C} f(v)\,. \tag{18}$$

(14) tritt auch selbst als Objekt auf (vgl. u. a. die Näherungsverfahren für (15) und (18) in diesem Punkt).

Auf die Probleme (15)—(18) führen u. a. folgende Aufgaben (Die in Klammern angeführten Arbeiten enthalten vor allem Näherungsverfahren zur Lösung der entsprechenden Aufgabe.).

1. Nichtklassische Variationsprobleme.
Hier wird z. B. (18) mit

$$f(u) = \int_G F(x, t(u))\, dG\,,$$

$t \equiv \{t_\alpha : |\alpha| \leqq l\}$, $t(u)(x) \equiv \{D^\alpha u(x) : |\alpha| \leqq l\}$, $x \in G \subset R_n$, $u(x) : G \to R_1$ und $C \subset \mathring{W}_p^l(G)$, betrachtet.

2. Das Dirichletsche Randwertproblem für nichtlineare partielle Differentialgleichungen bzw. Variationsungleichungen z. B. in der Form

$$\sum_{|\alpha| \leqq l} \int_G A_\alpha(x, t(u))\, D^\alpha(v - u)\, dG \geqq 0\,, \qquad v \in C \subset \mathring{W}_p^l(G)\,,$$

mit Differentialoperatoren in divergenter Form

$$\bar{A}u = \sum_{|\alpha| \leqq l} (-1)^{|\alpha|} D^\alpha A_\alpha(x, t(u))\,,$$

$A_\alpha : G \times R_N \to R_1$, $|\alpha| \leqq l$ (N bezeichnet die Dimension von t) (vgl. [14—16]).
Die Variationsungleichungen werden durch Ungleichungen für u auf G bzw. dem Rand von G (vgl. [33]) erzeugt oder als notwendige und hinreichende Bedingungen für die unter Punkt 1 angeführten nichtklassischen Variationsprobleme ($\partial F / \partial t_\alpha = A_\alpha$) benutzt.

3.a. Probleme der optimalen Steuerung mit fixierter Zeit und freiem rechten Ende, d. i. (18) mit

$$f(u) = \int_0^T F(x, u, t)\, dt + G(x(T))$$

und den linearen gewöhnlichen Differentialgleichungen

$$x'(t) = A(t)\, x(t) + B(t)\, u(t)\,, \qquad x(0) = d \in H_1$$

(vgl. [22]). Hier ist $A(t) \in (H_1 \to H_1)$, $B(t) \in (H_0 \to H_1)$ und $C \subset L_2(0, T; H_0)$ (H_0 und H_1 bezeichnen Hilberträume).

3.b. Probleme der optimalen Steuerung mit festem rechten Ende $x(T) = c$ (vgl. [23]).

3c. Steuerungsprobleme im Banachraum, d. i. z. B. (18) mit $f(u) = F(x[u], u)$ und der Operatorgleichung $T(x[u], u) = 0$ (vgl. [25, 30] und [31] in diesen Proceedings). Durch die Operatorgleichung wird die Steuerung mit gemischten und Randwertproblemen für pseudoparabolische bzw. elliptische partielle Differentialgleichungen und mit dem Goursat-Problem erfaßt.

3d. Probleme der Steuerung mit parabolischen partiellen Differentialgleichungen [24].

4. Abstrakte Differentialgleichungen und -ungleichungen im Hilbertraum (vgl. [28] und [29] (in diesen Proceedings). Sie führen auf (16) bzw. (15).

3.2. Wir gehen zunächst auf die Existenz von Lösungen der Variationsungleichung

$$(g - Tu, u - v) \geqq f(u) - f(v)\,, \qquad v \in C \subset B\,, \tag{14}$$

ein.

Definitionen. 1. T heißt monoton, wenn $(Tu - Tv, u - v) \geqq 0$ für $u, v \in B$; strikt monoton, wenn T monoton ist und $(Tu - Tv, u - v) = 0$ nur für $u = v$ gilt; gleichmäßig monoton, wenn eine reellwertige strikt wachsende stetige Funktion $\delta(r)$, $r \in R_+$ mit $\delta(0) = 0$ derart existiert, daß

$$(Tu - Tv, u - v) \geqq \delta(||u - v||) \quad \text{für} \quad u, v \in B\,;$$

stark monoton, wenn $\delta(r) = c_0 r^2$ mit $c_0 > 0$; φ-monoton, wenn eine reellwertige, stetige, strikt wachsende Funktion $\varphi(r)$ mit $\varphi(r) \to +\infty$ für $r \to +\infty$ derart existiert, daß

$$(Tu - Tv, u - v) \geqq (\varphi(||u||) - \varphi(||v||))\, (||u|| - ||v||)\,, \qquad u, v \in B\,.$$

2. Das Funktional $f \in (B \to R_1)$ heißt konvex, wenn

$$f_a(x, y) \equiv f(ax + (1 - a)\, y) - af(x) - (1 - a)\, f(y) \leqq 0\,, \quad a \in (0, 1), x, y \in B;$$

strikt konvex, wenn f konvex ist und $f_a(x, y) = 0$ nur für $x = y$ eintreten kann; gleichmäßig konvex, wenn eine Funktion $\delta(r)$ wie in Definition 1 existiert mit

$$f_a(x, y) \leqq -a(1 - a)\, \delta(||x - y||) \quad \text{für} \quad a \in (0, 1), x, y \in B;$$

stark konvex, wenn $\delta(r) = cr^2$ und $c > 0$; φ-konvex, wenn eine Funktion $\varphi(r)$ wie in Definition 1 existiert mit

$$f_a(x, y) \leqq -a(1 - a)\, (\varphi(||x||) - \varphi(||y||))\, (||x|| - ||y||)\,, \quad a \in (0, 1), x, y, \in B\,.$$

Bemerkung 11. Zwischen den Abbildungen und Funktionen in den Definitionen 1 und 2 bestehen enge Beziehungen. So ist z. B. der Gradient eines konvexen (strikt, gleichmäßig, stark bzw. φ-konvexen) stetigen Funktionals monoton (strikt, gleichmäßig, stark bzw. φ-monoton) (vgl. z. B. [19]). Die Umkehrung dieser Aussage ist nicht in allen Fällen in dieser Allgemeinheit bekannt.

Definition 3. $T \in (B \to B^*)$ heißt hemistetig, wenn für $u, v, w \in B$ $d(t) \equiv (T(u + tv), w)$ als Funktion von $t \in [0, 1]$ stetig ist.

Satz 3. *Sei B reflexiv, f konvex und unterhalbstetig, T monoton und hemistetig und außerdem eine der folgenden Voraussetzungen* (i) *oder* (ii) *erfüllt:*

(i) *C ist beschränkt.*
(ii) *Es gibt ein $w \in C$ mit $f(w) < +\infty$ und*

$$[(Tu, u - v) + f(u)]/||u|| \to +\infty \quad \text{für} \quad ||u|| \to \infty, \; u \in C\,.$$

Dann besitzt (14) *wenigstens (für strikt monotone T oder strikt konvexe f genau) eine Lösung.*

Beweis: vgl. z. B. [3, 4] (für $C = B$), [7, 33] und [27].

Bemerkung 12. Es seien die Voraussetzungen von Satz 3 erfüllt außer (i) und (ii). Dabei sei f endlich. Dann sind die Bedingungen

(iii) $$(Tu, u - w)/||u|| \to +\infty \quad \text{für} \quad ||u|| \to \infty, \; u \in C\,.$$

und

(iv) Es gilt $$f(u)/||u|| \to +\infty \quad \text{für} \quad ||u|| \to \infty, \; u \in C$$

jede für sich hinreichend für die Erfüllung von (ii). Gleichmäßig (φ-) monotone T erfüllen (iii) und stetige gleichmäßig (φ-) konvexe Funktionale f (iv) (vgl. [19]).

3.3. Wir gehen nun zu Näherungsverfahren für (14) in der Form von Folgen nichtlinearer Variationsungleichungen einfacheren Typs über.

Sei (B_n) eine Folge von Unterräumen $B_n \subset B$ mit $B_n \subset B_m$ für $n \leqq m$.

Definition 4. Wir sagen, daß die Folge (S_n) von Abbildungen $S_n \in (B \to B^*)$ approximationsstetig bezüglich (B_n) gegen $S \in (B \to B^*)$ konvergiert, wenn für jede Folge $(w_n) \subset B$ mit $w_n \in B_n$ und $||w_n - w|| \to 0$ gilt $||S_n w_n - Sw|| \to 0$.

Bemerkung 13. Jede stetige Abbildung S konvergiert bezüglich jeder Folge (B_n) approximationsstetig gegen sich selbst. Sei weiter $\varepsilon_n > 0$ und $\varepsilon_n \to 0$ für $n \to \infty$ und $S, S_0 \in (B \to B^*)$ stetig. Dann konvergiert (S_n) mit $S_n = S + \varepsilon_n S_0$ approximationsstetig bezüglich jeder Folge (B_n) gegen S. Weiter mögen Projektionsoperatoren Q_n von B auf B_n existieren mit $||Q_n|| \leqq k$ und $Q_n y \to y$ für $n \to \infty$ und $y \in B^*$. Außerdem konvergiere S approximationsstetig gegen sich selbst. Dann konvergiert $(Q_n S)$ approximationsstetig bezüglich (B_n) gegen S [19].

Sei weiter $C_n = C \cap B_n$, $C_1 \neq \varnothing$, $g_n \in B^*$ und (ε_n) eine Zahlenfolge mit $\varepsilon_n > 0$ und $\varepsilon_n \to 0$ für $n \to \infty$.

Definition 5. Seien $u_0 \in B$ und $(u_n), (v_n)$ zwei Elementenfolgen aus B mit $u_n, v_n \in C_n \cdot u_0$ heißt $(u_n, v_n, \varepsilon_n, C_n, g_n - T_n, f)$-approximierbar, wenn

1. $||v_n - u_0|| \to 0$ für $n \to \infty$.
2. $\overline{\lim\limits_{i \to +\infty}} [(g_{n_i} - T_{n_i}v_{n_i}, u_{n_i} - v_{n_i}) + f(v_{n_i}) - f(u_{n_i})]/\varepsilon_{n_i} ||u_{n_i}|| \leqq 0$ für jede Teilfolge (u_{n_i}) von (u_n) mit $||u_{n_i}|| \to +\infty$ für $i \to \infty$.
3. $\overline{\lim\limits_{n \to \infty}} [(g_n - T_nv_n, u_n - v_n) + f(v_n) - f(u_n)]/\varepsilon_n \leqq 0$ für den Fall der Beschränktheit der Folge (u_n).

Bemerkung 14. Sei u_0 Lösung von (1) und $(v_n) \subset B$ eine Folge mit $v_n \in C_n$, $||v_n - u_0||/\varepsilon_n \to 0$, $||g_n - g||/\varepsilon_n \to 0$, $||Tu_0 - T_nv_n||/\varepsilon_n \to 0$ und $|f(u_0) - f(v_n)|/\varepsilon_n \to 0$ für $n \to \infty$. Dann ist u_0 $(u_n, v_n, \varepsilon_n, C_n, g_n - T_n, f)$-approximierbar für jede Folge (u_n) aus B mit $u_n \in C_n$ [19]. Für (18) reichen folgende Voraussetzungen: Es existiert eine Folge (v_n) mit $v_n \in C_n$, $||v_n - u_0|| \to 0$ und $|f(v_n) - f(u_0)|/\varepsilon_n \to 0$ für $n \to \infty$.
Die in dieser Bemerkung formulierten hinreichenden Bedingungen für die $(u_n, v_n, \varepsilon_n, C_n, g_n - T_n, f)$-Approximierbarkeit von u_0 sind z. B. erfüllt, wenn u_0 Regularitätseigenschaften genügt (vgl. [2, 37]) und Sätze der Approximationstheorie über die Geschwindigkeit der Approximation von u_0 über Elemente aus B_n bekannt sind [12, 15, 17]. Im Fall des Projektionsverfahrens (z. B. des Galerkin-Verfahrens für (15) und des Ritz-Verfahrens für (18) sind die Bedingungen 2 und 3 in Definition 5 automatisch erfüllt (vgl. die weiteren Ausführungen). Für $B_n = B$, $n = 1, 2, \ldots$, ist u_0 immer $(u_n, v_n, \varepsilon_n, C_n, g_n - T_n, f)$-approximierbar. Bezüglich Probleme der optimalen Steuerung vgl. [19, 25].

Definition 6. $v \in C \subset B$ heißt (C_n)-approximierbar, wenn eine Folge (v_n) mit $v_n \in C_n$ und $||v_n - v|| \to 0$ für $n \to \infty$ existiert. C heißt (C_n)-approximierbar, wenn jedes $v \in C$ (C_n)-approximierbar ist. Bezüglich Beispiele vgl. [15, 17, 18].

Wir betrachten nun mit $T_n, S, S_n \in (B \to B^*)$, $h \in (B \to R_1)$, $g_n, z, z_n \in B^*$, $A_n = T_n + \varepsilon_n S_n$, $f_n = f + \varepsilon_n h$ und $a_n = g_n + \varepsilon_n z_n$ die Variationsungleichungen

$$(a_n - A_nu, u - v) \geqq f_n(u) - f_n(v)\,, \qquad v \in C_n,\ n = 1, 2, \ldots, \tag{19n}$$

und

$$(z - Su, u - v) \geqq h(u) - h(v)\,, \qquad v \in K\,, \tag{20}$$

mit $K = \{x \in C: x \text{ ist Lösung von (14)}\}$.

Bemerkung 15. (19n) ist gleichzeitig eine Methode der projezierten Strafvariationsungleichungen für (20) und eine Methode der projezierten Regularisierung für (14). Speziell enthält sie für $B_n = B$ verschiedene Strafverfahren (z. B. die Methoden der Straffunktionale und der Strafoperatoren) für (20) und verschiedene Regularisierungsverfahren (z. B. die Tychonoffsche und die elliptische Regularisierung) für (14) und für $C = K$ Projektionsverfahren für Variationsungleichungen (z. B. Galerkin- und Ritzverfahren). Darüber hinaus sind eine ganze Reihe neuer Näherungsverfahren in (19n) enthalten. Einige spezielle von ihnen wurden bereits in [11, 15, 17, 18] angegeben. Für gewisse Fälle vgl. auch [36].

Satz 4. *Sei B reflexiv, $C \subset B$ nicht leer, konvex und abgeschlossen, die Abbildungen T, T_n, S und S_n monoton und hemistetig, die Funktionale f und h konvex und unterhalbstetig und* (14) *besitze wenigstens eine Lösung. Weiter sei eine der beiden folgenden Voraussetzungen* (i) *oder* (ii) *erfüllt:*

(i) *S ist gleichmäßig monoton (φ-monoton und B gleichmäßig konvex) und $S_n = S$, $z_n = z$ für jedes n.*

(ii) *h ist gleichmäßig konvex (φ-konvex und B gleichmäßig konvex).*

Dann besitzt für jedes n die Variationsungleichung (19n) *genau eine Lösung u_n. Wenn außerdem im Fall $T \not\equiv g$ C (C_n)-approximierbar ist, die einzige Lösung u_0 von* (20) *$(u_n, v_n, \varepsilon_n, C_n, g_n - T_n, f)$-approximierbar ist, f und h stetig sind, (S_n) bzw. (T_n) approximationsstetig gegen S bzw. T konvergieren und $||g_n - g|| \to 0$ und $||z_n - z|| \to 0$ gilt, dann konvergiert (u_n) stark in B genau u_0, d. i. gegen die Lösung von* (14), *die gleichzeitig die einzige Lösung von* (20) *ist.*

Beweis: vgl. [19]. Bezüglich spezieller Verfahren siehe [15, 17, 18] und für die speziellen Probleme (15)—(18) und spezielle Verfahren die Arbeiten [9—12, 15].

Bemerkung 16. Mit Satz 4 haben wir gleichzeitig Näherungsverfahren für eindeutig wie für mehrdeutig lösbare Variationsungleichungen gewonnen.

Wir gehen nun zum Fall der Projektionsverfahren über.

Korollar 3. *Sei B reflexiv, $C = K \subset B$ konvex und abgeschlossen, S_n monoton und hemistetig und h konvex und stetig. Weiter mögen S und (S_n) der Bedingung* (i) *oder h der Bedingung* (ii) *von Satz* 4 genügen. *Dann besitzt für jedes n die Variationsungleichung*

$$(z_n - S_n u, u - v) \geqq h(u) - h(v)\,, \qquad v \in C_n\,, \tag{21}$$

genau eine Lösung u_n. Wenn weiter die einzige Lösung u_0 von (20) *(C_n)-approximierbar ist, (S_n) bezüglich (B_n) gegen (S) konvergiert und $||z_n - z|| \to 0$ für $n \to \infty$ gilt, dann konvergiert (u_n) stark in B gegen u_0.*

Bemerkung 17. [19] enthält neben dem Konvergenzbeweis auch Aussagen über die Konvergenzgeschwindigkeit von (21). Für spezielle Fälle vgl. dazu [15, 17, 18].

3.4. Ausgehend von Satz 4 und Korollar 3 geben wir nun auf der Basis von Punkt 1 Näherungsverfahren für (14) in der Form von Iterationsverfahren bei Folgen einfacherer Variationsungleichungen an. Dazu müssen wir uns zunächst auf den Hilbertraum H beschränken.

Sei $K \subset H$ konvex und abgeschlossen und $g \in (B \to R_1)$ endlich, konvex, und unterhalbstetig. Dann bezeichnet $P(K, g)\,u$ für $u \in H$ die einzige Lösung des Minimumproblems

$$\tfrac{1}{2}\,||u - v||^2 + g(v) = \min_{v \in K}$$

bzw. der Variationsungleichung

$$(u - w, w - v) \geqq g(w) - g(v)\,, \qquad v \in C\,.$$

Satz 5. *In den Voraussetzungen von Satz* 4 *sei B ein Hilbertraum und A_n für jedes n stark monoton mit der Konstanten $c(A_n)$ und Lipschitz-stetig mit der Kon-*

stanten $L(A_n)$. Dann konvergiert das Projektions-Iterationsverfahren

$$x_i = \big(P(C_i, t_i f_i)\, P_i W_i\big)^{n(i)}\, x_{i-1}\,, \qquad x_0 \in H\,, \tag{22}$$

mit $W_i = I - t_i(A_i - a_i)$, $0 < t_i < 2c(A_i)/L(A_i)^2$, $n(i) = [e_0/(1 - L(W_i))]$, $e_0 > 0$ *und* $L(W_i) = \sqrt{1 - 2t_i c(A_i) + t_i^2 L(A_i)^2}$ *gegen die einzige Lösung* u_0 *von* (20).

Beweis: vgl. [19].

Bemerkung 18. [19] enthält auch Stabilitätsaussagen für (22). Für spezielle Probleme und spezielle Verfahren (22) vgl. [9—12, 15, 17, 18].

19. A_n ist stark monoton, wenn wenigstens eine der Abbildungen T_n oder S_n stark monoton ist. Dabei gilt $c(A_n) \leqq c(T_n) + \varepsilon_n\, c(S_n)$. Wenn T_n und S_n Lipschitz-stetig sind, dann ist A_n Lipschitz-stetig mit $L(A_n) \geqq L(T_n) + \varepsilon_n L(S_n)$.

Korollar 4. *Wenn in den Voraussetzungen von Korollar* 3 B *ein Hilbertraum und* S_n *für jedes* n *stark monoton und Lipschitz-stetig ist, dann konvergiert das Projektions-Iterationsverfahren*

$$x_i = \big(P(C_i, t_i h)\, P_i W_i\big)^{n(i)}\, x_{i-1}\,, \qquad x_0 \in H\,, \tag{23}$$

mit $W_i = I - t_i(S_i - z_i)$, $0 < t_i < 2c(S_i)/L(S_i)^2$, $n(i) = [e_0/(1 - L(W_i))] = \sqrt{1 - 2t_i c(S_i) + t_i^2/L(S_i)^2}$ *in* B *gegen die einzige Lösung* u_0 *von* (20). *Dabei kann für* $c(T_i) = c > 0$ *und* $L(T_i) = L$ *für jedes* i *in* (23) $n(i) = 1$ *gesetzt werden.*

Beweis: vgl. [19].

4. Projektions-Iterationsverfahren im Banach-Raum

Die Verfahren (22) und (23) in Punkt 3 basieren auf dem allgemeinen Iterationsverfahren von Punkt 1. Sie sind einerseits nur für den Hilbertraum begründet worden, andererseits spielte dabei aber keine Rolle, ob die eingehenden Abbildungen Gradienten von Funktionalen bzw. Potentialoperatoren sind. Wir wollen nun für die Probleme (15) und (18) Projektions-Iterationsverfahren in allgemeineren Banachräumen als H angeben. Dabei benutzen wir aber die Eigenschaft der Abbildungen, Gradienten von Funktionalen zu sein.

Zunächst müssen wir jedoch einige Hilfsmittel anführen.

Definition 7. Sei $F\colon R_+ \to R_+$ stetig, strikt wachsend, $F(0) = 0$ und $F(r) \to +\infty$ für $r \to +\infty$. Der Operator $J_F \in (B \to B^*)$ heißt duale Abbildung bezüglich F, wenn gilt

$$(J_F u, u) = \|J_F u\|_* \, \|u\| \quad \text{und} \quad \|J_F u\|_* = F(\|u\|)\,, \qquad u \in B\,.$$

Für $F(r) = r$ setzen wir hier $J_F = J$.

Bemerkung 20. Für jedes F existiert wenigstens eine duale Abbildung J_F bezüglich F. J_F ist φ-monoton mit $\varphi = F$, für strikt konvexe B^* eindeutig erklärt, für strikt konvexe B strikt monoton und für reflexive B und strikt konvexe B^* demistetig. Sind beide B und B^* strikt konvex, dann ist $J_F(u)$ das Stützfunktional von $G(\|\cdot\|) = \int_0^{\|\cdot\|} F(s)\, ds$ in u. Sei weiter B reflexiv und B

und B^* strikt konvex. Dann ist die Gleichung

$$J_F u = g$$

für jedes $g \in B^*$ eindeutig lösbar. Dabei gilt $(J_F)^{-1} = J_{F^{-1}} \in (B^* \to B)$ [33].

Definition 8. Die Elementenfolge (e_i) aus B heißt Basis von B, wenn jedes Element $x \in B$ eindeutig in der Form

$$x = \sum_{i=1}^{+\infty} a_i e_i , \qquad a_i \in R_1 ,$$

dargestellt werden kann.

Bemerkung 21. Wenn B eine Basis besitzt, dann ist B separabel. Weiter existiert eine Folge (f_k) von Funktionalen $f_k \in B^*$, daß $f_k(x) = a_k$ und $f_i(e_j) = \delta_{ij}$. Der Operator

$$P_n x = \sum_{i=1}^{n} f_i(x)\, e_i$$

ist ein Projektionsoperator von B auf die lineare Hülle B_n der ersten n Elemente $e_1, \ldots, e_n$ mit

$$||P_n|| \leqq \alpha , \qquad n = 1, 2, \ldots .$$

Dabei gilt $||P_n x - x|| \to 0$ für $n \to \infty$ und $x \in B$ (d. h. B ist (B_n)-approximierbar). Der Operator

$$Q_n y = \sum_{i=1}^{n} (y, e_i)\, f_i , \qquad y \in B ,$$

ist ein Projektionsoperator von B auf die lineare Hülle der Elemente $f_1, \ldots, f_n$ und es gilt

$$||Q_n|| \leqq \beta , \qquad n = 1, 2, \ldots,$$

und $Q_n y \to y$ in B^* [35]. Man sieht, daß $R(Q_n) = B_n^*$ gesetzt werden kann.

Definition 9. Die Folge (B_n) von Unterräumen $B_n \subset B$ heißt zulässig, wenn $B_n \subset B_m$ für $n \leqq m$, für jedes n ein Projektionsoperator von B auf B_n mit

$$||P_n|| \leqq K$$

und ein Projektionsoperator Q_n von B^* auf B_n^* existiert.

Bemerkung 22. Wenn B eine Basis besitzt, dann ist die Folge (B_n) der linearen Hüllen B_n der ersten n Elemente der Basis zulässig.

Sei B reflexiv, B und B^* strikt konvex und $C \subset B$ konvex und abgeschlossen. Dann existiert der nichtlineare Projektionsoperator $P(C)$ $\left(||u - P(C)\, u|| \leqq ||u - v||,\ v \in C\right)$. Dabei ist $w = P(C)u$ äquivalent der Variationsungleichung

$$\left(J(u - w), v - w\right) \leqq 0 , \qquad w \in C .$$

Satz 6. *B sei reflexiv, B und B^* strikt konvex, $C \subset B$ nicht leer, konvex und abgeschlossen, f besitze den Lipschitz-stetigen Gradienten $T \in (B \to B^*)$ und sei gleichmäßig konvex (bzw. φ-konvex und dabei B gleichmäßig konvex), die Folge (B_n) sei zulässig und die einzige Lösung u_0 von*

$$f(u) = \min_{v \in C} f(v) \tag{18}$$

bzw. von

$$(Tu, v - u) \geqq 0\,, \qquad v \in C\,, \tag{15}$$

(C_n)-approximierbar. Dann konvergiert das Projektions-Iterationsverfahren

$$x_{n+1} = P(C_n)\,(x_n - \alpha_n P_n J^{-1} P_n^* Q_n T x_n)\,, \qquad x_0 \in B\,, \tag{24}$$

mit $0 < \varepsilon_1 \leqq \alpha_n < (2/L(T)K^2) - \varepsilon_2$, $\varepsilon_1, \varepsilon_2 > 0$ *fixiert, gegen* u_0.

Beweis. Vgl. [19].

Bemerkung 23. (24) wurde für Hilberträume und $C = B$ in [9] begründet. Weiter enthält (24) für $B_n = B$, $n = 1, 2, \ldots$, das projezierte Gradientenverfahren

$$x_{n+1} = P(C)\,(x_n - J^{-1}Tx_n)\,, \qquad x_0 \in B\,. \tag{25}$$

Für Hilberträume und beschränkte Mengen C wurde die Konvergenz von (25) z. B. in [32] bewiesen.

Literatur

[1] Brezis, H., Équations et inéquations non linéaires dans les espaces vectoriels en dualité. Ann. Inst. Fourier, Grenoble **18** (1968) 115–175.

[2] Brezis, H., and Stampacchia, G., Sur la regularite de la solution d'inequations elliptiques. Bull. Soc. Mat. de France **96** (1968) 153–180.

[3] Browder, F. E., On the unification of the calcules of variations and the theory of monotone nonlinear operators in Banach-spaces. Proc. Nat. Acad. Scien. **56** (1966) 419–425.

[4] —, Existence and approximation of solutions of nonlinear variational inequalities. Proc. Nat. Acad. Scien. **56** (1966) 1080–1086.

[5] Dalezki, J. L., u. Krein, M. G., Устойчивость решений дифференциальных уравнений в банаховом пространстве. Москва 1970.

[6] Gajewski, H., Iterationsverfahren bei Folgen kontraktiver Operatoren. Mber. Dt. Akad. Wiss. **12** (1970) 1–8.

[7] Hartmann, P. H., and Stampacchia, G., On some nonlinear elliptic differential functional equations. Acta Math. **115** (1966) 271–310.

[8] Hille, E., and Phillips, R. S., Functional Analysis and semi-groups. Providence (1957).

[9] Kluge, R., Ein Projektions-Iterationsverfahren bei Fixpunktproblemen und Gleichungen mit monotonen Operatoren. Mber. Dt. Akad. Wiss. **11** (1969) 599–609.

[10] —, Stark konvergierende numerische Verfahren zur Konstruktion von Fixpunkten pseudokontraktiver Abbildungen. Mber. Dt. Akad. Wiss. **12** (1970) 85–97.

[11] —, Zur approximativen Lösung nichtlinearer Variationsungleichungen. Mber. Dt. Akad. Wiss. **12** (1970) 120–134.

[12] —, Zur Konvergenz und asymptotischen Stabilität einiger Näherungsverfahren bei mehrdeutig lösbaren nichtlinearen Operatorgleichungen. Mber. Dt. Akad. Wiss. **12** (1970) 237–249.

[14] —, Ein Projektions-Iterationsverfahren zur approximativen Lösung von Minimumproblemen mit Nebenbedingungen und Variationsungleichungen mit nichtlinearen Potentialoperatoren. Mber. Dt. Akad. Wiss. **12** (1970) 721–734.

[15] —, Zur approximativen Lösung nichtlinearer Variationsungleichungen. Dissertation B. Berlin 1970.

[16] —, Zur approximativen Lösung konvexer Minimumprobleme mit Nebenbedingungen und nichtlinearer Variationsungleichungen über Folgen linearer algebraischer Gleichungssysteme. Math. Nachr. **48** (1971) 341–352.

[17] —, Näherungsverfahren zur approximativen Lösung nichtlinearer Variationsungleichungen. Math. Nachr. **51** (1971) 343—356.
[18] —, Folgen und Iterationsverfahren bei Folgen nichtlinearer Variationsungleichungen mit kontraktiven Operatoren. (Manuskript).
[19] —, Variationsungleichungen über Lösungsmengen von Variationsungleichungen.
[20] —, Zur Existenz und numerischen Approximation von Lösungen abstrakter Differentialgleichungen erster Ordnung mit Bochner-integrablen rechten Seiten. Math. Nachr. **53** (1972) 263—276.
[21] —, Über eine Klasse abstrakter Differentialgleichungen beliebiger Ordnung im Banach-Raum. Mber. Dt. Akad. Wiss. **14**, H. 10—12 (1972) 731—740.
[22] —, Näherungsverfahren für ein Problem der optimalen Steuerung mit abstrakten Differentialgleichungen erster Ordnung. Math. Nachr. **54** (1972) 229—248.
[23] —, Näherungsverfahren für ein Problem der optimalen Steuerung im Hilbert-Raum mit festem rechten Ende.
[24] —, Zur optimalen Steuerung mit parabolischen partiellen Differentialgleichungen.
[25] —, Zur optimalen Steuerung mit Variationsungleichungen.
[26] Kluge, R., und Bruckner, G., Iterationsverfahren bei Folgen von Abbildungen und nichtlineare Probleme mit monotonen Operatoren. Math. Nachr. **54** (1972) 335—353.
[27] —, Iterationsverfahren für einige nichtlineare Probleme mit Nebenbedingungen. Math. Nachr. **56** (1973) 345—369.
[28] —, Über einige Klassen nichtlinearer Differentialgleichungen und -ungleichungen im Hilbert-Raum. Math. Nachr. (im Druck).
[29] —, Gewöhnliche nichtlineare Differentialgleichungen und Differentialungleichungen im Hilbert-Raum. Diese Proceedings 147—153.
[30] Kluge, R., und Krauß, E., Näherungsverfahren für Probleme der Steuerung mit Operatorgleichungen.
[31] Krauß, E., Zur Steuerung mit Operatorgleichungen. Diese Proceedings.
[32] Lewitin, E. S., u. Poljak, B. T., Методы минимизации при налиъии ограничений. Ж. выч. мат.и мат. физ. **6** (1966) 787—823.
[33] Lions, J. L., Quelques méthodes de résolution des problèmes aux limites non linéaires. Dunod-Gauthier-Villars. Paris 1969.
[34] —, О неравенствах в частных производных. УМН XXVI (1971) 205—263.
[35] Люстерник, Л. А., Соболев, В. И., Элементы функционального анализа. Наука. Москва 1965.
[36] Sibony, M., Sur l'approximation d'équations et inéquations aux derivées partielles non linéaires de type monotone. J. Math. Anal. Appl. **34** (1971) 502—564.
[37] Stampacchia, G., Regularity of solutions of some variational inequalities. Proceedings of Symposia in Pure Mathematics. Vol. 18, **1** (1970) 271—281.

Anschrift des Autors:
Zentralinstitut für Mathematik
und Mechanik der AdW der DDR,
Berlin 108, Mohrenstr. 39.

Gewöhnliche nichtlineare Differentialgleichungen und Differentialungleichungen im Hilbertraum und monotone Operatoren

REINHARD KLUGE und GOTTFRIED BRUCKNER, Berlin

Wir lösen hier Anfangswertprobleme für nichtlineare Differentialgleichungen und -ungleichungen im Hilbertraum mittels Aussagen der Theorie nichtlinearer Gleichungen und Ungleichungen mit monotonen Operatoren [2—9]. Neben Existenz- und Einzigkeitsaussagen für Lösungen werden numerische Verfahren (Iterations- und Projektions-Interationsverfahren) zu ihrer näherungsweisen Bestimmung begründet. Die dabei erzielten Ergebnisse bezüglich Differentialgleichungen verallgemeinern einige der auf dem Kontraktionsprinzip basierenden Resultate der Arbeiten [10, 11] eines der Autoren und sind z. B. auf Rand-Anfangswertprobleme für pseudoparabolische partielle Differentialgleichungen und auf Probleme der optimalen Steuerung [12] anwendbar. Die Beweise der hier angegebenen Sätze und weitere Ergebnisse finden sich in [1].

1. Sei H ein Hilbertraum mit dem Skalarprodukt $(.,.)$, der Norm $\|\cdot\| = \sqrt{(.,.)}$, dem Nullelement 0 und dem identischen Operator I. Weiter bezeichnet $(H \to H)$ die Menge der Abbildungen von ganz H in H.

Auf dem Hilbertraum $H_0 \equiv L_2(0, T; H)$ mit dem Skalarprodukt

$$(u, v)_0 = \int_0^T \big(u(t), v(t)\big)\, dt$$

und der Norm

$$\|u\|_0 = \sqrt{(u, u)_0}\,, \qquad u, v \in H_0\,,$$

erklären wir ein weiteres Skalarprodukt

$$(u, v)_a = \int_0^T e^{-2at} \big(u(t), v(t)\big)\, dt$$

mit $a \geqq 0$ fixiert. Die mit dem Skalarprodukt $(.,.)_a$ versehene Menge H_0 ist ein Hilbertraum H_a $\big(H_a \equiv L_2(e^{-2at}; 0, T; H)\big)$.

Den Banachraum aller auf $[0, T]$ definierten abstrakten Funktionen mit der Norm

$$\|u\|_c = \max_{t \in [0,T]} \|u(t)\|$$

bezeichnen wir mit $C(0, T; H)$.

Die abstrakte Funktion $u(t)$ nennen wir differenzierbar in $t_0 \in [0, T]$, wenn ein Element $u'(t_0) \in H$ derart existiert, daß

$$\left\| \frac{u(t) - u(t_0)}{t - t_0} - u'(t_0) \right\| \to 0 \quad \text{für} \quad t \to t_0, \quad t \in [0, T].$$

$u'(t_0) = \left.\frac{du(t)}{dt}\right|_{t=t_0}$ heißt Ableitung von $u(t)$ in t_0.

Wir untersuchen folgende Probleme:

(P_1). Das Anfangswertproblem $A(0)\,x(0) = x_0 \in H$ für die abstrakte Differentialgleichung

$$\frac{d}{dt} A(t)\,x(t) = f(t, x(t)) \tag{1}$$

mit $A(t), f(t, .) \in (H \to H)$ für $t \in [0, T]$.

(P_2). Das Anfangswertproblem $x(0) = x_0 \in H$ für die abstrakte Differentialgleichung

$$A(t) \frac{dx(t)}{dt} = f(t, x(t)). \tag{2}$$

(P_3). Das Anfangswertproblem $A_1(0)\,x(0) = x_0 \in H$ für die abstrakte Differentialgleichung

$$A_2(t) \frac{d}{dt} A_1(t)\,x(t) = f(t, x(t)) \tag{3}$$

mit $A_2(t), A_1(t) \in (H \to H)$ für $t \in [0, T]$.

(P_4). Gesucht ist eine abstrakte Funktion $x\colon [0, T] \to H$ derart, daß für f. a. $t \in [0, T]$

$$\left(A(t)\,x(t) - x_0 - \int_0^t f(s, x(s))\,ds\,,\ v_t - x(t)\right) \geqq 0\,, \quad v_t \in K_5 \subset H\,, \tag{4}$$

mit der konvexen, abgeschlossenen Teilmenge $K_t \subset H$.

(P_5). Gesucht ist eine abstrakte für f. a. t differenzierbare Funktion $x(t)$: $[0, T] \to H$ derart, daß $x(0) = x_0 \in H$ und für f. a. $t \in [0, T]$

$$\left(A(t) \frac{d}{dt} x(t) - f(t, x(t))\,,\ v_t - \frac{d}{dt} x(t)\right) \geqq 0\,, \quad v_t \in K_t\,.$$

Wir formulieren Bedingungen an die Operatorenschar $D(t) \in (H \to H)$.

$(D)_1$. (Caratheodory-Bedingung). Für jedes $x \in H$ ist $D(t)\,x$ stark meßbar und für f. a. $t \in [0, T]$ $D(t)$ stetig.

$(D)_2$. Für $x(t) \in L_2(0, T; H)$ gilt $D(t)\,x(t) \in L_2(0, T; H)$ $\left(D \in (H_0 \to H_0)\right)$.

$(D)_3$. Für f. a. $t \in [0, T]$ und eine feste Zahl L_D gilt $\|D(t)\,x - D(t)\,y\| \leqq L_D \|x - y\|$ für $x, y \in H$ (Lipschitz-Bedingung).

$(D)_4$. Für f. a. $t \in [0, T]$ und eine feste Zahl $C_D > 0$ gilt $(D(t)\,x - D(t)\,y, x - y) \geqq C_D \|x - y\|^2$ für $x, y \in H$ (Monotonie-Bedingung).

2. Wir kommen zur Lösung der Aufgabe (P_1).

Lemma. *$A(t) \in (H \to H)$ genüge $(A)_1$, $(A)_2$ und $(A)_4$ und $f(t, .) \in (H \to H)$ genüge $(f)_2$ und $(f)_3$. Dann ist der Operator $B \in (H_a \to H_a)$*

$$(Bx)(t) \equiv A(t)\, x(t) - x_0 - \int_0^t f(s, x(s))\, ds\,, \qquad x_0 \in H\,,$$

stetig und für $a > \frac{T}{2} L_f^2/C_A^2$ stark monoton mit der Monotoniekonstanten $C_B = C_A - L_f \sqrt{T/2a}$.

Beweis. Wegen $(A)_1$ und $(A)_2$ ist A in H_a stetig (Krasnoselski) und wegen $(A)_4$ stark monoton mit der Monotoniekonstanten C_A. Wir untersuchen die Abbildung

$$(f_0 x)(t) \equiv \int_0^t f(s, x(s))\, ds$$

in H_a auf Lipschitz-Stetigkeit. Es gilt

$$\begin{aligned}
\|f_0 x - f_0 y\|_a^2 &= \int_0^T e^{-2at} \left\| \int_0^t f(s, x(s))\, ds - \int_0^t f(s, y(s))\, ds \right\|^2 dt \\
&\leqq \int_0^T e^{-2at} \left[L_f \int_0^t \|x(s) - y(s)\|\, ds \right]^2 dt \\
&= \int_0^T L_f^2 \left[\int_0^t e^{-at} e^{a(s-s)} \|x(s) - y(s)\|\, ds \right]^2 dt \\
&\leqq L_f^2 \int_0^T \left[\left(\int_0^t e^{2a(s-t)}\, ds \right) \left(\int_0^t e^{-2as} \|x(s) - y(s)\|^2\, ds \right) \right] dt \\
&\leqq L_f^2 \|x - y\|_a^2 \cdot \frac{1}{2a} \cdot T(1 - e^{-2aT})\,,
\end{aligned}$$

d. h. $\|f_0 x - f_0 y\|_a \leqq L_f \sqrt{T/2a}\, \|x - y\|_a = C_0 \|x - y\|_a$.

Folglich ist B stetig. Wir finden weiter

$$\begin{aligned}
(Bx - By, x - y)_a &= (Ax - Ay, x - y)_a - (f_0 x - f_0 y, x - y)_a \\
&\geqq C_A \|x - y\|_a^2 - C_0 \|x - y\|_a^2 = (C_A - C_0) \|x - y\|_a^2\,, \qquad x, y \in H_a\,.
\end{aligned}$$

Für $a > \frac{T}{2} L_f^2/C_A^2$ gilt $C_A - C_0 > 0$. Damit ist B stark monoton, q. e. d.

Definition 1. *$x(t)\colon [0, T] \to H$ heißt Lösung von (P_1), wenn $x \in L_2(0, T; H)$, für f. a. $t \in [0, T]$ $\frac{d}{dt}\big(A(t)\, x(t)\big)$ existiert, $A(0)\, x(0) = x_0$ ist und für f. a. t (1) erfüllt ist.*

Aus dem eben bewiesenen Lemma und dem Existenzsatz aus der Theorie der monotonen Operatoren [2] folgt

Satz 1. *Die Scharen $A(t)$ und $f(t, .)$ mögen den Voraussetzungen des Lemmas genügen. Dann besitzt (P_1) genau eine Lösung x, und es gilt $A(t)\,x(t) \in C(0, T; H)$ und $\frac{d}{dt}\big(A(t)\,x(t)\big) \in L_2(0, T; H)$.*

Zur approximativen Lösung von (P_1) geben wir der Kürze halber nur ein Projektions-Iterationsverfahren an.

Definition. Der Hilbertraum H habe die Eigenschaft (V), wenn eine Folge (H_n) von Unterräumen $H_n \subset H$ existiert mit $H_n \subset H_m$ für $n \leqq m$ und $\|P_n h - h\| \to 0$ für $n \to \infty$ und $h \in H$ (P_n bezeichnet den Operator der orthogonalen Projektion von H auf H_n).

Satz 1a. *Die Scharen $A(t)$ und $f(t, .)$ mögen den Voraussetzungen des Lemmas genügen. Außerdem seien $(A)_3$ und (V) erfüllt. Dann konvergiert das Projektions-Iterationsverfahren*

$$w_i = W_i\, w_{i-1}\,, \qquad i = 1, 2, \dots, \qquad w_0 \in L_2(0, T; H)\,,$$

mit

$$W_i = P_i - \frac{C_A - L_f\sqrt{T/2a}}{(L_A + L_f\sqrt{T/2a})^2}\left(P_i A(t) - P_i x_0 - \int_0^t P_i f(s, \cdot\,(s))\, ds\right)$$

in $L_2(0, T; H)$ gegen die einzige Lösung des Problems (P_1).

3. Die Lösung der Aufgaben (P_2) und (P_3) vollzieht sich analog zur Lösung von (P_1) in Punkt 2. Wir begnügen uns mit der Angabe der Ergebnisse.

Definition 2. $x(t)\colon [0, T] \to H$ heißt Lösung von (P_2), wenn $x \in C(0, T; H)$, $x(0) = x_0$ gilt, $x'(t) \in L_2(0, T; H)$ existiert und für f. a. t (2) erfüllt ist.

Satz 2. *$A(t) \in (H \to H)$ genüge $(A)_1$, $(A)_2$ und $(A)_4$ und $f(t, .) \in (H \to H)$ bilde Elemente aus $C(0, T; H)$ in Elemente aus $L_2(0, T; H)$ ab und genüge $(f)_3$. Dann besitzt (P_2) genau eine Lösung.*

Satz 2a. *Voraussetzungen wie zu Satz 2. Zusätzlich genüge $A(t)$ der Eigenschaft $(A)_3$ und H der Bedingung (V). Dann konvergiert das Projektions-Iterationsverfahren*

$$z_i(t) = P_i z_{i-1}(t) - \frac{C_A - L_f\sqrt{T/2a}}{(L_A + L_f\sqrt{T/2a})^2}\left[P_i A(t)\, z_{i-1}(t) - P_i f\Big(t, x_0 + \int_0^t z_{i-1}(s)\, ds\Big)\right]$$

in $L_2(0, T; H)$ gegen die Ableitung x' der einzigen Lösung x von (P_2). Dabei konvergiert die Folge

$$y_i(t) = P_i x_0 + \int_0^t z_i(s)\, ds$$

in $C(0, T; H)$ gegen x.

Definition 3. $x(t) \in L_2(0, T; H)$ heißt Lösung von (P_3), falls $A_1(t)\,x(t) \in C(0, T; H)$, $\frac{dA_1(t)\,x(t)}{dt}$ für f. a. t existiert, $A_1(0)\,x(0) = x_0 \in H$ und für f. a. $t \in [0, T]$ (3) erfüllt ist.

Satz 3. *$A_i(t)$, $i = 1, 2$, genüge $(A_i)_j$, $j = 1, 2, 4$, und f $(f)_1$ und $(f)_3$. Dann hat (P_3) genau eine Lösung x mit* $\frac{dA_1(t)\,x(t)}{dt} \in L_2(0, T; H)$.

Satz 3a. *Voraussetzungen wie zu Satz 3. Zusätzlich seien $(A_1)_3$ und $(A_2)_3$ erfüllt. Dann konvergiert das Iterationsverfahren*

$$v_i(t) = v_{i-1}(t) - s[A_2(t)\, v_{i-1}(t) - f(t, x_{i-1}(t))]\,, \qquad i = 1, 2, \ldots$$

mit

$$0 < s < 2\left(C_{A_2} - \frac{L_f}{C_{A_1}}\sqrt{T/2a}\right)\bigg/\left(L_{A_2} + \frac{L_f}{C_{A_1}}\sqrt{T/2a}\right)^2$$

und der einzigen Lösung $x_{i-1}(t)$ des Problems

$$\frac{d}{dt} A_1(t)\, x(t) = v_{i-1}(t)\,, \qquad A_1(0)\, x(0) = x_0\,,$$

in $L_2(0, T; H)$ gegen die einzige Lösung $v \in L_2(0, T; H)$ von

$$A_2(t)\, v(t) = f\left(t, A_1(t)^{-1}\left(x_0 + \int_0^t v(s)\, ds\right)\right).$$

Dabei konvergiert die entsprechende Folge $x_i(t)$ in $L_2(0, T; H)$ gegen die einzige Lösung von (P_3) und die Folge $\frac{d}{dt} A_1(t)\, x_i(t)$ in $L_2(0, T; H)$ gegen $\frac{d}{dt} A_1(t)\, x(t)$.

4. Wir kommen zur Lösung von (P_4) und (P_5).

$\overline{H}$ bezeichne einen reellen Hilbertraum. Sei $K \subset \overline{H}$ abgeschlossen und konvex. Wir führen den Operator $P(K)$ der Projektion auf K ein. Mit $P(K)\, h$, $h \in \overline{H}$, bezeichnen wir die eindeutig bestimmte Lösung des Minimumproblems

$$\min_{v \in K} \frac{1}{2} \|h - v\|^2$$

bzw. der äquivalenten Variationsungleichung

$$(h - u, u - v) \geqq 0\,, \qquad v \in K\,.$$

$P(K) \in (\overline{H} \to K)$ ist Lipschitz-stetig mit der Lipschitzkonstanten 1 (vgl. z. B. [7]).

Bemerkung a. $K_t \subset H$ sei für f. a. $t \in [0, T]$ konvex und abgeschlossen. Weiter genüge die Operatorenschar $Q(t) \equiv P(K_t)$ der Voraussetzung $(Q)_2$. Dann ist die Menge

$$K_0 \equiv \{x \in H_a\colon x(t) \in K_t \quad \text{für f. a.} \quad t \in [0, T]\}$$

eine konvexe und abgeschlossene Teilmenge von H_a, und es gilt für $y \in H_a$

$$(P(K_0)\, y)\,(t) = P(K_t)\, y(t) \qquad \text{für f. a. } t \in [0, T]\,.$$

Bemerkung b. Sei $K_t \neq \emptyset$ für $t \in [0, T]$ und

$$t \to K_t$$

eine stetige Abbildung des $\mathcal{L}^*$-Raumes $[0, T]$ in den $\mathcal{L}^*$-Raum der konvexen, abgeschlossenen Teilmengen von H (vgl. Kuratowski [13] und Mocso [14]). Dann genügt die Operatorenschar $Q(t) \equiv P(K_t)$, $t \in [0, T]$, der Bedingung $(Q)_2$.

Bemerkung c. Sei $H = R_n$, $a_i(t)$, $b_i(t)$ stetige reelle Funktionen für $i = 1, 2, \ldots, n$, $a_i(t) \leqq b_i(t)$. Sei weiter

$$K_t \equiv \{(d_1, \ldots, d_n): a_i(t) \leqq d_i \leqq b_i(t)\,, \quad i = 1, \ldots, n\} \neq \emptyset\,.$$

Dann ist $t \to K_t$ eine Abbildung im Sinne von Bemerkung b.

Nach diesen Vorbereitungen verfahren wir wieder analog zu der Lösung von (P_1) (Punkt 2).

Definition 4. $x(t)$: $[0, T] \to H$ heißt Lösung von (P_4), wenn $x \in L_2(0, T; H)$ und für f. a. $t \in [0, T]$ (4) erfüllt ist.

Mit Hilfe des Lemmas und Bemerkung a beweist man analog zu Satz 1 den

Satz 4. *Seien $K_t \subset H$ für f. a. $t \in [0, T]$ konvex und abgeschlossen. Die Projektionsoperatoren $Q(t) \equiv P(K_t)$ mögen der Bedingung $(Q)_2$ genügen. Weiter seien die Voraussetzungen des Lemmas erfüllt. Dann existiert genau eine Lösung der Aufgabe (P_4).*

Satz 4a. *Voraussetzungen wie zu Satz 4. Zusätzlich genüge die Schar $A(t)$ der Bedingung $(A)_3$ und H genüge (V). Weiter sei $K_0 \cap (H_1)_a \neq \emptyset$, und es existiere eine Folge $v_n \in K_0 \cap (H_n)_a$ mit $v_n \to x$ in $L_2(0, T; H)$. Dann konvergiert das Projektions-Iterationsverfahren*

$$x_i(t) = P(K_t)\, P_t \left[x_{i-1}(t) - \frac{C_B}{L_B^2} \{(Bx_{i-1})(t)\}\right], \qquad x_0 \in L_2(0, T; H)\,,$$

in $L_2(0, T; H)$ gegen die einzige Lösung x von (P_4). Dabei ist

$$(Bw)(t) \equiv A(t)\, w(t) - x_0 - \int_0^t f\big(s, w(s)\big)\, ds\,.$$

Definition 5. $x(t)$: $[0, T] \to H$ heißt Lösung von (P_5), wenn $x \in L_2(0, T; H)$, $\frac{dx(t)}{dt}$ existiert für f. a. t und $\frac{d}{dt}\, x(t) \in L_2(0, T; H)$, $x(0) = x_0$ und für f. a. t (5) erfüllt ist.

Satz 5. *$K_t \subset H$ sei für f. a. $t \in [0, T]$ konvex und abgeschlossen und die Operatoren $Q(t) \equiv P(K_t)$ mögen $(Q)_2$ genügen. Weiter genüge $A(t) \in (H \to H)$ $(A)_1$, $(A)_2$ und $(A)_4$ und $f(t, .) \in (H \to H)$ bilde Elemente aus $C(0, T; H)$ in Elemente aus $L_2(0, T; H)$ ab und genüge $(f)_3$. Dann existiert genau eine Lösung der Aufgabe (P_5).*

Satz 5a. *Voraussetzungen wie zu Satz 5. Zusätzlich genüge $A(t)$ der Bedingung $(A)_3$ und H der Bedingung (V). Weiter sei $K_0 \cap (H_1)_a \neq \emptyset$, und es existiere eine Folge $v_n \in K_0 \cap (H_n)_a$ mit $v_n \to w$ in $L_2(0, T; H)$; dabei ist w die einzige Lösung von*

$$\big((Dw)(t)\,,\ v_t - w(t)\big) \geqq 0\,, \quad v_t \in K_t\,, \quad \text{für f. a. } t\,,$$

mit

$$(Dw)(t) \equiv A(t)\, w(t) - f\left(t, x_0 + \int_0^t w(s)\, ds\right).$$

Dann konvergiert das Projektions-Iterationsverfahren

$$w_i(t) = P(K_t)\,P_i\left[w_{i-1}(t) - \frac{C_D}{L_D^2}\{(Dw_{i-1})(t)\}\right], \qquad w_0 \in L_2(0, T; H),$$

in $L_2(0, T; H)$ *gegen* w. *Weiter konvergiert die Folge*

$$y_n(t) = P_n x_0 + \int_0^t w_n(s)\,ds$$

in $C(0, T; H)$ *gegen* x, *der einzigen Lösung von* (P_5).

Literatur

[1] Kluge, R. und Bruckner, G., Über einige Klassen nichtlinearer Differentialgleichungen und -ungleichungen im Hilbert-Raum. Math. Nachr. (im Druck).
[2] Browder, F. E., Problèmes non linéaires, Montréal 1966.
[3] —, On the unification of the calculus of variations and the theory of monotone nonlinear operators. Proc. Nat. Acad. Sci. USA **56** (1966) 419—425.
[4] Lions, J. L., Quelques méthodes de résolution des problémes aux limites non linéaires, Paris 1969.
[5] Kluge, R., Zur approximativen Lösung nichtlinearer Variationsungleichungen. M.ber. Dt. Akad. Wiss. **12** (1970) 120—134.
[6] —, Zur Konvergenz und asymptotischen Stabilität einiger Näherungsverfahren bei mehrdeutig lösbaren nichtlinearen Operatorgleichungen. M.ber. Dt. Akad. Wiss. **12** (1970) 237—249.
[7] —. Dissertation B, Berlin 1970.
[8] —, Näherungsverfahren zur approximativen Lösung nichtlinearer Variationsungleichungen. Math. Nachr. **51** (1971) 343—356.
[9] — Näherungsverfahren für einige nichtlineare Probleme. Diese Proceedings 133—146.
[10] —, Zur Existenz und numerischen Approximation von Lösungen abstrakter Differentialgleichungen erster Ordnung mit Bochner-integrablen rechten Seiten. Math. Nachr. **53** (1972) 263—276.
[11] —, Über eine Klasse abstrakter Differentialgleichungen beliebiger Ordnung im Banach-Raum. Mber. Dt. Akad. Wiss. **14** (1972) 10—12.
[12] —, Näherungsverfahren für ein Problem der optimalen Steuerung mit abstrakten Differentialgleichungen erster Ordnung. Math. Nachr. **54** (1972) 229—248.
[13] Kuratowski, K., Topology, Bd. I, Warszawa 1966.
[14] Mosco, U., Convergence of convex sets and solutions of variational inequalities, Advances in Math. **3** (1969) 510—585.

Anschrift der Autoren:
ZIMM der AdW der DDR
108 Berlin, Mohrenstr. 39

Normal solvability and solvability of nonlinear equations

JOSEF KOLOMÝ, Prag

Introduction

This paper is divided according to its contents into three parts. Part 1 is devoted to the normal solvability of linear equations in normed linear spaces. Some known or less known results (see references in part 1) are slightly extended for linear densely defined (in general nonclosed) operators acting in normed spaces. Such slight extension admits also the result of Katětov [1] giving a connection between normal solvability and relative openness of linear continuous operators. The proofs of the known results and their simple extensions are given here for the sake of completeness. In Part 2 we sketch very briefly the known results concerning the normal solvability of nonlinear equations. Part 3 is devoted to the solvability of nonlinear equations and it is continuation of our paper [2]. This part deals with the surjectivity and existence theorems. Using the concept of A-feeble local contraction (compare [2], [3]) which is a generalization and extension of the local contraction of Edelstein [4] and generalized contraction of Kirk [5], [6], we derive, as a simple consequence of our general theorem, a fixed point theorem for feeble local contraction in Hilbert spaces.

1. Normal solvability (linear case)

Let X, Y be normed linear spaces (NLS), X^*, Y^* their dual spaces, T a linear operator having domain $D(T)$ a subspace of X (i.e. $D(T) \subset\subset X$) and range a subspace $R(T)$ of Y. The spaces X, Y are assumed complete only when specifically stated. Usually we assume that $\overline{D(T)} = X$. By T^* we denote a konjugate operator to T with domain $D(T^*) \subset\subset Y^*$ and range $R(T^*) \subset\subset X^*$. We shall assume that $D(T^*) \neq (0)$. Recall that $T: D(T) \to Y$ is said to be closed if its graph is closed in $X \times Y$, i.e. if $(u_n) \in D(T)$, $u_n \to u$ and $T(u_n) \to v$, then $u \in D(T)$ and $v = T(u)$. The null space of T is denoted by $N(T)$ and $T^{-1}(0) = N(T) = \{u \in D(T): T(u) = 0\}$, $\langle .,. \rangle$ is the pairing between X and X^* (Y and Y^*, resp.). For a set $M \subset X$, $M^\perp$ is an annihilator of M (for $A \subset X^*$, ${}^\perp A$ denotes its annihilator), i.e. $M^\perp = \{x^* \in X^*: x^*(M) = 0\}$.

Definition 1. Let X, Y be NLS, $M \subset X$, $F: M \to Y$ a mapping. Then F is said to be relatively open if for each $G \subset M$ which is open in M, $F(G)$ is open in $F(M)$.

We introduce slight extensions of some known necessary and sufficient conditions under which a linear map is relatively open (compare [7, § 4.2],

[8, Thm. 2.3], [1]). The proofs are given here for the sake of completeness and they use the same arguments as in [7], [1]).

Proposition 1. *Let X, Y be* NLS, $D(T) \subset\subset X$, $T: D(T) \to Y$ *a linear mapping. Then each of the following two conditions is necessary and sufficient for T to be relatively open:* (i) *There exists a constant $m > 0$ so that for each $v \in R(T)$ there is an element $u \in D(T)$ so that $T(u) = v$ and $||u|| \leqq m||v||$* (ii) *For each $v_n \in R(T)$, $v \in R(T)$, $v_n \to v = T(u)$, $u \in D(T)$ there exist $u_n \in D(T)$ so that $T(u_n) = v_n$ and $u_n \to u$ as $n \to \infty$.*

Proof. (It is given here for the sake of completeness.) Let T be relatively open. Then $C_\beta(0) \subset T(B_1(0))$ for some $\beta > 0$, where $B_1(0) = \{u \in D(T): ||u|| < 1\}$, $C_\beta(0) = \{v \in R(T): ||v|| < \beta\}$. Taking $0 < \beta' < \beta$, we set $m = (\beta')^{-1}$. If $v \in R(T)$, $v \neq 0$, then $\beta' v ||v||^{-1} \in C_\beta(0)$ and hence there exists $u_1 \in B_1(0)$ so that $T(u_1) = \beta' \frac{v}{||v||}$. Setting $u = \frac{||v||}{\beta'} u_1$, we have that $u \in D(T)$ and $T(u) = v$. Moreover, $||u|| \leqq m||v||$. If $v = 0$, we put $u = 0$ and our ondition is obviously satisfied.

Suppose that the condition (i) is satisfied. Let $\emptyset \neq G \subset D(T)$ be relatively open, $v_0 \in T(G)$, $v_0 = T(u_0)$, $u_0 \in G$ and suppose that $\delta > 0$ is such that $u \in D(T)$, $||u - u_0|| < \delta \Rightarrow u \in G$. We prove that $R(T)$ contains with its point v_0 also the ball $V = \left\{v \in R(T): ||v - v_0|| < \frac{\delta}{m}\right\}$. Let $v \in V$, then $v - v_0 \in R(T)$ and hence by our condition (i) there exists $z \in D(T)$ so that $v - v_0 = T(z)$ and $||z|| \leqq m||v - v_0|| < \delta$. Hence $u_0 + z \in D(T)$ and moreover, $u_0 + z \in G$. Then $T(u_0 + z) = T(u_0) + T(z) = v_0 + v - v_0 = v$.

Proof of (ii). Suppose T is relatively open and that $v_n \in R(T)$, $u_0 \in D(T)$, $v_0 = T(u_0)$, $v_n \to v_0$. Then $v_n - v_0 \to 0$ and in view of (i) there exist $z_n \in D(T)$ so that $T(z_n) = v_n - v_0$ and $z_n \to 0$. Setting $u_n = u_0 + z_n$, we have that $u_n \in D(T)$, $u_n \to u_0$ and $T(u_n) = T(u_0 + z_n) = v_n$.

Now, suppose that the condition (ii) is satisfied, but T is not relatively open. According to (i) there exists $v_n \in R(T)$ (obviously $v_n \neq 0$) so that for each $u \in D(T)$ satisfying $T(u) = v_n$ we have that $||u|| > n||v_n||$. Setting $v_n' = \frac{1}{\sqrt{n}} \frac{v_n}{||v_n||}$, we have that $v_n' \in R(T)$, $v_n' \to 0$. Then for each $u \in D(T)$ with $T(u) = v_n'$ we obtain that $||u|| > \sqrt{n}$. As $v_n' \to 0 = T(0)$, by our condition there exists $u_n' \in D(T)$ so that $T(u_n') = v_n'$ and $u_n' \to 0$, a contradiction.

Proposition 2 ([8], [9]). Let X, Y be Banach spaces, $D(T) \subset\subset X$, $T: D(T) \to Y$ linear injective and closed. Then $R(T)$ is closed $\Leftrightarrow T^{-1}$ is continuous.

Indeed if $v_n \in R(T)$, $v_n \to v$, then $v_n = T(u_n)$ for some $u_n \in D(T)$. As T^{-1} is continuous, (u_n) is a Cauchy sequence and therefore $u_n \to u_0 \in X$. Since T is closed, $u_0 \in D(T)$ and $T(u_0) = v$. Conversely, if $R(T)$ is closed, then a closed mapping T^{-1} is continuous by the closed graph theorem.

Let X be a Banach space, $D(T) \subset\subset X$, $T: D(T) \to Y$ a linear closed mapping. Then $X/N(T)$ is a Banach space ($N(T)$ is closed since T is closed) with the norm

$$||[u]|| = \inf\{||u - v||: v \in N(T)\} = \inf\{||q||: q \in [u]\},$$

where $[u] = \{u + v: v \in N(T)\}$. Let $\hat{T}$ be a mapping induced by T. Recall that $\hat{T}: D(\hat{T}) \to Y$ is defined by $\hat{T}([u]) = T(u)$, where $[u] \in D(\hat{T}) = D(T)/N(T)$. Then $\hat{T}$ is injective linear and closed with $R(\hat{T}) = R(T)$. (See [9].) Indeed, if $[u_n] \to [u]$ $[u_n] \in D(\hat{T})$ and $\hat{T}([u_n]) \to v$, then there is a sequence $v_n \in N(T)$ so that $u_n - v_n \to u$. Since $T(u_n - v_n) = \hat{T}([u_n]) \to v$ and T is closed, $u \in D(T)$ and $T(u) = v$. Hence $[u] \in D(\hat{T})$ and $\hat{T}([u]) = v$.

Proposition 3. *Let X, Y be* NLS, $D(T) \subset\subset X$, $T: D(T) \to Y$ *a linear and closed. We draw the following conclusions:*

(i) *(compare [8, Thm. 2.3]). If X, Y are complete, then $R(T)$ is closed $\Leftrightarrow$ T is relatively open.*

(ii) *If X is complete and $R(T)$ is of the 2nd category in itself, then T is relatively open.*

Proof. (i) (compare [8, § 1]). Suppose that $R(T)$ is closed. Then $R(\hat{T})$ is closed and $(\hat{T})^{-1}$ is continuous by proposition 2. Hence there exists a constant $k > 0$ so that $\|[u]\| \leqq k\|\hat{T}([u])\|$ for each $[u] \in D(\hat{T})$. Let $v_0 \in R(T) = R(\hat{T})$, $v_0 \neq 0$. Since $\hat{T}$ is injective there exists a unique $[u_0] \in D(\hat{T})$ so that $\hat{T}([u_0]) = v_0$. For $\varepsilon_0 = k\|\hat{T}([u_0])\| > 0$ there is $z_0 \in N(T)$ so that $\|[u_0]\| \leqq \|u_0 - z_0\| \leqq \|[u_0]\| + \varepsilon_0$. Setting $x_0 = u_0 - z_0$, we have that $x_0 \in D(T)$ and

$$\|x_0\| \leqq 2k\|\hat{T}([u_0])\| = 2k\|v_0\| = m\|v_0\| ,$$

where $m = 2k > 0$. Moreover, $T(x_0) = T(u_0 - z_0) = T(u_0) = \hat{T}([u_0]) = v_0$. For $v_0 = 0$ we set $x_0 = 0$. By proposition 1 T is relatively open. Conversely, suppose that T is relatively open. According to proposition 1 there is a constant $m > 0$ so that for each $v \in R(T)$ there exists $u \in D(T)$ such that $T(u) = v$ and $\|u\| \leqq m\|v\|$. Then there is a unique $[u] \in D(\hat{T})$ so that $\hat{T}([u]) = v$ and moreover, $\|[u]\| \leqq m\|\hat{T}([u])\|$. Hence $(\hat{T})^{-1}$ is continuous and by proposition 2 $R(\hat{T})$ and therefore $R(T)$ is closed.

The assertion (ii), which is a generalization of one part of (i), follows at once from the interior mapping principle [7, § 4.2].

Let X, Y be Banach spaces, $T: D(T) \to Y$ a linear mapping. Then any two of the following three statements imply the other: T closed, $R(T)$ closed, $\gamma(T) > 0$, where

$$\gamma(T) = \inf_{u \in D(T)} \frac{\|T(u)\|}{\operatorname{dist}(u, N(T))}, \qquad \left(\frac{0}{0} \overset{\text{def}}{\equiv} +\infty\right),$$

is so called minimum modulus of T (see [9, Chapt. IV]).

The concept of normal solvability has been introduced by F. Hausdorff [10]. Keeping his and Katětov's [1] terminology we proceed with the following.

Definition 2. Let X, Y be NLS, $D(T) \subset\subset X$, $\overline{D(T)} = X$, $T: D(T) \to Y$ a linear mapping. T is said to be normally solvable of the 1st type (or the equation $T(u) = v$ is normally solvable) if the following condition is true: $v \in R(T)$ if and only if for each $v^* \in D(T^*)$, $T^*v^* = 0 \Rightarrow \langle v^*, v \rangle = 0$.

A mapping T is said to be normally solvable of the 2nd type provided the following condition is valid: $u^* \in R(T^*)$ if and only if for each $u \in D(T)$, $T(u) = 0 \Rightarrow \langle u^*, u\rangle = 0$.

Otherwise stated a mapping T is called normally solvable of the 1st type (normally solvable of the 2nd type, respectively) if $R(T) = {}^{\perp}N(T^*)$, ($R(T^*) = N(T)^{\perp}$, respectively).

Remark 1. If $v \in R(T)$ and $v^* \in D(T^*)$, $T^*v^* = 0$, then there is $u \in D(T)$ so that $T(u) = v$ and $\langle v^*, v\rangle = \langle v^*, T(u)\rangle = \langle T^*v^*, u\rangle = 0$. Similarly, if $u^* \in R(T^*)$, $u \in N(T)$, then $u^* = T^*v^*$ for some $v^* \in D(T^*)$ and $\langle u^*, u\rangle = \langle T^*v^*, u\rangle = \langle v^*, T(u)\rangle = 0$. To prove that T is normally solvable of the 1st type (normally solvable of the 2nd type) it is sufficient to show the validity of the implication: if for each $v^* \in D(T^*)$, $T^*v^* = 0 \Rightarrow \langle v^*, v\rangle = 0$, then $v \in R(T)$ (if for each $u \in N(T) \Rightarrow \langle u^*, u\rangle = 0$, then $u^* \in R(T^*)$.

The following theorem is usually proved in textbooks under the conditions that either a) X, Y are Banach spaces and T is closed with $D(T) \subset\subset X$, or b) X, Y are NLS and $T: X \to Y$ is continuous. In fact, next result with its proof is contained in [11, Chapt. VII, § 5].

Theorem 1. *Let X, Y be* NLS, *$D(T) \subset\subset X$, $\overline{D(T)} = X$, $T: D(T) \to Y$ a linear mapping. Then T is normally solvable of the 1st type $\Leftrightarrow R(T)$ is closed.*

Proof. Suppose that T is normally solvable of the 1st type. Let $v_n \in R(T)$, $v_n \to v$, $v \in Y$, then $v_n = T(u_n)$ for some $u_n \in D(T)$. If $v^* \in D(T^*)$, $T^*v^* = 0$, then $\langle v^*, v_n\rangle = \langle v^*, T(u_n)\rangle = \langle T^*v^*, u_n\rangle = 0$. Moreover, $\langle v^*, v_n\rangle \to \langle v^*, v\rangle$ and therefore $\langle v^*, v\rangle = 0$. By our assumption $v \in R(T)$.

Assume that $R(T)$ is closed. In view of remark 1 it is sufficient to prove that if $v_0 \notin R(T)$, then there is $v^* \in D(T^*)$, $T^*v^* = 0 \Rightarrow \langle v^*, v_0\rangle \neq 0$. Let $v_0 \in Y - R(T)$, then $\operatorname{dist}(v_0, R(T)) > 0$ and by the Hahn-Banach theorem there exists $v^* \in Y^*$ so that $\langle v^*, v_0\rangle \neq 0$ and $v \in R(T) \Rightarrow \langle v^*, v\rangle = 0$. Hence $v^* \in D(T^*)$ and $u \in D(T) \Rightarrow \langle T^*v^*, u\rangle = \langle v^*, T(u)\rangle = 0$. Therefore $T^*v^* = 0$ and $\langle v^*, v_0\rangle \neq 0$ and this proves the theorem.

The next theorem is a slight extension of Katětov's result [1]. Instead of T is linear and continuous with $D(T) = X$ we assume here that T is only linear with $\overline{D(T)} = X$. The proof is based on proposition 1, 3 and it is almost the same as in [1].

Theorem 2. *Let X, Y be* NLS, *$D(T) \subset\subset X$, $\overline{D(T)} = X$, $T: D(T) \to Y$ a linear mapping. Then T is normally solvable of the 2nd type $\Leftrightarrow T$ is relatively open.*

Proof. Suppose that T is relatively open, $v^* \in X^*$ is so that for each $u \in D(T)$, $T(u) = 0 \Rightarrow \langle v^*, u\rangle = 0$. We prove that $v^* \in R(T^*)$. If $u_1, u_2 \in D(T)$, $T(u_1) = T(u_2)$, then $\langle v^*, u_1\rangle = \langle v^*, u_2\rangle$. For each $v = T(u)$, where $u \in D(T)$, we set $w^*(v) = \langle v^*, u\rangle$. Then w^* is a linear and continuous functional on $R(T)$. Indeed, if $v_n \in R(T)$, $v_0 \in R(T)$ and $v_n \to v_0$, $u_0 \in D(T)$, $v_0 = T(u_0)$ (let us remark that$w^*(v)$, where $v \in R(T)$, is equal to the same value $\langle v^*, u\rangle$ for each $u \in T^{-1}(v) = \{u \in D(T): T(u) = v\}$), in virtue of proposition 1 (ii) there exists $u_n \in D(T)$ so that $v_n = T(u_n)$ and $u_n \to u_0$ as $n \to \infty$. We have $w^*(v_n) = \langle v^*, u_n\rangle \to \langle v^*, u_0\rangle = w^*(v_0)$, since $v^* \in X^*$. By Hahn-Banach theorem there is

$z^* \in Y^*$ so that $z^*/R(T) = w^*$. Therefore $z^* \in D(T^*)$ and $\langle T^*(z^*), u\rangle = \langle z^*, T(u)\rangle = \langle w^*, T(u)\rangle = \langle v^*, u\rangle$ for each $u \in D(T)$. Hence $v^* = T^*z^*$ as $\overline{D(T)} = X$, $v^* \in R(T^*)$ and thus T is normally solvable of the 2nd type.

Suppose that T is normally solvable of the 2nd type. Let $v_n^* \in R(T^*)$, $v^* \in X^*$ be such that $v_n^* \to v^*$. If $u \in D(T)$, $T(u) = 0$, then $\langle v_n^*, u\rangle = \langle T^*u_n^*, u\rangle = \langle u_n^*, T(u)\rangle = 0$ for some $u_n^* \in D(T^*)$, $(n = 1, 2, \ldots)$. Hence $\langle v^*, u\rangle = 0$ for each $u \in D(T)$ for which $T(u) = 0$. Since T is normally solvable of the 2nd type, $v^* \in R(T^*)$. We have proved that $R(T^*)$ is closed in X^*. By proposition 3 (i) (let us remark that T^* is always closed [9, Chapt. II] and X^*, Y^* are complete) the mapping T^* is relatively open. We shall prove that T has also the same property.

Suppose on contrary that T is not relatively open. By proposition 1 (i) there exist $v_n = T(a_n)$, $a_n \in D(T)$ so that for each $u \in D(T)$ so that $v_n = T(u)$ we have $\|u\| > n\|v_n\|$. One can assume that $\|v_n\| = 1$, $(n = 1, 2, \ldots)$. Obviously $\operatorname{dist}(a_n, N(T)) \geqq n$, where $N(T) = \{u \in D(T)\colon T(u) = 0\}$. By the Hahn-Banach theorem there exist $\tilde{v}_n^* \in X^*$ so that $\tilde{v}_n^*(N(T)) = 0$, $\|\tilde{v}_n^*\| = 1$ and $\langle \tilde{v}_n^*, a_n\rangle = \operatorname{dist}(a_n, N(T))$, $(n = 1, 2, \ldots)$. Setting

$$v_n^* = \frac{\tilde{v}_n^*}{\operatorname{dist}(a_n, N(T))}, \qquad (n = 1, 2, \ldots),$$

we obtain that $v_n^*(N(T)) = 0$, $\|v_n^*\| \leqq \frac{1}{n}$ and $\langle v_n^*, a_n\rangle = 1$. Because $v_n^*(N(T)) = 0$ and T is normally solvable of the 2nd type, $v_n^* \in R(T^*)$. Moreover, $v_n^* \to 0$ as $n \to \infty$ and T^* being relatively open there exist $u_n^* \in D(T^*)$ so that $v_n^* = T^*(u_n^*)$ and $u_n^* \to 0$ as $n \to \infty$ (proposition 1 (ii)). Hence

$$|\langle v_n^*, a_n\rangle| = |\langle T^*u_n^*, a_n\rangle| = |\langle u_n^*, T(a_n)\rangle|$$
$$= |\langle u_n^*, v_n\rangle| = \|u_n^*\|\,\|v_n\| = \|u_n^*\| \to 0$$

a contradiction with the fact that $\langle v_n^*, a_n\rangle = 1$. Theorem is proved.

Corollary 1. *Let X, Y be* NLS, $D(T) \subset\subset X$, $\overline{D(T)} = X$, $T\colon D(T) \to Y$ *a linear relatively open mapping. Then T^* is also relatively open.*

Indeed, if T is relatively open, then by theorem 2 T is normally solvable of the 2nd type and hence $R(T^*)$ is closed in X^* (see proof of the second part of Thm. 2). By proposition 3 (i) T^* is relatively open.

The converse of Corollary 1 is not true in general. But in case X is complete and $T\colon X \to Y$ is linear and continuous, then T is relatively open $\Leftrightarrow T^*$ is relatively open (see [1]).

Let us also mention that the assumption $R(T^*)$ is closed does not imply that $R(T)$ is closed in normed linear space. Example: X is a NLS, $N \subset\subset X$ a subspace which is not closed in X, T = identity map of N onto N. Then $R(T^*)$ is closed, but $R(T)$ is not closed.

Using the closed range theorem [11, VII, § 5] and proposition 3 (i) applied to mappings $T\colon D(T) \to Y$, $D(T) \subset\subset X$, $T^*\colon D(T^*) \to X^*$, $D(T^*) \subset\subset Y^*$, we obtain

Theorem 3. *Let X, Y be Banach spaces, $D(T) \subset\subset X$, $\overline{D(T)} = X$, $T: D(T) \to Y$ a linear closed mapping. Then the following assertions are equivalent:* 1) $R(T)$ *is closed,* 2) $R(T^*)$ *is closed;* 3) T *is normally solvable of the* 1st *type;* 4) T *is normally solvable of the* 2nd *type;* 5) T *is relatively open;* 6) T^* *is relatively open.*

Example. Let X, Y be NLS, X complete, $T: X \to Y$ a linear continuous mapping so that $R(T^*)$ is of the 2nd category in itself. Then T is simultaneously normally solvable of the 1st and the 2nd types.

Indeed, by proposition 3 (ii) T^* is relatively open and hence T is also relatively open [1]. But the image $T(X) = R(T)$ of a complete space X under a linear continuous and relatively open mapping T is complete and hence closed. By theorems 1, 2 T is simultaneously normally solvable of the 1st and 2nd types.

Example. Let X, Y be NLS, X complete, $X \subset\subset Y$, $U: X \to Y$ a linear precompact (i.e. totally bounded) mapping. For each $u \in X$ we wet $T(u) = u - U(u)$. It is wellknown that T is relatively open and hence $T(X) = R(T)$ is complete and hence closed. In view of Theorems 1, 2 T is simultaneously normally solvable of the 1st and 2nd types. These assertions are known as the Fredholm theorems.

For some other examples with T closed (closable) having the closed range $R(T)$, see for instance [9]. The perturbation theory in connection with normal solvability and related topics is developed in Gochberg-Krein [12], Goldberg [9], Kato [13], Krein [8], Nashed [14], Przeworska-Rollewicz [15] (see also the papers cited in these literatures).

2. Normal solvability (nonlinear case)

The concept of normal solvability for nonlinear operators has been introduced and intensively studied by Pochoshajew [16—19].

Let X, Y be Banach spaces, Y uniformly convex, $F: X \to Y$ a Gâteaux differentiable mapping on X. Consider the equation (1)

$$F(u) = v\,. \tag{1}$$

Following Pochoshajew F (or the equation (1)) is said to be normally solvable if the following conditions are satisfied: (a) for each $v \in Y$ there exists a sequence $(v_n) \in Y$ so that $v_n \to v$ and for each $v_n \in Y$ there exists a point $u_n \in X$ such that $\|F(u) - v_n\|$ assumes its minimum value on X at u_n; (ii) for each (v_n), where (v_n) has the property described in (i), and $F(u_n) - v_n \in (N[F'(u_n)]^*)^{\perp}$, then $v \in F(X)$.

Pochoshajew's theorem asserts that if X, Y are Banach spaces, Y is uniformly convex and $F: X \to Y$ is Gâteaux differentiable on X, then F is normally solvable $\Leftrightarrow F(X)$ is closed.

One can sum up some other Pochoshajew's results as follows: Let F be a Gâteaux differentiable mapping on a Banach space X and assume that one of the following three conditions is fulfilled:

(i) Y is reflexive, $F(X)$ is weakly closed and $N(F'[(u)]^*) = 0$ for each $u \in X$.

(ii) Y is uniformly convex Banach space, $F(X)$ is closed and $N([F'(u)]^*) = 0$ for each $u \in X$.

(iii) $D(F)$ is a linear set in X, Y is uniformly convex Banach space, $F: D(F) \to Y$, $F(D(F))$ is closed and $\overline{F'(u)\,(D(F))} = Y$ for each $u \in D(F)$.

Then the image of F is all of Y.

Several surjectivity theorems for Fréchet differentiable weakly continuous mapping having weakly closed range and satisfying the condition $||[F'(u)]^*v|| \geqq k(u)\,||v||$, where $k(u) > 0$ for each $u \in X$ and $v \in Y^*$ have been established by Katschurowski [20], see also [21]. The assertion (iii) of Pochoshajew has been extended for arbitrary Banach spaces by Sabreiko and Krasnosjelski [22], while Browder [23—26] has considerably sharpened, generalized and extended the results of Pochoshajew and Katschurowski by use of different methods.

Using the results of Wulbert [27], Swaminathan [28] extends the Pochoshajev's theorem for range space having the property (E) instead of uniformly convex spaces. Recall that a Banach space X has property (E) (see Edelstein [29]) if for a given closed se $S \subset X$ the set of all points $u \in X$ which a nearest point in S (i.e. such that $||u - s|| = \inf\{||u - v||: v \in S\}$ is dense in X).

In [2] we have derived some surjectivity, existence and fixed point theorems for nonlinear maps. Some of these results are related to those of Pochoshajew, but in comparison with his method the concept of differentiability of map is not used. Our proof method is based on local approximation of a given map by a suitable operator, upon the variational properties of some functionals and the results of Edelstein [29] and Wulbert [27].

3. Solvability of nonlinear equations

Let X, Y be real normed linear spaces (NLS), X^*, Y^* their adjoint spaces. We use the symbols "$\to$", "$\xrightarrow{\omega}$" to denote the strong and weak convergences, respectively. Recall that a point $p \in X$ is said to be an internal point of a subset $W \subset X$ if for each $u \in X$ there exists $\varepsilon > 0$ so that $|\delta| \leqq \varepsilon \Rightarrow p + \delta u \in W$. The set of all internal points of W is denoted by $\mathrm{Int}_a\, W$. By $B_R(0)$ we denote the open ball centred about 0 and with the radius $R > 0$. For $V \subset X$, ∂V is the boundary of V and $\overline{V}$ denotes its closure in X. Identity map of X is denoted by id. A mapping $F: X \to Y$ is said to be: (a) p-positively homogeneous if $F(tu) = t^pF(u)$ for each $u \in X$ and $t \geqq 0$; (b) weakly continuous at $u_0 \in X$ if $u_n \xrightarrow{\omega} u_0 \Rightarrow F(u_n) \xrightarrow{\omega} F(u_0)$: (c) demicontinuous at u_0, if $u_n \to u_0 \Rightarrow F(u_n) \xrightarrow{\omega} F(u_0)$. Recall that a functional f is said to be weakly lower-semicontinuous at $u_0 \in X$ if $u \xrightarrow{\omega} u_0 \Rightarrow f(u_0) \leqq \lim_{n\to\infty} f(u_n)$.

Theorem 4. *Let X, Y be NLS, $F: X \to Y$ a given mapping. Suppose that for each $u_0 \in X$ (fixed) there exist mappings G_{u_0}: $X \to Y$ and D_{u_0} of X onto Y so that*

$$||F(v) - F(u_0) - G_{u_0}(v - u_0)|| \leqq \alpha\, ||D_{u_0}(v - u_0)||$$

for each $v \in X$ and some $\alpha > 0$. Assume further that $||D_u(z) - G_u(z)|| \leqq \varepsilon\, ||D_u(z)||$ for each $u \in X$, $z \in X$ and some ε, $0 \leqq \varepsilon < 1 - \alpha$. If either (i) *$X$ is reflexive,*

F is weakly continuous on X and $||F(u)|| \to +\infty$ *as* $||u|| \to +\infty$*: or* (ii) *X is complete, F is continuous and* $||F(u) - F(v)|| \geqq \psi(||u - v||)$, $u, v \in X$ *where* $\psi\colon R^+ \to R^+$ *is so that* $\psi(\varrho) \to 0_+ \Rightarrow \varrho \to 0_+$ *then* $F(X) = Y$.

Proof. Let $v_0 \in Y$ be arbitrary (but fixed). Set $Q(u) = F(u) - v_0$, $u \in X$. Let us put $q = \inf\{||Q(u)||\colon u \in X\}$; we prove that $q = 0$. Assume that $q > 0$ and let ε_0 such that $\varepsilon_0 < \frac{q - (\alpha + \varepsilon)\,q}{\alpha + \varepsilon}$, where α, ε are from hypotheses of our theorem, $\alpha + \varepsilon < 1$. Then there exists a point $u_0 \in X$ so that $q \leqq ||Q(u_0)|| < q + \varepsilon_0$. For the given element u_0 there are the mappings $G_{u_0}\colon X \to Y$, D_{u_0} from X onto Y so that

$$||F(v) - F(u_0) - G_{u_0}(v - u_0)|| \leqq \alpha\, ||D_{u_0}(v - u_0)||\,; \tag{a}$$

$$||D_{u_0}(v) - G_{u_0}(v)|| \leqq \varepsilon\, ||D_{u_0}(v)|| \tag{b}$$

for each $v \in X$. Since D_{u_0} is onto, there exists $\tilde{u} \in X$ so that $D_{u_0}(\tilde{u}) = -Q(u_0)$. Setting in (a), (b) instead v the element $u_0 + \tilde{u}$, we obtain

$$\begin{aligned} ||F(u_0 + \tilde{u}) - v_0|| &\leqq ||F(u_0 + \tilde{u}) - F(u_0) - G_{u_0}(\tilde{u})|| \\ &\quad + ||F(u_0) - v_0 + D_{u_0}(\tilde{u})|| + ||G_{u_0}(\tilde{u}) - D_{u_0}(\tilde{u})|| \\ &\leqq \alpha\, ||D_{u_0}(\tilde{u})|| + \varepsilon\, ||D_{u_0}(\tilde{u})|| = (\alpha + \varepsilon)\, ||D_{u_0}(\tilde{u})|| \\ &= (\alpha + \varepsilon)\, ||Q(u_0)|| < (\alpha + \varepsilon)\,(q + \varepsilon_0) < q\,, \end{aligned}$$

a contradiction that $q = \inf\{||Q(u)||\colon u \in X\}$. Hence $q = 0$. Therefore there exists $(u_n) \in X$ so that $\lim\limits_{n \to \infty} Q(u_n) = 0$ as $n \to \infty$. Assume (i) we have that $||Q(u)|| \to +\infty$ as $||u|| \to +\infty$, and that Q is weakly continuous. Then (u_n) is bounded and hence in view of reflexivity of X, one can assume that $u_n \xrightarrow{\omega} u_0 \in X$. By weak continuity of Q, $Q(u_n) \xrightarrow{\omega} Q(u_0)$. Therefore $Q(u_0) = 0$ and $F(u_0) = v_0$. Assuming (ii), we infer that $||Q(u) - Q(v)|| \geqq \psi(||u - v||)$; $u, v \in X$. This inequality implies that (u_n) is a Cauchy sequence and X being complete, $u_n \to w_0 \in X$. Continuity of Q gives that $Q(u_n) \to Q(w_0)$ and hence we get again that $Q(w_0) = 0$, i.e. $F(w_0) = v_0$. This proves our theorem.

Setting in Theorem 4 $D_u = G_u$ for each $z \in X$, $\varepsilon = 0$, we obtain the following

Corollary 2. *Let X, Y be* NLS, $F\colon X \to Y$ *a given mapping. Suppose that for each (fixed)* $u_0 \in X$ *there exists a mapping* G_{u_0} *of X onto Y so that* $||F(u) - F(u_0) - G_{u_0}(u - u_0)|| \leqq \alpha\, ||G_{u_0}(u - u_0)||$ *for each* $u \in X$ *and some* $\alpha \in [0, 1)$. *If either* (i) *X is reflexive, F is weakly continuous and* $||F(u)|| \to +\infty$ *as* $||u|| \to +\infty$, *or* (ii) *X is complete, F is continuous and* $||F(u) - F(v)|| \geqq \psi(||u - v||)$, $u, v \in X$ *where* $\psi\colon R^+ \to R^+$ *is so that* $\psi(\varrho) \to 0_+ \Rightarrow \varrho \to 0_+$, *then* $F(X) = Y$.

If $F\colon X \to Y$ is a Fréchet-differentiable mapping on X, $G_{u_0} = F'(u_0)$, $u_0 \in X$ ($F'(u_0)$ denotes the Fréchet derivative of F at u_0) and $||w(u_0, u - u_0)|| \leqq \frac{\alpha}{2} \times ||F(u) - F(u_0)||$ for each $u_0 \in X$ and all $u \in X$ ($w(u_0, u - u_0)$ is the remainder of $F'(u_0)$), then

$$||F(u) - F(u_0) - F'(u_0)(u - u_0)|| \leqq \alpha\, ||F'(u_0)\,(u - u_0)||$$

for each fixed $u_0 \in X$ and all $u \in X$.

Indeed,

$$\|F(u) - F(u_0) - F'(u_0)(u - u_0)\| = \|\omega(u_0, u - u_0)\|$$
$$\leqq \alpha\left(\|F(u) - F(u_0)\| - \|\omega(u_0, u - u_0)\|\right)$$
$$\leqq \alpha\|F'(u_0)(u - u_0)\| .$$

If we put in corollary 2 $F = \Phi + A$, $G_u = A$ we get

Corollary 3. *Let X, Y be* NLS, *$\Phi: X \to Y$, A a linear continuous mapping of X onto Y so that $\|\Phi(u) - \Phi(v)\| \leqq \alpha \|A(u - v)\|$ for each $u, v \in X$ and some $\alpha \in [0, 1)$. If either* (i) *X is reflexive, Φ is weakly continuous and $\|F(u)\| \to +\infty$ as $\|u\| \to +\infty$, or* (ii) *X is complete, Φ is continuous and $u, v \in X \Rightarrow \|F(u) - F(v)\| \geqq \psi(\|u - v\|)$ for some $\psi: R^+ \to R^+$ so that $\psi(\varrho) \to 0_+ \Rightarrow \varrho \to 0_+$, then $F(X) = Y$, where $F = \Phi + A$.*

These results are related to that of our paper [2], the conditions of thm. 4 and corollaries 2, 3 are different. We put stronger assumptions on a given mapping F (here we do not employ its local properties and the special properties of the mapping D_u) but we give less restrictive requirements on the given space Y (i.e. reflexivity or (E) property of Y). Let us remark that assumption (i) of thm. 4 implies that $F(X)$ is weakly closed. Indeed, if $(v_n) \in F(X)$, $v_n \xrightarrow{\omega} v$, then $F(u_n) = v_n$, $u_n \in X$ and (u_n) is bounded. Since X is reflexive, one may assume that $u_n \xrightarrow{\omega} u_0 \in X$. As F is weakly continuous, $F(u_n) \xrightarrow{\omega} F(u_0)$ and hence $v = F(u_0) \in F(X)$. Of course, the assumptions (ii) of thm. 4 guarantee that $F(X)$ is closed.

In next we shall use a slightly generalized lemma of Wainberg [30, Lemma 9.2].

Lemma 1. *Let X be a real* NLS, *f a functional defined on the ball $B_{R+r}(0)$, $(r > 0)$. Assume that f has the Gâteaux derivative $f'(u)$ on $\partial B_R(0)$ and that $\langle f'(u), u\rangle > 0$ for each $u \in \partial B_R(0)$. Then all points u_0 for which $f(u_0) = \inf \times \{f(u): u \in \overline{B_R(0)}\}$ must lie in $B_R(0)$.*

Proof. (See [30, Chapt. III.]) Suppose that $f(u_0) = \inf\{f(u): u \in \overline{B_R(0)}\}$ for some $u_0 \in \partial B_R(0)$. For $0 \leqq t \leqq 1$ we set $\varphi(t) = f(u_0 + t(-u_0))$. Then $\varphi(t) \geqq \varphi(0)$ for sufficiently small $t > 0$ and therefore $\varphi'_+(0) \geqq 0$. Since

$$\varphi'_+(0) = \lim_{t\to 0_+} \frac{1}{t}\left[f(u_0 + t(-u_0)) - f(u_0)\right] = \langle f'(u_0), -u_0\rangle = -\langle f'(u_0), u_0\rangle < 0 ,$$

a contradiction.

Theorem 5. *Let X be a real Hilbert space, $F: B_{R+r}(0) \to X$ (for some $r > 0$) a mapping which is weakly continuous on $\overline{B_R(0)}$, Gâteaux differentiable on $\partial B_R(0)$ and such that $F(u) \neq 0$ for each $u \in \partial B_R(0)$. Assume that for each fixed point $u_0 \in B_R(0)$ there exist a p_{u_0}-positively homogeneous mapping G_{u_0} of X onto Y, a set $V(u_0)$ and a constant $\alpha_{u_0} \in [0, 1)$ so that $u_0 \in \mathrm{Int}_a V(u_0)$, $V(u_0) \subset B_R(0)$ and that*

$$\|F(u) - F(u_0) - G_{u_0}(u - u_0\| \leqq \alpha_{u_0} \|G_{u_0}(u - u_0)\|$$

holds for each $u \in V(u_0)$. If $\langle [F'(u)]^* F(u), u\rangle > 0$ for each $u \in \partial B_R(0)$, then there is a point $u^* \in B_R(0)$ so that $F(u^*) = 0$.

Proof. Put $f(u) = ||F(u)||$ for $u \in \overline{B_R(0)}$. Then $f'(u) = (||F(u)||)' \circ F'(u)$ for each $u \in \partial B_R(0)$. Hence (here f' denotes the Gâteaux derivative of f)

$$\langle f'(u), u\rangle = \left\langle \frac{F(u)}{||F(u)||}, F'(u)\, u\right\rangle = \frac{1}{||F(u)||} \langle [F'(u)]^* F(u), u\rangle > 0$$

for each $u \in \partial B_R(0)$ by our hypothesis. Furthermore, f is weakly lower-semicontinuous on weakly compact and weakly closed set $\overline{B_R(0)}$. Therefore there exists $u^* \in \overline{B_R(0)}$ so that $f(u^*) = \inf\{f(u): u \in \overline{B_R(0)}\}$. By lemma 1 $u^* \in B_R(0)$. We prove that $F(u^*) = 0$. Assume on contrary that $F(u^*) \neq 0$. By our hypothesis for a given element $u^* \in B_R(0)$ there exists a p-positively homogeneous mapping G_{u^*} of X onto Y, a set $V(u^*)$ and a constant $\alpha_{u^*} \in [0, 1)$ so that $u^* \in \mathrm{Int}_a\, V(u^*)$, $V(u^*) \subset B_R(0)$ and that

$$||F(u) - F(u^*) - G_{u^*}(u - u^*)|| \leqq \alpha_{u^*} ||G_{u^*}(u - u^*)|| \,.$$

Since $G_{u^*}: X \to Y$ is onto, there exists a point $\tilde{u} \in X$ so that $G_{u^*}(\tilde{u}) = -F(u^*)$. Then $u^* + t_0 \tilde{u} \in V(u^*)$ for sufficiently small t_0, $(0 < t_0 < 1)$ and

$$||F(u^* + t_0\tilde{u}) - F(u^*) - G_{u^*}(t_0\tilde{u})|| \leqq \alpha_{u^*} ||G_{u^*}(t_0\tilde{u})|| = \alpha_{u^*} t_0^{p_{u^*}} ||G_{u^*}(\tilde{u})|| \,,$$

$$||F(u^*) + G_{u^*}(t_0\tilde{u})|| = ||-G_{u^*}(\tilde{u}) + G_{u^*}(t_0\tilde{u})||$$
$$= ||G_{u^*}(\tilde{u}) - G_{u^*}(t_0\tilde{u})|| = (1 - t_0^{p_{u^*}})\, ||G_{u^*}(\tilde{u})|| \,.$$

Employing these inequalities, we obtain that

$$||F(u^* + t_0\tilde{u})|| \leqq ||F(u^* + t_0\tilde{u}) - F(u^*) - G_{u^*}(t_0\tilde{u})|| + ||F(u^*) + G_{u^*}(t_0\tilde{u})||$$
$$\leqq [\alpha_{u^*} t_0^{p_{u^*}} + (1 - t_0^{p_{u^*}})]\, ||G_{u^*}(\tilde{u})|| < ||G_{u^*}(\tilde{u})|| = ||F(u^*)|| \,,$$

a contradiction that $||F(u^*)|| = \inf\{||F(u)||: u \in \overline{B_R(0)}\}$. Hence $F(u^*) = 0$, where $u^* \in B_R(0)$. theorem is proved.

The assumption of theorem 5 (see also corollaries 4, 5) one can replace by the following assumptions: F is demicontinuous on $\overline{B_R(0)}$ and $f(u) = ||F(u)||$ is quasiconvex on $\overline{B_R(0)}$ (i.e. $f(\lambda\mu + (1 - \lambda)\, v) \leqq \max\,(f(u), f(v))$ for each $u, v \in \overline{B_R(0)}$ and $\lambda \in [0, 1]$).

Using the Hess [31] variational principle, one can obtain an existence theorem for mapping $F(u) = f'(u)$, where f is a continuously differentiable functional and $F: X \to X^*$ satisfies the following (P)-condition [31]: if $(u_n) \in X$ X is a NLS, and $u_n \xrightarrow{\omega} u \in X$, then $\lim\limits_{n\to\infty} \langle F(u_n), u_n - u\rangle \geqq 0$.

Now we derive some conclusions of Theorem 5. First of all we recall here the concept of A-feeble local contraction in slightly more general form than in [2], [3]. This notion generalizes and extends simultaneously the concept of local contraction of Edelstein [4] and generalized contraction of Kirk [5], [6], [34].

Definition 3. Let (X, ϱ), (Y, σ) be linear metric spaces, $\emptyset \neq \dot{M} \subset X$ and open subset of X, $A: X \to Y$ a mapping. A mapping $\Phi: M \to Y$ is said to be an A-feeble local contraction on M [an A-local contraction on M] if for each point $u_0 \in M$ there exist a set $V(u_0)$ a constant α_{u_0} $(0 \leqq \alpha_{u_0} < 1)$ so that $u_0 \in \mathrm{Int}_a\, V(u_0)$, $V(u_0) \subset M$ and $\varrho(\Phi(u), \Phi(u_0)) \leqq \alpha_{u_0} \sigma(0, A(u - u_0))$ holds for each $v \in V(u_0)$ [$\varrho(\Phi(u), \Phi(v)) \leqq \alpha_{u_0} \sigma(0, A(u - w))$ for each $v, w \in V(u_0)$].

In case $X = Y$, $A = \mathrm{id}$, Φ is called a feeble local contraction on M [a local contraction on M].

Corollary 4. *Let X be a real Hilbert space, Φ: $B_{R+r}(0) \to X$, $(r > 0)$ a mapping which is weakly continuous on $\overline{B_R(0)}$ and Gâteaux differentiable $\partial B_R(0)$. Suppose Φ is an A-feeble local contraction on $B_R(0)$, where A is linear continuous operator from X onto X. If $\langle [F'(u)]^* F(u), u \rangle > 0$ and $F(u) \neq 0$ for each $u \in \partial B_R(0)$, where $F = \Phi + A$, then there is $u^* \in B_R(0)$ so that $F(u^*) = 0$.*

Setting $A = -\,\mathrm{id}$, we obtain the following fixed-point theorem.

Corollary 5. *Let X be a Hilbert space, Φ: $B_{R+r}(0) \to X$, $(r > 0)$ a mapping which is weakly continuous on $\overline{B_R(0)}$ and Gâteuax differentiable on $\partial B_R(0)$. Assume Φ is a feeble local contraction on $B_R(0)$. If $\langle [G'(u)]^* G(u), u \rangle > 0$ and $G(u) \neq 0$ for each $u \in \partial B_R(0)$, where $G = \Phi - \mathrm{id}$, then there is a point $u^* \in B_R(0)$ so that $\Phi(u^*) = u^*$.*

Remarks. Fixed-point theorems and surjectivity theorems for A-feeble local contractions in normed linear spaces under different conditions have been established in our papers [2], [32], [33]. The result of Edelstein [4] says: If X is a complete ε-chainable metric space, f: $X \to X$ an (ε, λ)-uniformly locally contractive mapping, then there is a unique point $u^* \in X$ so that $f(u^*) = u^*$. Fixed point theorems for generalized contractive mappings have been developed by Kirk [5], [6], [34].

Let us remark that the classical Banach's contraction principle has been further extended by many authors, see for instance Rakotsch [35], Nadler, Jr. [36], Wong [37], Wong [38] (cf. also the references quoted in these papers). Some fixed point theorems based on variational arguments have been derived by Belluce-Kirk [39], Kirk [40], Montagnana-Vignoli [41] and in our papers [2], [32]. The concept of quasi-convex functionals in extrema problems has been used by Poljak [42], [43]. This concept was introduced and studied in fixed-point theory and solvability of nonlinear equations independently by Montagnana-Vignoli [41] and in our papers [32], [33], [2].

References

[1] Katětov, M., O normovaných vektorových prostorach. Über normierte Vektorräume (Zusammenfassung). Mitteilungen der Tsch. Akad. der Wiss. 53, Nummer 45 (1943) 1–30.

[2] Kolomý, J., Normal solvability, solvability and fixed-point theorems Colloq. Math.

[3] —, Invariance of domain theorem for weakly continuous mappings. Math. Nachr.

[4] Edelstein, M., An extension of Banach's contraction principle. Proc. Am. Math. Soc. 12 (1961) 7–10.

[5] Kirk, W. A., Mappings of generalized constractive type. Journ. Math. Anal. Appl. 32 (1970) 567–572.

[6] —, On nonlinear mappings of strongly semicontractive type. Journ. Math. Anal. Appl. 27 (1969) 409–412.

[7] Taylor, A. E., Introduction to functional analysis. John Wiley & Sons, Inc., New York (1958).

[8] Krein, S. G., Линейные уравнения в банаховом пространстве. Москва (1971).

[9] Goldberg, S., Unbounded linear operators. MacGraw-Hill Co., New York (1966).

[10] Hausdorff, F., Zur Theorie der linearen metrischen Räume. J. Reine Angew. Math. **167** (1932) 294–311.

[11] Yosida, K., Functional analysis. Springer-Verlag, Berlin (1965). [Russian translation, Moscow 1967].

[12] Gochberg, I. Z., u. Krein, M. G., Основные положения о дефектных числах, корневых числах и индексах линейных операторов. УМН **12**, 2 [74] (1957) 43—118.

[13] Kato, T., Perturbation theory of linear operators. Springer-Verlag, Berlin (1966).

[14] Nashed, M. Z., Generalized inverses, normal solvability and iteration for singular operator equations. Nonlinear functional analysis and applications. Edited by Louis B. Rall Academic Press, New York, London (1971).

[15] Przeworska-Rollewicz, D., and Rollewicz, S., Equations in linear spaces. PWN Warszawa (1968).

[16] Pochoshajew, S. I., О нормальной разрешимости нелинейных уравнений. ДАН СССР **184**, 1 (1969) 40—43.

[17] —, О нелинейных операторах, имеющих слабо замкнутую область значений, и квазилинейных эллиптических уравнениях. Матем. сб. **78**, 2 (1969) 237—259.

[18] —, Нормальная разрешимость нелинейных уравнений в равномерно выпуклых банаховых пространствах. Функц. анализ и его прилож. **3**, 2 (1969) 80—84.

[19] —, О слабо нелинейных гиперболических системах. Труды Моск. матем. о-ва **23** (1970) 61—76.

[20] Katschurowski, R. I., О разрешимости некоторых классов нелинейных уравнений. ДАН СССР **189**, 1 (1969) 35—37.

[21] —, Обобщение теорем Фредгольма и теорем о линейных операторах с замкнутой областью значений на некоторые классы нелинейных операторов. ДАН СССР **197** (1971) 520—523.

[22] Sabroiko, P. P., u. Krasnosjelski, M. A., О разрешимости нелинейных операторных уравнений. Функц. анализ и его прилож. **5** (1971) 42—44.

[23] Browder, F. E., On the Fredholm alternative for nonlinear operators. Bull. Amer. Math. Soc. **76** (1970) 993–998.

[24] –, Normal solvability for nonlinear mappings into Banach spaces. Bull. Amer. Math. Soc. **77** (1971) 73–77.

[25] –, Normal solvability and the Fredholm alternative for mappings into infinite dimensional manifolds. Journ. Funct. Anal. **8** (1971) 250–274.

[26] –, Normal solvability and Φ-accretive mappings of Banach spaces. Bull. Amer. Math. Soc. **78** (1972) 186–192.

[27] Wulbert, D. E., Uniqueness and differential characterization from manifolds of functions. Amer. Journ. of Math. XCIII (1971) 350–366.

[28] Swaminanthan, S., On the closed range theorem for nonlinear operators. Proc. of the Third Prague Topol. Sympos. (1971), Academia Prague (1972), 417–418.

[29] Edelstein, M., On nearest points of sets in uniformly convex Banach spaces. Journ. London Math. Soc. **43** (1968) 375–377.

[30] Wainberg, M. M., Вариационный метод монотонных операторов в теории нелинейных уравнений. Москва 1972.

[31] Hess, P., A variational approach to a class of nonlinear eigenvalue problem. Proc. Amer. Math. Soc. **29** (1971) 272–271.

[32] Kolomý, J., Some mapping and fixed-point theorems. Proc. of the Third Prague Topol. Sympos. (1971), Academia Prague (1972), 245–248.

[33] –, Some mapping theorems and solvability of nonlinear equations in Banach space. Comment. Math. Univ. Carolinae **12** (1971) 413–417.

[34] Kirk, W. A., Fixed point theorems for nonlinear nonexpansive and generalized contraction mappings, Pac. Journ. Math. **38** (1971) 89–94.
[35] Rakotsch, E., A note on contractive mappings. Proc. Amer. Math. Soc. **13** (1962) 459–465.
[36] Nadler, S. B., Jr., Multivalued contraction mappings. Pac. Journ. Math. **30** (1969) 475–488.
[37] Wong, Ch. S., Fixed point theorems for nonexpansive mappings. Journ. Math. Anal. Appl. **37** (1972) 142–156.
[38] Wong, James S. W., Mappings of contractive type on abstract spaces. Journ. Math. Anal. Appl. **37** (1972) 331–340.
[39] Belluce, L. P., and Kirk, W. A., Some fixed point theorems in metric and Banach spaces. Canad. Math. Bull. **12** (1969) 481–491.
[40] Kirk, W. A., Fixed point theorems for nonexpansive mappings. Nonlinear functional analysis, editor F. E. Browder, Proc. of Symp. in Pure Math. Vol. 18, part. 1. Amer. Math. Soc., Providence, Rhode Island, (1970) 162–168.
[41] Montagnana, M., and Vignoli, A., On quasiconvex mappings and fixed point theorems. Boll. Unione Mat. Ital. **4** (1971) 870–878.
[42] Poljak, B. T., Теоремы существования и сходимость минимизирующих последовательностей для задач на экстремум при наличии ограничений. ДАН СССР 166 (1966) 287—290.
[43] —, Один общий метод решения экстремальных задач. ДАН СССР **174** (1967) 33—36.

Anschrift des Autors:

Mathematical Institute of Charles University
Prague 8-Karlín
Sokolovská 83

Zur Steuerung mit Operatorgleichungen

ECKEHARD KRAUSS, Berlin

B_1 und B_2 seien Banachräume, B_2 reflexiv, $C \subseteq B_2$ nicht leer, konvex und abgeschlossen und F ein endliches reellwertiges Funktional auf $B_1 \times B_2$.

Wir betrachten das folgende Minimumproblem: Gesucht wird eine Lösung von

$$f(u_0) = \min_{u \in C \subseteq B_2} f(u) \tag{0}$$

mit $f(u) = F(x, u)$, wobei u und x durch die den Prozeß beschreibende Operatorgleichung

$$T(x, u) = \theta\,, \qquad T \in (B_1 \times B_2 \to B_1^*)\,, \tag{I}$$

miteinander verbunden sind.

Die Schreibweise $f(u)$ unterstreicht, daß der „Steuerparameter" u die unabhängige Variable ist, während „die Phasenkoordinate" x als Lösung von (I) gefunden wird. Wir schreiben in diesem Fall auch $x[u]$. u_0 nennen wir optimalen „Steuerparameter".

Neben (I) betrachten wir das Fixpunktproblem

$$x[u] = A(x[u], u)\,, \qquad A \in (B_1 \times B_2 \to B_1)\,. \tag{II}$$

Durch (0), (I) werden bei geeigneten Monotonievoraussetzungen an $T(\cdot, u)$ u. a. Probleme der Steuerung mit elliptischen partiellen Differentialgleichungen und durch (0), (II) bei Kontraktionsvoraussetzungen an $A(\cdot, u)$ u. a. Probleme der Steuerung mit Integralgleichungen, mit gewöhnlichen Differentialgleichungen in B_1 und mit dem Goursat-Problem erfaßt.

Für Hilberträume B_1 lassen sich die Probleme (I) und (II) ineinander überführen. Damit sind hier getroffene Aussagen bezüglich (0), (II) unter diesen Voraussetzungen auch auf (0), (I) anwendbar.

In [3] hat R. Kluge Näherungsverfahren für die Steuerung mit linearen gewöhnlichen Differentialgleichungen im Hilbertraum und damit u. a. mit gemischten Problemen bei pseudo-parabolischen partiellen Differentialgleichungen angegeben. Dabei stützt er sich auf in dem Übersichtsartikel [2] angeführte Näherungsverfahren für nichtlineare Variationsungleichungen und Minimumprobleme mit Nebenbedingungen. Wir wollen an dieser Stelle die Probleme (0), (I) und (0), (II) auf Existenz von optimalen „Steuerparametern" untersuchen und Näherungsverfahren zu ihrer approximativen Bestimmung angeben. Dabei benutzen wir die Idee der Linearisierung von (II) durch Iterations- und Projektions-Iterationsverfahren (vgl. [2]), auf deren Basis „diagonalisierte"

Näherungsverfahren angegeben werden können, die nicht den Gradienten des Ausgangsproblems (0), (II), sondern einfachere Ausdrücke enthalten. Die Beweise der hier angeführten Aussagen finden sich in [1].

1. Zur Existenz von optimalen „Steuerparametern"

Für den Existenzbeweis benutzen wir die Sätze von Weierstraß-Poljak [4]. Zunächst jedoch wird die schwache (Folgen-) Unterhalbstetigkeit von f untersucht. Dabei bezeichne $\xrightarrow{B_i}$ die starke und $\rightharpoonup^{B_i}$ die schwache Konvergenz in B_i.

Definitionen: Eine Abbildung $S \in (B_1 \to B_1^*)$ heißt monoton, wenn $(Sx - Sy, x - y)_1 \geqq 0$ für $x, y \in B_1$ ($(w, z)_1$ bezeichnet dabei den Wert des Funktionals $w \in B_1^*$ auf $z \in B_1$). Sie heißt gleichmäßig monoton, wenn eine stetige, streng monoton wachsende Funktion $\delta(r)$: $[0, \infty) \to [0, \infty$, mit $\delta(0) = 0$ derart existiert, daß

$$(Sx - Sy, x - y)_1 \geqq \delta(||x - y||)\, ||x - y|| \quad \text{für} \quad x, y \in B_1 \,.$$

S heißt hemistetig, wenn $d(t) \equiv (S(x + ty), z)_1$ für alle $x, y, z \in B_1$ als Funktion von $t \in [0, 1]$ stetig ist. Eine Abbildung heißt verstärkt stetig, wenn sie schwach konvergente in stark konvergente Folgen überführt.

Satz 1. *Voraussetzungen:*

1. *B_1 sei reflexiv.*
2. *$T(\cdot, u)$ sei für jedes $u \in B_2$ hemistetig und gleichmäßig monoton mit der Funktion $\delta_u(r)$, wobei $\delta_u(r) \geqq \delta_M(r)$ für $u \in M$ bei beliebiger beschränkter Menge $M \subseteq B_2$ mit der stetigen, streng monoton wachsenden Funktion $\delta_M(r)$: $[0, \infty) \to [0, \infty)$, $\delta_M(0) = 0$.*
3. *$T(x, \cdot)$ sei für $x \in B_1$ verstärkt stetig.*
4. *Mit $x_n \xrightarrow{B_1} x$ und $u_n \rightharpoonup^{B_2} u$ sei $\varliminf_{n\to\infty} F(x_n, u_n) \geqq F(x, u)$.*

Behauptung: f in (0), (I) ist schwach unterhalbstetig.

Beweis. Wegen Voraussetzung 4 reicht es zu zeigen, daß die Lösung $x[u]$ von (I) verstärkt stetig von u abhängt (Unter den Voraussetzungen des Satzes existiert für jedes $u \in B_2$ genau eine Lösung $x[u]$ von (I).). Nach Voraussetzung 3 gilt mit $u_n \rightharpoonup^{B_2} u$

$$T(x[u], u_n) \to T(x[u], u) = \theta = T(x[u_n], u_n)\,,$$

also

$$\begin{aligned}&(T(x[u], u) - T(x[u], u_n), x[u_n] - x[u])_1\\ &= (T(x[u_n], u_n) - T(x[u], u_n), x[u_n] - x[u])_1\\ &\geqq \delta_{u_n}(||x[u_n] - x[u]||)\, ||x[u_n] - x[u]||\\ &\geqq \delta_M(||x[u_n] - x[u]||)\, ||x[u_n] - x[u]||\end{aligned}$$

oder

$$||x[u_n] - x[u]|| \leqq \delta_M^{-1}(||T(x[u], u) - T(x[u], u_n)||) \to 0$$

für $n \to \infty$, q. e. d.

Im folgenden wird Voraussetzung 3 von Satz 1 abgeschwächt.

Satz 2. *Voraussetzungen:*

1. *Es seien die Bedingungen* 1 *und* 2 von *Satz* 1 *erfüllt.*
2. $T(x, \cdot)$ *sei für* $x \in B_1$ *stetig.*
3. *Für* $x_n \xrightarrow{B_1} x$ *und* $u_n \xrightarrow{B_2} u$ *sei* $\underline{\lim}_{n\to\infty} F(x_n, u_n) \geqq F(x, u)$.
4. f *sei quasikonvex.*

Behauptung: f *in* (0), (I) *ist schwach unterhalbstetig.*

Beweis. vgl. [1].

Bemerkungen. Satz 2 hat für uns gegenüber Satz 1 den Nachteil, daß wir Näherungsverfahren auf seiner Grundlage nur für lineare Steuerungsaufgaben begründen können. Die Voraussetzung 4 in Satz 2 ist im Fall eines linearen Operators T erfüllt, wenn man $F(\lambda x_1 + (1 - \lambda) x_2, \lambda u_1 + (1 - \lambda) u_2) \leqq \max \{F(x_1, u_1), F(x_2, u_2)\}$, $\lambda \in [0, 1]$, fordert. Für nichtlineare T vgl. [1].

Satz 3. *In den Voraussetzungen 2—4 von Satz 2 (oder den Voraussetzungen 3 und 4 von Satz 1) an die Abbildung* $A \in (B_1 \times B_2 \to B_1)$ *und die Funktionale* F *und* f *existiere eine natürliche Zahl* k *und für jede beschränkte Menge* $M \subseteq B_2$ *eine Zahl* $q(M) < 1$ *derart, daß* $A(\cdot, u)^k$ *Lipschitz-stetig ist mit der Konstanten* $q(u)$ *und* $q(u) \leqq q(M)$ *für* $u \in M$. *Dann ist* f *in der Aufgabe* (0), (II) *schwach unterhalbstetig.*

Beweis, vgl. [1].

Nun gilt

Satz: *Es seien die Voraussetzungen einer der Sätze* 1—3 *erfüllt. Wenn dann eine der folgenden Bedingungen* a)—c)

a) C *ist beschränkt.*
b) $F(x, u_n) \to +\infty$ *gleichmäßig in* x *für* $||u_n|| \to \infty$.
c) f *ist gleichmäßig konvex.*

erfüllt ist, dann besitzt (0), (I) *bzw.* (0), (II) *wenigstens (im Fall* c) *genau) eine Lösung.*

2. Zur Approximation über Folgen von Problemen

Wir gehen nun zur näherungsweisen Lösung von (0), (II) über. Zu diesem Zweck wenden wir auf (II) Iterations- und Projektions-Iterationsverfahren an. Auf diese Weise entstehen Folgen von einfacheren Steuerungsproblemen. Die hier getroffenen Aussagen werden im letzten Punkt der Arbeit zur Begründung konstruktiver Näherungsverfahren für (0), (II) herangezogen. Es gilt

Satz 4. *Voraussetzungen:*

1. $B_1 = H_1$ *sei ein Hilbertraum,* H_1^n *eine Folge linearer Unterräume* H_1^n *von* H_1 *mit* $H_1^n \subseteq H_1^m$ *für* $n \leqq m$ *und* $\overline{\bigcup_n H_1^n} = H_1$ *und* P_n *der Operator der orthogonalen Projektion von* H_1 *auf* H_1^n.

2. $A: H_1 \times B_2 \to H_1$ *genüge* $\|A(x, u) - A(y, u)\| \leqq q(u)\|x - y\|$ *mit* $q(u) \leqq q(M) < 1$ *für jedes* u *aus einer beliebigen beschränkten Menge* $M \subseteq B_2$.
3. $A(x, \cdot)$ *sei für* $x \in H_1$ *verstärkt stetig.*
4. $F(\cdot, u)$ *sei für* $u \in B_2$ *stetig. Weiter folge aus* $x_n \xrightarrow{H_1} x$ *und* $u_n \xrightarrow{H_1} u$

$$\varliminf_{n\to\infty} F(x_n, u_n) \geqq F(x, u)$$

5. *Entweder sei* C *beschränkt, oder es gelte* $F(x, u_n) \to +\infty$ *gleichmäßig in* x *für* $\|u_n\| \to \infty$.

Behauptungen:

1. *Für jedes* n *existiert eine Lösung* u_n *des Minimumproblems*

$$f_n(u) = \min_{v \in C} f_n(v) \tag{1n}$$

mit $f_n(v) = F(x_n[v], v)$;

$$x_n[v] = P_n A(x_{n-1}[v], v)\,, \quad n = 1, 2, \dots, \quad \text{und} \quad x_0[v] \equiv x_0 \in H_1\,.$$

2. *Die Menge der Minimalstellen der Probleme* (1n) *ist beschränkt. Der schwache Grenzwert jeder schwach konvergenten Teilfolge einer Folge* u_n *von Minimalstellen* u_n *der* (1n) *ist eine Lösung von* (0), (II).
3. *Wenn zusätzlich gilt*

$$f_n\left(\frac{v_1 + v_2}{2}\right) \leqq \frac{f_n(v_1) + f_n(v_2)}{2} - c\|v_1 - v_2\|^2 \tag{2}$$

mit $c > 0$, $v_1, v_2 \in B_2$ *und* $n = 1, 2, \dots$ *und* (0), (II) *genau eine Lösung* u_0 *besitzt, dann besitzt* (1n) *für jedes* n *genau eine Lösung* u_n, *und* u_n *konvergiert stark in* B_2 *gegen* u_0.

Weiter gilt

Satz 5. *Voraussetzungen:*

1. $A(\cdot, u)^n$ *sei für jedes* n *Lipschitz-stetig mit der von* u *unabhängigen Lipschitz-Konstanten* q_n. *Dabei gelte* $q_n \to 0$ *für* $n \to \infty$ (B_1 *und* B_2 *sind in den folgenden Sätzen, wenn nicht anders hervorgehoben, Banachräume;* B_2 *reflexiv*).
2. $A(x, \cdot)$ *sei verstärkt stetig und Lipschitz-stetig für alle* $x \in B_1$ *mit einer von* x *unabhängigen Lipschitzkonstanten.*
3. *Es gilt* $|F(x, u) - F(y, u)| \leqq K\|x - y\|$ *für* $x, y \in B_1$, $u \in B_2$. *Außerdem sei* $F(x, \cdot)$ *schwach unterhalbstetig für* $x \in B_1$.
4. *Alle* f_n *in Satz 5 genügen* (2) *und* C *sei beschränkt.*

Behauptung: Für jedes n *besitzt das Minimumproblem*

$$f_n(u) = \min_{v \in C} f_n(v) \tag{2n}$$

mit $f_n(v) = F(x_n[v], v)$, $x_n[v] = A(x_{n-1}[v], v)$, $n = 1, 2, \dots$ *und* $x_0[v] \equiv x_0 \in B_1$ *genau eine Lösung* u_n. *Weiter besitzt das Problem* (0), (II) *genau eine Lösung* u *und* u_n *konvergiert stark in* B_2 *gegen* u.

Der folgende Satz zeigt, daß wir für lineare Steuerungsaufgaben die verstärkte Stetigkeit von A im 2. Argument fallenlassen können:

Satz 6. *Voraussetzungen:*

1. *A sei ein linearer, stetiger Operator auf $B_1 \times B_2$ mit $\|A(x, 0)^n\| \leqq q_n\|x\|$ und $q_n \to 0$ für $n \to \infty$.*
2. *$F(\cdot, u)$ sei oberhalbstetig für $u \in B_2$ und aus $x_n \xrightarrow{B_1} x$ und $u_n \xrightarrow{B_2} u$ folge $\underline{\lim}_{n\to\infty} F(x_n, u_n) \geqq F(x, u)$.*
3. *Sei* a) *C beschränkt*

   oder

   b) *$F(x, u_n) \to +\infty$ für $\|u_n\| \to \infty$ gleichmäßig in x.*

Behauptungen:

1. *Für jedes n existiert eine Lösung von $(2n)$.*
2. *Die Menge der Lösungen u_n der Aufgaben $(2n)$ ist beschränkt.*
3. *Der Grenzwert jeder schwach konvergenten Teilfolge einer Folge u_n von Minimumstellen u_n der Aufgabe $(2n)$ ist eine Lösung von (0), (II).*
4. *Gilt außerdem noch*

$$F\left(\frac{x_1 + x_2}{2}, \frac{u_1 + u_2}{2}\right) \leqq \frac{F(x_1, u_1) + F(x_2, u_2)}{2} - c\|u_1 - u_2\|^2$$

$c > 0$, so konvergieren die eindeutig bestimmten Minimumstellen u_n von $(2n)$ stark gegen die einzige Lösung von (0), (II).

Bemerkung: Der 2. Teil der Voraussetzung 2 ist erfüllt, wenn F den folgenden Bedingungen genügt:

a) $F(\lambda x_1 + (1 - \lambda) x_2, \lambda u_1 + (1 - \lambda) u_2) \leqq \max \{F(x_1, u_1), F(x_2, u_2)\}$ für $x_1, x_2 \in B_1$, $u_1, u_2 \in B_2$, $\lambda \in [0, 1]$.

b) Aus $x_n \xrightarrow{B_1} x$ und $u_n \xrightarrow{B_2} u$ folge $\underline{\lim}_{n\to\infty} F(x_n, u_n) \geqq F(x, u)$.

Satz 7. *Voraussetzungen:*

1. *C sei beschränkt.*
2. *A sei ein linearer stetiger Operator auf $B_1 \times B_2$ mit $\|A(x, 0)^n\| \leqq q_n\|x\|$, wobei $q_n \to 0$ für $n \to \infty$.*
3. *Für $u_n \xrightarrow{B_2} u$ gelte $\underline{\lim}\, F(x, u_n) \geqq F(x, u)$, und es sei $|F(x, u) - F(y, u)| \leqq K\|x - y\|$ für $x, y \in B_1$, $u \in B_2$*
4. *Es ist $F\left(\frac{x_1 + x_2}{2}, \frac{u_1 + u_2}{2}\right) \leqq \frac{F(x_1, u_1) + F(x_2, u_2)}{2} - c\|u_1 - u_2\|^2$ mit $c > 0$.*

Behauptung: Dann konvergieren die eindeutig bestimmten Minimalstellen von $(2n)$ in B_2 stark gegen die einzige Lösung von (0), (II).

3. Gradientenverfahren

In diesem Punkt seien H_1 und H_2 separable Hilberträume, $A \in (H_1 \to H_1)$, $B \in (H_2 \to H_1)$ und $C \subseteq H_2$ beschränkt, konvex und abgeschlossen. Wir betrachten (0), (II) in der speziellen Form

$$(0),\,(\mathrm{III}) \qquad f(u_0) = \min_{u \in C} f(u)$$

mit $f(u) = F(x, u)$ und

$$((\mathrm{III})) \qquad x = Ax + Bu\,.$$

Die folgenden Sätze 8–10 enthalten Aussagen über die Konvergenz gewisser „iterierter" Verfahren gegenüber denen von Punkt 2. Dabei bezeichnet $P(K)$ den Operator der Projektion von H_2 auf die konvexe und abgeschlossene Menge $K \subseteq H_2$.

Satz 8. *Voraussetzungen:*

1. $H_i^j (j = 1, 2)$ *sei eine Folge von Unterräumen von* H_j *mit* $H_i^j \subseteq H_k^j$ *für* $i \leqq k$ *und* $\overline{\bigcup_i H_i^j} = H_j$.

   Weiter sei $C_i = C \cap H_i^2$, $C_1 \neq \emptyset$.

2. *B sei linear und vollstetig und A sei Fréchet-differenzierbar mit*

$$||DA(x, \cdot)||_{(H_1 \to H_1)} \leqq q < 1 \qquad \text{für} \quad x \in H_1$$

und

$$||DA(x, \cdot) - DA(y, \cdot)||_{(H_1 \to H_1)} \leqq a||x - y||_{H_1} \qquad \text{für} \quad x, y \in H_1\,.$$

3. *F sei Fréchet-differenzierbar und*

$$||F_x(x_1, u_1) - F_x(x_2, u_2)||_{(H_1 \to R)} \leqq f_{11}||x_1 - x_2||_{H_1} + f_{12}||u_1 - u_2||_{H_2}$$
$$||F_u(x_1, u_1) - F_u(x_2, u_2)||_{(H_2 \to R)} \leqq f_{21}||x_2 - x_1||_{H_1} + f_{22}||u_1 - u_2||_{H_2}$$
$$||F_x(x, u)||_{(H_1 \to R)} \leqq K_1 \qquad \text{für} \quad x \in H_1,\ u \in H_2$$

4. *Es sei*

$$F\left(x, \frac{u_1 + u_2}{2}\right) \leqq \frac{F(x, u_1) + F(x, u_2)}{2} - c||u_1 - u_2||^2$$

für alle $x \in H_1$; $u_1, u_2 \in H_2$ *mit* $d \underset{\mathrm{Df}}{=} c - L + f_{22} > 0$,

wobei

$$L = \begin{pmatrix} f_{11} & f_{12} \\ f_{21} & f_{22} \\ a & 0 \end{pmatrix} \begin{pmatrix} \dfrac{B}{1-q} \\ 1 \end{pmatrix} \left(\frac{B}{1-q},\ 1,\ \frac{K_1 B}{(1-q)^2}\right)$$

(Matrizenmultiplikation).

Behauptungen:

1. *Das Problem* (0), (III) *besitzt genau eine Lösung* u_0.
2. *Die durch die Iterationsvorschrift*

$$u_{i+1} = P(C_i)\left(u_i - tP_i^2 f_i'(u_i)\right)\,, \qquad u_0 \in H_2\,, \qquad 0 < t < \frac{2d}{L^2}$$

definierte Folge $\{u_i\}$ *konvergiert stark gegen* u_0. *Die in der Iterationsvorschrift auftretenden Größen* $P_i^2 f_i'(u_i) \in H_i^2$ *berechnen sich dabei aus folgendem Rekursionsschema:*

$$(f_i'(u_i), h)_{H_2} = F_x(x_i[u_i], u_i)\, w_i(h) + F_u(x_i[u_i], u_i)\, h \qquad \text{für} \quad h \in H_i^2$$

mit

$$v_i = P_{i-1}^1(Av_{i-1} + Bu_{i-1}) \qquad i = 1, 2, \ldots, \qquad v_0 \in H_1$$

und

$$w_i(h) = P_{i-1}^1 DA(v_{i-1})\, w_{i-1}(h) + P_{i-1}^1 Bu_{i-1} \qquad i = 1, 2, \ldots$$

$$w_0(h) = \theta \in H_1 .$$

Satz 9. *Voraussetzungen:*

1. C_i *wie in Satz* 8.
2. A *und* B *seien linear und stetig mit* $\|A\| < 1$.
3. *Es sei Voraussetzung* 3 *von Satz* 8 *erfüllt.*
4. *Es gilt* $F\left(\frac{x_1 + x_2}{2}, \frac{u_1 + u_2}{2}\right) \leqq \frac{F(x_1, u_1) + F(x_2, u_2)}{2} - c\|u_1 - u_2\|^2$

für $x_1, x_2 \in H_1$ *und* $u_1, u_2 \in H_2$; $c > 0$.

Behauptungen:

1. *Das Problem* (0), (III) *besitzt genau eine Lösung* u_0.
2. *Das Iterationsverfahren*

$$u_{i+1} = P(C_i)\left(u_i - tP_i^2 f_i'(u_i)\right), \qquad i = 0, 1, 2, \ldots, \qquad u_0 \in H_2,$$

mit

$$0 < t < \frac{2c}{L^2}$$

und

$$L = \begin{pmatrix} f_{11} & f_{12} \\ f_{21} & f_{22} \end{pmatrix} \begin{pmatrix} \dfrac{\|B\|}{1 - \|A\|} \\ 1 \end{pmatrix} \begin{pmatrix} \dfrac{\|B\|}{1 - \|A\|} & 1 \end{pmatrix}$$

konvergiert stark gegen u_0. *Die* $P_i^2 f_i'(u_i) \in H_i^2$ *ergeben sich wie folgt:*

$$(f_i'(u_i), h)_{H_2} = F_x(x_i[u_i], u_i)\, x_i'[u_i](h) + F_u(x_i[u_i], u_i)\, h \qquad \text{für} \quad h \in H_i^2$$

mit

$$x_i[u] = Ax_{i-1}[u] + Bu \qquad i = 1, 2, \ldots, \qquad x_0[u] \equiv x_0 \in H_1$$

und

$$x_i'[u](h) = Ax_{i-1}'[u](h) + Bu \qquad i = 1, 2, \ldots; \quad x_0'[u](h) = \theta \in H_1$$

für alle $h \in H_2$.

Der nächste Satz zeigt, daß man unter gewissen Voraussetzungen die Aussagen der Sätze 8 und 9 kombinieren kann:

Satz 10. *Voraussetzungen:*

1. *Es seien die Voraussetzungen* 1 *und* 3 *von Satz* 8 *erfüllt.*

2. *A und B seien linear, A stetig mit* $\|A\| < 1$ *und B vollstetig.*
3. *Wie Voraussetzung 4 in Satz 9.*

Behauptungen:

1. *Das Problem* (0), (III) *besitzt genau eine Lösung* u_0.
2. *Das Iterationsverfahren*

$$u_{i+1} = P(C_i)\left(u_i - tP_i^2 f_i'(u_i)\right), \qquad i = 0, 1, 2, \dots, \qquad u_0 \in H_2,$$

konvergiert für $0 < t < \frac{2c}{L^2}$ (L wie in Satz 9) *stark gegen diese Lösung. Die* $P_i^2 f_i'(u_i) \in H_i^2$ *ergeben sich aus*

$$f_i'\left((u_i), h\right)_{H_2} = F_x\left(x_i[u_i], u_i\right) x_i'[u_i](h) + F_u\left(x_i[u_i], u_i\right) h$$

für alle $h \in H_i^2$ *mit*

$$x_i[u] = P_i^1 A x_{i-1}[u] + P_i^1 Bu, \qquad i = 1, 2, \dots, \qquad x_0[u] \equiv x_0 \in H_1$$

und

$$x_i'[u](h) = P_{i-1}^1 A x_{i-1}'[u](h) + P_{i-1}^1 Bu, \qquad i = 1, 2, \dots$$

$$x_0[u](h) = \theta \in H_1 \quad \text{für alle} \quad h \in H_2.$$

Literatur

[1] Krauß, E., Über ein Steuerungsproblem im Banachraum. Manuskript.

[2] Kluge, R., Näherungsverfahren zur approximativen Lösung nichtlinearer Variationsungleichungen. Math. Nachr. **51** (1971) 343–356.

[3] —, Näherungsverfahren für ein Problem der optimalen Steuerung mit abstrakten Differentialgleichungen erster Ordnung. Math. Nachr. **54** (1972) 229–248.

[4] Poljak, B. T., Теоремы существования и сходимость минимизирующих последовательностей для задач на экстремум при наличии ограничений. ДАН СССР **166** (1966) 287—290.

Anschrift des Autors:

ZIMM der AdW der DDR
108 Berlin, Mohrenstr. 39

Differenzierbare Zweige von Eigenfunktionen im Bifurkationselement

A. LANGENBACH, B. WEGNER und K. WIEDEMANN, Berlin

Im reellen Hilbertraum H mit dem Skalarprodukt $(\cdot, \cdot)$ und der Norm $\|\cdot\|$ betrachten wir die Gleichung

$$Au - \lambda Ku = \Theta \tag{1.1}$$

für $u \in H$ und reelle Parameterwerte $\lambda \in R$; Θ Nullelement in H. Hierbei sei $A \in (H \to H)$ ein nichtlinearer Operator mit

$$A\Theta = \Theta$$

und $K \in [H \to H]$, wobei $[H \to H]$ den Raum der linearen beschränkten Operatoren von H in H bezeichne. Für jedes $\lambda \in R$ besitzt (1.1) dann die Lösung Θ, während für gewisse λ nichttriviale Lösungen existieren mögen. Solche Lösungen nennen wir Eigenfunktionen von (1.1), die zugehörigen λ Eigenwerte. Weiter heißt $\lambda_0 \in R$ Bifurkationselement von (1.1), wenn eine Folge von Eigenwerten $\{\lambda_n\}$, $\lim\limits_{n\to\infty} \lambda_n = \lambda_0$, mit Eigenfunktionen u_n, $\lim\limits_{n\to\infty} u_n = \Theta$ existiert.

Eine stetige Abbildung u eines Intervalls $I \subseteq R$ in H nennen wir Weg in H. Ein Weg $u(t)$, $t \in I$ heißt in $t_* \in I$ differenzierbar, wenn ein $u'(t_*) \in H$ existiert mit

$$\lim_{\delta t \to 0} \left\| \frac{u(t_* + \delta t) - u(t_*)}{\delta t} - u'(t_*) \right\| = 0$$

für jede Folge $\{t_* + \delta t\} \subseteq I$.

Einen Weg $u(t)$, $t \in I$ nennen wir Trajektorie von (1.1), wenn es eine stetige Funktion $\lambda(t)$, $t \in I$ gibt, so daß die Gleichung

$$Au(t) - \lambda(t)\, Ku(t) = \Theta\,, \qquad t \in I$$

erfüllt ist. Punkte $u(t)$ einer Trajektorie von (1.1) mit $\|u(t)\| > 0$ sind offenbar Eigenfunktionen der Gleichung (1.1). Unser Ziel ist es, differenzierbare Trajektorien $u(t)$, $t \in (-\infty, t_0)$ von (1.1) derart nachzuweisen, daß $\|u(t)\| > 0$ für $t \in (-\infty, t_0)$ und $\lim\limits_{t\to-\infty} u(t) = \Theta$. Dann ist der Wert $\lambda_0 = \lim\limits_{t\to-\infty} \lambda(t)$ Bifurkationselement von (1.1) und die Trajektorie $u(t)$, $t \in (-\infty, t_0)$ ein differenzierbarer Zweig von Eigenfunktionen im Bifurkationselement λ_0.

Zunächst gehen wir von folgenden Bedingungen aus:

I: Der Operator A sei auf H Fréchet-differenzierbar, d. h. in jedem $u \in H$ bestehe für jedes $v \in H$ die Zerlegung

$$A(u + v) - Au = A'(u)\, v + \omega_A(u, v) \tag{1.2}$$

mit $A'(u) \in [H \to H]$ und $\omega_A(u, v) = o(\|v\|)$ für $\|v\| \to 0$.

II: Die linearisierte Gleichung

$$A'(\Theta)\,u - \lambda K u = \Theta \tag{1.3}$$

besitze einen einfachen Eigenwert $\lambda_0 \neq 0$ zur normierten Eigenfunktion $\varphi \in H$.

Bezeichne $\Phi = \mathfrak{L}\{\varphi\}$ und $E = H \ominus \Phi$, dann hat jedes $u \in H$ die Darstellung $u = a\varphi + e$ mit $a \in R$, $e \in E$. Gleichung (1.1) ist äquivalent dem System

$$P_\Phi\big(A(a\varphi + e) - \lambda K(a\varphi + e)\big) = \Theta\,, \tag{1.4}$$

$$P_E\big(A(a\varphi + e) - \lambda K(a\varphi + e)\big) = \Theta\,, \tag{1.5}$$

wobei P_Φ und P_E die orthogonalen Projektoren auf Φ und E bedeuten. Wir verfahren nun folgendermaßen:

Mit der Bezeichnung $\tau = \lambda - \lambda_0$ fassen wir (1.5) als eine vom Parameter $(a, \tau) \in R^2$ abhängige Gleichung auf und nehmen zunächst an, daß sie in einer gewissen Umgebung $G \subseteq R^2$ des Nullelements $\mathfrak{O} \in R^2$ eine eindeutig bestimmte, stetige Lösung $e(a, \tau) \in E$, $(a, \tau) \in G$, mit $e(0, 0) = \Theta$ besitzt. Diese Lösung setzen wir in (1.4) ein und gewinnen Lösungen von (1.4) aus Trajektorien eines autonomen Differentialgleichungssystems in der Umgebung eines singulären Punktes.

Autonomes Differentialgleichungssystem zur Lösung von (1.4)

Sei $u(a, \tau) = a\varphi + e(a, \tau)$ und $f(a, \tau) = \big(Au(a, \tau) - (\tau + \lambda_0)\,Ku(a, \tau), \varphi\big)$. Gleichung (1.4) betrachten wir nun in der Form

$$f(a, \tau) = 0 \tag{1.6}$$

und stellen an $f(a, \tau)$ folgende Forderungen:

(i) In G mögen stetige partielle Ableitungen von $f(a, \tau)$ nach a und τ existieren.

(ii) Diese besitzen die Zerlegungen

$$\frac{\partial f}{\partial \tau}(a, \tau) = -\,ac - r(a, \tau)\,, \quad \frac{\partial f}{\partial a}(a, \tau) = -\,\tau c + s(a, \tau)$$

mit $c = \text{const}$, $c > 0$ und den Wachstumsschranken

$$r(a, \tau)\,,\ s(a, \tau) = o(|a| + |\tau|) \quad \text{für} \quad (|a| + |\tau|) \to 0\,.$$ [1]

(iii) Die Funktionen $r(a, \tau)$, $s(a, \tau)$, $(a, \tau) \in G$ seien in G Lipschitz-stetig.

Besitzt nun (1.6) eine differenzierbare Trajektorie $\big(a(t), \tau(t)\big)$, $t \in I \subseteq R$, so gilt auf dieser unter Bedingung (i) mit

$$\frac{d}{dt} f\big(a(t), \tau(t)\big) \equiv 0 \tag{1.7}$$

[1]) Im allgemeinen Fall genügt es, von den Zerlegungen $\frac{\partial f}{\partial \tau}(a, \tau) = -\,ac - r(a, \tau)$, $\frac{\partial f}{\partial a}(a, \tau) = -\,\tau c + ac_1 + s(a, \tau)$ mit $c_1 = \text{const}$ auszugehen. Das im folgenden gebildete autonome Syystem (1.9) wird dann durch die Transformation $x = \tau - \frac{c_1}{2c}\,a$ in die Form $a' = ac + \bar{r}(a, x)$, $x' = -\,xc + \bar{s}(a, x)$ gebracht.

auch

$$\frac{\partial}{\partial a} f(a, \tau) \frac{da}{dt} + \frac{\partial}{\partial \tau} f(a, \tau) \frac{d\tau}{dt} = 0 .$$

Setzen wir nun

$$\frac{da}{dt} = - \frac{\partial}{\partial \tau} f(a, \tau) , \qquad \frac{d\tau}{dt} = \frac{\partial}{\partial a} f(a, \tau) , \tag{1.8}$$

dann genügt jede Lösung von (1.8) wegen

$$\tau' a' - a' \tau' = 0$$

auch (1.7). Mit (i) und (ii) erhält (1.8) die Gestalt

$$\frac{da}{dt} = ac + r(a, \tau) , \qquad \frac{d\tau}{dt} = - \tau c + s(a, \tau) . \tag{1.9}$$

Lemma. *Unter den Bedingungen* (i), (ii), (iii) *existiert eine Trajektorie* $(a(t), \tau(t))$, $t \in (-\infty, t_0)$ *von* (1.9) *mit* $a(t) > 0$ *für* $t > -\infty$ und $\lim_{t \to -\infty} (a(t), \tau(t)) = \mathfrak{D}$.[1]

Beweis. Wir transformieren $x = -t$ und zeigen, daß das System

$$\frac{da}{dx} = - ac - r(a, \tau) , \qquad \frac{d\tau}{dx} = \tau c - s(a, \tau) \tag{1.10}$$

eine Trajektorie $(a(x), \tau(x))$, $x \in (t_0, \infty)$ besitzt mit $a(x) > 0$ für $x < \infty$ und $\lim_{x \to \infty} (a(x), \tau(x)) = \mathfrak{D}$.

Dazu betrachten wir in der $a\tau$-Ebene die Funktionen $w(a, \tau) = \tau - \alpha a$, $\overline{w}(a, \tau) = \tau + \alpha a, \alpha > 0$ beliebig, aber fixiert und bilden ein Dreieck $D = [\mathfrak{D}, A, B] \subseteq G$ durch die Geraden $w(a, \tau) = 0$, $\overline{w}(a, \tau) = 0$, $a = \eta$, wobei $\eta > 0$ noch hinreichend klein zu wählen ist (vgl. Abb. 1).

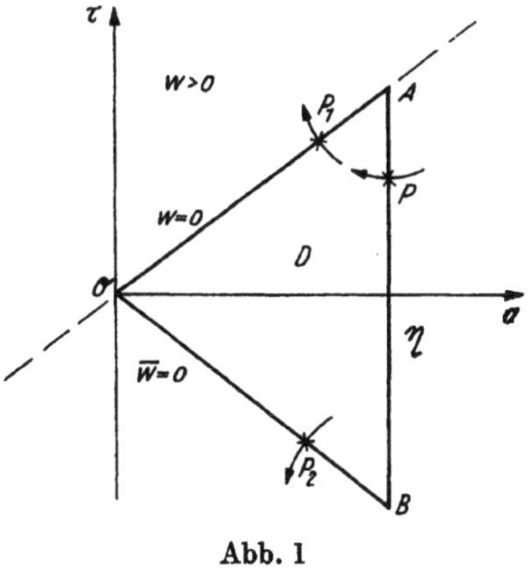

Abb. 1

[1]) Unter den Bedingungen (i), (ii), (iii) ist der singuläre Punkt (0, 0) von (1.9) Sattelpunkt, s. [1]; das Lemma stellt ein bekanntes Ergebnis aus der Theorie der Differentialgleichungen dar.

Sei P_1 Punkt auf $(\mathfrak{O}, A)$ und $\varphi(x, P_1)$ Trajektorie von (1.10) durch P_1. Wegen (ii) gilt bei geeignetem η für die zeitliche Änderung von $w(a, \tau)$ auf der Trajektorie $\varphi(x, P_1)$

$$\frac{d}{dx} w(a, \tau) = \tau' - \alpha a' = \tau c - s(a, \tau) + \alpha a c + \alpha r(a, \tau)$$

und im Punkt P_1

$$\frac{d}{dx} w(a, \alpha a) = 2\alpha a c - s(a, \alpha a) + \alpha r(a, \alpha a) > 0 \quad \text{für} \quad a \leqq \eta\,.$$

Daher läuft die Trajektorie $\varphi(x, P_1)$ in die Richtung, in der die Funktion w wächst, d. h. in die Halbebene $w > 0$.

Analog zeigen wir, daß eine Trajektorie von (1.10) durch einen Punkt $P_2 \in (\mathfrak{O}, B)$ in die Halbebene $\overline{w} < 0$ hineinläuft.

Weiter gilt für Trajektorien $\varphi(x, P)$, die in Punkten $P \in (A, B)$ beginnen, bei hinreichend kleinem η

$$\frac{da}{dx} = -ac - r(a, \tau) < 0\,,$$

d. h. die Trajektorien $\varphi(x, P)$ verlaufen in D von rechts nach links.

Damit läuft durch jeden Punkt $P \in (A, B)$ eine Trajektorie $\varphi(x, P)$, die D durch $(\mathfrak{O}, A)$ oder $(\mathfrak{O}, B)$ verläßt oder sich $\mathfrak{O}$ asymptotisch nähert. Der Punkt $\mathfrak{O}$ kann von keiner Trajektorie $\varphi(x, P)$ geschnitten werden, da er selbst eine Trajektorie des Systems (1.10) darstellt. Auf (A, B) definieren wir eine Abbildung ψ durch

$$\psi(P) = \begin{cases} \text{Schnittpunkt von } \varphi(x, P) & \text{mit } (\mathfrak{O}, A) \cup (\mathfrak{O}, B) \\ \mathfrak{O}\,, & \text{sonst}\,. \end{cases}$$

Da die rechten Seiten von (1.10) nach (iii) Lipschitz-stetig sind, hängen die Lösungen von (1.10) stetig von ihren Anfangswerten ab, vgl. [1]. Deshalb ist die Abbildung ψ stetig und die Urbilder $\psi^{-1}(\mathfrak{O}, A)$ bzw. $\psi^{-1}(\mathfrak{O}, B)$ der offenen Mengen $(\mathfrak{O}, A)$ bzw. $(\mathfrak{O}, B)$ sind offen in (A, B). Auf Grund der — durch die Lipschitz-Stetigkeit gesicherten — Eindeutigkeit der Lösungen folgt

$$\psi^{-1}(\mathfrak{O}, A) \cap \psi^{-1}(\mathfrak{O}, B) = \emptyset\,.$$

Für das zusammenhängende Intervall (A, B) ist somit

$$\psi^{-1}(\mathfrak{O}, A) \cup \psi^{-1}(\mathfrak{O}, B) \neq (A, B)\,.$$

Also existiert ein Punkt $P_0 \in (A, B)$, so daß für die Trajektorie $\varphi(x, P_0) = (a(x), \tau(x))$ gilt:

$$\lim_{x \to \infty} (a(x), \tau(x)) = \mathfrak{O} \quad \text{und} \quad a(x) > 0 \quad \text{für} \quad x < \infty\,; \qquad \text{q. e. d.}$$

Satz 1. *Der Operator A sei demistetig. Weiter existiere eine eindeutige, stetige Lösung $e(a, \tau)$, $(a, \tau) \in G$ von* (1.5) *mit $e(0, 0) = \Theta$. Die Funktion $f(a, \tau)$ $(a, \tau) \in G$ aus* (1.6) *genüge der Bedingung* (i) *und das System* (1.8) *besitze eine Lösung*

$(a(t), \tau(t))$, $t \in (-\infty, t_0)$ *mit* $\lim\limits_{t \to -\infty} (a(t), \tau(t)) = (0, 0)$ *und* $a(t) > 0$ *für* $t > -\infty$. *Dann liefern diese Lösungen eine Trajektorie*

$$u(t) = a(t)\,\varphi + e(a(t), \tau(t)),\ t \in (-\infty, t_0)$$

der Gleichung (1.1), *wobei gilt:* $||u(t)|| > 0$ *für* $t > -\infty$ *und* $\lim\limits_{t \to -\infty} u(t) = \Theta$.

Beweis. Als Lösung von (1.8) genügt $(a(t), \tau(t))$ auch der Beziehung (1.7):

$$\frac{d}{dt}\,(Au(t) - (\tau(t) + \lambda_0)\,Ku(t), \varphi) = 0\,, \qquad t \in (-\infty, t_0)\,;$$

d. h. $(Au(t) - \lambda(t)\,Ku(t), \varphi) = C_1 = \text{const}$, $t \in (-\infty, t_0)$ mit $\lambda(t) = \tau(t) + \lambda_0$. Die Stetigkeit von $e(a, \tau)$, $(a, \tau) \in G$ zieht die Stetigkeit von $u(t) = a(t)\,\varphi + e(a(t), \tau(t))$, $t \in (-\infty, t_0)$ nach sich. Aus der Demistetigkeit von A, aus $u(0, 0) = e(0, 0) = \Theta$ und $\lim\limits_{t \to -\infty} a(t) = 0$ folgt weiter $\lim\limits_{t \to -\infty} u(t) = \Theta$ und $\lim\limits_{t \to -\infty} (Au(t) - \lambda(t)\,Ku(t), \varphi) = 0$, also $C_1 = 0$. Daher gilt

$$P_\Phi(Au(t) - \lambda(t)\,Ku(t)) = \Theta\,, \qquad t \in (-\infty, t_0)\,.$$

Nach Konstruktion von $e(a, \tau)$ genügt $u(t)$ auch der Gleichung

$$P_E(Au(t) - \lambda(t)\,Ku(t)) = \Theta\,, \qquad t \in (-\infty, t_0)\,.$$

Damit ist $u(t)$ Trajektorie von (1.1) mit

$$||u(t)||^2 = ||a(t)\,\varphi||^2 + ||e(a(t), \tau(t))||^2 \geqq a^2(t) > 0; \qquad \text{q. e. d.}$$

Korollar. *Der Operator* A *sei stetig. Gleichung* (1.5) *besitze eine eindeutige, stetige Lösung* $e(a, \tau)$, $(a, \tau) \in G$ *mit* $e(0, 0) = \Theta$. *Die Bedingungen* (i), (ii), (iii) *des Lemmas seien erfüllt. Dann existiert eine Trajektorie* $u(t) = a(t)\,\varphi + e(a(t), \tau(t))$, $t \in (-\infty, t_0)$ *von* (1.1) *mit* $||u(t)|| > 0$ *für* $t > -\infty$ *und* $\lim\limits_{t \to -\infty} u(t) = \Theta$.

Beweis. Auf Grund des Lemmas sind alle Voraussetzungen von Satz 1 nachweisbar.

Existenz von Lösungen der Gleichung (1.5)

Wir wenden uns der Gleichung

$$P_E(Au - (\tau + \lambda_0)\,Ku) = P(e, a, \tau) = \Theta \tag{1.5}$$

zu und geben zwei Sätze über die Existenz von Lösungen an. Zunächst betrachten wir Bedingung

III: Es existiere eine Umgebung G_1 von $\mathfrak{O} \in R^2$, so daß für alle $(a, \tau) \in G_1$ die Operatoren $P(\cdot, a, \tau) \in (E \to E)$ auf E gleichmäßig stark monoton sind, d. h. für alle $e_1, e_2 \in E$, $(a, \tau) \in G_1$ gibt es eine Konstante $\gamma > 0$ mit

$$(P(e_1, a, \tau) - P(e_2, a, \tau), e_1 - e_2) \geqq \gamma ||e_1 - e_2||^2\,. \tag{1.11}$$

Satz 2. *Sei der Operator A auf H stetig und Bedingung* III *erfüllt. Dann gibt es eine eindeutig bestimmte, stetige Abbildung $e \in (G_1 \to E)$ mit $P(e(a,\tau), a, \tau) = \Theta$ für $(a,\tau) \in G_1$ und $e(0,0) = \Theta$.*

Beweis. Nach Voraussetzung ist für jedes $(a,\tau) \in G_1$ der Operator $P(\cdot, a, \tau)$ auf E stark monoton und stetig. Daher existiert für jedes $(a,\tau) \in G_1$ ein eindeutig bestimmtes $e(a,\tau)$ mit $P(e(a,\tau), a, \tau) = \Theta$, vgl. [2].

Seien nun $(a_1,\tau_1), (a_2,\tau_2) \in G_1$ und $e_1 = e(a_1,\tau_1), e_2 = e(a_2,\tau_2)$ Punkte der Lösung von (1.5). Dann folgt aus (1.11)

$$\begin{aligned}\Theta &= \left(P(e_1, a_1, \tau_1) - P(e_2, a_2, \tau_2), e_1 - e_2\right)\\ &= \left(P(e_1, a_1, \tau_1) - P(e_2, a_1, \tau_1), e_1 - e_2\right)\\ &\quad + \left(P(e_2, a_1, \tau_1) - P(e_2, a_2, \tau_2), e_1 - e_2\right)\\ &\geqq \gamma\|e_1 - e_2\|^2 - \|P(e_2, a_1, \tau_1) - P(e_2, a_2, \tau_2)\| \, \|e_1 - e_2\|\end{aligned}$$

und nach Division durch $\|e_1 - e_2\|$

$$\|e(a_1,\tau_1) - e(a_2,\tau_2)\| \leqq \frac{1}{\gamma} \|P(e(a_2,\tau_2), a_1, \tau_1) - P(e(a_2,\tau_2), a_2, \tau_2)\| . \quad (1.12)$$

Für $(a_1,\tau_1) \to (a_2,\tau_2)$ ergibt sich aus (1.12) und der Stetigkeit von $P(v, \cdot, \cdot) \in (R^2 \to E)$ für jedes $v \in E$ die Stetigkeit von $e(a,\tau)$, $(a,\tau) \in G_1$.

Wegen $P(\Theta, 0, 0) = P_E(A(0\varphi + \Theta) - \lambda_0 K\Theta) = \Theta$ und der Eindeutigkeit von $e(a,\tau)$ finden wir $e(0,0) = \Theta$; q. e. d.

Um (1.5) als Fixpunktgleichung zu formulieren, verlangen wir

IV: Der Operator A sei in Θ Fréchet-differenzierbar, und es gelte

$$A'(\Theta) = I , \qquad I \in [H \to H] \qquad \text{identischer Operator.}$$

V: Auf E besitze der Operator $I - \lambda_0 K$ ein beschränktes Inverses

$$S = (I - \lambda_0 K)^{-1} \in [E \to E] .$$

VI: Die Operatoren $K \in [H \to H]$ und $A'(u) \in [H \to H]$, $u \in H$ seien selbstadjungiert.

Setzen wir $F = A - I \in (H \to H)$, dann läßt sich (1.5) unter den Bedingungen II, IV, V und VI bezüglich K umformen zu

$$\begin{aligned}\Theta &= P_E(u + Fu - (\tau + \lambda_0) Ku)\\ &= e + P_E Fu - (\tau + \lambda_0) Ke = (I - \lambda_0 K) e + P_E Fu - \tau Ke .\end{aligned}$$

Nach Anwendung von S erhalten wir

$$\Theta = e + SP_E F (a\varphi + e) - \tau SKe$$

und mit $Q(e, a, \tau) = - SP_E F (a\varphi + e) + \tau SKe$

$$e - Q(e, a, \tau) = \Theta . \quad (1.13)$$

Satz 3. *Sei A auf H stetig, und es gelten die Bedingungen* II, IV, V *und* VI *bezüglich K; dazu*

VII: *Es existieren eine Kugel $K(\Theta, r) = \{e \in E, \|e\| \leqq r\}$ in E und eine Umgebung G_1 von $\mathfrak{D} \in R^2$ derart, daß für alle $(a,\tau) \in G_1$ die Operatoren $Q(\cdot, a, \tau)$ die*

Kugel $K(\Theta, r)$ in sich abbilden und auf $K(\Theta, r)$ strikt kontraktiv sind mit einer einheitlichen Konstante $\alpha < 1$:

$$||Q(e_1, a, \tau) - Q(e_2, a, \tau)|| \leqq \alpha ||e_1 - e_2|| , \qquad e_1, e_2 \in K(\Theta, r) , \qquad (a, \tau) \in G_1 .$$

Dann gibt es eine eindeutige, stetige Abbildung $e \in (G_1 \to E)$ mit $P(e(a, \tau), a, \tau) = \Theta$, $(a, \tau) \in G_1$ und $e(0, 0) = \Theta$.

Beweis. Für jedes $(a, \tau) \in G_1$ ergibt sich die Existenz der eindeutigen Lösung $e(a, \tau)$ von (1.13) und damit von (1.5) aus dem Banachschen Fixpunktsatz [3]. Weiter folgt die Stetigkeit von $e(a, \tau)$, $(a, \tau) \in G_1$ aus der Gleichmäßigkeit der Kontraktionskonstante und aus der Stetigkeit von $Q(v, \cdot, \cdot) \in (R^2 \to E)$ für $v \in E$, die durch die Stetigkeit von A gesichert ist. Wieder erhalten wir aus $A\Theta = \Theta$ auch $e(0, 0) = \Theta$; q. e. d.

Eigenschaften der Lösung $e(a, \tau)$ von (1.5)

Im folgenden sollen unter gewissen Bedingungen Eigenschaften von $e(a, \tau)$ derart nachgewiesen werden, daß die Voraussetzungen (i), (ii), (iii) des Lemmas erfüllt sind. Dazu geben wir die

Definition. Die Abbildung $e \in (G_1 \to E)$ heißt Fréchet-differenzierbar in G_1, wenn in jedem $(a, \tau) \in G_1$ für jedes $(a + \delta a, \tau + \delta\tau) \in G_1$ Elemente $\frac{\partial e}{\partial a}(a, \tau)$, $\frac{\partial e}{\partial \tau}(a, \tau) \in E$ existieren, so daß die Zerlegung

$$e(a + \delta a, \tau + \delta\tau) - e(a, \tau) = \delta a \frac{\partial e}{\partial a}(a, \tau) + \delta\tau \frac{\partial e}{\partial \tau}(a, \tau) + \omega_e(a, \tau, \delta a, \delta\tau)$$

besteht mit

$$||\omega_e(a, \tau, \delta a, \delta\tau)|| = o(|\delta a| + |\delta\tau|) \quad \text{für} \quad (|\delta a| + |\delta\tau|) \to 0 .$$

Satz 4. *Seien die Bedingungen* I *und* III *erfüllt. Dann ist die Abbildung $e \in (G_1 \to E)$ in G_1 Fréchet-differenzierbar.*

Beweis. Wegen I besitzt in jedem $(e, a, \tau) \in E \times R^2$ der Operator P für jedes $(e + \delta e, a + \delta a, \tau + \delta\tau) \in E \times R^2$ die Zerlegung

$$\begin{aligned} &P(e + \delta e, a + \delta a, \tau + \delta\tau) - P(e, a, \tau) \\ &= P_E A'(a\varphi + e)(\delta a\varphi + \delta e) - (\tau + \lambda_0) P_E K(\delta a\varphi + \delta e) - \delta\tau\, P_E K(a\varphi + e) \\ &\quad + P_E \omega_A(a\varphi + e, \delta a\varphi + \delta e) - \delta\tau\, P_E K(\delta a\varphi + \delta e) \\ &= P_E(A'(a\varphi + e) - (\tau + \lambda_0) K)\, \delta e + \delta a\, P_E(A'(a\varphi + e) - (\tau + \lambda_0) K)\, \varphi \\ &\quad - \delta\tau\, P_E K(a\varphi + e) + \omega_1(e, a, \tau, \delta e, \delta a, \delta\tau) \end{aligned} \tag{1.14}$$

mit

$$||\omega_1(e, a, \tau, \delta e, \delta a, \delta\tau)|| = o(||\delta e|| + |\delta a| + |\delta\tau|) \tag{1.15}$$

$$\text{für} \quad (||\delta e|| + |\delta a| + |\delta\tau|) \to 0 .$$

Nach [4] ist wegen der Zerlegung (1.14) und Voraussetzung III für jedes $(e, a, \tau) \in E \times G_1$ die Einschränkung $D(e, a, \tau) \in [E \to E]$ von $P_E\big(A'(a\varphi + e) - (\tau + \lambda_0) K\big)$ auf E stark monoton und eindeutig umkehrbar, wobei gilt

$$||D^{-1}(e, a, \tau)|| \leqq \frac{1}{\gamma}, \qquad (e, a, \tau) \in E \times G_1 . \tag{1.16}$$

Seien nun $(a + \delta a, \tau + \delta\tau)$, $(a, \tau) \in G_1$ beliebig. Für die Punkte $e + \delta e = e(a + \delta a, \tau + \delta\tau)$, $e = e(a, \tau)$ der Lösung von (1.5) finden wir mit der Zerlegung (1.14)

$$\begin{aligned}\Theta = {} & D\big(e(a, \tau), a, \tau\big)\, \delta e + \delta a\, P_E\big(A'(u(a, \tau)) - (\tau + \lambda_0) K\big)\, \varphi \\ & - \delta\tau\, P_E K u(a, \tau) + \omega_1\big(e(a, \tau), a, \tau, \delta e, \delta a, \delta\tau\big)\end{aligned}$$

und nach Anwendung von $D^{-1}\big(e(a, \tau), a, \tau\big)$

$$\begin{aligned}e(a + \delta a, \tau + \delta\tau) - e(a, \tau) = {} & -\delta a\, D^{-1}\big(e(a, \tau), a, \tau\big)\, P_E\big(A'(u(a, \tau)) - (\tau + \lambda_0) K\big)\, \varphi \\ & + \delta\tau\, D^{-1}\big(e(a, \tau), a, \tau\big)\, P_E K u(a, \tau) \\ & + D^{-1}\big(e(a, \tau), a, \tau\big)\, \omega_1\big(e(a, \tau), a, \tau, \delta e, \delta a, \delta\tau\big) . \end{aligned} \tag{1.17}$$

Zur Abschätzung des Restgliedes

$$\begin{aligned}\omega_e(a, \tau), \delta a, \delta\tau) = {} & D^{-1}\big(e(a, \tau), a, \tau\big) \\ & \times \omega_1\big(e(a, \tau), a, \tau, e(a + \delta a, \tau + \delta\tau) - e(a, \tau), \delta a, \delta\tau\big)\end{aligned}$$

beachten wir, daß auf Grund von (1.12) und Zerlegung (1.14) gilt

$$\begin{aligned}& \frac{||e(a + \delta a, \tau + \delta\tau) - e(a, \tau)||}{|\delta a| + |\delta\tau|} \\ & \leqq \frac{1}{\gamma}\left\{||\big(A'(u(a,\tau)) - (\tau + \lambda_0) K\big)\varphi|| + ||Ku(a, \tau)|| + \frac{||\omega_1\big(e(a, \tau), a, \tau, \Theta, \delta a, \delta\tau\big)||}{|\delta a| + |\delta\tau|}\right\}.\end{aligned} \tag{1.18}$$

Daher sind für $(|\delta a| + |\delta\tau|) \to 0$ die Ausdrücke

$$\frac{||e(a + \delta a, \tau + \delta\tau) - e(a, \tau)||}{|\delta a| + |\delta\tau|}$$

beschränkt.

Mit (1.15) erhalten wir wegen der Stetigkeit von $e(a, \tau)$, $(a, \tau) \in G_1$ aus

$$\begin{aligned}& \frac{||\omega_e(a, \tau, \delta a, \delta\tau)||}{|\delta a| + |\delta\tau|} \\ & \leqq \frac{1}{\gamma}\, \frac{||\omega_1\big(e(a, \tau), a, \tau, \delta e, \delta a, \delta\tau\big)||}{||e(a + \delta a, \tau + \delta\tau) - e(a, \tau)|| + |\delta a| + |\delta\tau|} \left(\frac{||\delta e||}{|\delta a| + |\delta\tau|} + 1\right)\end{aligned}$$

die Abschätzung

$$||\omega_e(a, \tau, \delta a, \delta\tau)|| = o(|\delta a| + |\delta\tau|) \quad \text{für} \quad (|\delta a| + |\delta\tau|) \to 0 .$$

Mithin entnehmen wir (1.17) die Fréchet-Differenzierbarkeit der Abbildung $e \in (G_1 \to E)$ in G_1 und die Darstellungen

$$\frac{\partial e}{\partial a}(a,\tau) = -D^{-1}(e(a,\tau),a,\tau)\,P_E(A'(u(a,\tau)) - (\tau + \lambda_0)\,K)\,\varphi\,, \tag{1.19}$$

$$\frac{\partial e}{\partial \tau}(a,\tau) = D^{-1}(e(a,\tau),a,\tau)\,P_E K u(a,\tau)\,, \tag{1.20}$$

q. e. d.

Satz 5. *Seien die Voraussetzungen* I, II, IV, V, VI, VII *erfüllt. Dann gibt es eine Umgebung* G_2 *von* $\mathfrak{O} \in R^2$, $G_2 \subseteq G_1$, *so daß die Abbildung* $e \in (G_1 \to E)$ *in* G_2 *Fréchet-differenzierbar ist.*

Beweis. Wieder liefert Bedingung I die Zerlegung:

$$\begin{aligned} &Q(e + \delta e, a + \delta a, \tau + \delta\tau) - Q(e,a,\tau) \\ &= -\,(SP_E F'(a\varphi + e) - \tau SK)\,\delta e - \delta a\, SP_E F'(a\varphi + e)\,\varphi \\ &\quad + \delta\tau\, SKe + \omega_2(e,a,\tau,\delta e,\delta a,\delta\tau) \end{aligned} \tag{1.21}$$

mit

$$\|\omega_2(e,a,\tau,\delta e,\delta a,\delta\tau)\| = o(\|\delta e\| + |\delta a| + |\delta\tau|)$$

für $(\|\delta e\| + |\delta a| + |\delta\tau|) \to 0$.

Sei nun $0 < r' < r$. Dann sind für alle $(e,a,\tau) \in K(\Theta, r') \times G_1$ die Einschränkungen $R(e,a,\tau)$ der Operatoren $SP_E F'(a\varphi + e) - \tau SK$ auf E gleichmäßig strikt kontraktiv auf E. Denn für beliebiges $v \in E$ und $(e,a,\tau) \in K(\Theta, r') \times G_1$ ist für $t > 0$ und t hinreichend klein $e + tv \in K(\Theta, r)$. Mithin gilt nach VII

$$\begin{aligned} \alpha\|v\| &\geqq \lim_{t\to 0} \frac{1}{t}\,\|Q(e + tv, a,\tau) - Q(e,a,\tau)\| \\ &\geqq \lim_{t\to 0} \left|\,\|SP_E F'(a\varphi + e)\,v - \tau SKv\| - \frac{\|\omega_2(e,a,\tau,tv,0,0)\|}{\|tv\|}\,\|v\|\,\right| \\ &= \|R(e,a,\tau)\,v\|\,. \end{aligned}$$

Somit sind auf E die Operatoren $I - R(e,a,\tau) \in [E \to E]$, $(e,a,\tau) \in K(\Theta, r') \times G_1$ stark monoton mit der Monotoniekonstante $1 - \alpha$, also existieren beschränkte Inverse, wobei gilt

$$\|(I - R(e,a,\tau))^{-1}\| \leqq \frac{1}{1-\alpha}\,.$$

Wählen wir nun die Umgebung $G_2 \subseteq G_1$, $\mathfrak{O} \in G_2$ so, daß für $(a,\tau) \in G_2$ gilt $e(a,\tau) \in K(\Theta, r')$. (Das ist wegen $e(0,0) = \Theta$ und der Stetigkeit von $e(a,\tau)$, $(a,\tau) \in G_1$ möglich.) Jetzt verfahren wir wie in Satz 4: Für $(a + \delta a, \tau + \delta\tau)$, $(a,\tau) \in G_2$ und die Punkte $e(a + \delta a, \tau + \delta\tau)$, $e(a,\tau)$ der Lösung von (1.13) setzen wir in die Differenz

$$\begin{aligned} \Theta = {}& e(a + \delta a, \tau + \delta\tau) - Q(e(a + \delta a, \tau + \delta\tau), a + \delta a, \tau + \delta\tau) \\ & - (e(a,\tau) - Q(e(a,\tau),a,\tau)) \end{aligned}$$

die Zerlegung (1.21) ein, wenden darauf $(I - R(a(e,\tau), a, \tau))^{-1}$ an und erhalten

$$\begin{aligned} & e(a+\delta a, \tau+\delta\tau) - e(a,\tau) \\ &= -\delta a(I - R(e(a,\tau), a, \tau))^{-1} SP_E F'(a\varphi + e) \\ &\quad + \delta\tau(I - R(e(a,\tau), a, \tau))^{-1} SKe(a,\tau) \\ &\quad + (I - R(e(a,\tau), a, \tau))^{-1} \omega_2(e(a,\tau), a, \tau, e(a+\delta a, \tau+\delta\tau) - e(a,\tau), \delta a, \delta\tau). \end{aligned} \tag{1.22}$$

Da für jedes $(a,\tau) \in G_1$ der Operator $I - Q(\cdot, a, \tau)$ auf $K(\Theta, r)$ stark monoton ist mit der Konstante $1 - \alpha$, gilt analog zur Ungleichung (1.12) die Beziehung

$$\begin{aligned} & \|e(a+\delta a, \tau+\delta\tau) - e(a,\tau)\| \\ & \leqq \frac{1}{1-\alpha} \|Q(e(a,\tau), a+\delta a, \tau+\delta\tau) - Q(e(a,\tau), a, \tau)\| . \end{aligned}$$

Hieraus folgt mit Hilfe von (1.21) die Beschränktheit der Ausdrücke

$$\frac{\|e(a+\delta a, \tau+\delta\tau) - e(a,\tau)\|}{|\delta a| + |\delta\tau|} \quad \text{für} \quad (|\delta a| + |\delta\tau|) \to 0 .$$

Aus (1.22) ergibt sich damit durch dieselben Abschätzungen für das Restglied wie oben die Fréchet-Differenzierbarkeit von $e(a,\tau)$ für $(a,\tau) \in G_2$; q. e. d.

Satz 6. *Es gelten die Bedingungen* I, III *und*

VIII: *Die Fréchet-Ableitung* $A' \in (H \to [H \to H])$ *von* A *genüge auf jeder beschränkten Menge von* H *einer Lipschitz-Bedingung.*

Dann sind auf jedem Kompaktum $k \subseteq G_1$ *die Abbildungen* $e, \partial e/\partial a, \partial e/\partial\tau \in (G_1 \to E)$ *Lipschitz-stetig.*

Beweis. Auf dem Kompaktum k ist die stetige Funktion $u(a,\tau) = a\varphi + e(a,\tau)$, $(a,\tau) \in G_1$ beschränkt, $\|u(a,\tau)\| \leqq r_k$. Wegen Bedingung VIII existiert dann eine Konstante $C_2 > 0$ mit

$$\|A'(u_1) - A'(u_2)\| \leqq C_2\|u_1 - u_2\|; \qquad u_1, u_2 \in K(\Theta, r_k) \subseteq H .$$

Daher sind für alle $(a,\tau) \in k$ die Operatoren $A'(u(a,\tau)) - (\tau + \lambda_0) K$ gleichmäßig beschränkt:

$$\|A'(u(a,\tau)) - (\tau+\lambda_0) K\| \leqq C_2\|u(a,\tau)\| + \|A'(\Theta)\| + |\tau+\lambda_0| \, \|K\| \leqq C_3$$

mit $C_3 = \text{const}$, $C_3 > 0$.

Das hat der Darstellungen (1.19), (1.20) wegen die Beschränktheit der partiellen Fréchet-Ableitungen $\partial e/\partial a$, $\partial e/\partial\tau$ über k zur Folge:

$$\begin{aligned} \left\|\frac{\partial e}{\partial a}(a,\tau)\right\| &\leqq \|D^{-1}(e(a,\tau), a, \tau)\| \, \|P_E(A'(u(a,\tau)) - (\tau+\lambda_0) K)\| \leqq \frac{C_3}{\gamma} \leqq C_4 , \\ \left\|\frac{\partial e}{\partial \tau}(a,\tau)\right\| &\leqq \|D^{-1}(e(a,\tau), a, \tau)\| \, \|P_E K u(a,\tau)\| \leqq \frac{\|K\|}{\gamma} r_k \leqq C_4 \end{aligned}$$

mit $C_4 = \text{const}$, $C_4 > 0$. Für $(a_1, \tau_1), (a_2, \tau_2) \in k$ folgt deshalb nach dem Mittelwertsatz [5]:

$$||e(a_1, \tau_1) - e(a_2, \tau_2)|| \leqq |a_1 - a_2|\, C_4 + |\tau_1 - \tau_2|\, C_4 . \tag{1.23}$$

Hieraus erhalten wir auch eine Lipschitz-Bedingung für $\partial e/\partial a$ und $\partial e/\partial \tau$. Denn wenden wir auf Gleichung (1.19) $D\big(e(a, \tau), a, \tau\big)$ an, so ergibt sich

$$\begin{aligned} & D\big(e(a_1, \tau_1), a_1, \tau_1\big) \frac{\partial e}{\partial a}(a_1, \tau_1) - D\big(e(a_2, \tau_2), a_2, \tau_2\big) \frac{\partial e}{\partial a}(a_2, \tau_2) \\ &= D\big(e(a_1, \tau_1), a_1, \tau_1\big) \frac{\partial e}{\partial a}(a_1, \tau_1) - D\big(e(a_1, \tau_1), a_1, \tau_1\big) \frac{\partial e}{\partial a}(a_2, \tau_2) \\ &\quad + D\big(e(a_1, \tau_1), a_1, \tau_1\big) \frac{\partial e}{\partial a}(a_2, \tau_2) - D\big(e(a_2, \tau_2), a_2, \tau_2\big) \frac{\partial e}{\partial a}(a_2, \tau_2) \\ &= - P_E\big(A'(u(a_1, \tau_1)) - (\tau_1 + \lambda_0) K\big) \varphi + P_E\big(A'(u(a_2, \tau_2)) - (\tau_2 + \lambda_0) K\big) \varphi \end{aligned}$$

und nach Anwendung von $D^{-1}\big(e(a_1, \tau_1), a_1, \tau_1\big)$

$$\begin{aligned} & \left\| \frac{\partial e}{\partial a}(a_1, \tau_1) - \frac{\partial e}{\partial a}(a_2, \tau_2) \right\| \leqq ||D^{-1}\big(e(a_1, \tau_1), a_1, \tau_1\big)|| \\ & \times \left\{ \left\| P_E[A'(u(a_1, \tau_1)) - (\tau_1 + \lambda_0) K - A'(u(a_2, \tau_2)) + (\tau_2 + \lambda_0) K] \frac{\partial e}{\partial a}(a_2, \tau_2) \right\| \right. \\ & \left. + ||P_E\big(A'(u(a_1, \tau_1)) - A'(u(a_2, \tau_2))\big) \varphi|| + ||P_E(\tau_1 - \tau_2) K\varphi|| \right\} \\ & \leqq \frac{1}{\gamma} \{(C_2\, C_4 + C_2)||\, u(a_1, \tau_1) - u(a_2, \tau_2)|| + |\tau_1 - \tau_2|\; ||K||\, (C_4 + 1)\} . \end{aligned}$$

Mit (1.23) ist folglich $\partial e/\partial a$ auf k Lipschitz-stetig. Ebenso berechnen wir aus (1.20) die Lipschitz-Stetigkeit von $\partial e/\partial \tau$; q.e.d.

Analoge Betrachtungen führen uns zu

Satz 7. *Unter den Bedingungen* I, II, IV, V, VI, VII *und* VIII *sind auf jedem Kompaktum* $k \subseteq G_2$ *die Abbildungen* $e, \partial e/\partial a, \partial e/\partial \tau \in (G_2 \to E)$ *Lipschitz-stetig.*

Satz 8. *Seien die Bedingungen* I, II, III *und* VIII *erfüllt. Dann besteht für* $(|a| + |\tau|) \to 0$ *das Wachstumsverhalten* $||e(a, \tau)|| = o(|a| + |\tau|)$

$$\left\| \frac{\partial e}{\partial a}(a, \tau) \right\|, \left\| \frac{\partial e}{\partial \tau}(a, \tau) \right\| = O(|a| + |\tau|) .$$

Beweis. Mit $e(0, 0) = \Theta$ und $A'(\Theta)\, \varphi = \lambda_0\, K\varphi$ erhalten wir aus (1.12)

$$\frac{||e(a, \tau)||}{|a| + |\tau|} \leqq \frac{1}{\gamma} \left\{ ||A'(\Theta)\, \varphi - \lambda_0 K\varphi|| + ||K\Theta|| + \frac{||\omega_1(\Theta, 0, 0, \Theta, a, \tau)||}{|a| + |\tau|} \right\}$$

und wegen (1.15)

$$||e(a, \tau)|| = o(|a| + |\tau|) \qquad \text{für} \quad (|a| + |\tau|) \to 0 .$$

Aus Formel (1.20) entnehmen wir

$$\left\|\frac{\partial e}{\partial \tau}(a,\tau)\right\| \leqq \|D^{-1}(e(a,\tau),a,\tau)\,P_E K(a\varphi + e(a,\tau))\| \leqq \frac{\|K\|}{\gamma}(|a| + \|e(a,\tau)\|)\,.$$

Dann liefert uns die Schranke für $e(a,\tau)$

$$\frac{\left\|\frac{\partial e}{\partial \tau}(a,\tau)\right\|}{|a|+|\tau|} \leqq \frac{\|K\|}{\gamma}\left(1 + \frac{\|e(a,\tau)\|}{|a|+|\tau|}\right) \leqq \text{const} \quad \text{für} \quad (|a|+|\tau|) \to 0.$$

Aus (1.19) folgt

$$\left\|\frac{\partial e}{\partial a}(a,\tau)\right\| \leqq \frac{1}{\gamma}\|(A'(u(a,\tau)) - (\tau+\lambda_0)K)\,\varphi\|$$

$$\leqq \frac{1}{\gamma}\{\|A'(u(a,\tau)) - A'(\Theta)\| + |\tau|\,\|K\|\}\,.$$

Nach Voraussetzung VIII gibt es eine Konstante $C_5 > 0$, so daß gilt

$$\frac{\left\|\frac{\partial e}{\partial a}(a,\tau)\right\|}{|a|+|\tau|} \leqq \frac{1}{\gamma}\,\frac{C_5\|u(a,\tau)\| + |\tau|\,\|K\|}{|a|+|\tau|} \leqq \frac{1}{\gamma}\left\{C_5 + \frac{C_5\|e(a,\tau)\|}{|a|+|\tau|} + \|K\|\right\}$$

$$\leqq \text{const} \quad \text{für} \quad (|a|+|\tau|) \to 0; \quad \text{q. e. d.}$$

Dieselben Überlegungen liefern

Satz 9. *Unter den Bedingungen* I, II, IV, V, VI, VII *und* VIII *gilt für* $(|a|+|\tau|) \to 0$

$$\|e(a,\tau)\| = o(|a|+|\tau|)$$

$$\left\|\frac{\partial e}{\partial \tau}(a,\tau)\right\|,\ \left\|\frac{\partial e}{\partial a}(a,\tau)\right\| = O(|a|+|\tau|)\,.$$ [1]

Nachweis der Voraussetzungen (i), (ii), (iii) des Lemmas

Gehen wir von weiteren Bedingungen aus:

IX: Sei $c = (K\varphi, \varphi) > 0$.

X: Für (a,τ) aus dem Existenzbereich der Lösung $e(a,\tau)$ von (1.5) gelte:

$$(A'(u(a,\tau))\,\varphi - A'(\Theta)\,\varphi, \varphi) = o(|a|+|\tau|) \quad \text{für} \quad (|a|+|\tau|) \to 0\,.$$

Satz 10. *Seien die Forderungen* I, II, III, VI, VIII, IX *und* X *erfüllt. Auf jeder beschränkten Umgebung G von $\mathfrak{O} \in R^2$, $G \subseteq G_1$ besitzt dann* (1.5) *eine eindeutige, stetige Lösung $e(a,\tau)$, $(a,\tau) \in G$ mit $e(0,0) = \Theta$ derart, daß die Voraussetzungen* (i), (ii), (iii) *nachweisbar sind.*

[1]) Unter den Voraussetzungen von Satz 9 läßt sich sogar zeigen: $\left\|\frac{\partial e}{\partial \tau}(a,\tau)\right\| = o(|a|+|\tau|)$ für $(|a|+|\tau|) \to 0$.

Beweis. Sei G beschränkte Umgebung von $\mathfrak{O}$ und $\overline{G} \subseteq G_1$. Die Existenz der eindeutigen, stetigen Lösung $e(a, \tau), (a, \tau) \in G_1$ mit $e(0, 0) = \Theta$ folgt aus Satz 2. Sei weiter $P_1(u, \tau) = Au - (\tau + \lambda_0) Ku$, $u \in H$, $\tau \in R$. Dann ist nach Voraussetzung I die Abbildung $P_1 \in (H \times R \to H)$ Fréchet-differenzierbar. Auf Grund von Satz 4 ist mit $e \in (G_1 \to H)$ auch die Abbildung $u \in (G_1 \to H)$ Fréchet-differenzierbar, $u(a, \tau) = a\varphi + e(a, \tau)$. Daher ist in G_1, s. [5], die zusammengesetzte Abbildung $\mathfrak{F} \in (G_1 \to H)$, $\mathfrak{F}(a, \tau) = P_1(u(a, \tau), \tau)$, und somit auch f, $f(a, \tau) = (\mathfrak{F}(a, \tau), \varphi)$, Fréchet-differenzierbar. Die partiellen Fréchet-Ableitungen von f haben dann die Form

$$\frac{\partial f}{\partial a}(a, \tau) = \left([A'(u(a, \tau)) - (\tau + \lambda_0) K]\left[\varphi + \frac{\partial e}{\partial a}(a, \tau)\right], \varphi\right)$$

$$= -\tau(K\varphi, \varphi) + s(a, \tau) = -\tau c + s(a, \tau), \tag{1.24}$$

$$\frac{\partial f}{\partial \tau}(a, \tau) = \left([A'(u(a, \tau)) - (\tau + \lambda_0) K] \frac{\partial e}{\partial \tau}(a, \tau), \varphi\right) - (Ku(a, \tau), \varphi)$$

$$= -a(K\varphi, \varphi) - r(a, \tau) = -ac - r(a, \tau) \tag{1.25}$$

mit

$$s(a, \tau) = \left([A'(u(a, \tau)) - (\tau + \lambda_0) K] \frac{\partial e}{\partial a}(a, \tau), \varphi\right)$$

$$+ (A'(u(a, \tau)) \varphi - A'(\Theta) \varphi, \varphi), \tag{1.26}$$

$$r(a, \tau) = -\left([A'(u(a, \tau)) - (\tau + \lambda_0) K] \frac{\partial e}{\partial \tau}(a, \tau), \varphi\right) + (Ke(a, \tau), \varphi). \tag{1.27}$$

Wegen (1.24), (1.25) besitzen $\partial f/\partial a$ und $\partial f/\partial \tau$ die in (ii) geforderten Zerlegungen. Zeigen wir nun die Lipschitz-Stetigkeit von $s(a, \tau)$, $r(a, \tau)$ in G. Nach Satz 6 sind über dem Kompaktum $\overline{G}$ die Abbildungen $e(a, \tau)$, $u(a, \tau)$, $\frac{\partial e}{\partial a}(a, \tau)$, $\frac{\partial e}{\partial \tau}(a, \tau)$, $(a, \tau) \in G_1$ Lipschitz-stetig und beschränkt, $\|e\|$, $\|u\|$, $\|\partial e/\partial a\|$, $\|\partial e/\partial \tau\| \leqq r_G$. Weiter ist wegen Bedingung VIII die Abbildung $P_2 \in (G_1 \to [H \to H])$, $P_2(a, \tau) = A'(u(a, \tau)) - (\tau + \lambda_0) K$, über G Lipschitz-stetig, und damit sind es auch die zusammengesetzten Abbildungen $P_2(a, \tau) \frac{\partial e}{\partial a}(a, \tau)$ und $P_2(a, \tau) \frac{\partial e}{\partial \tau}(a, \tau)$, $(a, \tau) \in G_1$.

Diese Aussagen ziehen dann unmittelbar die Lipschitz-Stetigkeit der einzelnen Summanden in (1.26) und (1.27) und daher auch der Funktionen $s(a, \tau)$, $r(a, \tau)$, $(a, \tau) \in G_1$ in G nach sich. Auf Grund von (1.24) und (1.25) sind somit insbesondere die partiellen Ableitungen $\partial f/\partial a$ und $\partial f/\partial \tau$ in G stetig und folglich die Forderungen (i) und (iii) bestätigt. Zum Nachweis von (ii) genügt es wegen Bedingung IX, die Wachstumsschranken für $r(a, \tau)$ und $s(a, \tau)$ zu berechnen. Diese ergeben sich aus den Schranken für die einzelnen Glieder in (1.26) und (1.27):

Wegen der Stetigkeit von $u(a, \tau)$ über G_1, der Bedingungen I und II und $u(0, 0) = \Theta$ gaben wir

$$\lim_{(|a| + |\tau|) \to 0} (A'(u(a, \tau)) \varphi - (\tau + \lambda_0) K\varphi) = A'(\Theta) \varphi - \lambda_0 K\varphi = \Theta.$$

Da nach Satz 8 $\left\|\frac{\partial e}{\partial a}(a,\tau)\right\| = O(|a| + |\tau|)$ für $(|a| + |\tau|) \to 0$ ist, läßt sich mit VI abschätzen

$$\frac{\left|\left([A'(u(a,\tau)) - (\tau+\lambda_0) K] \frac{\partial e}{\partial a}(a,\tau), \varphi\right)\right|}{|a| + |\tau|}$$

$$= \frac{\left|\left(\frac{\partial e}{\partial a}(a,\tau), A'(u(a,\tau))\,\varphi - (\tau+\lambda_0) K\varphi\right)\right|}{|a| + |\tau|}$$

$$\leqq \frac{\left\|\frac{\partial e}{\partial a}(a,\tau)\right\|}{|a| + |\tau|} \, \|A'(u(a,\tau))\,\varphi - (\tau+\lambda_0) K\varphi\| \to 0 \quad \text{für} \quad (|a| + |\tau|) \to 0;$$

d. h. $\left([A'(u(a,\tau)) - (\tau+\lambda_0) K] \frac{\partial e}{\partial a}(a,\tau), \varphi\right) = o(|a| + |\tau|)$. Ebenso berechnen wir aus $\left\|\frac{\partial e}{\partial \tau}(a,\tau)\right\| = O(|a| + |\tau|)$ auch

$$\left([A'(u(a,\tau)) - (\tau+\lambda_0) K] \cdot \frac{\partial e}{\partial \tau}(a,\tau), \varphi\right) = o(|a| + |\tau|)\,.$$

Da $K \in [H \to H]$ folgt mit Satz 8 aus $\|e(a,\tau)\| = o(|a| + |\tau|)$ auch $(Ke(a,\tau), \varphi) = o(|a| + |\tau|)$. Voraussetzung X sichert für das restliche Glied $(A'(u(a,\tau))\,\varphi - A'(\Theta)\,\varphi, \varphi) = o(|a| + |\tau|)$. Damit ist Satz 10 bewiesen.

Ist (1.5) der Fixpunktgleichung (1.13) äquivalent, dann lautet die entsprechende Aussage:

Satz 11. *Unter den Voraussetzungen* I, II, IV, V, VI, VII, VIII, IX *und* X *existiert eine Umgebung* G *von* $\mathfrak{O} \in R^2$, *so daß* (1.5) *eine eindeutige, stetige Lösung* $e(a,\tau)$, $(a,\tau) \in G$ *mit* $e(0,0) = \Theta$ *besitzt und die Bedingungen* (i), (ii), (iii) *nachweisbar sind.*

Lösung des Bifurkationsproblems

Zusammenfassend gelangen wir zu folgendem Ergebnis:

Satz 12. *Seien die Forderungen* I, II, III, VI, VIII, IX, X *von Satz* 10 *oder* I, II, IV, V, VI, VII, VIII, IX, X *von Satz* 11 *erfüllt.*

Dann ist λ_0 *Bifurkationspunkt von* (1.1). *In* λ_0 *beginnt ein differenzierbarer Zweig* $u(t)$, $t \in (-\infty, t_0)$ *von Eigenfunktionen von* (1.1).

Beweis. Nach Satz 10 oder Satz 11 sind unter Bedingung I alle Voraussetzungen des Korollars zu Satz 1 erfüllt. Daher ist $u(t) = a(t)\,\varphi + e(a(t), \tau(t))$, $t \in (-\infty, t_0)$ eine Trajektorie von Eigenfunktionen von (1.1), die wegen

$\lim_{t \to -\infty} u(t) = \Theta$ und $\lim_{t \to -\infty} \lambda(t) = \lim_{t \to -\infty} (\tau(t) + \lambda_0) = \lambda_0$ im Bifurkationspunkt λ_0 abzweigt.

Die Differenzierbarkeit von $u(t)$ ergibt sich daraus, daß $(a(t), \tau(t))$, $t \in (-\infty, t_0)$ als Lösung von (1.8) differenzierbar und die Abbildung $e \in (G_1 \to E)$ nach Satz 4 oder Satz 5 in G Fréchet-differenzierbar ist.

Beispiel

Bifurkation und differenzierbare Zweige von Eigenfunktionen der nichtlinear gestörten Potentialgleichung

Auf einem beschränkten Gebiet $\Omega \subseteq R^2$, dessen Rand die Anwendung des Gaußschen Satzes gestatte und für das die Sobolewschen Einbettungssätze gültig seien, betrachten wir das Randwertproblem

$$-\Delta u + f(u) - \lambda u = 0 \quad \text{in} \quad \Omega \tag{2.1}$$

$$u = 0 \quad \text{auf} \quad \partial\Omega \tag{2.2}$$

mit $\Delta = \frac{\partial^2}{\partial x^2} + \frac{\partial^2}{\partial y^2}$, für $u \in \dot{C}^2(\overline{\Omega})$ und $\lambda \in R$.

$\dot{C}^2(\overline{\Omega})$ bezeichne den Raum der in $\overline{\Omega} = \Omega \cup \partial\Omega$ zweimal stetig differenzierbaren Funktionen, die auf $\partial\Omega$ verschwinden. Die nichtlineare Funktion $f \in C^2(R)$ genüge den Abschätzungen

$$\left.\begin{aligned} |f(t)| &\leqq M|t|^{\nu_0}, & \nu_0 &\geqq 1, \\ |f'(t)| &\leqq M|t|^{\nu_1}, & \nu_1 &\geqq 2, \\ |f''(t)| &\leqq M|t|^{\nu_2}, & \nu_2 &\geqq 0, \\ f'(t) &\geqq 0 & & \end{aligned}\right\} \tag{2.3}$$

für eine geeignete Konstante $M > 0$ und natürliche Zahlen ν_0, ν_1, ν_2.

(Z. B. erfüllt die Funktion $f(t) = t^3$ die Bedingungen (2.3).) Zum Hilbertraum H_0 gelangen wir durch Vervollständigung von $\dot{C}^2(\overline{\Omega})$ in der durch das Skalarprodukt

$$(u, v)_0 = \int_\Omega \left\{\frac{\partial u}{\partial x}\frac{\partial v}{\partial x} + \frac{\partial u}{\partial y}\frac{\partial v}{\partial y}\right\} d\Omega = \int_\Omega T[u, v]\, d\Omega \tag{2.4}$$

erzeugten Metrik.

Aus der Friedrichsschen Ungleichung folgt

$$\|u\|_{L_2}^2 \leqq a^2 \|u\|_0^2, \quad u \in H_0 \tag{2.5}$$

mit $0 < a = \text{const}$. Damit erweist sich H_0 als Unterraum des Sobolew-Raumes $W_2^1(\Omega)$, wobei die Norm $\|\cdot\|_0$ der Norm $\|\cdot\|_{2,1}$ in $W_2^1(\Omega)$,

$$\|u\|_{2,1}^2 = \int_\Omega u^2\, d\Omega + \int_\Omega T[u, u]\, d\Omega\,,$$

äquivalent ist. Nach den Sobolewschen Einbettungssätzen, s. [6], ist dann der Einbettungsoperator $E \in (H_0 \to L_q(\Omega))$ für jedes natürliche $q > 1$ vollstetig und beschränkt. Es gibt also Konstanten $M_q > 0$ mit

$$||u||_{L_q} \leqq M_q ||u||_0 , \qquad u \in H_0 . \tag{2.6}$$

Nach dem Gaußschen Satz finden wir

$$\int_\Omega - \Delta uv \, d\Omega = \int_\Omega T[u, v] \, d\Omega \qquad \text{für} \quad u, v \in \dot{C}^2(\bar{\Omega})$$

und verallgemeinern (2.1), (2.2) zu der Aufgabe

$$\int_\Omega T[u, v] \, d\Omega + \int_\Omega f(u) \, v \, d\Omega - \lambda \int_\Omega uv \, d\Omega = 0 , \qquad v \in H_0 . \tag{2.7}$$

Dann sind wegen (2.3) für jedes $u \in H_0$ die Formen

$$a(u, v) = \int_\Omega (T[u, v] + f(u) \, v) \, d\Omega , \tag{2.8}$$

$$b(u, v) = \int_\Omega uv \, d\Omega \tag{2.9}$$

in v lineare beschränkte Funktionale über H_0. Diese erzeugen nach dem Rieszschen Satz über die Darstellung linearer beschränkter Funktionale Operatoren $A_0, K_0 \in (H_0 \to H_0)$ durch

$$(A_0 u, v)_0 = \int_\Omega (T[u, v] + f(u) \, v) \, d\Omega , \tag{2.10}$$

$$(K_0 u, v)_0 = \int_\Omega uv \, d\Omega . \tag{2.11}$$

Wegen $|(K_0 u, v)| \leqq ||u||_{L_2} \, ||v||_{L_2} \leqq a^2 ||u||_0 \, ||v||_0$ ist K_0 ein linearer beschränkter Operator, $K_0 \in [H_0 \to H_0]$. (2.7) erhält dann die Form der Operatorgleichung

$$A_0 u - \lambda K_0 u = \Theta , \qquad u \in H_0 , \qquad \lambda \in R . \tag{2.12}$$

(2.12) hat die Gestalt von Gleichung (1.1), wobei mit (2.3) aus $f(0) = 0$ auch $A_0 \Theta = \Theta$ folgt. Bestätigen wir nun die Bedingungen I, II, III, VI, VIII, IX und X von Satz 10.

I. Für $u, v, w \in H_0$ betrachten wir die Differenz

$$(A_0(u + v) - A_0 u, w)_0 = \int_\Omega \{ T[v, w] + (f(u + v) - f(u)) \, w \} \, d\Omega .$$

Da auf Grund von Voraussetzung (2.3) $f(u) \, w \in L_1(\Omega)$, ist für den Integranden in fast allen $(x, y) \in \Omega$ die Taylor-Formel anwendbar. Für $0 \leqq \vartheta \leqq 1$ folgt daher

$$(A_0(u + v) - A_0 u, w) = \int_\Omega \left\{ T[v, w] + f'(u) \, vw + \frac{1}{2} f''(u + \vartheta v) \, v^2 w \right\} d\Omega .$$

Bilden wir für $v, w \in H_0$ bei beliebig fixiertem $u \in H_0$

$$c(u, v, w) = \int_\Omega \{ T[v, w] + f'(u) \, vw \} \, d\Omega ; \tag{2.13}$$

wegen (2.3), (2.6) und

$$|c(u, v, w)| \leqq ||v||_0 \, ||w||_0 + M||u||^{r_1}_{L_{2r_1}} ||v||_{L_4} \, ||w||_{L_4} \leqq k_0 ||v||_0 \, ||w||_0 \tag{2.14}$$

mit $k_0 = \text{const}$ ist $c(u, v, w)$ ein in w lineares beschränktes Funktional über H_0. Es definiert durch

$$\left(A_0'(u) \, v, w\right) = \int_\Omega \{T[v, w] + f'(u) \, vw\} \, d\Omega \tag{2.15}$$

für jedes $u \in H_0$ einen linearen Operator $A_0'(u)$, wobei infolge (2.14) gilt $A_0'(u) \in [H_0 \to H_0]$. (2.3) liefert weiter die Abschätzung

$$\left|\int_\Omega \frac{1}{2} f''(u + \vartheta v) \, v^2 w \, d\Omega\right| \leqq k_1 ||u + \vartheta v||_0^{r_2} \, ||v||_0^2 \, ||w||_0 ,$$

$k_1 = \text{const}$, und damit

$$||A_0(u + v) - A_0 u - A_0'(u) \, v|| \leqq k_1 ||u + \vartheta v||_0 \, ||v||_0^2 = o(||v||) \qquad \text{für} \quad ||v|| \to 0;$$

d. h. $A_0'(u)$ ist die Fréchet-Ableitung von A_0 in u.

VI. (2.11) und (2.15) entnehmen wir unmittelbar die Selbstadjungiertheit von $K_0 \in [H_0 \to H_0]$ und $A_0'(u) \in [H_0 \to H_0]$, $u \in H_0$.

II. Offenbar ist K_0 positiv. Weiter ist der stetige Operator K_0 vollstetig. Denn wegen der vollstetigen Einbettung $E \in \left(H_0 \to L_2(\Omega)\right)$ enthält jede in H_0 beschränkte Menge N eine in $L_2(\Omega)$ konvergente Folge $\{u_n\}$. Mit (2.11) und (2.5) ergibt sich dann aus

$$|(K_0 u_n - K_0 u_m, v)_0| \leqq a ||u_n - u_m||_{L_2} \, ||v||_0 \qquad \text{für alle} \quad v \in H_0$$

$||K_0 u_n - K_0 u_m||_0 \to 0$ für $n, m \to \infty$; d. h. K_0 ist kompakt in H_0. Der lineare selbstadjungierte vollstetige Operator K_0 besitzt einen betragsmäßig größten Eigenwert μ_0, [3], wobei wegen der Positivität von K_0 gilt: $\mu_0 > 0$ und

$$\mu_0 = \sup_{u \in H_0 \setminus \{\Theta\}} \frac{(K_0 u, u)_0}{||u||_0^2} .$$

(2.15) zeigt $A_0'(\Theta) = I$. Deshalb ist $\lambda_0 = 1/\mu_0$ kleinster Eigenwert der linearisierten Gleichung

$$A_0'(\Theta) \, u - \lambda K_0 u = u - \lambda K_0 u = \Theta . \tag{2.16}$$

Da der erste Eigenwert des $-\Delta$-Operators einfach ist, gehen wir im weiteren davon aus, daß λ_0 einfacher Eigenwert von (2.16) zur normierten Eigenfunktion φ sei.

III. Bilden wir wieder $E_0 = H_0 \ominus \Phi$, $\Phi = \mathfrak{L}\{\varphi\}$. Für den nächsten Eigenwert $\lambda_1 > \lambda_0$ von (2.16) erhalten wir aus

$$\lambda_1 = \inf_{e \in E_0 \setminus \{\Theta\}} \frac{(e, e)}{(K_0 e, e)}, \qquad -(K_0 e, e) \geqq -\frac{1}{\lambda_1}(e, e) , \qquad e \in E_0 . \tag{2.17}$$

Sei $k < \lambda_1$ beliebig und $\gamma(k) = \min\left(1, 1 - \dfrac{k}{\lambda_1}\right)$; dann ist $\gamma(k) > 0$. Für jedes $\lambda \in (-\infty, k]$ und jedes $u \in H_0$ ist der Operator $P_E\left(A_0'(u) - \lambda K_0\right) \in [E_0 \to E_0]$

positiv definit. Denn mit $f'(t) \geqq 0$, $f'(0) = 0$ und (2.17) gelangen wir zu

$$\begin{aligned}(A_0'(u)\,e - \lambda K_0 e, e) &= (A_0'(u)\,e - A_0'(\Theta)\,e, e) + (A_0'(\Theta)\,e - \lambda K_0 e, e) \\ &= \int_\Omega (f'(u) - f'(0))\,e^2\,d\Omega + (e, e) - \lambda(K_0 e, e) \\ &\geqq (e, e) - \lambda(K_0 e, e) \geqq \gamma(k)\,\|e\|^2 .\end{aligned}$$

Deshalb läßt sich unter Bedingung I nachweisen, daß für alle $(a, \tau) \in R \times (-\infty, k - \lambda_0)$ die Operatoren $P_0(\cdot, a, \tau) = P_E(A_0(a\varphi + \cdot) - (\tau + \lambda_0) K_0(a\varphi + \cdot))$ auf E_0 stark monoton sind mit der einheitlichen Monotoniekonstante $\gamma(k) > 0$:

$$\begin{aligned}&(P_0(e_1, a, \tau) - P_0(e_2, a, \tau), e_1 - e_2) \\ &= \int_0^1 \frac{d}{ds}(A_0(a\varphi + e_2 + s(e_1 - e_2)), e_1 - e_2)\,ds - (\tau + \lambda_0)\,(K_0(e_1 - e_2), e_1 - e_2) \\ &= \int_0^1 (A_0'(a\varphi + e_2 + s(e_1 - e_2))\,(e_1 - e_2) - \lambda K_0(e_1 - e_2), e_1 - e_2)\,ds \geqq \gamma(k)\,\|e\|_0^2 .\end{aligned}$$

VIII. Sei N beschränkte Teilmenge von H_0. Dann existiert eine Konstante $k_2 > 0$, so daß für $u_1, u_2 \in N$; $v, w \in H_0$ wegen (2.15) und (2.3) für $0 \leqq \vartheta \leqq 1$ folgt

$$\begin{aligned}|(A_0'(u_1)\,v - A_0'(u_2)\,v, w)| &= \left|\int_\Omega \{f'(u_1)\,vw - f'(u_2)\,vw\}\,d\Omega\right| \\ &= \left|\int_\Omega f''(u_2 + \vartheta\,(u_1 - u_2))\,(u_1 - u_2)\,vw\,d\Omega\right| \\ &\leqq k_2 \|u_1 - u_2\|_0\,\|v\|_0\,\|w\|_0 ;\end{aligned}$$

d. h. die Abbildung $A_0' \in (H_0 \to [H_0 \to H_0])$ genügt auf jeder beschränkten Teilmenge von H_0 einer Lipschitz-Bedingung.

IX. Aus K_0 poitiv und $\varphi \neq \Theta$ folgt $c = (K_0\varphi, \varphi) > 0$.

X. Sei (a, τ) Element des Existenzbereiches der Lösung $e(a, \tau)$ von $P_E(A_0 u(a, \tau) - \lambda K_0 u(a, \tau)) = \Theta$. Dann besteht nach Satz 8 auf Grund der Voraussetzungen I, II, III, VIII die Schranke $\|e(a, \tau)\| = o(|a| + |\tau|)$ für $(|a| + |\tau|) \to 0$. Wegen (2.3) und (2.6) gibt es eine Konstante $k_3 > 0$ mit

$$\begin{aligned}(A_0'(u(a, \tau))\,\varphi - A_0'(\Theta)\,\varphi, \varphi) &= \int_\Omega f'(u(a, \tau))\,\varphi^2\,d\Omega \\ &\leqq M \|u(a, \tau)\|_{L_{2\nu_1}}^{\nu_1} \|\varphi\|_{L_4}^2 \leqq k_3 \|u(a, \tau)\|_0^{\nu_1} .\end{aligned}$$

Aus $\nu_1 \geqq 2$ und $\|u(a, \tau)\|^2 = |a|^2 + \|e(a, \tau)\|^2$ erhalten wir deshalb

$$(A_0'(u(a, \tau))\,\varphi - A_0'(\Theta)\,\varphi, \varphi) = o(|a| + |\tau|) .$$

Damit sind die Forderungen I, II, III, VI, IX und X nachgewiesen.

Es existiert also nach Satz 12 *eine Trajektorie* $u(t)$, $t \in (-\infty, t_0)$, *die einen differenzierbaren Zweig von Eigenfunktionen der verallgemeinerten Gleichung* (2.12) *im Bifurkationselement* λ_0 *darstellt.*

Beispiel

Bifurkation und differenzierbare Zweige von Eigenfunktionen in der Theorie der Kármánschen Platten

Die im unbelasteten Zustand ebene Mittelfläche einer Platte nimmt beim Ausbeulen unter Last eine gewölbte Form an. Die Ausbeulung wird dann durch eine Funktion $u(x_1, x_2)$, $(x_1, x_2) \in \Omega$, auf dem beschränkten Gebiet $\Omega \subseteq R^2$ beschrieben. In der Darstellung Th. v. Kármán's genügt diese Funktion einem System von partiellen Differentialgleichungen, s. [7],

$$\Delta^2 u = -\alpha L(u, F) + \frac{1}{D} q(x_1, x_2), \qquad \Delta^2 F = \beta L(u, u) \quad \text{in } \Omega. \tag{3.1}$$

Hierin stellt q die senkrecht zur Mittelfläche wirkende Last dar

$$L(u, v) = 2 \frac{\partial^2 u}{\partial x_1 \partial x_2} \frac{\partial^2 v}{\partial x_1 \partial x_2} - \frac{\partial^2 u}{\partial x_1^2} \frac{\partial^2 v}{\partial x_2^2} - \frac{\partial^2 v}{\partial x_1^2} \frac{\partial^2 u}{\partial x_2^2}, \tag{3.2}$$

$\alpha = h/D$, $\beta = E/2$ und D sind positive Konstanten, h bezeichnet die Plattendicke, D die Steifigkeit und E den Youngmodul des elastischen Materials. Schließlich ist $F(x_1, x_2)$, $(x_1, x_2) \in \Omega$, das Potential der Plattenkräfte, die teilweise von Lasten parallel zur Mittelfläche herrühren.

Bei der Untersuchung des Beulproblems für eingespannte Platten setzen wir

$$q(x_1, x_2) \equiv 0. \tag{3.3}$$

Die äußere Last wirkt also parallel zur Mittelfläche und sei durch das Spannungspotential

$$v_0 = \mu z, \tag{3.4}$$

mit der hinreichend glatten Funktion $z(x_1, x_2)$ vorgegeben. Zur Einführung von Operatoren gehen wir zunächst von dem Raum $C^4(\bar{\Omega})$ der in $\bar{\Omega}$ viermal stetig differenzierbaren Funktionen aus. Wir verlangen $z \in C^4(\bar{\Omega})$ und setzen $F = v + v_0$. Alsdann nehmen wir die Existenz einer Lösung der Gleichung (3.1) an, in der die Wölbungsfunktion u und der Anteil v des Potentials F Elemente des Raumes $\mathring{C}^4(\bar{\Omega})$ sind,

$$\mathring{C}^4(\bar{\Omega}) = \left\{ u \in C^4(\bar{\Omega});\ u(x), \frac{\partial u}{\partial x_1}(x), \frac{\partial u}{\partial x_2}(x) = 0 \quad x \in \partial\Omega \right\}.$$

Zur Vereinfachung der Schreibweise nehmen wir noch an, daß z Lösung der Gleichung (3.1) unter der Annahme $u(x_1, x_2) = 0$, also

$$\Delta^2 z = 0 \quad \text{in } \Omega. \tag{3.5}$$

z ist damit Lösung des Problems des ebenen Spannungszustandes der linearen Elastizitätstheorie. Die Lösungen dieses Problems sind gut bekannt. Wir erhalten somit für das Beulproblem der Kármánschen Plattentheorie das folgende Gleichungssystem:

$$\Delta^2 u = \lambda L(u, z) - \alpha L(u, v), \qquad \Delta^2 v = \beta L(u, u), \qquad u, v \in \mathring{C}^4(\bar{\Omega}). \tag{3.6}$$

Wir erwarten: für $|\lambda| = \alpha|\mu|$ hinreichend klein besitzt das System (3.6) die einzige Lösung $u, v = \Theta$, also

$$u(x_1, x_2) \equiv 0\,, \qquad F(x_1, x_2) = \mu z(x_1, x_2)\,. \tag{3.7}$$

Die äußeren Kräfte, die parallel zur Mittelfläche wirken, erzeugen lediglich einen ebenen Spannungszustand. Für hinreichend große $|\lambda|$ mag es neben dieser trivialen Lösung auch nichttriviale Lösungen von (3.6) geben. Sie beschreiben einen Beulzustand. Anliegen der Beultheorie ist die Berechnung kritischer Beullasten (hier dargestellt durch den reellen Parameter λ), bei denen ein Ausbeulen möglich wird, und die Beschreibung des Beulverhaltens nach dem Überschreiten der kritischen Last.

Das Beulproblem als nichtlineares Eigenwertproblem

Beginnen wir mit der Bemerkung, daß man den Differentialausdruck $L(u, v)$ für $u, v \in \mathring{C}^4$ in Divergenzform schreiben kann:

$$L(u, v) = -\frac{\partial}{\partial x_1}\left\{\frac{\partial^2 v}{\partial x_2^2}\frac{\partial u}{\partial x_1} - \frac{\partial^2 v}{\partial x_1\,\partial x_2}\frac{\partial u}{\partial x_2}\right\} - \frac{\partial}{\partial x_2}\left\{\frac{\partial^2 v}{\partial x_1^2}\frac{\partial u}{\partial x_2} - \frac{\partial^2 v}{\partial x_1\,\partial x_2}\frac{\partial u}{\partial x_1}\right\}. \tag{3.8}$$

Wir setzen nun voraus, daß Ω sowohl die Bedingungen des Gaußschen Satzes wie auch die Bedingungen der Sobolewschen Einbettungssätze erfüllt, und verjüngen das System (3.6) mit einem Element $h \in \mathring{C}^4$.

$$\int_\Omega \sum_{i,j=1}^{2} \frac{\partial^2 u}{\partial x_i\,\partial x_j}\frac{\partial^2 h}{\partial x_i\,\partial x_j}\,d\Omega$$

$$= \lambda \int_\Omega \left\{\frac{\partial^2 z}{\partial x_2^2}\frac{\partial u}{\partial x_1}\frac{\partial h}{\partial x_1} - \frac{\partial^2 z}{\partial x_1\,\partial x_2}\frac{\partial u}{\partial x_2}\frac{\partial h}{\partial x_1} + \frac{\partial^2 z}{\partial x_1^2}\frac{\partial u}{\partial x_2}\frac{\partial h}{\partial x_2} - \frac{\partial^2 z}{\partial x_1\,\partial x_2}\frac{\partial u}{\partial x_1}\frac{\partial h}{\partial x_2}\right\} d\Omega$$

$$-\alpha \int_\Omega \left\{\frac{\partial^2 v}{\partial x_2^2}\frac{\partial u}{\partial x_1}\frac{\partial h}{\partial x_1} - \frac{\partial^2 v}{\partial x_1\,\partial x_2}\frac{\partial u}{\partial x_2}\frac{\partial h}{\partial x_1} + \frac{\partial^2 v}{\partial x_1^2}\frac{\partial u}{\partial x_2}\frac{\partial h}{\partial x_2} - \frac{\partial^2 v}{\partial x_1\,\partial x_2}\frac{\partial u}{\partial x_1}\frac{\partial h}{\partial x_2}\right\} d\Omega\,,$$

$$\int_\Omega \sum_{i,j=1}^{2} \frac{\partial^2 v}{\partial x_i\,\partial x_j}\frac{\partial^2 h}{\partial x_i\,\partial x_j}\,d\Omega$$

$$= \beta \int_\Omega \left\{\frac{\partial^2 u}{\partial x_2^2}\frac{\partial u}{\partial x_1}\frac{\partial h}{\partial x_1} - \frac{\partial^2 u}{\partial x_1\,\partial x_2}\frac{\partial u}{\partial x_2}\frac{\partial h}{\partial x_1} + \frac{\partial^2 u}{\partial x_1^2}\frac{\partial u}{\partial x_2}\frac{\partial h}{\partial x_2} - \frac{\partial^2 u}{\partial x_1\,\partial x_2}\frac{\partial u}{\partial x_1}\frac{\partial h}{\partial x_2}\right\} d\Omega\,. \tag{3.9}$$

Für die weitere Behandlung des Systems (3.9) benötigen wir die Sobolew-Räume $W_4^1(\Omega)$ und $W_2^2(\Omega)$. Es gilt $\mathring{C}^4 \subseteq W_2^2 \cap W_4^1$; die Abschließung von $\mathring{C}^4$ in W_2^2 nennen wir H_0. H_0 ist Unterraum von W_2^2 und mit dem Skalarpunkt

$$(u, v) = \int_\Omega \sum_{i,j=1}^{2} \frac{\partial^2 u}{\partial x_i\,\partial x_j}\frac{\partial^2 v}{\partial x_i\,\partial x_j}\,d\Omega \qquad u, v \in H_0 \tag{3.10}$$

Hilbertraum. Auf H_0 ist die durch das Skalarprodukt (3.10) definierte Norm $\|\cdot\|$ der Norm des Raumes W_2^2 äquivalent. Auf W_4^1 wählen wir die Norm

$$||u||_{4,1}^4 = \int_\Omega \left\{ u^4 + \sum_{i,j=1}^{2} \left(\frac{\partial u}{\partial x_i}\right)^4 \right\} d\Omega . \tag{3.11}$$

Auf Grund der Sobolewschen Einbettungssätze ist der Einbettungsoperator $E \in (H_0 \to W_4^1)$ beschränkt und vollstetig, vgl. [6]. In den Gleichungen (3.9) kommt das Funktional

$$\Psi(u, w)\, h = \int_\Omega \left\{ \frac{\partial^2 w}{\partial x_2^2} \frac{\partial u}{\partial x_1} \frac{\partial h}{\partial x_1} - \frac{\partial^2 w}{\partial x_1 \partial x_2} \frac{\partial u}{\partial x_2} \frac{\partial h}{\partial x_1} + \frac{\partial^2 w}{\partial x_1^2} \frac{\partial u}{\partial x_2} \frac{\partial h}{\partial x_2} - \frac{\partial^2 w}{\partial x_1 \partial x_2} \frac{\partial u}{\partial x_1} \frac{\partial h}{\partial x_2} \right\} d\Omega$$

vor. Für fixierte $u \in W_4^1$, $w \in W_2^2$ ist $\Psi(u, w)$ ein lineares beschränktes Funktional in H_0. Nach dem Rieszschen Satz erklären wir den Operator $B \in (W_4^1 \times W_2^2 \to H_0)$,

$$\big(B(u, w), h\big) = \Psi(u, w)\, h . \tag{3.12}$$

Der Operator B ist bilinear. Weiter gilt für die Einschränkung $B_0 \in (H_0 \times H_0 \to H_0)$ des Operators B die Abschätzung

$$||B_0(u, w)|| \leqq \nu ||w||\, ||u|| \qquad w, u \in H_0 \tag{3.13}$$

mit einer Konstante $\nu > 0$.

Das System (3.9) schreiben wir nun in der Form

$$u = \lambda B(u, z) - \alpha B_0(u, v) , \qquad v = \beta B_0(u, u) , \tag{3.14}$$

woraus wir durch Einsetzen unmittelbar zur Operatorgleichung

$$u = \lambda B(u, z) - \alpha\beta B_0\big(u, B_0(u, u)\big) \tag{3.15}$$

kommen. Diese Gleichung bestimmt die Beulfunktion u; wir betrachten sie für vorgegebene $z \in W_2^2$ und suchen Eigenwerte $\lambda \in R$ zu Eigenfunktionen $u \in H_0$. Die Operatorgleichung (3.15) betrachten wir als die Beulgleichung der Kármánschen Plattentheorie für eingespannte Platten.

Einige Eigenschaften der Operatoren der Beulgleichung (3.15)

Mit $A_0 u = u + \alpha\beta B_0\big(u, B_0(u, u)\big)$ und $K_0 u = B(u, z)$ hat Gleichung (3.15) für jedes fixierte $z \in W_2^2$ die Gestalt von Gleichung (1.1), $A_0 \Theta = \Theta$ folgt unmittelbar aus der Bilinearität von B_0. Für die Operatoren der Gleichung (3.15) weisen wir nun die Forderungen I, II, IV, V, VI, VII, IX und X nach und greifen dabei auf die in [7] hergeleiteten Eigenschaften zurück. Nach [7], Lemma 4 ist der Operator $F_0 \in (H_0 \to H_0)$, $F_0 u = B_0\big(u, B_0(u, u)\big)$, auf H_0 Fréchet-differenzierbar und $F_0'(u)\, v = 2B_0\big(u, B_0(u, v)\big) + B_0\big(v, B_0(u, u)\big)$.

Damit ist I und insbesondere IV erfüllt, da wegen $F_0'(\Theta)\, v = \Theta$, für alle $v \in H_0$, $A_0'(\Theta) = \mathrm{I}$ gilt. Die linearisierte Gleichung (1.3) hat dann die Form

$$u = \lambda B(u, z) . \tag{3.16}$$

Da die charakteristischen Zahlen des Operators $B(\cdot, z)$ vom Element $z \in W_2^2$ abhängen, welches die Belastung der Platte repräsentiert, schränken wir die zulässigen $z \in W_2^2$ ein. Wir nennen nunmehr ein $z \in W_2^2$ zulässig, wenn die folgenden Bedingungen erfüllt sind:

(i_z) $B(\cdot, z)$ ist positiv
(ii_z) $B(\cdot, z)$ besitzt eine einfache charakteristische Zahl λ_0.

Das Element $z \in W_2^2$ in Gleichung (3.15) sei zulässig. $\varphi \in H_0$ sei die normierte Eigenfunktion der Gleichung (3.16) zum einfachen Eigenwert λ_0. Dann ist Bedingung II gültig und wegen $(B(\varphi, z), \varphi) = c > 0$ auch Bedingung IX.

Sei $\Phi = \mathfrak{L}\{\varphi\}$, $E_0 = H_0 \ominus \Phi$ und für $u \in H_0$ $u = a\varphi + e$. Die restlichen Forderungen V, VI, VII, VIII und X ergeben sich dann aus folgenden Resultaten in [7]:

Der Operator $B(\cdot, z) \in (H_0 \to H_0)$ ist für jedes $z \in W_2^2$ vollstetig, linear und selbstadjungiert; daher existiert auf E_0 das beschränkte Inverse $S_0 = (I - \lambda_0 B(\cdot, z))^{-1} \in [E_0 \to E_0]$. In jedem $u \in H_0$ ist $F_0'(u) \in [H_0 \to H_0]$ selbstadjungiert. Es existieren eine Umgebung $G \subseteq R^2$ von $\mathfrak{O} \in R^2$ und eine Kugel $K(\Theta, r) \subseteq E_0$ derart, daß der Operator $Q_0(\cdot, a, \tau)$, $Q_0(e, a, \tau) = \tau B(e, z) - \alpha\beta S_0 P_E F_0 (a\varphi + e)$, für jedes $(a, \tau) \in G$ die Kugel $K(\Theta, r)$ in sich abbildet und auf $K(\Theta, r)$ strikt kontraktiv ist mit einer für alle $(a, \tau) \in G$ einheitlichen Konstante $\delta < 1$.

Auf jeder beschränkten Teilmenge von H_0 ist die Abbildung $F_0' \in (H_0 \to [H_0 \to H_0])$ Lipschitz-stetig.

Für (a, τ) aus dem Existenzbereich der Lösung $e(a, \tau)$ der Gleichung $P_E(A_0 - \lambda K_0)(a\varphi + e(a, \tau)) = \Theta$ gilt

$$\left(A_0'(u(a, \tau))\,\varphi - A_0'(\Theta)\,\varphi, \varphi\right) = \left(F_0'(a\varphi + e(a, \tau))\,\varphi, \varphi\right) = o(|a| + |\tau|)$$

für $(|a| + |\tau|) \to 0$.

Mit Satz 12 erhalten wir folgendes Ergebnis:

Sei $z \in W_2^2$ zulässig. Dann beginnt in jeder einfachen charakteristischen Zahl des Operators $B(\cdot, z)$ eine differenzierbare Trajektorie von Eigenfunktionen der Beulgleichung (3.15).

Literatur

[1] Pontrjagin, L. S., Gewöhnliche Differentialgleichungen, Berlin (1965).
[2] Katschurowski, R. I., Нелинейные монотонные операторы в банаховых пространствах. УМН **23** (1968) 121—168.
[3] Kantorowitsch, L. W., und Akilow, G. P., Funktionalanalysis in normierten Räumen, Berlin 1964.
[4] Langenbach, A., Über implizite Operatorfunktionen im Hilbert-Raum. Mber. Dt. Akad. Wiss. **13** (1971) 648—655.
[5] Dieudonné, J., Grundzüge der modernen Analysis, Berlin 1971.
[6] Smirnow, W. I., Lehrgang der höheren Mathematik, Teil V, Berlin (1962).
[7] Langenbach, A., Bifurkation und differenzierbare Zweige von Eigenfunktionen in der Theorie der Kármánschen Platten. Math. Nachr. (im Druck).

Anschrift der Autoren:

Humboldt-Universität, Berlin,
Sektion Mathematik

Über die Differenzierbarkeit der Lösungen gewisser quasilinearer elliptischer Differentialgleichungen am Rande des Gebiets

MANFRED MÜLLER, Berlin

1. Im beschränkten Gebiet $\Omega \subset R_n$ mit dem Rand Γ betrachten wir das Randwertproblem

$$\sum_{|\alpha|,|\beta|=2} a_{\alpha\beta} D^\beta[f(T^2(u))\, D^\alpha u] = g\,, \tag{1}$$

$$D^\alpha u|_\Gamma = 0\,, \quad |\alpha| \leqq 1\,. \tag{2}$$

Hierbei bedeutet

$$T^2(u) = \sum_{|\alpha|,|\beta|=2} a_{\alpha\beta}\, D^\alpha u\, D^\beta u\,,$$

Voraussetzungen:

I. $f \in C^1([0, +\infty))$.

II. Es existieren Konstanten $\varkappa$, N derart, daß für alle $\xi \in R_1$ gilt:

$$0 < \varkappa \leqq \frac{d}{d\xi} f(\xi^2)\, \xi \leqq N\,.$$

III. a) $a_{\alpha\beta} \in R_1$, $\quad a_{\alpha\beta} = a_{\beta\alpha}$ für alle $|\alpha|, |\beta| = 2$.

b) Für jeden Vektor $\lambda = (\lambda_\alpha)$, $|\alpha| = 2$, gilt die Ungleichung

$$\sum_{|\alpha,|\beta|=2} a_{\alpha\beta}\lambda_\alpha\lambda_\beta \geqq b_0 \sum_{|\alpha|=2} \lambda_\alpha^2\,, \quad b_0 > 0\,.$$

IV. $g \in L_2(\Omega)$.

2. Sei $\mathring{W}_2^2(\Omega)$ die Abschließung der Menge $\dot{C}^\infty(\Omega)$ aller in Ω finiten, beliebig oft stetig differenzierbaren Funktionen in der Norm

$$\|\varphi\|^2_{W_2^2(\Omega)} = \int_\Omega \sum_{|\alpha|\leqq 2} (D^\alpha\varphi)^2\, dx\,.$$

Definition 1. Die Funktion $u \in \mathring{W}_2^2(\Omega)$ heißt verallgemeinerte Lösung des Randwertproblems (1), (2), wenn für jedes $\varphi \in \mathring{W}_2^2(\Omega)$ die Integralidentität

$$\int_\Omega f(T^2(u))\, T^2(u, \varphi)\, dx = \int_\Omega g\varphi\, dx \tag{3}$$

erfüllt ist.

3. In [4] ist mit Hilfe der Ergebnisse von A. Langenbach [3] gezeigt, daß folgender Satz gilt:

Satz 1. *Unter den Voraussetzungen* I—IV *existiert für das Randwertproblem* (1), (2) *genau eine verallgemeinerte Lösung* $u \in \mathring{W}_2^2(\Omega)$.

Es fragt sich nun, ob man für die verallgemeinerte Lösung $u \in \mathring{W}_2^2(\Omega)$ die Existenz von Ableitungen höherer Ordnung nachweisen kann. Diese Frage läßt sich positiv beantworten, wenn es sich um Ableitungen 3. Ordnung und deren Zugehörigkeit zu $L_2(\Omega')$ für jedes echte Teilgebiet $\Omega' \subset \Omega$ handelt (vgl. [4] sowie [5]):

Satz 2. *Unter den Voraussetzungen* I—IV *gehört die verallgemeinerte Lösung u des Randwertproblems* (1), (2) *zu* $W_2^3(\Omega')$ *für jedes echte Teilgebiet* $\Omega' \subset \Omega$, *und es gilt die Abschätzung*

$$\|u\|_{W_2^3(\Omega')} \leqq C \|g\|_{L_2(\Omega)}$$

mit einer von $d(\Omega', \Gamma)$ *abhängigen Konstanten C.*

4. Gegenstand der vorliegenden Arbeit ist es, das Verhalten der Ableitungen 3. Ordnung am Rand Γ des Gebiets Ω zu untersuchen und eine Aussage über deren Zugehörigkeit zu L_2 im ganzen Gebiet Ω zu gewinnen. Genauer: wir beweisen

Satz 3. *Wenn die Voraussetzungen* I.—IV. *erfüllt sind und der Rand* Γ *dreimal stetig differenzierbar ist, gehört die verallgemeinerte Lösung des Randwertproblems* (1), (2) *zu* $W_2^3(\Omega)$, *und es gilt die Abschätzung*

$$\|u\|_{W_2^3(\Omega)} \leqq C \|g\|_{L_2(\Omega)} ,$$

wobei die Konstante C nicht von u abhängt.

Der Vorbereitung des Beweises dienen die folgenden Bemerkungen über die Differenzierbarkeit des Randes Γ sowie über Abschätzungen für lineare elliptische Differentialoperatoren (vgl. [2], [1]).

5. Definition 2. Der Rand Γ heißt l-mal stetig differenzierbar, wenn gilt:
a) Es existiert eine Zahl $\sigma > 0$ derart, daß der Durchschnitt $\Gamma \cap K_\sigma(M_0)$ für jeden beliebigen Punkt $M_0 \in \Gamma$ eine zusammenhängende Fläche darstellt. $K_\sigma(M_0)$ bedeutet dabei eine Kugel mit dem Radius σ und dem Mittelpunkt in M_0.
b) Der Durchschnitt $\Gamma \cap K_\sigma(M_0)$ genügt in einem cartesischen Ortskoordinatensystem $(y_1, \ldots, y_n)$ mit dem Ursprung in M_0 der Gleichung

$$y_n = \omega(y_1, \ldots, y_{n-1}) ,$$

wobei die Funktion $w(y_1, \ldots, y_{n-1})$ l-mal stetig differenzierbar ist in dem durch Projektion von $\Gamma \cap K_\sigma(M_0)$ auf die Hyperebene $y_n = 0$ entstehenden Gebiet.

Definition 3. Der Rand Γ heißt rektifizierbar in einer gewissen Umgebung des Punktes $M_0 \in \Gamma$, wenn in dieser Umgebung U Funktionen

$$z_i = z_i(x_1, \ldots, x_n) , \qquad i = 1, \ldots, n ,$$

mit folgenden Eigenschaften existieren:
a) Das Bild des Durchschnitts $\Gamma \cap U$ liegt in der Hyperebene: $z_n = 0$.

b) Der Durchschnitt $\Omega \cap U$ wird in den Halbraum $z_n > 0$ abgebildet.
c) Die Funktionen $z_i(x_1, \ldots, x_n)$, $i = 1, \ldots, n$, sind differenzierbar und die Funktionaldeterminante $\dfrac{D(z)}{D(x)}$ ist in $\Omega \cap U$ verschieden von Null.

Im Falle eines l-mal stetig differenzierbaren Randes Γ lassen sich Koordinaten $(z_1, \ldots, z_n)$ mit den angegebenen Eigenschaften folgendermaßen einführen:

Zuerst transformiert man $(x_1, \ldots, x_n)$ auf Ortskoordinaten $(y_1, \ldots, y_n)$ mit dem Ursprung in $M_0 = (x_1^{(0)}, \ldots, x_n^{(0)})$:

$$y_k = \sum_{m=1}^{n} c_{km}(x_m - x_m^{(0)}) , \qquad k = 1, \ldots, n .$$

Dabei gilt offensichtlich $\frac{D(y)}{D(x)} = ||c_{km}|| = \text{const} \neq 0$.

Danach führt man in $K_\sigma(M_0)$ neue Koordinaten $(z_1, \ldots, z_n)$ mit Hilfe der Formeln ein

$$\left.\begin{aligned} z_i &= y_i , \qquad i = 1, \ldots, n-1 , \\ z_n &= y_n - \omega(y_1, \ldots, y_{n-1}) . \end{aligned}\right\} \tag{4}$$

Die Funktionen $z_i = z_i(y_1(x_1, \ldots, x_n), \ldots, y_n(x_1, \ldots, x_n))$, $i = 1, \ldots, n$, sind l-mal stetig differenzierbar. Außerdem ist $\frac{D(z)}{D(y)} = 1$, so daß gilt $\frac{D(z)}{D(x)} = \text{const} \neq 0$.

Der Zylinder

$$Z = \left\{ (z_1, \ldots, z_n) \,\middle|\, \sum_{i=1}^{n-1} z_i^2 \leqq \frac{\varrho}{2}, \ 0 \leqq z_n \leqq \tau \right\}$$

gehört dann für hinreichend kleine ϱ, τ zu $\bar{\Omega}$.

6. Für jede Funktion $v \in L_2(\Omega)$ gilt die Beziehung

$$||v||_{L_2(\Omega)} = \sup_{w \in L_2(\Omega)} \frac{\left| \int_\Omega vw \, dx \right|}{||w||_{L_2(\Omega)}} . \tag{5}$$

Das Gleichheitszeichen bleibt erhalten, wenn das Supremum nicht über ganz $L_2(\Omega)$, sondern nur über eine dichte Teilmenge genommen wird.

7. Wir betrachten den Operator

$$L_0 = (-1)^m \sum_{|\alpha| = m} D^{2\alpha} .$$

Lemma 1. *Für beliebiges $u \in \dot{C}^\infty(\Omega)$ gelten die beiden Abschätzungen*

$$||u||_{W_2^{2m}(\Omega)} \leqq C||L_0 u||_{L_2(\Omega)} , \tag{6}$$

$$||u||_{W_2^{2m}(\Omega)} \leqq C||(L_0 + I)\, u||_{L_2(\Omega)} , \tag{7}$$

wobei die Konstante C von u unabhängig ist.

Beweis: Die Abschätzung (6) folgt unmittelbar aus den Ergebnissen von [1], wenn man berücksichtigt, daß L_0 stark elliptisch ist.

Für die Abschätzung (7) betrachten wir den Ausdruck

$$||(L_0 + \lambda I)\, u||_{L_2(\Omega)}^2 = ||L_0 u||_{L_2(\Omega)}^2 + 2\lambda(L_0 u, u) + \lambda^2 ||u||_{L_2(\Omega)}^2 .$$

Wegen

$$(L_0 u, u) = (-1)^m \int_\Omega \sum_{|\alpha| = 2m} D^{2\alpha} u \cdot u \, dx = \int_\Omega \sum_{|\alpha| = m} (D^\alpha u)^2 \, dx = \sum_{|\alpha| = m} ||D^\alpha u||_{L_2(\Omega)}^2$$

erhalten wir

$$||(L_0 + \lambda I)\, u||_{L_2(\Omega)}^2 = ||L_0 u||_{L_2(\Omega)}^2 + 2\lambda \sum_{|\alpha| = m} ||D^\alpha u||_{L_2(\Omega)}^2 + \lambda^2 ||u||_{L_2(\Omega)}^2 .$$

Für $\lambda = 1$ folgt hieraus

$$||(L_0 + I)\, u||^2_{L_2(\Omega)} \geqq ||L_0 u||^2_{L_2(\Omega)}\,,$$

was zusammen mit (6) die Abschätzung (7) liefert.

8. Lemma 2. *Wenigstens eine der Mengen* $\{L_0\varphi\}$, $\{(L_0 + I)\,\varphi\}$, $\varphi \in \dot{C}^\infty(\Omega)$, *ist dicht in* $L_2(\Omega)$.

Beweis: Falls entgegen der Behauptung keine der beiden Mengen dicht ist in $L_2(\Omega)$, so existiert wenigstens eine Funktion $w \in L_2(\Omega)$, $w \neq 0$, die orthogonal zu allen Funktionen von $\{L_0\varphi\}$ und $\{(L_0 + I)\,\varphi\}$ ist. Das heißt

$$\int_\Omega w L_0 \varphi\, dx = 0 \qquad \text{und} \qquad \int_\Omega w(L_0 + I)\,\varphi\, dx = 0\,.$$

Die Differenz dieser beiden Gleichungen ergibt

$$\int_\Omega w\varphi\, dx = 0$$

für jede Funktion $\varphi \in \dot{C}^\infty(\Omega)$, was aber $w = 0$ zur Folge hätte.

9. Aus Lemma 2 folgern wir, daß wenigstens eine der beiden Beziehungen gilt:

$$||v||_{L_2(\Omega)} = \sup_{\varphi \in \dot{C}^\infty(\Omega)} \frac{\left|\int_\Omega v L_0 \varphi\, dx\right|}{||L_0\varphi||_{L_2(\Omega)}}\,,$$

$$||v||_{L_2(\Omega)} = \sup_{\varphi \in \dot{C}^\infty(\Omega)} \frac{\left|\int_\Omega v(L_0 + I)\,\varphi\, dx\right|}{||(L_0 + I)\,\varphi||_{L_2(\Omega)}}\,.$$

In jedem Falle ist dann die Abschätzung richtig

$$||v||_{L_2(\Omega)} \leqq \sup_{\varphi \in \dot{C}^\infty(\Omega)} \frac{\left|\int_\Omega v L_0 \varphi\, dx\right|}{||L_0\varphi||_{L_2(\Omega)}} + \sup_{\varphi \in \dot{C}^\infty(\Omega)} \frac{\left|\int_\Omega v(L_0 + I)\,\varphi\, dx\right|}{||(L_0 + I)\,\varphi||_{L_2(\Omega)}}\,,$$

woraus wir unter Verwendung von (6) und (7) erhalten

$$||v||_{L_2(\Omega)} \leqq C \sup_{\varphi \in \dot{C}^\infty(\Omega)} \frac{\left|\int_\Omega v L_0 \varphi\, dx\right|}{||\varphi||_{W_2^{2m}(\Omega)}} + C \sup_{\varphi \in \dot{C}^\infty(\Omega)} \frac{\left|\int_\Omega v(L_0 + I)\,\varphi\, dx\right|}{||\varphi||_{W_2^{2m}(\Omega)}}\,.$$

Die Integrale der Zähler lassen sich folgendermaßen abschätzen:

$$\left|\int_\Omega v L_0 \varphi\, dx\right| = \left|(-1)^m \int_\Omega v \sum_{|\alpha| = m} D^{2\alpha}\varphi\, dx\right| \leqq \sum_{|\alpha| = m} \left|\int_\Omega v D^{2\alpha}\varphi\, dx\right|,$$

$$\left|\int_\Omega v(L_0 + I)\,\varphi\, dx\right| \leqq \left|\int_\Omega v L_0\varphi\, dx\right| + \left|\int_\Omega v\varphi\, dx\right| \leqq \sum_{|\alpha| \leqq m} \left|\int_\Omega v\, D^{2\alpha}\varphi\, dx\right|.$$

Damit bekommen wir endgültig die Abschätzung

$$||v||_{L_2(\Omega)} \leqq C \sup_{\varphi \in \dot{C}^\infty(\Omega)} \frac{\sum_{|\alpha| \leqq m} \left|\int_\Omega v\, D^{2\alpha}\varphi\, dx\right|}{||\varphi||_{W_2^{2m}(\Omega)}}\,, \tag{8}$$

wo die Konstante C von v nicht abhängt.

10. Zur Bezeichnung der Differenzenquotienten einer in Ω definierten Funktion $u(x)$ verwenden wir die Schreibweise

$$D^1_{\pm\Delta_i}u(x) = \frac{1}{\pm\Delta x_i}\left(\overset{\pm i}{u(x)} - u(x)\right)$$
$$= \frac{1}{\pm\Delta x_i}\left(u(x_1, \dots, x_i \pm \Delta x_i, \dots, x_n) - u(x_1, \dots, x_i, \dots, x_n)\right),$$
$$i = 1, \dots, n .$$

Wenn $u, v \in L_2(\Omega)$ und der Träger Ω' von v echtes Teilgebiet von Ω ist, so gilt für $\Delta x_i < d(\Omega', \Gamma)$ die Formel

$$\int_\Omega u D^1_{\pm\Delta_i} v\, dx = -\int_{\Omega'} D^1_{\mp\Delta_i} u\, v\, dx, \qquad i = 1, \dots, n . \tag{9}$$

Wenn $u \in W^l_p(\Omega)$, $p > 1$, so gilt für alle $|\alpha| \leqq l - 1$ und $\Delta x_i < d(\Omega', \Gamma)$ die Abschätzung

$$\int_{\Omega'} |D^1_{\pm\Delta_i} D^\alpha u|^p\, dx \leqq ||u||^p_{W^l_p(\Omega)} . \tag{10}$$

Die Existenz von verallgemeinerten Ableitungen einer gegebenen Funktion kann man nachweisen, indem man die gleichmäßige Beschränktheit der Integrale über die Differenzenquotienten beweist. Nach [2] gilt nämlich das folgende

Lemma 3. *Seien $u \in L_p(\Omega)$, $p > 1$, Ω' echtes Teilgebiet von Ω und*

$$\int_{\Omega'} |D^1_{\Delta_i} u|^p\, dx \leqq C_1$$

gleichmäßig für alle $\Delta x_i < c_i$.

Dann besitzt die Funktion u die verallgemeinerte Ableitung $D^1_i u$, die dem Raum $L_2(\Omega')$ angehört.

Außerdem gilt für jede Nullfolge $\{\Delta x^{(k)}_i\}$, $k = 1, 2, \dots$,

$$D^1_{\Delta_i} u \xrightarrow[k\to\infty]{} D^1_i u \qquad \textit{in } L_p(\Omega')$$

und

$$\int_{\Omega'} |D^1_i u|^p\, dx \leqq C_1 .$$

11. Beweis von Satz 3: Sei $M_0 \in \Gamma$ beliebig, aber fixiert. Wir gehen aus von der Integralidentität (3) und nehmen an, daß der Übergang zu Ortskoordinaten $(y_1, \dots, y_n)$ mit dem Ursprung in M_0 bereits vollzogen ist. Wir haben also die Identität

$$\int_\Omega f\big(T^2(u)\big)\, T^2(u, \varphi)\, dy = \int_\Omega g\varphi\, dy \tag{11}$$

für jedes $\varphi \in \mathring{W}^2_2(\Omega)$.

Wir betrachten die Identität (11) speziell für alle $\varphi \in \mathring{W}^2_2(\Omega_0)$, wobei Ω_0 das Urbild des Zylinders Z vermittels der Transformation (4) darstellt.

Als nächstes gehen wir zu den Koordinaten $(z_1, \dots, z_n)$ über. Dabei gilt $\frac{D(z)}{D(y)} = 1$. Wir berechnen $T^2(u, \varphi)$ in den z-Koordinaten

$$D_i^1 u(y_1, \dots, y_n) = \sum_{l=1}^{n} D_l^1 u(z_1, \dots, z_n)\, D_i^1 z_l(y_1, \dots, y_n)\,,$$

$$D_j^1 D_i^1 u(y_1, \dots, y_n) = \sum_{l,m=1}^{n} D_m^1 D_l^1 u(z_1, \dots, z_n)\, D_i^1 z_m(y_1, \dots, y_n)\, D_i^1 z_l(y_1, \dots, y_n)$$
$$+ \sum_{l_1=1}^{n} D_{l_1}^1 u(z_1, \dots, z_n)\, D_j^1 D_i^1 z_{l_1}(y_1, \dots, y_n)\,.$$

Bezeichnen α_i, α_j bzw. β_p, β_q diejenigen Komponenten der Multiindizes α und β, $|\alpha|, |\beta| = 2$, die von Null verschieden sind, so ergibt sich

$$T^2(u, \varphi) = \sum_{|\alpha|, |\beta| = 2} d_{\alpha\beta} \left(\sum_{l,m=1}^{n} D_m^1 D_l^1 u\, D_i^1 z_l\, D_j^1 z_m + \sum_{l_1=1}^{n} D_{l_1}^1 u\, D_j^1 D_i^1 z_{l_1} \right)$$
$$\times \left(\sum_{r,s=1}^{n} D_s^1 D_r^1 \varphi\, D_p^1 z_r\, D_q^1 z_s + \sum_{r_1=1}^{n} D_{r_1}^1 \varphi\, D_q^1 D_p^1 z_{r_1} \right)$$
$$= \sum_{|\gamma|, |\delta| = 2} b_{\gamma\delta}\, D^\gamma u\, D^\delta \varphi + \sum_{\substack{|\gamma| = 2 \\ |\delta_1| = 1}} c_{\gamma\delta_1}\, D^{\delta_1} u\, D^\gamma \varphi + \sum_{\substack{|\gamma| = 2 \\ |\delta_1| = 1}} c_{\gamma\delta_1} D^{\delta_1} \varphi\, D^\gamma u$$
$$+ \sum_{|\gamma_1|, |\delta_1| = 1} d_{\gamma_1\delta_1}\, D^{\gamma_1} u\, D^{\delta_1} \varphi \underset{\text{Def.}}{\equiv} \hat{T}^2(u, \varphi)\,.$$

Wir vermerken, daß in den Koeffizienten $b_{\gamma\delta}$, $c_{\gamma\delta_1}$, $d_{\gamma_1\delta_1}$ Ableitungen 2. Ordnung der Funktionen z_i, $i = 1, \dots, n$, vorkommen. Wenn man berücksichtigt, daß im Verlaufe des Beweises von den Koeffizienten Differenzenquotienten 1. Ordnung zu berechnen sind, wird hier sichtbar, daß $\Gamma \in C^3$ gefordert werden muß.

Die Koeffizienten $b_{\gamma\delta}$, $c_{\gamma\delta_1}$, $d_{\gamma_1\delta_1}$ hängen offensichtlich von $y_1, \dots, y_n$ und damit von $z_1, \dots, z_n$ ab.

Es läßt sich zeigen, daß det $|b_{\gamma\delta}| = 1$, so daß die Matrix $||b_{\gamma\delta}||$ eine positiv definite quadratische Form erzeugt:

$$\sum_{|\gamma|, |\delta| = 2} b_{\gamma\delta}\, \lambda_\gamma \lambda_\delta \geqq b_1 \sum_{|\gamma| = 2} \lambda_\gamma^2\,.$$

Da in der Form $\hat{T}^2(u, \varphi)$ die drei letzten Summen Ableitungen niederer Ordnung enthalten, kommt es vor allem auf die Summe

$$\sum_{|\gamma|, |\delta| = 2} b_{\gamma\delta} D^\gamma u\, D^\delta \varphi$$

an. Der Einfachheit und Übersichtlichkeit halber nehmen wir deshalb an, daß die Integralidentität (11) in den Koordinaten $(z_1, \dots, z_n)$ dasselbe Aussehen hat wie in den Ortskoordinaten $(y_1, \dots, y_n)$. D. h., es gilt:

$$\int_Z f\big(T^2(u)\big)\, T^2(u, \varphi)\, dz = \int_Z g\varphi\, dz \tag{12}$$

für jede Funktion $\varphi \in \mathring{W}_2^2(Z)$.

Wir führen folgende Bezeichnungen ein:

$$\overline{G}_1 = \left\{ z \in Z \,\middle|\, \sum_{i=1}^{n-1} z_i^2 \leqq \frac{\varrho}{8}, \quad 0 \leqq z_n \leqq \frac{\tau}{4} \right\},$$

$$\overline{G}_2 = \left\{ z \in Z \,\middle|\, \sum_{i=1}^{n-1} z_i^2 \leqq \frac{\varrho}{4}, \quad 0 \leqq z_n \leqq \frac{\tau}{2} \right\}.$$

Sei $\psi \in \mathring{W}_2^2(G_2)$ beliebig. Nach Fortsetzung durch Null auf ganz Z gilt auch $\psi \in \mathring{W}_2^2(Z)$. Die Funktion

$$\varphi = D^1_{-\Delta_i}\psi\,, \qquad i = 1, \dots, n-1\,,$$

gehört dann für hinreichend kleines Δz_i zu $\mathring{W}_2^2(Z)$ und kann in der Identität (12) als Vergleichsfunktion genommen werden:

$$\int_Z \sum_{|\alpha|,|\beta|=2} a_{\alpha\beta} f\big(T^2(u)\big)\, D^\alpha u\, D^\beta D^1_{-\Delta_i}\psi\, dz = \int_Z g\, D^1_{-\Delta_i}\psi\, dz\,.$$

Die linke Seite formen wir mit Hilfe von Formel (9) um

$$\int_Z \sum_{|\alpha|,|\beta|=2} a_{\alpha\beta} D^1_{\Delta_i}[f\big(T^2(u)\big)\, D^\alpha u]\, D^\beta \psi\, dz = -\int_Z g\, D^1_{-\Delta_i}\psi\, dz\,.$$

Sei $\zeta = \zeta(z)$ eine Schnittfunktion für das Gebiet G_1 mit folgenden Eigenschaften:

1. $\zeta \in C^\infty(R_n)$,
2. $0 \leqq \zeta(z) \leqq 1$,
3. $\zeta(z) \equiv 1$ in $\overline{G}_1$,
4. $\zeta(z) \equiv 0$ in $\overline{Z} \setminus \overline{G}_2$,
5. $|D^\alpha\zeta| \leqq \dfrac{C_\alpha}{a^{|\alpha|}}$ mit $a = \min(\varrho, \tau)$.

Sei nun speziell $\psi = D^1_{\Delta_i}u \cdot \zeta^4$.

Unter Verwendung der positiven Definitheit der durch $\|a_{\alpha\beta}\|$ erzeugten quadratischen Form, der Eigenschaften der Funktion f und der Zugehörigkeit von u zu $W_2^2(Z)$ erhalten wir für alle hinreichend kleinen Δz_i die Abschätzung (vgl. z. B. [4])

$$\int_{G_1} \sum_{|\alpha|=2} (D^\alpha D^1_{\Delta_i}u)^2\, dz \leqq C\|g\|^2_{L_2(\Omega)}\,.$$

Nach Lemma 3 folgt aus dieser Abschätzung zunächst

$$D_i^1 u \in W_2^2(G_1)\,, \qquad i = 1, \dots, n-1\,,$$

und

$$\sum_{|\alpha|=2} \|D_i^1 D^\alpha u\|_{L_2(G_1)} \leqq C\|g\|_{L_2(\Omega)}\,, \qquad i = 1, \dots, n-1\,. \tag{13}$$

Es bleibt zu zeigen, daß auch die Ableitung $D_n^1 u$ zu W_2^2 gehört. Dazu weisen wir die gleichmäßige Beschränktheit von $D^1_{\Delta_n} D_n^2 u$ in der L_2-Norm nach.

Wir bezeichnen noch

$$\overline{G} = \left\{ z \in Z \,\middle|\, \sum_{i=1}^{n-1} z_i^2 \leqq \frac{\varrho}{8}, \quad 0 \leqq z_n \leqq \frac{\tau}{8} \right\}$$

und betrachten die Integralidentität (12) für alle Funktionen $\varphi \in \mathring{W}_2^2(G_1)$

$$\int_{G_1} f(T^2(u))\, T^2(u, \varphi)\, dz = \int_{G_1} g\varphi\, dz\,. \tag{14}$$

Sei $\psi \in \dot{C}^\infty(G)$ beliebig. Setzen wir ψ durch Null außerhalb von G fort, so gehört die Funktion $D^1_{-\Delta_n}\psi$ für $\Delta z_n < \frac{\tau}{8}$ zum Raum $\mathring{W}_2^2(G_1)$ und kann demzufolge als Vergleichsfunktion in die Integralidentität (14) eingesetzt werden

$$\int_{G_1} \sum_{|\alpha|, |\beta| = 2} a_{\alpha\beta} f(T^2(u))\, D^\alpha u\, D^\beta D^1_{-\Delta_n}\psi\, dz = \int_{G_1} g\, D^1_{-\Delta_n}\psi\, dz\,.$$

Für $A \in L_2(G_1)$ gilt die Formel

$$\int_{G_1} A(z)\, D^1_{-\Delta_n}\psi\, dz = -\int_G D^1_{\Delta_n} A(z)\, \psi\, dz\,. \tag{15}$$

In der Tat, wir haben

$$\int_{G_1} A(z)\, D^1_{-\Delta_n}\psi(z)\, dz = -\frac{1}{\Delta z_n}\int_{G_1} A(z)\left(\overset{-n}{\psi}(z) - \psi(z)\right) dz$$

$$= -\frac{1}{\Delta z_n}\int_{G_1} A\overset{-n}{\psi}\, dz + \frac{1}{\Delta z_n}\int_{G_1} A\psi\, dz = -\frac{1}{\Delta z_n}\int_G \overset{+n}{A}\psi\, dz + \frac{1}{\Delta z_n}\int_G A\psi\, dz$$

$$= -\int_G D^1_{\Delta_n} A(z)\, \psi(z)\, dz\,.$$

Sei $\mu = (0, \ldots, 0, 2)$. Dann ist offensichtlich

$$\sum_{|\beta|=2} A_\beta = A_\mu + \sum_{\substack{|\beta|=2\\ \beta \neq \mu}}' A_\beta\,.$$

Entsprechend schreiben wir die Integralidentität in der Form

$$\int_{G_1} \sum_{|\alpha|=2} a_{\alpha\mu} f(T^2(u))\, D^\alpha u\, D^\mu D^1_{-\Delta_n}\psi\, dz$$

$$= -\int_{G_1} \sum_{\substack{|\beta|=2\\ \beta\neq\mu}} \left\{\sum_{|\alpha|=2} a_{\alpha\beta} f(T^2(u))\, D^\alpha u\right\} D^\beta D^1_{-\Delta_n}\psi\, dz + \int_{G_1} g D^1_{-\Delta_n}\psi\, dz\,.$$

Genauso trennen wir auch die Summe über $|\alpha| = 2$ auf der linken Seite

$$\int_{G_1} a_{\mu\mu} f(T^2(u))\, D^\mu u\, D^\mu\, D^1_{-\Delta_n}\psi\, dz = -\int_{G_1} \sum_{\substack{|\alpha|=2\\ \alpha\neq\mu}} a_{\alpha\mu} f(T^2(u))\, D^\alpha u\, D^\mu D^1_{-\Delta_n}\psi\, dz$$

$$-\int_{G_1} \sum_{\substack{|\beta|=2\\ \beta\neq\mu}} \left\{\sum_{|\alpha|=2} a_{\alpha\beta} f(T^2(u))\, D^\alpha u\right\} D^\beta D^1_{-\Delta_n}\psi\, dz + \int_{G_1} g D^1_{-\Delta_n}\psi\, dz\,.$$

Nach Anwendung der Formel (15) erhalten wir hieraus

$$\int_G a_{\mu\mu} D^1_{\Delta_n}[f(T^2(u))\, D^\mu u]\, D^\mu\psi\, dz = -\int_G \sum_{\substack{|\alpha|=2\\ \alpha\neq\mu}} a_{\alpha\mu} D^1_{\Delta_n}[f(T^2(u))\, D^\alpha u]\, D^\mu\psi\, dz$$

$$+ \int_{G_1} \sum_{\substack{|\beta|=2\\ \beta\neq\mu}} \left\{\sum_{|\alpha|=2} a_{\alpha\beta} f(T^2(u))\, D^\alpha u\right\} D^\beta D^1_{-\Delta_n}\psi\, dz - \int_{G_1} g D^1_{-\Delta_n}\psi\, dz\,.$$

Seien $\beta = \gamma + \delta$, $|\gamma|, |\delta| = 1$ und $\sigma = (0, \ldots, 0, 1)$.

Dann liefert die partielle Integration des zweiten Integrals der rechten Seite

$$\int_{G_1} \sum_{\substack{|\beta|=2\\ \beta\neq\mu}} \left\{ \sum_{|\alpha|=2} a_{\alpha\beta} f(T^2(u))\, D^\alpha u \right\} D^\beta D^1_{-\Delta_n}\psi\, dz$$

$$= \int_{G_1} \sum_{\substack{|\beta|=2\\ \delta\neq\sigma}} \left\{ \sum_{|\alpha|=2} a_{\alpha\beta} f(T^2(u))\, D^\alpha u \right\} D^{\gamma+\delta} D^1_{-\Delta_n}\psi\, dz$$

$$= -\int_{G_1} \sum_{\substack{|\beta|=2\\ \delta\neq\sigma}} \left\{ \sum_{|\alpha|=2} a_{\alpha\beta} D^\sigma[f(T^2(u))\, D^\alpha u \right\} D^\gamma D^1_{-\Delta_n}\psi\, dz\,.$$

Damit ergibt sich unter Verwendung von (10), II. und III. die Abschätzung

$$b_0 \left| \int_G D^1_{\Delta_n}[f(T^2(u))\, D^\mu u]\, D^\mu\psi\, dz \right| \leqq \left| \int_G \sum_{\substack{|\alpha|=2\\ \alpha\neq\mu}} a_{\alpha\mu} D^1_{\Delta_n}[f(T^2(u))\, D^\alpha u]\, D^\mu \psi\, dz \right|$$

$$+ \left| \int_{G_1} \sum_{\substack{|\beta|=2\\ \delta\neq\sigma}} \left\{ \sum_{|\alpha|=2} a_{\alpha\beta} D^\delta[f(T^2(u))\, D^\alpha u] \right\} D^\gamma D^1_{-\Delta_n}\psi\, dz \right| + \left| \int_{G_1} g\, D^1_{-\Delta_n}\psi\, dz \right|$$

$$\leqq C \sum_{\substack{|\alpha|=2\\ \alpha\neq\mu}} \|D^1_n D^\alpha u\|_{L_2(G_1)} \|D^\mu\psi\|_{L_2(G_1)} + C \sum_{\substack{|\alpha|=2\\ |\gamma|=1\\ \delta\neq\sigma}} \|D^\delta D^\alpha u\|_{L_2(G_1)} \sum_{|\gamma|=1} \left[\int_{G_1} (D^1_{-\Delta_n} D^\gamma\psi)^2\, dz \right]^{1/2}$$

$$+ \|g\|_{L_2(G_1)} \left[\int_{G_1} (D^1_{-\Delta_n}\psi)^2\, dz \right]^{1/2}.$$

Für die Ableitungen $D^1_n D^\alpha u$, $\alpha \neq \mu$, und $D^\delta D^\alpha u$, $\delta \neq \sigma$, haben wir aber bereits die Abschätzung (13) bewiesen, so daß wir jetzt folgendes Zwischenresultat erhalten

$$\left| \int_G D^1_{\Delta_n}[f(T^2(u))\, D^\mu u]\, D^\mu\psi\, dz \right|$$

$$\leqq C \|g\|_{L_2(\Omega)} \left\{ \|D^\mu\psi\|_{L_2(G)} + \sum_{|\gamma|\leqq 1} \left[\int_{G_1} (D^1_{-\Delta_n} D^\gamma\psi)^2\, dz \right]^{1/2} \right\}.$$

Für $D^\gamma\psi$ gilt aber die Darstellung

$$D^1_{-\Delta_n} D^\gamma\psi(z_n) = -\frac{1}{\Delta z_n} [D^\gamma\psi(z_1, \ldots, z_{n-1}, z_n - \Delta z_n) - D^\gamma\psi(z_1, \ldots, z_{n-1}, z_n)]$$

$$= -\frac{1}{\Delta z_n} [D^\gamma\psi(z_n - \Delta z_n) - D^\gamma\psi(z_n)] = -\frac{1}{\Delta z_n} \int_0^1 \frac{d}{dt} D^\gamma\psi(z_n - t\Delta z_n)\, dt$$

$$= \int_0^1 D^1_n D^\gamma\psi(z_n - t\Delta z_n)\, dt\,,$$

woraus folgt

$$\int_{G_1} (D^1_{-\Delta_n} D^\gamma\psi)^2\, dz = \int_{G_1} \left(\int_0^1 D^1_n D^\gamma\psi(z_n - t\Delta z_n)\, dt \right)^2 dz$$

$$\leqq \int_{G_1} \int_0^1 (D^1_n D^\gamma\psi(z_n - t\Delta z_n))^2\, dt\, dz = \int_0^1 \int_{G_1} (D^1_n D^\gamma\psi(z_n - t\Delta z_n))^2\, dz\, dt$$

$$\leqq \int_0^1 \int_{G_1} (D^1_n D^\gamma\psi)^2\, dz\, dt = \int_{G_1} (D^1_n D^\gamma\psi)^2\, dz = \int_G (D^1_n D^\gamma\psi)^2\, dz\,.$$

Damit gewinnen wir für jede Funktion $\psi \in \dot{C}^\infty(G)$ die Abschätzung

$$\left|\int_G D^1_{\Delta_n}[f(T^2(u))\, D^\mu u]\, D^\mu \psi\, dz\right| \leqq C\|g\|_{L_2(\Omega)}\, \|\psi\|_{W_2^2(G)}\,. \tag{16}$$

Für $|\alpha| = 2$, $\alpha \neq \mu$ lassen sich die Integrale

$$\int_G D^1_{\Delta_n}[f(T^2(u))\, D^\mu u]\, D^\alpha \psi\, dz$$

durch Anwendung der Formel (9) und durch partielle Integration direkt abschätzen. Sei $\alpha = \gamma + \delta$, $|\gamma|, |\delta| = 1$. Wegen $\alpha \neq \mu$ ist wenigstens einer der Multiindizes γ, δ verschieden von $\sigma = (0, \ldots, 0, 1)$. Sei z. B. $\delta \neq \sigma$. Wir haben dann für jede Funktion $\psi \in \dot{C}^\infty(G)$ und $|\alpha| = 2$, $\alpha \neq \mu$:

$$\left|\int_G D^1_{\Delta_n}[f(T^2(u))\, D^\mu u]\, D^\alpha \psi\, dz\right| = \left|\int_{G_1} f(T^2(u))\, D^\mu u\, D^1_{-\Delta_n} D^{\gamma+\delta} \psi\, dz\right|$$
$$= \left|\int_{G_1} D^\delta[(f(T^2(u))\, D^\mu u]\, D^1_{-\Delta_n} D^\gamma \psi\, dz\right| \leqq \|D^\delta D^\mu u\|_{L_2(G_1)} \left[\int_{G_1} (D^1_{-\Delta_n} D^\gamma \psi)^2\, dz\right]^{1/2}$$
$$\leqq C\|g\|_{L_2(\Omega)}\, \|\psi\|_{W_2^2(G)}\,. \tag{17}$$

Schließlich bekommen wir für $|\alpha| \leqq 1$ nach Anwendung von Formel (9)

$$\left|\int_G D^1_{\Delta_n}[f(T^2(u))\, D^\mu u]\, D^\alpha \psi\, dz\right| = \left|\int_G f(T^2(u))\, D^\mu u\, D^1_{-\Delta_n} D^\alpha \psi\, dz\right|$$
$$\leqq C\|D^\mu u\|_{L_2(G_1)} \left[\int_{G_1} (D^1_{-\Delta_n} D^\alpha \psi)^2\, dz\right]^{1/2} \leqq C\|g\|_{L_2(\Omega)}\, \|\psi\|_{W_2^2(G)}\,. \tag{18}$$

Die Abschätzungen (16), (17), (18) ergeben zusammen

$$\sum_{|\alpha| \leqq 2} \left|\int_G D^1_{\Delta_n}[f(T^2(u))\, D^\mu u]\, D^\alpha \psi\, dz\right| \leqq C\|g\|_{L_2(\Omega)}\, \|\psi\|_{W_2^2(G)}\,,$$

woraus dann für jedes $\Phi \in \dot{C}^\infty(G)$ folgt

$$\sum_{|\alpha| \leqq 2} \left|\int_G D^1_{\Delta_n}[f(T^2(u))\, D^\mu u]\, D^{2\alpha} \Phi\, dz\right| \leqq C\|g\|_{L_2(\Omega)}\, \|\Phi\|_{W_2^4(G)}\,. \tag{19}$$

Mit Hilfe von (8) und (19) können wir nun zunächst die Funktionen $D^1_{\Delta_n}[f(T^2(u)) \times D^\mu u]$ in der $L_2(G)$-Norm gleichmäßig bezüglich Δz_n abschätzen

$$\|D^1_{\Delta_n}[f(T^2(u))\, D^\mu u]\|_{L_2(G)}$$
$$\leqq C \sup_{\varphi \in \dot{C}^\infty(G)} \frac{\sum_{|\alpha| \leqq 2} \left|\int_G D^1_{\Delta_n}[f(T^2(u))\, D^\mu u]\, D^{2\alpha} \varphi\, dz\right|}{\|\varphi\|_{W_2^4(G)}}$$
$$\leqq C \sup_{\varphi \in \dot{C}^\infty(G)} \frac{C\|g\|_{L_2(\Omega)}\, \|\varphi\|_{W_2^4(G)}}{\|\varphi\|_{W_2^4(G)}} = C\|g\|_{L_2(\Omega)}\,. \tag{20}$$

Sei $u_t = u + t\left(\overset{+n}{u} - u\right)$. Dann gilt auf Grund von II. für fast alle $z \in G$ die Ungleichung

$$\sum_{|\alpha|, |\beta| = 2} a_{\alpha\beta} D^1_{\Delta_n}[f(T^2(u))\, D^\alpha u]\, D^\beta D^1_{\Delta_n} u$$
$$= \sum_{|\alpha|, |\beta| = 2} a_{\alpha\beta} \int_0^1 [f(T^2(u_t))\, D^\alpha D^1_{\Delta_n} u + 2\, f'(T^2(u_t))\, T^2(u_t, D^1_{\Delta_n} u)\, D^\alpha u_t]\, dt\, D^\beta D^1_{\Delta_n} u$$
$$= \int_0^1 [f(T^2(u_t))\, T^2(D^1_{\Delta_n} u) + 2f'(T^2(u_t))\, (T^2(u_t, D^1_{\Delta_n} u))^2]\, dt \geqq \varkappa T^2(D^1_{\Delta_n} u)\,.$$

Nach Integration über G ergibt sich

$$\varkappa \int_G T^2(D^1_{\Delta_n} u)\, dz \leqq \int_G \sum_{|\alpha|,|\beta|=2} a_{\alpha\beta} D^1_{\Delta_n}[f(T^2(u))\, D^\alpha u]\, D^\beta D^1_{\Delta_n} u\, dz$$

$$\leqq C \left\{ \int_G \sum_{|\alpha|=2} [D^1_{\Delta_n}[f(T^2(u))\, D^\alpha u]]^2\, dz \right\}^{1/2} \left\{ \int_G \sum_{|\beta|=2} (D^\beta D^1_{\Delta_n} u)^2\, dz \right\}^{1/2}$$

und damit

$$\int_G \sum_{|\alpha|=2} (D^\alpha D^1_{\Delta_n} u)^2\, dz \leqq C \sum_{|\alpha|=2} \int_G [D^1_{\Delta_n}[f(T^2(u))\, D^\alpha u]]^2\, dz$$

$$= C \| D^1_{\Delta_n}[f(T^2(u))\, D^\mu u] \|^2_{L_2(G)} + C \sum_{\substack{|\alpha|=2 \\ \alpha \neq \mu}} \| D^1_{\Delta_n}[f(T^2(u))\, D^\alpha u] \|^2_{L_2(G)} .$$

Hieraus erhalten wir endlich unter Berücksichtigung von (13) und (20) für alle hinreichend kleinen Δz_n die Abschätzung

$$\int_G (D^1_{\Delta_n} D^2_n u)^2\, dz \leqq C \| g \|^2_{L_2(\Omega)} .$$

Nach Lemma 3 folgt aus dieser Abschätzung

$$D^3_n u \in L_2(G)$$

und

$$\| D^3_n u \|_{L_2(G)} \leqq C \| g \|_{L_2(\Omega)} .$$

Insgesamt gilt also $u \in W^3_2(G)$ und

$$\| u \|_{W^3_2(G)} \leqq C \| g \|_{L_2(\Omega)} .$$

In den Koordinaten $(y_1, \dots, y_n)$ bedeutet das $u \in W^3_2(D)$ und

$$\| u \|_{W^3_2(D)} \leqq C \| g \|_{L_2(\Omega)} ,$$

wobei D das Urbild von G bezeichnet.

Offensichtlich existiert ein $h_0 = h(M_0) > 0$, so daß $K_{h_0}(M_0) \cap \Omega \subset D$. Die Mengen $K_h(M) \cap \Gamma$, $M \in \Gamma$, bilden eine (offene) Überdeckung von Γ, aus der wir eine endliche Überdeckung $H_i = K_{h_i}(M_i) \cap \Gamma$, $i = 1, \dots, N$, auswählen. Die Vereinigung von Gebieten

$$\bigcup_{i=1}^N B_i , \qquad B_i = K_{h_i}(M_i) \cap \Omega$$

enthält dann einen Randstreifen Ω_δ positiver Breite. Für jedes Gebiet B_i, $i = 1, \dots, N$, haben wir $u \in W^3_2(B_i)$ und

$$\| u \|_{W^3_2(B_i)} \leqq C \| g \|_{L_2(\Omega)} , \qquad i = 1, \dots, N .$$

Nach Aufsummierung über alle $i = 1, \dots, N$ bekommen wir für den ganzen Randstreifen Ω_δ die Beziehungen $u \in W^3_2(\Omega_\delta)$ und

$$\| u \|_{W^3_2(\Omega_\delta)} \leqq C \| g \|_{L_2(\Omega)} .$$

Berücksichtigen wir zudem, daß nach Satz 2 für jedes echte Teilgebiet $\Omega' \subset \Omega$ die Eigenschaften $u \in W^3_2(\Omega')$ und

$$\| u \|_{W^3_2(\Omega')} \leqq C \| g \|_{L_2(\Omega)}$$

bereits bekannt sind, kommen wir zu dem Endresultat $u \in W^3_2(\Omega)$ und

$$\| u \|_{W^3_2(\Omega)} \leqq C \| g \|_{L_2(\Omega)} ,$$

wobei C nur von den Eigenschaften des Gebiets Ω sowie von den Schranken für die Funktion f und die Koeffizienten $a_{\alpha\beta}$ abhängt.

Damit ist der Satz bewiesen.

12. Bemerkung. Man erkennt sofort, daß die Beweismethode nicht an eine bestimmte Ordnung der Differentialgleichung (1) gebunden ist. Die Behauptung des Satzes gilt in entsprechender Weise auch für das Randwertproblem

$$\sum_{|\alpha|,|\beta|=m} a_{\alpha\beta} D^{\beta}[f(T^2(u))\, D^{\alpha} u] = g\,,$$

$$D^{\alpha} u|_{\Gamma} = 0\,, \qquad |\alpha| \leqq m-1\,.$$

Der Rand muß dann allerdings $(m+1)$-mal stetig differenzierbar sein.

Der Autor dankt Herrn I. W. Skrypnik, Donezk, für einige wertvolle Hinweise zum Beweis des Satzes.

Literatur

[1] Agmon, S., Douglis, A., and Nirenberg, L., Estimates near the boundary for solutions of elliptic partial equations satisfying general boundary conditions I. Comm. on pure and applied mathematics, Vol. XII (1959) 623–727.

[2] Ladyshenskaja, O. A., u. Uralzewa, N. N., Линейные и квазилинейные уравнения эллиптического типа. Москва (1964).

[3] Langenbach, A., Über Gleichungen mit Potentialoperatoren und Minimalfolgen nichtquadratischer Funktionale. Math. Nachrichten **32** (1966) 9–24.

[4] Müller, M., Das Differenzenverfahren für gewisse quasilineare elliptische Differentialgleichungen I. Math. Nachrichten **46** (1970) 341–361.

[5] Wischik, M. I., Краевые задачи для квазилинейных сильно эллиптических систем уравнений, имеющих дивергентную форму. ДАН СССР **138**, 3 (1961) 518—521.

Anschrift des Autors:

Humboldt-Universität, Berlin
Sektion Mathematik

On a regularization of an undamped problem in nonlinear plate vibration

J. NAUMANN, Berlin

1. Introduction

In this paper we shall present a method of a regularization of a system of nonlinear partial differential equations governing undamped vibrations of a thin elastic plate.

Let Ω be a bounded domain in the x,y-plane (representing the shape of the plate) with boundary $\partial\Omega$, and let $0 < T < +\infty$ be fixed. We then consider the equations

$$u'' + a_1\Delta^2 u = [v, u] + f \qquad \text{in} \quad \Omega \times [0, T] \tag{1.1}$$

$$a_2\Delta^2 v = -[u, u] \qquad \text{in} \quad \Omega \times [0, T]\,, \tag{1.2}$$

subject to the boundary conditions

$$\left.\begin{matrix} u = u_n = 0\,, & v = v_n = 0 \\ \text{on } \partial\Omega\,, & \text{for all} \quad t \in [0, T] \end{matrix}\right\} \tag{1.3}$$

and to the initial conditions

$$u(0) = u_0\,, \qquad u'(0) = u_1 \qquad \text{for all} \quad (x, y) \in \Omega\,. \tag{1.4}$$

Here Δ^2 denotes the biharmonic operator, $[\,,\,]$ in (1.1), (1.2) is defined to be

$$[\varphi, \psi] = \varphi_{xx}\psi_{yy} + \varphi_{yy}\psi_{xx} - 2\varphi_{xy}\psi_{xy}\,,$$

while n in (1.3) means the outer normal with respect to Ω. Further, a_1 and a_2 in (1.1) and (1.2), respectively, are fixed positive constants. Then, the system (1.1), (1.2) together with data (1.3), (1.4) describes the nonlinear undamped vibration of a thin elastic plate that is clamped along its edge and subjected to an external force as well as prescribed initial conditions.

The existence of a solution of problem (1.1)—(1.4) (in a certain generalized sense precisely formulated below) has been proved i.e. by Lions [1] and Worowitch [7]. Furthermore, for the corresponding problem with the allowance of rotational inertia (i.e. (1.1) involves the additional term $-\alpha\,\Delta u''$ where $\alpha = \text{const} > 0$), Morosow [4] and Lions [1] have proved the existence and uniqueness of the solution.

We note that the investigation of problem (1.1)—(1.4) involves considerable mathematical difficulties. In fact, the problem of uniqueness for (1.1)—(1.4) seems to be an open question (cf. [1], p. 147) and, related to this one, no continuous dependence of the solution with respect to data can be expected.

In what follows, we replace problem (1.1)—(1.4) by the following one:

$$u_\varepsilon'' + \varepsilon \Delta^2 u_\varepsilon' + a_1 \Delta^2 u_\varepsilon = [v_\varepsilon, u_\varepsilon] + f \quad \text{in} \quad \Omega \times [0, T] \qquad (1.1)_\varepsilon$$

$$a_2 \Delta^2 v_\varepsilon = -[u_\varepsilon, u_\varepsilon] \quad \text{in} \quad \Omega \times [0, T], \qquad (1.2)_\varepsilon$$

$$\left. \begin{array}{l} u_\varepsilon = \dfrac{\partial u_\varepsilon}{\partial n} = 0 , \qquad v_\varepsilon = \dfrac{\partial v_\varepsilon}{\partial n} = 0 \\ \text{on } \partial\Omega , \qquad \text{for all } \; t \in [0, T] , \end{array} \right\} \qquad (1.3)_\varepsilon$$

$$u_\varepsilon(0) = u_0 , \qquad u_\varepsilon'(0) = u_1 \quad \text{for all} \quad (x, y) \in \Omega \qquad (1.4)_\varepsilon$$

where the notations above are preserved (especially, u_0, u_1, f in $(1.1)_\varepsilon$—(1.4) are identical with those of (1.1)—(1.4), respectively). Here ε denotes a real positive number. We remark that our procedure is similar so that in the linear case dicussed in Lions-Magenes [2] though the object of the present paper is somewhat else.

The aim of this paper is to prove that the solution of $(1.1)_\varepsilon$—$(1.4)_\varepsilon$ approximates in a certain sense that of (1.1)—(1.4). By this, the solution of problem $(1.1)_\varepsilon$—$(1.4)_\varepsilon$ continuously depends on the data such that its uniqueness and a regularity result with respect to t can be derived. Thus, $(1.1)_\varepsilon$—$(1.4)_\varepsilon$ presents a type of regularization of (1.1)—(1.4).

Section 2 contains the terminology used as well as some preliminaries being necessary for all that follows. Our main results are established in the third section. Theorem 1 yields the existence of a solution of $(1.1)_\varepsilon$—$(1.4)_\varepsilon$ in which the proof is based upon the Faedo-Galerkin method (cf. [1] and, for a more general approach, [3]). For the solution obtained, the following theorem states a regularity result with respect to t as well as the continuous dependence with respect to data. This basic tool for this is an energy identity (Strauß [6]). Theorem 3 then gives the precise sense in which the solution of $(1.1)_\varepsilon$—$(1.4)_\varepsilon$ approximates that of (1.1)—(1.4) for letting $\varepsilon \to 0$. We complete our paper with some remarks collected in section 4. An important observation (remark 1) consists in our deriving an inequality of energy type for the solution of problem (1.1)—(1.4). This result is obtained by a careful study of compactness properties of [,].

2. Definitions. Preliminaries

Let Ω be a bounded domain in the x,y-plane whose boundary $\partial\Omega$ is of type $\mathfrak{N}^{(0),1}$. [1])

By $L^p(\Omega)$ we denote the space of all real functions which are integrable of power p ($1 \leq p < +\infty$) on Ω with respect to the Lebesgue measure $dx\,dy$. For $u, v \in L^2(\Omega)$ we especially put

$$(u, v) = \int_\Omega uv \, dx\,dy , \qquad |u| = (u, u)^{1/2} .$$

[1]) Without particularly referring to it, furthermore we use some wellknown facts which can be found on the following textbooks: Nečas, J.: *Les méthodes directes en théorie des equations elliptiques.* — Prague, 1967; Treves, F.: *Topological vector spaces, distributions and kernels.* — New York, London, 1967.

Using the usual notation

$$D^\alpha = \frac{\partial^{|\alpha|}}{\partial x^{\alpha_1}\, \partial y^{\alpha_2}} \quad \text{where} \quad \alpha = (\alpha_1, \alpha_2)\,, \quad |\alpha| = \alpha_1 + \alpha_2$$

we define for an integer $m \geqq 1$:

$$W^{m,p}(\Omega) = \{u \mid u \in L^p(\Omega)\,, \quad D^\alpha u \in L^p(\Omega) \quad \text{for } |\alpha| \leqq m\}$$

(the derivatives are taken in the sense of distributions). $W^{m,p}(\Omega)$ is a Banach space with respect to the norm

$$||u||_{m,p} = \left\{\sum_{|\alpha| \leqq m} \int_\Omega |D^\alpha u|^p \, dx\, dy\right\}^{1/p}.$$

By $W_0^{m,p}(\Omega)$ will be denoted the closure of $\mathcal{D}(\Omega)$ with respect to the norm $||\ ||_{m,p}$ where $\mathcal{D}(\Omega)$ is the set of all infinitely differentiable functions with compact support in Ω.

In what follows we make a full use of the space $W_0^{2,2}(\Omega)$ which we assume to be equipped with the scalar product

$$((u, v)) = \int_\Omega (u_{xx}v_{xx} + 2u_{xy}v_{xy} + u_{yy}v_{yy})\, dx\, dy\,.$$

The norms $||\ ||_{2,2}$ and $||\ || = ((\ ,\ ))^{1/2}$ are equivalent on $W_0^{2,2}(\Omega)$.

We denote by $W^{-2,2}(\Omega)$ the space of all distributions f in Ω which are representable in the form $f = \sum_{|\alpha| \leqq 2} D^\alpha f_\alpha$ where $f_\alpha \in L^2(\Omega)$ for all $|\alpha| \leqq 2$ and the derivatives are to be understood in the sense of distributions. Providing $W^{-2,2}(\Omega)$ with the usual infimum-norm, there is a linear isometry of $W^{-2,2}(\Omega)$ onto $\left(W_0^{2,2}(\Omega)\right)'$ (where the latter is furnished with the dual norm).

Since $[u, v] \in L^1(\Omega)$ for any $u, v \in W_0^{2,2}(\Omega)$, $[u, v]$ can be identified with a distribution in Ω. On the other hand, according to Sobolew's embedding theorem, v may be regarded as being continuous on $\overline{\Omega}$. It is then easy to see that

$$[u, v] = (u_{xx}v)_{yy} + (-2u_{xy}v)_{xy} + (u_{yy}v)_{xx}$$

holds in the sense of distributions which implies $[u, v] \in W^{-2,2}(\Omega)$. Denoting by $\langle F, u\rangle$ the value of $F \in \left(W_0^{2,2}(\Omega)\right)'$ in the point $u \in W_0^{2,2}(\Omega)$ and identifying $[u, v]$ with its isometric image in $\left(W_0^{2,2}(\Omega)\right)'$, we have

$$\langle [u, v], \varphi\rangle = \int_\Omega (u_{xx}\varphi_{yy} - 2u_{xy}\varphi_{xy} + u_{yy}\varphi_{xx})\, v\, dx\, dy$$

for all $\varphi \in W_0^{2,2}(\Omega)$. From this one we get the obvious symmetry relation

$$\langle [u, v], \varphi\rangle = \langle [\varphi, v], u\rangle \tag{2.1}$$

for arbitrary $u, v, \varphi \in W_0^{2,2}(\Omega)$.

We now derive an estimate which is useful for some considerations below. By the aid of integration by parts one obtains

$$\int_\Omega (u_{xx}\varphi_{yy} - 2u_{xy}\varphi_{xy} + u_{yy}\varphi_{xx})\, v\, dx\, dy$$

$$= \int_\Omega [(u_{xy}v_y - u_{yy}v_x)\, \varphi_x + (u_{xy}v_x - u_{xx}v_y)\, \varphi_y]\, dx\, dy\,.$$

Hölder's inequality and Sobolew's embedding theorem imply

$$|\langle [u, v], \varphi \rangle| \leqq \text{const}\, ||u||\, ||v||_{1,4}\, ||\varphi||_{1,4}$$
$$\leqq (\text{const}\, ||u||\, ||v||_{1,4})\, ||\varphi|| \,.$$

Denoting by $|| \; ||_{-2,2}$ the dual norm on $(W_0^{2,2}(\Omega))'$, we can conclude

$$||[u, v]||_{-2,2} \leqq \text{const}\, ||u||\, ||v||_{1,4} \,. \tag{2.2}$$

We finally consider the boundary value problem

$$\Delta^2 u = h \quad \text{in } \Omega\,, \qquad u = u_n = 0 \quad \text{on } \partial\Omega\,. \tag{*}$$

This problem possesses for each $h \in W^{-2,2}(\Omega)$ a uniquely determined solution u belonging to $W_0^{2,2}(\Omega)$; moreover, it holds $||u|| \leqq \text{const}\, ||h||_{-2,2}$. Thus, introducing for the boundary value problem (*) Green's operator G which assigns to each $h \in W^{-2,2}(\Omega)$ the solution u of (*), one obtains

$$||G(h)|| \leqq \text{const}\, ||h||_{-2,2} \,. \tag{2.3}$$

Since $\Delta^2 u \in W^{-2,2}(\Omega)$ for any $u \in W_0^{2,2}(\Omega)$, as above, identifying $\Delta^2 u$ with its isometric image in $(W_0^{2,2}(\Omega))'$ we have

$$\langle \Delta^2 u, v \rangle = \big((u, v)\big) \qquad \forall\, u, v \in W_0^{2,2}(\Omega)\,. \tag{2.4}$$

To complete our notations, let X be a Banach space with norm $|| \; ||_X$. We then denote by $L^p(0, T; X)$ $(1 \leqq p < +\infty)$ the space (of classes) of functions f valued in X and strongly measurable on $(0, T)$ (with respect to the Lebesgue measure on $(0, T)$) such that

$$||f||_{L^p(0,T;X)} = \left\{ \int_0^T ||f(t)||_X^p \, dt \right\}^{1/p} < +\infty \,.$$

For $p = \infty$ we replace this norm by the following one:

$$||f||_{L^\infty(0,T;X)} = \sup_{t \in (0,T)} \text{ess}\, ||f(t)||_X \,.$$

Provided with this norm, $L^p(0, T; X)$ $(1 \leqq p \leqq +\infty)$ is a Banach space.
Furthermore, from Sobolew's embedding theorem we have

$W_0^{2,2}(\Omega) \subset W^{1,4}(\Omega) \subset L^2(\Omega)$; the embedding $W_0^{2,2}(\Omega) \to W^{1,4}(\Omega)$ is compact.

Defining now the space

$$W = \{ u \mid u \in L^4\big(0, T; W_0^{2,2}(\Omega)\big)\,, \qquad u' \in L^4\big(0, T; L^2(\Omega)\big) \}$$

where the derivative is taken in the sense of distributions valued in $L^2(\Omega)$, the following embedding theorem holds (cf. [1]):

The embedding $W \to L^4\big(0, T; W^{1,4}(\Omega)\big)$ *is compact.* (2.5)

3. Statement of results

We now formulate the precise sense in which the solvability of $(1.1)_\varepsilon$—$(1.4)_\varepsilon$ is to be understood, and in which sense problem $(1.1)_\varepsilon$—$(1.4)_\varepsilon$ approximates problem (1.1)—(1.4).

Theorem 1. *Suppose we are given*

$$f \in L^2(0, T; L^2(\Omega)) . \tag{3.1}$$

$$u_0 \in W_0^{2,2}(\Omega) , \qquad u_1 \in L^2(\Omega) . \tag{3.2}$$

Then for each fixed $\varepsilon > 0$ there exists a pair u_ε, v_ε such that

$$u_\varepsilon \in L^\infty(0, T; W_0^{2,2}(\Omega)) , \tag{3.3}$$

$$u_\varepsilon' \in L^\infty(0, T; L^2(\Omega)) \cap L^2(0, T; W_0^{2,2}(\Omega)) , \tag{3.4}$$

$$v_\varepsilon \in L^\infty(0, T; W_0^{2,2}(\Omega)) \tag{3.5}$$

where the functions u_ε and v_ε satisfy the relation

$$- \int_0^T (u_\varepsilon', \varphi')\, dt + \varepsilon \int_0^T ((u_\varepsilon', \varphi))\, dt + a_1 \int_0^T ((u_\varepsilon, \varphi))\, dt$$

$$= \int_0^T \langle [u_\varepsilon, v_\varepsilon], \varphi\rangle\, dt + \int_0^T (f, \varphi)\, dt + (u_1, \varphi(0)) \tag{3.6}$$

for all

$$\varphi \in L^2(0, T; W_0^{2,2}(\Omega)) \qquad \textit{with} \quad \varphi' \in L^2(0, T; L^2(\Omega)) \textit{ and } \varphi(T) = 0$$

as well as the relation

$$a_2 \int_0^T ((v_\varepsilon, \psi))\, dt = - \int_0^T \langle [u_\varepsilon, u_\varepsilon\rangle, \psi\rangle\, dt \tag{3.7}$$

for all $\psi \in L^2(0, T; W_0^{2,2}(\Omega))$.

Remark. We briefly dicuss the interpretation of $(1.1)_\varepsilon$—$(1.4)_\varepsilon$ (completely the same remarks are valid for (1.1)—(1.4); see theorem 3 below).

First of all, it is easy see that (3.6) implies the existence of the second derivative of u_ε in the sense of distributions valued in $W^{-2,2}(\Omega)$:

$$u_\varepsilon'' = f + ([u_\varepsilon, v_\varepsilon] - \varepsilon\, \Delta^2 u_\varepsilon' - a_1\, \Delta^2 u_\varepsilon) ;$$

thus

$$u_\varepsilon'' \in L^2(0, T; L^2(\Omega)) + L^\infty(0, T; W^{-2,2}(\Omega)) .$$

On the other hand, the initial condition $u_\varepsilon(0) = u_0$ will be ensured by the construction of proof; by this, since (3.3) and (3.4) imply (eventually, after modification on a set of measure zero in $(0, T)$) that u_ε is continuous of $[0, T]$ in $L^2(\Omega)$, this condition is satisfied in the sense of $L^2(\Omega)$. Finally, since u_ε' (again, eventually, after modification on a set of measure zero in $(0, T)$) is continuous of $[0, T]$ into $W^{-2,2}(\Omega)$, equation (3.6) implies the satisfying of the second condition in $(1.4)_\varepsilon$ in the sense of $W^{-2,2}(\Omega)$.

To state our next result we introduce the space

$$\tilde{W} = \{u \mid u \in L^\infty(0, T; W_0^{2,2}(\Omega)), \qquad u' \in L^\infty(0, T; L^2(\Omega))\}$$

provided with the norm

$$||u||_{\tilde{W}} = ||u||_{L^\infty(0,T;W_0^{2,2}(\Omega))} + ||u'||_{L^\infty(0,T;L^2(\Omega))} .$$

Theorem 2. *Suppose that* (3.1), (3.2) *is satisfied, and let* $\varepsilon > 0$ *be fixed. Then:*

(i) *For* u_ε *according to theorem* 1 *(eventually, after modification on a set of measure zero in* $(0, T)$*) it holds that*

$$u_\varepsilon \in C^{(0)}([0, T]; W_0^{2,2}(\Omega)) , \qquad u_\varepsilon' \in C^{(0)}([0, T]; L^2(\Omega)) .$$

(ii) *The mapping*

$$(u_0, u_1, f) \mapsto u_\varepsilon \quad \textit{(according to theorem 1)}$$

is Lipschitzian of $W_0^{2,2}(\Omega) \times L^2(\Omega) \times L^2(0, T; L^2(\Omega))$ *into* $\tilde{W}$.

Corollary to theorem 2. *The solution* u_ε, v_ε *of problem* $(1.1)_\varepsilon$—$(1.4)_\varepsilon$ $(u_\varepsilon, v_\varepsilon$ *according to* (3.3), (3.5)$)$ *is uniquely determined.*

We now state the result concerning the passage to limit $\varepsilon \to 0$ in $(1.1)_\varepsilon$—$(1.4)_\varepsilon$.

Theorem 3. *Let be* $\varepsilon = \varepsilon_j \to 0$ *for* $j \to \infty$ $(\varepsilon_j > 0$ *for each* $j)$. *Denoting by* $u_j = u_{\varepsilon_j}$, $v_j = v_{\varepsilon_j}$ *the functions corresponding to* $\varepsilon = \varepsilon_j$ *according to theorem* 1, *then:*

There exist subsequences $\{u_\nu\} \subset \{u_j\}$ *and* $\{v_\nu\} \subset \{v_j\}$, *respectively, such that*

$$\left.\begin{aligned} &u_\nu \to u \ \textit{weakly* in}\ L^\infty(0, T; W_0^{2,2}(\Omega)) ,\\ &u_\nu' \to u' \ \textit{weakly* in}\ L^\infty(0, T; L^2(\Omega)) ,\\ &v_\nu \to v \ \textit{weakly in}\ L^2(0, T; W_0^{2,2}(\Omega)) \end{aligned}\right\} \tag{3.8}$$

where the functions u *and* v *satisfy the identity*

$$\begin{aligned} &-\int_0^T (u', \varphi')\, dt + a_1 \int_0^T ((u, \varphi))\, dt \\ &= \int_0^T \langle [u, v], \varphi\rangle\, dt + \int_0^T (f, \varphi)\, dt + (u_1, \varphi(0)) \end{aligned} \tag{3.9}$$

for all

$$\varphi \in L^2(0, T; W_0^{2,2}(\Omega)) \quad \textit{with} \quad \varphi' \in L^2(0, T; L^2(\Omega)) \quad \textit{and} \quad \varphi(T) = 0$$

as well as the identity

$$a_2 \int_0^T ((v, \psi))\, dt = -\int_0^T \langle [u, u], \psi\rangle\, dt \tag{3.10}$$

for all $\psi \in L^2(0, T; W_0^{2,2}(\Omega))$.

3.1. Proof of theorem 1

For the method of proof we essentially follow Lions [1] (since there is no possibility of confusion throughout this section and the next one, we drop the index ε).

1° Approximation solutions. Let $\{w_1, w_2, \dots, w_j, \dots\}$ be a base in $W_0^{2,2}(\Omega)$ (cf. [1]).

For each integer m we find two systems of real numbers $\{\xi_{jm}\}$, $\{\eta_{jm}\}$ ($j = 1, \dots, m$), respectively, such that

$$\sum_{j=1}^{m} \xi_{jm} w_j \to u_0 \quad \text{strongly in} \quad W_0^{2,2}(\Omega) \quad \text{for} \quad m \to \infty\,, \tag{3.11}$$

$$\sum_{j=1}^{m} \eta_{jm} w_j \to u_1 \quad \text{strongly in} \quad L^2(\Omega) \quad \text{for} \quad m \to \infty\,. \tag{3.12}$$

We define the approximation solution

$$u_m = u_m(t) = \sum_{j=1}^{m} g_{jm}(t)\, w_j$$

where the real functions $g_{jm} = g_{jm}(t)$ satisfy the initial value problem

$$\begin{aligned}&\big(u_m''(t), w_j\big) + \varepsilon\big((u_m(t), w_j)\big) + a_1\big((u_m(t), w_j)\big)\\ &+\frac{1}{C_2}\langle[G([u_m(t), u_m(t)]), u_m(t)], w_j\rangle = \big(f(t), w_j\big)\,, \qquad 1 \leqq j \leqq m\,,\end{aligned} \tag{3.13}$$

$$g_{jm}(0) = \xi_{jm}\,, \qquad g_{jm}'(0) = \eta_{jm}\,, \qquad 1 \leqq j \leqq m\,. \tag{3.14}$$

Observing the estimates (2.2) and (2.3), one obtains for each m the existence of a solution $\{g_{jm}\}$ ($j = 1, \dots, m$) of initial value problem (3.13), (3.14). This solution exists on a certain interval $[0, t_m]$ where $t_m > 0$.

2° A-priori-estimates. Multiplying (3.13) with g_{jm}' and summing on j, one obtains

$$\begin{aligned}&\frac{1}{2}\frac{d}{dt}\big(|u_m'(t)|^2 + a_1\|u_m(t)\|^2\big) + \varepsilon\|u_m'(t)\|^2\\ &+\frac{1}{a_2}\langle[G([u_m(t), u_m(t)]), u_m(t)], u_m'(t)\rangle = \big(f(t), u_m'(t)\big)\,.\end{aligned} \tag{3.15}$$

By the aid of (2.1) and (2.4) we can write the nonlinear term in the following manner:

$$\begin{aligned}&\langle[G([u_m, u_m]), u_m], u_m'\rangle = \langle[u_m', u_m], G([u_m, u_m])\rangle\\ &= \big((G([u_m', u_m]), G([u_m, u_m]))\big) = \frac{1}{4}\frac{d}{dt}\|G([u_m, u_m])\|^2\end{aligned}$$

(the last equality is guaranteed by symmetry of $[\ ,\ ]$). Thus, (3.15) can be rewritten as

$$\begin{aligned}&\frac{1}{2}\frac{d}{dt}\Big(|u_m'(t)|^2 + a_1\|u_m(t)\|^2 + \frac{1}{2a_2}\|G([u_m(t), u_m(t)])\|^2\Big)\\ &+ \varepsilon\|u_m'(t)\|^2 = \big(f(t), u_m'(t)\big)\,.\end{aligned} \tag{3.16}$$

The estimates (2.3) and (2.2) yield

$$||G([u_m(0), u_m(0)])|| \leqq \text{const } ||u_m(0)||^2 .$$

Using this one, (3.16) now implies

$$\tfrac{1}{2}\left(|u_m'(t)|^2 + a_1||u_m(t)||^2\right) + \varepsilon \int_0^t ||u_m'(\sigma)||^2 \, d\sigma$$

$$\leqq \text{const } \left(|u_m'(0)|^2 + ||u_m(0)||^2 + ||u_m(0)||^4\right) + \int_0^j (f, u_m') \, d\sigma . \tag{3.17}$$

Taking into account (3.11) and (3.12), one gets by the aid of Gronwall's lemma

$$|u_m'(t)|^2 + a_1||u_m(t)||^2 + \varepsilon \int_0^t ||u_m'(\sigma)||^2 \, d\sigma \leqq \text{const} \quad \text{for all} \quad t \in [0, T] . \tag{3.18}$$

We now define

$$v_m = v_m(t) = -\frac{1}{a_2} G\big([u_m(t), u_m(t)]\big) \quad \text{for all } t \in [0, T] . \tag{3.19}$$

The combining of (2.3) and (2.2) with (3.18) supplies

$$||v_m(t)|| \leqq \text{const} \quad \text{for all } t \in [0, T] . \tag{3.20}$$

From (3.18) and (3.20) we can conclude that the solution of (3.13), (3.14) in fact exists on the whole interval $[0, T]$. Moreover, it has been established

$$\left.\begin{aligned} \{u_m\} &\text{ remains bounded in } L^\infty\big(0, T; W_0^{2,2}(\Omega)\big) , \\ \{u_m'\} &\text{ remains bounded in } L^\infty\big(0, T; L^2(\Omega)\big) , \\ \{\varepsilon^{1/2} u_m'\} &\text{ remains bounded in } L^2\big(0, T; W_0^{2,2}(\Omega)\big) , \\ \{v_m\} &\text{ remains bounded in } L^\infty\big(0, T; W_0^{2,2}(\Omega)\big) . \end{aligned}\right\} \tag{3.21}$$

3° Passage to limit. By the boundedness (3.21), Alaouglu's theorem implies the existence of subsequences $\{u_\mu\} \subset \{u_m\}$ and $\{v_\mu\} \subset \{v_m\}$, respectively such that

$$\left.\begin{aligned} u_\mu &\to u \quad \text{weakly* in } L^\infty\big(0, T; W_0^{2,2}(\Omega)\big) , \\ u_\mu' &\to \chi_1 \quad \text{weakly* in } L^\infty\big(0, T; L^2(\Omega)\big) , \\ v_\mu &\to v \quad \text{weakly* in } L^\infty\big(0, T; W_0^{2,2}(\Omega)\big) . \end{aligned}\right\} \tag{3.22}$$

Obviously, we have $u' = \chi_1$.

By the first two boundedness properties in (3.21) and the embedding theorem (2.5), we may assume without loss of generality that

$$u_\mu \to u \quad \text{strongly in} \quad L^4\big(0, T; W^{1,4}(\Omega)\big) . \tag{3.23}$$

If we now define $\bar{v}(t) = -\frac{1}{a_2} G\big([u(t), u(t)]\big)$ almost everywhere on $(0, T)$ $\big(u$ according to (3.22)$\big)$ if follows by the aid of (2.3), (2.2), the boundedness in (2.1) and (3.23) that

$$\int_0^T ||v_\mu(t) - \bar{v}(t)||^4 \, dt = \frac{1}{a_2^4} \int_0^T ||G([u_\mu(t), u_\mu(t)]) - G([u(t), u(t)])||^4 \, dt$$

$$\leqq \text{const} \int_0^T ||u_\mu(t) - u(t)||_{1,4}^4 \, dt \xrightarrow[\mu \to \infty]{} 0$$

(in deriving this estimate the symmetry of [,] is used). By comparison with the last convergence in (3.22), we have

$$v_\mu \to v \quad \text{strongly in} \quad L^4(0, T; W_0^{2,2}(\Omega)) . \tag{3.24}$$

Using again the estimate (2.2) (as well as the symmetry of [,]), (3.23) and (3.24) yield

$$[u_\mu, v_\mu] \to [u, v] \quad \text{strongly in} \quad L^2(0, T; W^{-2,2}(\Omega)) . \tag{3.25}$$

Finally, since ε is fixed, we may assume, again without loss of generality, that

$$u'_\mu \to \chi_2 \quad \text{strongly in} \quad L^2(0, T; W_0^{-2,2}(\Omega)) \tag{3.26}$$

where, as above, $u' = \chi_2$.

Let be $\varphi_j \in C^{(1)}([0, T])$ such that $\varphi_j(T) = 0$ for $j = 1, \ldots, j_0$. We then define for $t \in [0, T]$

$$\varphi\colon t \mapsto \varphi(t) = \sum_{j=1}^{j_0} \varphi_j(t)\, w_j . \tag{3.27}$$

Setting $m = \mu$ in (3.13) one can conclude for $j_0 \leqq \mu$:

$$\begin{aligned} &- \int_0^T (u'_\mu, \varphi')\, dt + \varepsilon \int_0^T ((u'_\mu, \varphi))\, dt + a_1 \int_0^T ((u_\mu, \varphi))\, dt \\ &= \int_0^T \langle [u_\mu, v_\mu], \varphi\rangle\, dt + \int_0^T (f, \varphi)\, dt + (u_\mu(0), \varphi(0)) \end{aligned} \tag{3.28}$$

where the definition (3.19) is used.

By (3.22), (3.25), and (3.26), the tending $\mu \to \infty$ in (3.28) yields the identity (3.6) for all functions of type (3.27). Thus, we complete the proof of (3.6) by the remark that the set of all functions φ of type (3.27) is dense in the space

$$\{\varphi \mid \varphi \in L^2(0, T; W_0^{2,2}(\Omega)) \text{ with } \varphi' \in L^2(0, T; L^2(\Omega)) \text{ and } \varphi(T) = 0\} .$$

To finish the proof of theorem 1 we observe that (2.4) and the definition (3.19) provide

$$a_2 \int_0^T ((v_\mu, \psi))\, dt = - \int_0^T \langle [u_\mu, u_\mu], \psi\rangle\, dt$$

for any $\psi \in L^2(0, T; W_0^{2,2}(\Omega))$. Repeating essentially the same conclusions just carried out for proving the first identity, the proof of theorem 1 is complete.

3.2. Proof of theorem 2

The proof of theorem 2 is based upon the following result which in this form is due to Strauß (cf. [6]).

Proposition 1. *Let be*

$$w \in L^\infty(0, T; W_0^{2,2}(\Omega)) , \qquad w' \in L^\infty(0, T; L^2(\Omega)) \cap L^2(0, T; W_0^{2,2}(\Omega)) ,$$

and let w satisfy

$$w'' + a_1 \Delta^2 w = f_1 + f_2$$

where $f_1 \in L^1(0, T; L^2(\Omega))$, $f_2 \in L^2(0, T; W^{-2,2}(\Omega))$.

Then the identity

$$|w'(t)|^2 + a_1||w(t)||^2 = |w'(0)|^2 + a_1||w(0)||^2 + 2\int_0^t (f_1, w')\,d\sigma + 2\int_0^t \langle f_2, w'\rangle\,d\sigma$$

holds for all $t \in [0, T]$.

Proof of (i). Since the identity (3.6) implies the satisfying of $(1.1)_\varepsilon$—$(1.4)_\varepsilon$ in the sense of $L^2(0, T; W^{-2,2}(\Omega))$ we may write

$$u'' + a_1\,\Delta^2 u = f + \left(-\frac{1}{a_2}\,[G([u, u]), u] - \varepsilon\,\Delta^2 u'\right).$$

We now set

$$f_1 = f, \qquad f_2 = -\frac{1}{a_2}\,[G([u, u]), u] - \varepsilon\,\Delta^2 u'.$$

Taking into consideration (3.3), (3.4) one obtains from proposition 1

$$|u'(t)|^2 + a_1||u(t)||^2 = |u_1|^2 + a_1||u_0||^2 + 2\int_0^t (f, u')\,dt$$

$$-2\int_0^t \left\langle\frac{1}{a_2}\,[G([u, u]), u] + \varepsilon\,\Delta^2 u', u'\right\rangle d\sigma \tag{3.29}$$

for all $t \in [0, T]$.

Equality (3.29) means that the function

$$\chi : t \mapsto \chi(t) = |u'(t)|^2 + a_1||u(t)||^2$$

is continuous on $[0, T]$.

On the other hand, (3.3) and (3.4) imply the existence of a representative in the class u (we denote it again by u) such that

$$u \in C_w([0, T]; W_0^{2,2}(\Omega)) \tag{3.30}$$

(cf. [2], [6]), and since $u'' \in L^2(0, T; W^{-2,2}(\Omega))$, by concluding in the same way, one obtains the existence of a representative such that

$$u' \in C_w([0, T]; L^2(\Omega))\,. \tag{3.31}$$

Let be $\{t_n\} \subset [0, T]$, $t_n \to t_0$ for $n \to \infty$. We have

$$|u'(t_0) - u'(t_n)|^2 + a_1||u(t_0) - u(t_n)||^2$$
$$= \chi(t_0) + \chi(t_n) - 2\big(u'(t_0), u'(t_n)\big) - 2a_1\big((u(t_0), u(t_n))\big)\,.$$

The continuity of χ and (3.30), (3.31) now yield

$$|u'(t_0) - u'(t_n)|^2 + a_1||u(t_0) - u(t_n)||^2 \to 0\,.$$

Before passing to proof of (ii) we note an estimate which will be used below. Since

$$\langle[G([u, u]), u], u\rangle = \frac{1}{4}\frac{d}{dt}\,||G([u, u])||^2$$

(u according to (3.3), (3.4)), (3.29) provides for all $t \in [0, T]$

$$|u'(t)|^2 + a_1||u(t)||^2 + \int_0^t \varepsilon \, ||u'(\sigma)||^2 \, d\sigma$$

$$\leqq |u_1|^2 + a_1||u_0||^2 + \frac{1}{2a_2} ||G([u_0, u_0])||^2 + 2 \int_0^t (f, u') \, d\sigma .$$

By (2.3) and (2.2), for every $t \in [0, T]$,

$$|u'(t)|^2 + a_1||u(t)||^2 + \varepsilon \int_0^t ||u'(\sigma)||^2 \, d\sigma$$

$$\leqq |u_1|^2 + a_1||u_0||^2 + \text{const} \, ||u_0||^4 + 2 \int_0^t (f, u') \, d\sigma . \tag{3.32}$$

Thus, we especially have

$$||u(t)||^2 \leqq K(u_0, u_1, f) \qquad \forall \, t \in [0, T] \tag{3.33}$$

where

$$K(u_0, u_1, f) = \text{const} \left(|u_1|^2 + ||u_0||^2 + ||u_0||^4 + \int_0^T |f(\sigma)|^2 \, d\sigma \right)$$

and the constant does not depend on $||u_0||$, $||u_1||$, $||f||_{L^2(0, T; L^2(\Omega))}$.

Proof of (ii). Suppose besides (3.1), (3.2) we are given

$$\bar{f} \in L^2(0, T; W_0^{2,2}(\Omega)) ,$$

$$\bar{u}_0 \in W_0^{2,2}(\Omega) , \qquad \bar{u}_1 \in L^2(\Omega);$$

we then denote by $\bar{u}, \bar{v}$ the solution of $(1.1)_\varepsilon$—$(1.4)_\varepsilon$ (according to theorem 1) which corresponds with these data.

Setting $w = u - \bar{u}$, one obtains

$$w'' + a_1 \Delta^2 w = f - \bar{f} + \left\{ -\frac{1}{a_2} ([G([u, u]), u] - [G([\bar{u}, \bar{u}]), \bar{u}]) - \varepsilon \Delta^2 w' \right\}.$$

We can again apply proposition 1 which yields

$$|w'(t)|^2 + a_1||w(t)||^2 + 2\varepsilon \int_0^t ||w'(\sigma)||^2 \, d\sigma$$

$$= |u_1 - \bar{u}_1|^2 + a_1||u_0 - \bar{u}_0||^2 + 2 \int_0^t (f - \bar{f}, w') \, d\sigma$$

$$- \frac{1}{2a_2} \int_0^t \langle [G([u, u]), u] - [G([\bar{u}, \bar{u}]), \bar{u}], w' \rangle \, d\sigma$$

$$\text{for all } t \in [0, T] . \tag{3.34}$$

For the estimate of the nonlinear term we use (2.2) and (2.3):

$$||[G([u, u]), u] - [G([\bar{u}, \bar{u}]), \bar{u}]||_{-2,2}$$

$$\leqq \text{const} \, (||u||^2 + ||\bar{u}||^2) \, ||w|| .$$

Thus, by the aid of (3.33), from (3.34) we can conclude

$$|w'(t)|^2 + a_1||w(t)||^2 + 2\varepsilon \int_0^t ||w'(\sigma)||^2 \, d\sigma$$

$$\leqq |u_1 - \bar{u}_1|^2 + a_1||u_0 - \bar{u}_0||^2 + 2 \int_0^t (f - \bar{f}, w') \, d\sigma$$

$$+ \text{const} \, [K(u_0, u_1, f) + K(\bar{u}_0, \bar{u}_1, \bar{f})] \int_0^t ||w|| \, ||w'|| \, d\sigma$$

$$\leqq |u_1 - \bar{u}_1|^2 + a_1||u_0 - \bar{u}_0||^2 + \int_0^t |f - \bar{f}|^2 \, d\sigma$$

$$+ C_\varepsilon(u_0, u_1, f; \bar{u}_0, \bar{u}_1, \bar{f}) \int_0^t (|w'|^2 + a_1||w||^2) \, d\sigma$$

$$+ \varepsilon \int_0^t ||w'||^2 \, d\sigma \qquad \text{for all} \quad t \in [0, T]$$

where

$$C_\varepsilon(u_0, u_1, f; \bar{u}_0, \bar{u}_1, \bar{f}) = \max \left\{1, \frac{1}{\varepsilon} \cdot \text{const} \cdot [K(u_0, u_1, f) + K(\bar{u}_0, \bar{u}_1, \bar{f})]^2\right\}.$$

It is obvious that the constant in the expression on the right hand side neither depends on u_0, u_1, f nor on $\bar{u}_0, \bar{u}_1, \bar{f}$.

Therefore, it holds that

$$|w'(t)|^2 + a_1||w(t)||^2 \leqq |u_1 - \bar{u}_1|^2 + a_1||u_0 - \bar{u}_0||^2 + \int_0^T |f - \bar{f}|^2 \, d\sigma$$

$$+ C_\varepsilon(u_0, u_1, f; \bar{u}_0, \bar{u}_1, \bar{f}) \int_0^t (|w'|^2 + a_1||w||^2) \, d\sigma$$

Gronwall's lemma finally implies

$$|u(t) - \bar{u}(t)|^2 + a_1||u(t) - \bar{u}(t)||^2 \leqq \exp [T C_\varepsilon(u_0, u_1, f; \bar{u}_0, \bar{u}_1, \bar{f})]$$

$$\times \left(|u_1 - \bar{u}_1|^2 + a_1||u_0 - \bar{u}_0||^2 + \int_0^T |f - \bar{f}|^2 \, d\sigma\right) \quad \text{for all} \quad t \in [0, T]. \qquad (3.35)$$

The desired result now follows from (3.35).

The assertion of the corollary to theorem 2 follows in a straightforward manner from (3.35).

3.3. Proof of theorem 3

The estimate (3.32) provides

$\{u_j\}$ remains bounded in $L^\infty(0, T; W_0^{2,2}(\Omega))$,

$\{u_j'\}$ remains bounded in $L^\infty(0, T; L^2(\Omega))$,

$\{\varepsilon_j^{1/2} u_j\}$ remains bounded in $L^2(0, T; W(\Omega))$.

Setting $\psi = v_j$ in (3.7) and using (2.3) and (2.2), it follows

$$\int_0^T \|v_j(t)\|^2\,dt \leqq \text{const} \int_0^T \|u_j(t)\|^2\,\|v_j(t)\|\,dt$$

which implies

$$\{v_j\} \quad \text{remains bounded in } L^2(0, T;\, W_0^{2,2}(\Omega))\,.$$

Thus, there exist Subsequences $\{u_\nu\} \subset \{u_j\}$ and $\{v_\nu\} \subset \{v_j\}$ such that

$$\left.\begin{aligned} u_\nu &\to u \quad \text{weakly* in } L^\infty(0, T;\, W_0^{2,2}(\Omega))\,,\\ u_\nu' &\to \chi_0 \quad \text{weakly* in } L^\infty(0, T;\, L^2(\Omega))\,,\\ v_\nu &\to v \quad \text{weakly in } \ L^2(0, T;\, W_0^{2,2}(\Omega))\,. \end{aligned}\right\} \tag{3.36}$$

Obviously, $u' = \chi_0$.

The boundedness of $\{\varepsilon_j^{1/2}\, u_j\}$ in $L^2(0, T;\, W_0^{2,2}(\Omega))$ yields

$$\varepsilon_j \left| \int_0^T ((u_j', \psi))\,dt \right| \leqq \varepsilon_j^{1/2} \cdot \text{const} \cdot \int_0^T \|\psi\|^2\,dt \xrightarrow[j\to\infty]{} 0 \tag{3.37}$$

for any $\psi \in L^2(0, T;\, W_0^{2,2}(\Omega))$.

Furthermore, by embedding theorem (2.5), as in Section 3.1 it holds that (without loss of generality)

$$u_\nu \to u \quad \text{strongly in } L^4(0, T;\, W_0^{2,2}(\Omega))\,.$$

Setting $\hat{v}(t) = -\dfrac{1}{a_2}\, G([(u(t), u(t)])$ almost everywhere on $(0, T)$, we obtain

$$\hat{v}_\nu \to v \ \text{strongly in } L^4(0, T;\, W_0^{2,2}(\Omega))\,.$$

This one and (3.36) imply

$$v = \hat{v} \quad \text{in} \quad L^2(0, T;\, W_0^{2,2}(\Omega))\,.$$

Repeating now the same conclusions as those following (3.22) and observing (3.37), the passage to limit $j \to \infty$ in (3.6) and (3.7), respectively, provides (3.9) and (3.10), respectively.

4. Remarks

1. Turning back the to proof of theorem 2, we observe that the fact of fixedness of $\varepsilon > 0$ is essentially used. In particular, the definition of $C_\varepsilon(u_0, u_1, f;\ \bar{u}_0, \bar{u}_1, \bar{f})$ (cf. p. 222) immediately implies that $C_\varepsilon \to +\infty$ as $\varepsilon \to 0$. Therefore, from our conclusions above the well-posedness of (1.1)—(1.4) cannot be expected.

Disregarding this fact, *for the solution of* (1.1)—(1.4) *(in the sense of* (3.10)*), an inequality of energy type can be proved.*

Indeed, let (3.1) and (3.2) be satisfied. As in [1], we consider the initial value problem (cf. section 3.1)

$$\begin{aligned} \left(u_m''(t), w_j\right) + a_1\left((u_m(t), w_j)\right) + \frac{1}{a_2} \langle [G([u_m(t), u_m(t)]), u_m(t)], w_j\rangle \\ = \left(f(t), w_j\right)\,, \qquad 1 \leqq j \leqq m\,, \\ g_{jm}(0) = \xi_{jm}\,, \quad g_{jm}'(0) = \eta_{jm}\,, \qquad 1 \leqq j \leqq m \end{aligned}$$

where (3.11) and (3.12) may be satisfied.

This initial value problem possesses for each m a solution which in fact exists on the whole interval $[0, T]$ (cf. [1]). Then there exist subsequences $\{u_\mu\}$ and $\{v_\mu\}$ (see definition (3.19)), such that

$$\left.\begin{aligned} u_\mu &\to u \quad \text{weakly* in } L^\infty(0, T; W_0^{2,2}(\Omega)) , \\ u'_\mu &\to u' \quad \text{weakly* in } L^\infty(0, T; L^2(\Omega)) , \\ v_\mu &\to v \quad \text{weakly* in } L^\infty(0, T; W_0^{2,2}(\Omega)) \end{aligned}\right\} \tag{4.1}$$

(cf. [1]), and, by arguing in the same way as in section 3.1, we have

$$[u_\mu, v_\mu] \to [u, v] \text{ strongly in } L^2(0, T; W^{-2,2}(\Omega)) . \tag{4.2}$$

At this stage we remark that the pair u, v according to (4.1) is a solution of (1.1)–(1.4) in the sense of (3.9) and (3.10), respectively (cf. [1] for details, and sections 3.1, 3.3).

Repeating the same conclusions which led us to (3.16), we obtain after an integration

$$\begin{aligned} &\frac{1}{2}\left(|u'_m(t)|^2 + a_1||u_m(t)||^2 + \frac{1}{2a_2}||G([u_m(t), u_m(t)])||^2\right) \\ &= \frac{1}{2}\left(|u_m(0)|^2 + a_1||u_m(0)||^2 + \frac{1}{2a_2}||G([u_m(0), u_m(0)])||^2\right) \\ &\quad + \int_0^t (f(\sigma), u'_m(\sigma))\, d\sigma \end{aligned} \tag{4.3}$$

for all $t \in [0, T]$. Let $\theta \in C^{(0)}([0, T])$, $\theta \geqq 0$ be arbitrary. We multiply (4.3) by θ and integrate over $[0, T]$. The use of (3.11), (3.12) as well as the convergence properties (4.1) and (4.2) for $m = \mu \to \infty$ in (4.3) now imply

$$\begin{aligned} &\frac{1}{2}\int_0^T \left(|u'(t)|^2 + a_1||u(t)||^2 + \frac{1}{2a_2}||G([u(t), u(t)])||^2\right)\theta(t)\, dt \\ &\leqq \frac{1}{2}\liminf_{\mu\to\infty}\int_0^T \left(|u'_\mu(t)|^2 + a_1||u_\mu(t)||^2 + \frac{1}{2a_2}||G_\mu([u(t), u_\mu(t)])||^2\right)\theta(t)\, dt \\ &= \frac{1}{2}\left(|u_1|^2 + a_1||u_0||^2 + \frac{1}{2a_2}||G([u_0, u_0])||^2\right)\int_0^T \theta(t)\, dt \\ &\quad + \int_0^T \theta(t)\left[\int_0^t (f(\sigma), u'(\sigma)), d\sigma\right] dt , \end{aligned}$$

and since θ was arbitrary, we can conclude

$$\begin{aligned} &\frac{1}{2}\left(|u'(t)|^2 + a_1||u(t)||^2 + \frac{1}{2a_2}||G([u(t), u(t)])||^2\right. \\ &\leqq \frac{1}{2}\left(|u_1|^2 + a_1||u_0||^2 + \frac{1}{2a_2}||G([u_0, u_0])||^2\right) + \int_0^t (f, u')\, d\sigma \end{aligned} \tag{4.4}$$

for almost all $t \in (0, T)$ which presents desired inequality.

Obviously, (4.4) yields the estimate

$$|u'(t)|^2 + ||u(t)||^2 \leqq \text{const}\left(|u_1|^2 + ||u_0||^2 + ||u_0||^4 + \int_0^T |f(\sigma)|^2\right) d\sigma$$

for almost all $t \in (0, T)$.

2. The convergence properties (3.8) for $\{u_\nu\}$ and $\{u'_\nu\}$ imply

$$u_\nu(t_0) \rightarrow u(t_0) \text{ weakly in } L^2(\Omega) \text{ for all } t_0 \in [0, T]. \tag{4.5}$$

Furthermore, taking into account the first convergence in (3.22) and moreover (3.25), (3.36) we obtain by use of (3.6)

$$\int_0^T \langle u''_\nu, \varphi\rangle\, dt = \int_0^T (f, \varphi)\, dt - a_1 \int_0^T ((u_\nu, \varphi))\, dt - \int_0^T \langle [u_\nu, v_\nu], \varphi\rangle\, dt - \varepsilon_\nu \int_0^T ((u'_\nu, \varphi))\, dt$$

$$\rightarrow \int_0^T (f, \varphi)\, dt - a_1 \int_0^T ((u, \varphi))\, dt - \int_0^T \langle [u, v], \varphi\rangle\, dt = \int_0^T \langle u'', \varphi\rangle\, dt$$

for all φ of the form

$$\varphi = \tilde{\varphi} \otimes w \quad \text{with} \quad \tilde{\varphi} \in \mathcal{D}((0, T)), \quad w \in W_0^{2,2}(\Omega).$$

By wellknown density theorems, it follows that

$$u''_\nu \rightarrow u'' \text{ weakly in } L^2(0, T; W^{-2,2}(\Omega)).$$

Combining this with the second convergence in (3.8), we have

$$u'_\nu(t_0) \rightarrow u'(t_0) \text{ weakly in } W^{-2,2}(\Omega) \text{ for all } t_0 \in [0, T]. \tag{4.6}$$

Thus, for example, if for $(1.1)_\varepsilon$—$(1.4)_\varepsilon$ (in the sense of theorem 1) the existence of a periodic solution is proved, by (4.4) and (4.5) we can conclude the existence of a periodic solution for the undamped problem (1.1)—(1.4).

3. The convergence properties (3.8) can be sharpened if appropriate regularity properties (with respect to x, y) of the solution of problem (1.1)—(1.4) are known. For example, under certain regularity properties for u, u', and v, respectively, the convergence of the whole sequences $\{u_{\varepsilon_j}\}$, $\{u'_{\varepsilon_j}\}$, and $\{v_{\varepsilon_j}\}$ in a corresponding sense can be proved.

Moreover, *we remark that under sufficiently strong regularity properties (with respect to x, y) of the solution of* (1.1)—(1.4) *their uniqueness (and continuity in the sense of theorem* 2, (i)*) can be proved.* In fact, if e.g.

$$u(t) \in W^{2,4}(\Omega) \cap W_0^{2,2}(\Omega) \quad \text{a.e. on} \quad (0, T) \tag{4.7}$$

we may employ the result of Strauß (cf. section 3.2) to derive an equality of energy type. For this we only note that (4.7) implies

$$[u(t), v(t)] \in L^2(\Omega), \quad \text{a.e. on} \quad (0, T).$$

4. In similar way as above, the damped problem

$$u'' + a_0 u' + a_1 \Delta^2 u = [u, v] + f \quad \text{in } \Omega \times [0, T]$$

$$a_2 \Delta^2 v = -[u, u] \quad \text{in } \Omega \times [0, T]$$

$$u = u_n = 0, \quad v = v_n = 0 \quad \text{on } \partial\Omega, \quad \text{for all } t \in [0, T]$$

$$u(0) = u_0, \quad u'(0) = u_1 \quad \text{on } \Omega$$

($a_0 = \text{const} > 0$) can be replaced by a regularized system by adding $\varepsilon \Delta^2 u'$ in the first equation. The damping term $a_0 u'$ in fact does not improve the corresponding properties of problem (1.1)—(1.4) (see [5], also).

References

[1] Lions, J. L., Quelques méthodes des résolution des problèmes aux limites non linéaires. Dunod, Paris (1967).

[2] Lions, J. L. et Magenes, E.: Problèmes aux limites non homogènes et applications, vol. 1. — Dunod, Paris (1968) (russ. transl.: Moscow 1971).

[3] Lions, J. L., and Strauss, W. A., Some nonlinear evolution equations. Bull. Soc. math. France, vol. **93** (1965) 43—96.

[4] Morosow, N. F., О нелинейных колебаниях тонких пластин с учетом инерции вращения. ДАН СССР **176** (1967) 522—525.

[5] —, Исследование нелинейных колебаний тонких пластин с учетом затухания. Дифф. уравн. **3** (1967) 619—635.

[6] Strauss, W. A., On continuity of functions with values in various Banach spaces. Pacific J. Math., vol. **19** (1966) 543—551.

[7] Worowitsch, I. I., О некоторых прямых методах в нелинейной теории колебаний пологих оболочек. Изв. АН СССР сер. мат. **21** (1957) 747—784.

Anschrift des Autors:

Sektion Mathematik
Humboldt-Universität zu Berlin
108 Berlin

Mathematical models of elastic-inelastic materials

JINDŘICH NEČAS, Prague

Introduction. In the present discussion on the physical and mathematical foundations of the theory of inelastic behaviour of materials, several models have been suggested. We describe briefly three models:

1) The deformation theory of plasticity, where both existence and uniqueness theorems for the solution are proved by the variational method. Also, the natural numerical method, the secant modulus method, is convergent. This model is, in general, very far from the reality, especially if loading and unloading occurs in the considered body. See for example K. Washizu [1], S. G. Michlin [2], P. P. Mosolov, V. P. Mjasnikov [3].

2) The incremental strain theory, very popular and suggesting a natural numerical method of finite differences in time. There are some difficulties to choose an appropriate mathematical apparatus to define the basic problems. Up to this time, neither proofs of the existence nor the uniqueness are given. It can be, that this model, because of some its irregularity, does not allow a correct mathematical description. For details see K. Washizu [1] and J. Nečas [4].

3) The model with internal state variables, which can be considered as some regularisation of the second model. As a special case, it contains the first model and it seems be physically acceptable. A full mathematical background can be established relatively easily and natural numerical methods related to the theoretical considererations can be proposed. For details, see J. Kratochvíl, O. W. Dillon [5] and J. Kratochvíl, J. Nečas [6].

§ 1. The deformation theory of plasticity

This theory goes back to the works of Hencky and Mises. A modern approach can be found in L. M. Katschanov [7], [8], A. Langenbach [9], J. Nečas [10], S. G. Michlin [2], P. P. Mosolow, V. P. Mjasnikov [3].

From the physical point of view, this theory can be very far from the reality when loading and unloading occurs in the examined body.

In the next three paragraphes, let $u = (u_1, u_2, u_3)$ be the displacement vector, where $u \in [W^{1,2}(\Omega)]^3$. $W^{1,2}(\Omega)$ denotes the Sobolev space of L_2 functions with L_2 first derivatives in Ω, a bounded domain with Lipschitz boundary $\partial\Omega$. Let $\varepsilon_{ij} = \frac{1}{2}\left(\frac{\partial u_i}{\partial x_j} + \frac{\partial u_j}{\partial x_i}\right)$ be the infinitesimal strain tensor, σ_{ij} from $L_2(\Omega)$ be the symmetrical stress tensor. We shall suppose in the sequel that

$$\frac{\partial \sigma_{ij}}{\partial x_j} + F_i = 0 \tag{1.1}$$

in the sense of distributions, where F_i, from $L_2(\Omega)$, are the components of the body force. Put

$$\Gamma(u, v) = -\tfrac{2}{3}\, e_{ii}(u)\, e_{ii}(v) + 2e_{ij}(u)\, e_{ij}(v)\,. \tag{1.2}$$

The Hook's law, in this paragraphe, has the form

$$\sigma_{ij}(x) = \big(k(x) - \tfrac{2}{3}\,\mu(x, \Gamma(u,u)\,(x))\big)\, \delta_{ij} e_{kk}(u)\,(x) + 2\mu\big(x, \Gamma(u,u)\,(x)\big)\, e_{ij}(u)\,(x)\,. \tag{1.3}$$

In the next we shall use a clear abbreviation for (1.3) (and analogous formulas): $\sigma_{ij} = \big(k - \tfrac{2}{3}\,\mu(\Gamma(u, u))\big)\, \delta_{ij} e_{kk}(u) + 2\mu(\Gamma(u, u))\, e_{ij}(u)$. $k(x)$ is the bulk modulus, $k \in L_\infty(\Omega)$, $\mu(x, s)$ is the Lamé's coefficient, mesurable in x for s from the interval $0 \leqq s < \infty$; for almost all x from Ω once continuously differentiable in s and satisfying:

$$0 < \mu_1 \leqq \mu(x, s) \leqq \tfrac{3}{2}\, k(x)\,, \tag{1.4}$$

$$s\,\frac{\partial \mu}{\partial s}\,(x, s) \leqq c_1,\ \mu(x, s) + 2s\,\frac{\partial \mu}{\partial s}\,(x, s) \geqq c_2 > 0\,. \tag{1.5}$$

Let g_i, from $L_2(\Gamma_1)$, where $\Gamma_1 \subset \partial\Omega$, be the components of the vector of the boundary force. The principle of the virtual work gives

$$\int_\Omega [k e_{ii}(u)\, e_{ii}(v) + \mu(\Gamma(u, u))\, \Gamma(u, v) - v_i F_i]\, dx - \int_{\Gamma_1} v_i g_i dS = 0\,, \tag{1.6}$$

where v are taken from the space $V = \{v \mid v = 0 \text{ on } \partial\Omega \setminus \Gamma_1\}$. We suppose that $u = u^0$ on $\partial\Omega \setminus \Gamma_1$, u^0 beeing a given element from $[W^{1,2}]^3$. If $\Gamma_1 = \partial\Omega$, we suppose fulfilled the condition of the total equilibry:

$$\int_\Omega F\, dx + \int_{\partial\Omega} g\, dS = 0\,, \qquad \int_\Omega [F \times x]\, dx + \int_{\partial\Omega} [g \times x]\, dS = 0 \tag{1.7}$$

and we consider only such displacements that $\int_\Omega u\, dx = 0$, $\int_\Omega [u \times x]\, dx = 0$. Put

$$\Phi(u) = \int_\Omega \left[\frac{k}{2}\,(e_{ii}(u))^2 + \frac{1}{2}\int_0^{\Gamma(u,u)} \mu(\sigma)\, d\sigma - u_i F_i\right] dx - \int_{\Gamma_1} u_i g_i\, dS\,. \tag{1.8}$$

We see that the differential of Gateaux: $D\Phi(u, v) = \frac{d}{dt}\,\Phi(u + tv)]_{t=0} = \int_\Omega [k e_{ii}(u)\, e_{ii}(v) + \mu(\Gamma(u, u))\, \Gamma(u, v) - v_i F_i]\, dx - \int_{\Gamma_1} v_i g_i\, dS$.

The condition (1.4) gives the coercivity of the functional (1.8):

$$\lim_{\|v\| \to \infty} \Phi(u^0 + v) = \infty\,, \qquad v \in V\,, \tag{1.9}$$

and the condition (1.5) the convexity condition of the functional (1.8) in the form:

$$D\Phi(u + h, h) - D\Phi(u, u) \geqq c\|h\|^2\,, \quad h \in V\,. \tag{1.10}$$

By the well known theorems from the classical abstract calculus of variations, see for example M. M. Wainberg [11], we obtain easily:

Theorem 1.1. *Under the conditions* (1.4), (1.5) *there exists a unique* $u \in [W^{1,2}]^3$ *such that* $u - u^0 \in V$ *and* (1.6) *is valid. This solution minimises the functional* (1.8) *among the functions satisfying the condition* $u - u^0 \in V$.

For a numerical method, we can use the classical Galerkin's method and we obtain, because of the condition (1.10) a strong convergence to the solution.

We can use also the secant modulus method, see the paper J. Kačúr, J. Nečas, J. Polák, J. Souček [12]. Define the iteration as follows: $u^1 \in [W^{1,2}]^3$, $u^1 - u^0 \in V$ and if u^n is defined, let $u^{n+1} - u^0 \in V$ and let u^{n+1} solve the linear problem: for every $v \in V$

$$\int_\Omega [ke_{ii}(u^{n+1})\, e_{ii}(v) + \mu\big(\Gamma(u^n, u^n)\big)\, \Gamma(u^{n+1}, v) - v_i F_i]\, dx - \int_{\Gamma_1} v_i g_i\, dS = 0 . \quad (1.11)$$

We have

Theorem 1.2. *If* (1.4), (1.5) *are satisfied and* $\partial\mu/\partial s \leqq 0$, *then* $u^n \to u$, *to the solution of the problem.*

§ 2. The incremental strain theory

This theory was developped by sevaral authors, see for example R. Hill [13], W. T. Koiter [14]. We follow the paper of the author, see J. Nečas [4]. We consider in this paragraphe the displacements, strain and stress tensors as functions of the time, and only the traction problem.

Let σ be a vector with 9 components σ_{ij} such that $\sigma_{ij} = \sigma_{ji}$. We consider nonnegative functions $f(\sigma)$ and $h(\sigma)$, f once continuously differentiable, h continuous, such that

$$|h(\sigma)| \leqq c , \qquad \left|\frac{\partial f}{\partial \sigma_{ij}}(\sigma)\right| \leqq c . \quad (2.1)$$

Put $\chi_1(s) = 0$ for $0 \leqq s < p$, $\chi_1(s) = 1$ for $s \geqq p$, $\chi_2(s) = 1$ for $s = 0$, $\chi_2(s) = 0$ for $s > 0$. Put $r(s) = \frac{1}{2}(s + |s|)$.

Let S be the subspace of $[L_2]^9$ of elements σ such that $\sigma_{ij} = \sigma_{ji}$, with the scalar product

$$\int_\Omega \sigma_{ij}\, \tau_{ij}\, dx = (\sigma, \tau) . \quad (2.2)$$

Put $Q = \Omega \times (0, T)$. Let us look for the elements σ such that $\sigma_{ij} \in L_2(Q)$, $\partial\sigma_{ij}/\partial t \in L_2(Q)$ in the sense of distributions and $\sigma(x, 0) = 0$. It follows immediately that for almost all x from Ω, the function $f\big(\sigma(x,\tau)\big)$ is absolutely continuous in the interval $\langle 0, T\rangle$ and so we can define

$$n(\sigma)\,(x, t) = \max_{0 \leq \tau \leq t} f\big(\sigma(x, \tau)\big) . \quad (2.3)$$

We suppose that the Hook's law is given in rate independent increments: ($\dot{\varepsilon}_{ij} = \partial \varepsilon_{ij}/\partial t$)

$$\dot{\varepsilon}_{ij} = -\frac{\nu}{E}\dot{\sigma}_{kk}\,\delta_{ij} + \frac{1+\nu}{E}\dot{\sigma}_{ij} + \chi_1(f(\sigma))\,\chi_2(n(\sigma) - f(\sigma))\,. \tag{2.4}$$

$$h(\sigma)\frac{\partial f}{\partial \sigma_{ij}}(\sigma)\,r\left(\frac{\partial f}{\partial \sigma_{kl}}(\sigma)\,\dot{\sigma}_{kl}\right), \quad 0 < \nu < \frac{1}{2}, \quad E > 0\,.$$

We name the set $f(\sigma) = k$ the yield surface.

Let F_i, the components of the body force, be in $C^{(1)}(\langle 0, T\rangle, L_2(\Omega))$ (once continuously differentiable functions with values in the space $L_2(\Omega)$) and g_i, the components of the surface traction, be in $C^{(1)}(\langle 0, T\rangle, L_2(\partial\Omega))$. We suppose $F_i(x, 0) = 0$, $g_i(x, 0) = 0$ and for every $t \in \langle 0, T\rangle$ the condition (1.7).

Let K be the subspace of S generated by strain tensors

$$\frac{1}{2}\left(\frac{\partial v_i}{\partial x_j} + \frac{\partial v_j}{\partial x_i}\right), \qquad v_i \in W^{1,2}(\Omega)\,.$$

We are looking for such σ, that $\sigma_{ij}(x, 0) = 0$ and such that for $v \in [W^{1,2}]^3$:

$$0 = \int_\Omega [\dot{\sigma}_{ij}\varepsilon_{ij}(v) - \dot{F}_i v_i]\,dx - \int_{\partial\Omega} \dot{g}v\,dS = 0 \tag{2.5}$$

(what gives equilibry and boundary conditions).

Put $H = S \dot{-} K$. We suppose for $h \in H$

$$\int_\Omega \dot{\varepsilon}_{ij}h_{ij}\,dx = 0 \tag{2.6}$$

(compatibility conditions or Castigliano's principle). It follows from (2.4), (2.5) and (2.6) that

$$\dot{\sigma}(t) = G(\sigma(t), n(\sigma)(t)), \sigma(0) = 0\,. \tag{2.7}$$

In the class, where we consider the stress tensor σ_{ij}, the function r in (2.4) can be replaced by identity and $\dot{\sigma}(t)$ can be obtained as a solution of a linear traction boundary value problem in dependence of $\sigma(t)$ and $n(\sigma)(t)$. So (2.7) is an abstract differential equation with retarded argument. Because for G nor lipschitz, neither compatibility conditions are valid, the question on the solution of the equation (2.7) remains open.

§ 3. The model with internal state variables

The preceding model is intuitively a singular limit of the present model. It appears that this model is also a generalisation of the first one. The model with internal state variables is closely connected with the dislocation theory. For details see P. Perzyna, W. Wojno [15], J. Kratochvíl, O. W. Dillon [5], J. Kratochvíl, J. Nečas [6].

We consider only the traction problem.

We look for σ from $C(\langle 0, T\rangle, S)$ and we consider ε^e from $C(\langle 0, T\rangle, S)$ (elastic part of the strain tensor), ε^p from $C(\langle 0, T\rangle, S)$ (plastic part), F (body force) from $C(\langle 0, T\rangle, [L_2(\Omega)]^3)$, g (surface traction) from $C(\langle 0, T\rangle, [L_2(\partial\Omega)]^3)$, α from $C(\langle 0, T\rangle, [L_2(\Omega)]^n)$ (vector of internal state variables).

We suppose (equilibrium and boundary conditions)

$$\int_\Omega [\sigma_{ij}\varepsilon_{ij}(v) - F_i v_i]\,dx - \int_{\partial\Omega} g_i v_i\,dS = 0 \tag{3.1}$$

for every v from $[W^{1,2}]^3$.

We suppose that there exists a function $P(\sigma, \alpha)$ with Lipschitz like continuous first derivatives in $R_q \times R_n$, such that

$$\left|\frac{\partial^2 P}{\partial\sigma_{ij}\,\partial\alpha_k}\right| + \left|\frac{\partial^2 P}{\partial\alpha_k\,\partial\alpha_l}\right| \leqq c < \infty \tag{3.2}$$

and such that

$$c_1\xi_{ij}\xi_{ij} \leqq \frac{\partial^2 P}{\partial\sigma_{ij}\,\partial\sigma_{kl}}\,\xi_{ij}\xi_{kl} \leqq c_2\xi_{ij}\xi_{ij}\,. \tag{3.3}$$

Further, let $B_{ij}(\sigma, \alpha)$ and $C_i(\sigma, \alpha)$ be Lipschitz like continuous functions with derivatives bounded by a fixed constant. We suppose that the elements $\alpha_0 \in [L_2(\Omega)]^n$ and $\varepsilon_0^p \in S$ are prescribed.

We are looking for $\sigma, \alpha, \varepsilon^p, \varepsilon^e$ satisfying

$$\varepsilon^p(t) - \varepsilon_0^p = \int_0^t B(\sigma(\tau), \alpha(\tau))\,d\tau\,, \tag{3.4}$$

$$\alpha(t) - \alpha_0 = \int_0^t C(\sigma(\tau), \alpha(\tau))\,d\tau\,, \tag{3.5}$$

$$\varepsilon_{ij}^e(t) = \frac{\partial P}{\partial\sigma_{ij}}\,(\sigma(t), \alpha(t)) \tag{3.6}$$

and

$$\varepsilon^e + \varepsilon^p \in K \tag{3.7}$$

(the definition of K see § 2).

In the paper [6] is proved the following theorem:

Theorem 3.1. *There exists a unique solution of the problem* (3.1), (3.4)—(3.7).

Sketch of the proof: Let σ be in $C(\langle 0, \delta\rangle, S)$ with δ enough small. Put $M_{\sigma,\delta} = \{\sigma \in C(\langle 0, \delta\rangle, S)\}$. It follows, from (3.5) that the mapping $\sigma \mapsto \alpha$ is a contracting mapping. Also the mapping $\sigma \mapsto \varepsilon^p$ is contracting. This follows from (3.4). Let us look for $\omega(t)$, $t \in \langle 0, \delta\rangle$, satisfying (3.1) and

$$\int_\Omega \left[\frac{\partial P}{\partial\sigma_{ij}}\,(\omega(t), \alpha(\sigma)\,(t)) + \varepsilon_{ij}^p(\sigma)\,(t)\right] h_{ij}\,dx = 0\,, \tag{3.8}$$

$h \in H$ (see the notation of the § 2). But $\omega(t)$ is defined by (3.1) and (3.8) uniquely and we can find it by minimising the functional

$$\int_\Omega [P(\omega, \alpha(\sigma)\,(t)) + \varepsilon_{ij}^p(\sigma)\,\omega_{ij}]\,dx \tag{3.9}$$

in the space of functions satisfying (3.1). Coercivity condition as well the condition (1.10) follows from (3.2) and (3.3). But those conditions imply that the mapping $\sigma \mapsto \omega$ from $M_{\sigma,\delta}$ to itself is a contracting mapping, so we obtain a unique fixed point. By a finite number of steps, replacing the interval $\langle 0, \delta\rangle$ by $\langle k\delta, (k + 1)\,\delta\rangle$, we obtain the theorem.

References

[1] Washizu, K., Variational methods in elasticity and plasticity (1968).
[2] Michlin, S. G., Čislěnnaja realizacija variacionnych metodov, Moskva 1966.
[3] Mosolow, P. P., Mjasnikow, W. P., Вариационные методы в теории течений жестко-вязко-пластических сред, (1971).
[4] Nečas, J., On the formulation of the traction problem for the flow theory of plasticity. Aplikace matematiky (1973) Vol. 18, n 2, 119–127.
[5] Kratochvíl, J., and Dillon, O. W., Thermodynamics of Elastic-Plastic Materials as a Theory with Internal State Variables. Jour. Appl. Physics, Vol. **40**, No. 8 (1969) 3207–3218.
[6] Kratochvíl, J., and Nečas, J., On the Solution of the Traction Boundary Value Problem for Elastic-Inelastic Materials, to appear.
[7] Katschanow, L. M., Вариационные принципы для упруго- пластических сред, ПММ **6** (1942) 187—196.
[8] —, Основы теории пластичности, Москва 1969.
[9] Langenbach, A., Variationsmethoden in der nichtlinearen Elastizität und Plastizitätstheorie. Wiss. Z. Humboldt Univ. Berlin, IX (1959/1960).
[10] Nečas, J., Matematická teorie pružnosti, lecture notes, University Prague (1970).
[11] Wainberg, M. M., Вариационные методы исследования нелинейных операторов, Москва (1956).
[12] Kačúr, J., Nečas, J., Polák, J., and Souček, J., Convergence of a method for solving the magnetostatic field in nonlinear media. Aplikace matematiky **13**, n. 6 (1968) 456–465.
[13] Hill, R., Mathematical Theory of Plasticity, Oxford (1950).
[14] Koiter, W. T., General Theorems for Elastic-Plastic Solids (1960).
[15] Perzyna, P., and Wojno, W., Thermodynamics of a rate sensitive plastic material. Arch. Mechaniki Stosowanej **20** (1968) 499.

Anschrift des Autors:

Department of Mathematics
Charles University
18600 Prague 8 — Karlin

Zur starken Konvergenz des Galerkin-Verfahrens für Evolutionsgleichungen

REINER NÜRNBERG, Berlin

Sei B ein reflexiver Banachraum, B^* sein dualer Raum, B und B^* dual mit der Bilinearform $(.,.)$. Eine Abbildung $T \in (D(T) \to B^*)$, $D(T) \subset B$ ihr Definitionsbereich, heißt monoton, wenn

$$(Tu - Tv, u - v) \geqq 0 \qquad \text{für } u, v \in D(T),$$

strikt monoton, wenn $(Tu - Tv, u - v) = 0$ nur gilt für $u = v$; hemistetig, wenn $(T(tu + (1 - t) v), h)$ stetig in $t \in [0, 1]$ ist für $u, v \in D(T)$ und $h \in B$; koerzitiv, wenn $(Tu, u)/||u|| \to +\infty$ für $||u|| \to +\infty$.

Wir betrachten die Operatorgleichung

$$Tu = 0. \tag{1}$$

Dabei sei $T = L + A$ mit dem linearen Operator L und dem nichtlinearen Operator A, wobei $D(L)$ dicht in B und $D(A) = B$. Mit der Existenz von Lösungen der Gleichung (1) hat sich u. a. F.E. Browder befaßt (vgl. z. B. [1, 2]).

Wir beweisen hier im ersten Teil der Arbeit Aussagen über die starke Konvergenz des Galerkin-Verfahrens in der folgenden Form. Seien B_n abgeschlossene lineare Unterräume von B ($B_i \subset B_j$ für $i \leqq j$; nicht unbedingt endlich dimensional) mit

$$\bigcup_n B_n \text{ liegt dicht in } B.$$

Wir betrachten dann die folgenden Näherungsgleichungen

$$T_n u_n = 0 \tag{2}$$

mit $T_n = j_n^* \circ T \circ j_n$, wobei $j_n \in (B_n \to B)$ die identische Einbettung, j_n^* die duale dazu ist.

Vom Galerkin-Verfahren sprechen wir, wenn die Näherungsgleichung (2) für alle n eine eindeutige Lösung besitzt. Wir sagen, das Galerkin-Verfahren konvergiert, wenn die Näherungslösungen gegen die einzige Lösung von (1) konvergieren. ($\to$ kennzeichnet die starke und $\rightleftharpoons$ die schwache Konvergenz in B, wie auch in B^*).

Diese Ergebnisse werden dann im zweiten Teil der Arbeit auf nichtlineare Evolutionsgleichungen vom parabolischen Typ angewendet. Bemerkt sei, daß J. A. Dubinski in [3] Aussagen über die schwache Konvergenz des Galerkin-Verfahrens für Evolutionsgleichungen angegeben hat.

1.

In diesem Abschnitt werden wir Aussagen über die Konvergenz des Galerkin-Verfahrens in der eingeführten Form für die Operatorgleichung (1) beweisen. Zunächst untersuchen wir die schwache Konvergenz.

Satz 1. *Seien B und B^* strikt konvex und $T \in (D(T) \to B^*)$, $D(T) \subset B$, ein dichtdefinierter Operator der Form $T = L + A$ mit folgenden Eigenschaften:*

1. *$L \in (D(L) \to B^*)$, $D(L) \subset B$, ist linear, dichtdefiniert, abgeschlossen und monoton. L_n^* sei für alle n monoton ($L_n = j_n^* \circ L \circ j_n$). L und L^* seien die Abschließung der Einschränkung auf $\bigcup_n B_n \cap D(L)$ bzw. $\bigcup_n B_n \cap D(L^*)$.*
2. *$A \in (B \to B^*)$ ist hemistetig, strikt monoton und koerzitiv.*

Dann konvergiert das Galerkin-Verfahren schwach.

Beweis. B_n sind ebenfalls reflexive, strikt konvexe Banachräume. Die Abbildungen $T_n \in (D(T_n) \to B_n^*)$, $D(T_n) \subset B_n$, besitzen folgende Eigenschaften ($T_n = L_n + A_n$):

L_n ist linear, dichtdefiniert und abgeschlossen,
L_n und L_n^* sind monoton,
A_n ist hemistetig, strikt monoton und koerzitiv.

Damit ist die Existenz der Näherungslösungen gesichert (vgl. z. B. J. L. Lions [5]).

Da T bzw. T_n strikt monoton sind, sind die Lösungen jeweils eindeutig.

Wir betrachten die Näherungslösungen

$$u_n \in D(T_n) = B_n \cap D(L) \quad \text{mit} \quad T_n u_n = 0 .$$

Die Folge (u_n) ist beschränkt, da T koerzitiv ist:

$$0 = (T_n u_n, u_n) = (T u_n, u_n) \geqq c(\|u_n\|)\, \|u_n\| ,$$

wobei $c(r) \to +\infty$ für $r \to +\infty$.

Da im reflexiven Banachraum eine beschränkte Menge schwach kompakt ist, existiert ein $u_0 \in B$ mit

$$u_n \rightharpoonup u_0$$

(zunächst konvergiert nur eine Teilfolge, es wird sich aber zeigen, daß der Grenzwert von u_n eindeutig ist).

Die Folge (Au_n) ist ebenfalls beschränkt. Dies ergibt sich aus einem Resultat von Hess und F. E. Browder:

Da L monoton ist, gilt zunächst

$$(Au_n, u_n) \leqq 0 .$$

Aus der Beziehung

$$(Au_n, v) = -(Au_n - Av, u_n - v) + (Au_n, u_n) + (Av, v - u_n), \; v \in B$$

folgt

$$|(Au_n, v)| \leqq \|Av\| \, (\|v\| + K) ,$$

da A monoton und (u_n) beschränkt ist ($\|u_n\| \leqq K$).

Für einen Operator mit offenem Definitionsbereich folgt aus monoton und hemistetig auch demistetig, d. h. stetig aus der starken Topologie in die schwache (T. Kato). Deshalb ist A lokalbeschränkt und damit folgt aus der obigen Beziehung die *Behauptung:* (Au_n) ist beschränkt.

Da B^* reflexiv ist, existiert ein $x \in B^*$ mit

$$Au_n \rightharpoonup x$$

(zunächst wieder nur Teilfolge).

Analog zum Existenzbeweis von F. E. Browder [1] zeigen wir nun, daß $u_0 \in D(L)$ und $Tu_0 = 0$ gilt. Für $v \in B_n$ haben wir

$$0 = (T_n u_n, v) = (Tu_n, v) = (Lu_n, v) + (Au_n, v)$$

und damit für $v \in B_n \cap D(L^*)$

$$(u_n, L^*v) = (Lu_n, v) = -(Au_n, v)\,.$$

Wenn wir in dieser Gleichung zur Grenze übergehen, erhalten wir

$$(u_0, L^*v) = -(x, v)$$

für alle $v \in \bigcup_n B_n \cap D(L^*)$. Da L^* die Abschließung seiner Einschränkung auf $\bigcup_n B_n \cap D(L^*)$ ist, bleibt diese Beziehung auch für alle $v \in D(L^*)$ erhalten.

Daraus folgt $u_0 \in D(L^{**})$. Da L abgeschlossen ist, gilt $L = L^{**}$. Deshalb erhalten wir

$$u_0 \in D(L) \quad \text{und} \quad Lu_0 = -x\,.$$

Es bleibt zu zeigen, daß $Tu_0 = 0$ (dann gilt auch $x = Au_0$). Da T monoton ist, haben wir für $u \in B_n \cap D(L)$

$$0 \leqq (Tu_n - Tu, u_n - u) = -(Tu, u_n - u)$$

und wenn wir zur Grenze übergehen

$$-(Tu, u_0 - u) \geqq 0$$

für alle $u \in \bigcup_n B_n \cap D(L)$. Da L die Abschließung seiner Einschränkung auf $\bigcup_n B_n \cap D(L)$ ist, sowie A demistetig ist, gilt diese Ungleichung auch für alle $u \in D(L)$. Weil T hemistetig ist, folgt daraus $Tu_0 = 0$ (siehe F. E. Browder [1]) und damit ist der Satz bewiesen.

Für die starke Konvergenz des Galerkin-Verfahrens benötigen wir folgende Monotoniebedingungen an T:

T heißt *gleichmäßig monoton*, wenn eine stetige, strikt wachsende Funktion $c(r) \geqq 0$ für $r \geqq 0$ mit $c(0) = 0$ derart existiert, daß für alle $u, v \in D(T)$ gilt

$$(Tu - Tv, u - v) \geqq c(||u - v||)\,.$$

T heißt *d-monoton*, wenn eine reelle strikt wachsende Funktion $d(r)$ für $r \geqq 0$ mit $d(r) \to +\infty$ für $r \to +\infty$ derart existiert, daß für alle $u, v \in D(T)$ gilt

$$(Tu - Tv, u - z) \geqq \big(d(||u||) - d(||v||)\big)\,(||u|| - ||v||)\,.$$

Bemerkung. Wenn T gleichmäßig monoton oder d-monoton ist, dann ist T koerzitiv.

Satz 2. *Unter den Voraussetzungen von Satz 1 sei A gleichmäßig monoton (oder d-monoton und B gleichmäßig konvex) und es gelte für die Lösung der Gleichung* (1) $u_0 \in D(L^*)$.

Dann konvergiert das Galerkin-Verfahren stark.

Beweis. Als erstes zeigen wir $(Tu_n, u_0) \to 0$.

Für $v \in D(L^*)$ gilt

$$(Tu_n, v) = (Lu_n, v) + (Au_n, v) = (u_n, L^*v) + (Au_n, v) .$$

Da $u_n \rightharpoonup u_0$ und $Au_n \rightharpoonup Au_0$, konvergiert dieser Ausdruck

$$(u_n, L^*v) + (Au_n, v) \to (u_0, L^*v) + (Au_0, v) .$$

Dieser Wert ist gleich 0

$$(u_0, L^*v) + (Au_0, v) = (Lu_0, v) + (Au_0, v) = (Tu_0, v) = 0 ,$$

deshalb folgt nach Voraussetzung über die Lösung u_0

$$(Tu_n, u_0) \to 0 .$$

Wir zeigen nun $u_n \to u_0$. Es gilt

$$(Tu_0 - Tu_n, u_0 - u_n) = -(Tu_n, u_0 - u_n) = -(Tu_n, u_0) ,$$

da $Tu_0 = 0$ und $(Tu_n, u_n) = (T_n u_n, u_n) = 0$.

1. *Fall.* T sei gleichmäßig monoton. Dann gilt

$$c(||u_0 - u_n||) \leqq -(Tu_n, u_0) ,$$

und damit $u_n \to u_0$, da $r \to 0$ wenn $c(r) \to 0$.

2. *Fall.* T sei d-monoton. Dann gilt

$$\big(d(||u_0||) - d(||u_n||)\big) (||u_0|| - ||u_n||) \leqq -(Tu_n, u_0)$$

und deshalb $||u_n|| \to ||u_0||$, da $d(r)$ strikt wachsend ist. Weil B gleichmäßig konvex ist, folgt daraus und aus $u_n \rightharpoonup u_0$, daß $u_n \to u_0$. Damit ist die starke Konvergenz des Verfahrens in beiden Fällen bewiesen.

2.

Wir wollen nun diese Sätze auf Evolutionsgleichungen anwenden. Sei B ein reflexiver Banachraum, der stetig in einen Hilbertraum H eingebettet ist, B liege dicht in H. B und B^* seien strikt konvex. Wenn wir dann H mit seinem dualen identifizieren, erhalten wir

$$B \subset H \subset B^* , \quad \text{dual mit der Bilinearform } (.,.) .$$

Wir führen darüber hinaus die folgenden Räume ein

$$B_0 = L^p(0, T; B) , \quad H_0 = L^2(0, T; H) , \quad B_0^* = L^q(0, T; B^*) ,$$

wobei $\frac{1}{p} + \frac{1}{q} = 1$. Wenn $p \geqq 2$ ist, dann gilt ebenfalls die Beziehung

$$B_0 \subset H_0 \subset B_0^*, \quad \text{dual mit der Bilinearform } (.,.)_0 = \int_0^T (.,.)\, dt .$$

Sei weiter B separabel und

$$\bigcup_n B_n \quad \text{dicht in } B$$

mit den endlich dimensionalen linearen Unterräumen B_n ($B_i \subset B_j$ für $i \leqq j$). Dann gilt für $B_{0,n} = L^p(0, T; B_n)$ ebenfalls $\bigcup_n B_{0,n}$ dicht in B_0.

Wir betrachten die *Evolutionsgleichung*

$$u' + Au = f \in B_0^*, \quad u \in B_0 \quad \text{mit} \quad u(0) = u_0 \in H . \tag{1}$$

Dabei ist $u' = du/dt$ die Ableitung im Distributionensinn, d. h. wenn $u \in B_0$, gilt für $u' \in B_0^*$ die Definitionsgleichung

$$\int_0^T \bigl(u'(t), v\bigr)\, \varphi(t)\, dt = -\int_0^T \bigl(u(t), v\bigr)\, \varphi'(t)\, dt$$

für alle $\varphi \in \dot{C}^\infty\bigl((0, T)\bigr)$ (φ beliebig oft differenzierbar und mit kompakten Träger) und alle $v \in B$.

Wenn $u \in L^p(0, T; B)$ und $u' \in L^q(0, T; B^*)$, dann gilt $u \in C(0, T; H)$ und die Abbildung $u \to u(0)$ ist surjektiv auf H. Damit hat das Anfangswertproblem einen Sinn und wir können ohne Beschränkung der Allgemeinheit voraussetzen $u_0 = 0$: Denn es ist möglich, ein $w_0 \in B_0$ einzuführen mit $w_0' \in B_0^*$ und $w_0(0) = u_0$. Wenn wir dann $u - w_0 = w$ setzen, erhalten wir die zu (1) äquivalente Gleichung

$$w' + A(w + w_0) = f - w_0' , \qquad w(0) = 0 .$$

Ebenso setzen wir im weiteren voraus $f = 0 \in B_0^*$.

Wir betrachten weiterhin die Galerkin-Gleichungen

$$u_n' + A_n u_n = 0 \in B_{0,n}^* , \qquad u \in B_{0,n} \quad \text{mit} \quad u_n(0) = 0 . \tag{2}$$

Dabei ist $A_n = j_n^* \circ A \circ j_n$ und $j_n \in (B_{0,n} \to B_0)$ die identische Einbettung.

Über das Galerkin-Verfahren für die Gleichung (1) können wir die folgende Aussage formulieren.

Satz 3. *Sei $A \in (B_0 \to B_0^*)$ hemistetig und gleichmäßig monoton (bzw. d-monoton und B gleichmäßig konvex).*

Dann besitzt die Näherungsgleichung (2) *für alle n genau eine Lösung und das Galerkin-Verfahren konvergiert stark gegen die einzige Lösung von* (1).

Beweis. (In diesem Beweis zur Anwendung der Sätze 1 und 2 wird die Methode der Anwendung des Eixstenzbeweises auf die veränderten Bedingungen übertragen (vgl. F. E. Browder [1] oder R. I. Katschurowski [4]).)

Wir bemerken, daß der Übergang zur homogenen Gleichung mit homogenem Anfangswert nicht die geforderten Eigenschaften des Operators A ändert.

Zunächst definieren wir folgenden linearen Operator

$$L = \frac{d}{dt}, \qquad D(L) = \{u \in B_0 \colon u' \in B_0^* , \ u(0) = 0\} .$$

Damit ist unser Cauchy-Problem (1) äquivalent zu der Operatorgleichung

$$Lu + Au = 0 \in B_0^*, \quad u \in B_0 \tag{1'}$$

Um für die Lösung von (1)' die Beziehung $u_0 \in D(L^*)$ nachweisen zu können, die wir für den Nachweis der starken Konvergenz des Galerkin-Verfahrens benötigen, betten wir das Problem in den folgenden Raum ein

$$B_1 = L^p(R_1; B), \quad B_1^* = L^q(R_1; B^*),$$

$$\text{dual mit der Bilinearform } (.,.)_1 = \int_{R_1} (.,.)\, dt.$$

Weiterhin sei $B_{1,n} = L^p(R_1; B_n)$.

Wir definieren wieder einen linearen Operator

$$L_1 = \frac{d}{dt}, \quad D(L_1) = \{u \in B_1 : u' \in B_1^*\}.$$

Aus der Beziehung für partielle Integration erhalten wir $L_1^* = -L_1$ (damit gilt $D(L_1^*) = D(L_1)$) und außerdem gilt

$$(L_1 u, u)_1 = 0 \quad \text{für alle } u \in D(L_1).$$

Wir behaupten

L_1 ist die Abschließung der Einschränkung auf $\bigcup_n B_{1,n} \cap D(L_1)$.

Aus der Definition der verallgemeinerten Ableitung folgt

$$L_1 = (-L_1')^*,$$

wobei L_1' die Einschränkung von L_1 auf die Menge $\left\{\sum_{i=1}^{n} \varphi_i v_i : \varphi_i \in \dot{C}^\infty(R_1),\ v_i \in B\right\}$ ist. Daraus folgt $L_1^* = (-L_1')^{**}$, und da $L_1^* = -L_1$, erhalten wir

L_1 ist die Abschließung der Einschränkung auf $D(L_1')$.

Daraus ergibt sich unmittelbar die geforderte Eigenschaft, da $\bigcup_n B_n$ dicht in B liegt.

Wir erweitern ebenfalls die Abbildung A zu einer Abbildung $A_1 \in (B_1 \to B_1^*)$.

Es sei $J \in (B_1 \to B_1^*)$ die duale Abbildung bzgl. der Funktion $\Phi(r) = r^{p-1}$, d. h.

$$(Ju, u)_1 = ||Ju||_{1*}\, ||u||_1, \quad ||Ju||_{1*} = ||u||_1^{p-1}.$$

Da B_1^* strikt konvex ist, existiert diese Abbildung eindeutig und ist hemistetig und monoton. Wir definieren

$$(A_1 u)(t) = \begin{cases} (Au)(t) & \text{für } t \in [0, T] \\ (Ju)(t) & \text{sonst}. \end{cases}$$

Die Abbildung A ist hemistetig, strikt monoton und koerzitiv, und damit genügt $T_1 = L_1 + A_1$ Satz 1.

Sei nun u_{1*} die Lösung von $T_1 u = 0$. Wir behaupten $u_* = u_{1*}|_{[0,T]}$ (die Einschränkung der Funktion auf das Intervall $[0, T]$) ist die Lösung von (1). Dafür bleibt nur zu zeigen $u_*(0) = 0$.

Für $t < 0$ gilt (wobei χ die charakteristische Funktion bezeichnet)

$$\|u_{1*}(t)\|_H^2 = 2(L_1 u_{1*}, \chi_{(-\infty, t)]} u_{1*})_1 = -2(A_1 u_{1*}, \chi_{(-\infty, t]} u_{1*})_1$$
$$= -2 \int_{-\infty}^{t} \big((J u_{1*})(\tau), u_{1*}(\tau)\big)\, d\tau \leqq 0 .$$

Daraus folgt $u_{1*}(0) = 0$, da u_{1*} stetig in H ist.

Für die Näherungsgleichungen gilt auch: wenn $u_{1n} \in B_{1,n} \cap D(L_1)$ die Lösung von $T_{1,n} u_{1n} = 0$ bezeichnet, so ist $u_n = u_{1n}|_{[0,T]}$ Lösung von (2).

Da die Abbildung T_1 nicht gleichmäßig monoton bzw. d-monoton ist, müssen wir folgendes bemerken, um Satz 2 anwenden zu können

$$(T_1 u - T_1 v, u - v)_1 = (A_1 u - A_1 v, u - v)_1 ,$$

da $(L_1 u, u)_1 = 0$, und

$$(A_1 u - A_1 v, u - v)_1 \geqq (Au - Av, u - v)_0 ,$$

da die duale Abbildung J monoton ist.

Deshalb folgt aus $(T_1 u_{1n}, u_{1*})_1 \to 0$, daß $u_n \to u_*$. Damit ist der Satz bewiesen.

Literatur

[1] Browder, F. E., Problémes non-linéaires. Montreal Press 1966.

[2] —, Nonlinear Maximal Monotone Operators in Banach Space. Math. Ann. **175** (1968) 89–113.

[3] Dubinski, J. A., Квазилинейные эллиптические и параболические уравнения любого порядка. УМН **23**, 1 [139] (1968) 45—90.

[4] Katschurowski, R. I., Нелинейные монотонные операторы в банаховых пространствах. УМН **23**, 2 [140] (1968) 121—168.

[5] Lions, J. L., Quelques méthodes des résolution des problémes aux limites non linéaires. Dunod (1969).

Anschrift des Autors:

Sektion Mathematik
der Humboldt-Universität zu Berlin
108 Berlin, Unter den Linden 6

On the stability of the two-dimensional Janenko scheme with respect to boundary values

GISBERT STOYAN

Introduction

We consider the first boundary value problem for the two-dimensional equation of heatflow,

$$\frac{\partial u}{\partial t} = \sum_1^2 L_i u\,, \qquad x = (x_1, x_2) \in G\,, \qquad (x, t) \in Q = G \times (0, T]\,,$$
$$u = \varphi(x, t)\,, \qquad x \in \Gamma\,, \qquad 0 < t \leqq T\,,$$
$$u(x, 0) = 0\,, \qquad x \in G \cup \Gamma\,,$$

where L_i are the elliptic differential operators

$$L_i u = \frac{\partial}{\partial x_i}\left(k_i(x, t)\,\frac{\partial u}{\partial x_i}\right) + r_i(x, t)\,\frac{\partial u}{\partial x_i} - q_i(x, t)\,u\,.$$

The domain G (with boundary Γ) in which the boundary value problem is stated, is composed of a finite number of connected rectangles with sides arranged parallel to the coordinate axes.

The coefficients of the operators L_i are assumed to satisfy the following inequalities:

$$k_i(x, t) \geqq c_0 > 0\,, \qquad q_i(x, t) \geqq \frac{1}{c_0} \max_{(x, t) \in Q} r_i^2(x, t), \tag{2}$$

the latter of which, if initially not met, can be made true by passing over from $u(x, t)$ to $v(x, t) = e^{ct}u(x, t)$ with an appropriate constant c.

For the numerical solution of the boundary value problem (1) Janenko [1, 2] proposed a difference scheme, but, as can be seen in [2], without overcoming the difficulties connected with the formulation of the boundary conditions for the difference scheme.

In [3] has been solved the problem of finding such boundary conditions for the scheme of Janenko which do not reduce the order $O(\tau^2 + h^2)$ of the scheme (τ, h — mesh-widths in time and space direction) obtainable in the case of homogeneous boundary values.

For this purpose in [3] the scheme [2] was considered in a new setting, as an additive scheme in the sense of Samarski [4, 5, 6]. Using ideas and results of Frjasinow [7, 8], the boundary conditions for the auxiliary function $y^{j+1/2}$ (see (7) below) of the scheme were given the form

$$y^{j+1/2} = \psi^{j+1/2}\,, \qquad x \in \Gamma\,, \qquad \psi = \varphi + \tau U_1 \varphi + \tau^2 U_2 \varphi\,, \tag{3}$$

where U_i, $i = 1, 2$, are partial differential operators, e.g.

$$U_1\varphi = (-1)^i \left(\frac{1}{2}\frac{\partial\varphi}{\partial t} - L_i\varphi\right) \tag{4}$$

for x lying on a piece of Γ parallel to the x_i-axis, $i = 1, 2$. U_2 contains higher derivatives.

In this paper we shall derive an a-priori estimate for the solution of the Janenko scheme which shows the stability of the scheme with respect to boundary values and allows the troublesome term $\tau^2 U_2$ in (3) to be dispensed with and to replace $\tau U_1\varphi$ by appropriate finite differences of φ.

Notations, formulation of the Janenko scheme

Let be ω_h a grid in G with mesh-width h_i in x_i-direction and γ the discrete boundary. For functions defined on ω_h we introduce the scalarproduct

$$(y, z) = \sum_{\omega_h} yzh_1h_2$$

and the norm $||y|| = (y, y)^{1/2}$.

Let be $\omega_\tau = \{n\tau,\ n = 0, 1, \ldots, N,\ N\tau = T\}$. For functions depending on t we use the notations

$$y^j = y(x, j\tau)\,, \quad y^{j+1/2} = y(x, (j + 1/2)\,\tau)\,, \quad y_t^j = (y^{j+1/2} - y^{j-1/2})/\tau\,.$$

We employ standard approximations Λ_i of order h_i^2 for the operators L_i (see e.g. [6]),

$$L_i y = (a_i y_{\bar{x}_i})_{x_i} + b_i y_{\mathring{x}_i} - d_i y\,, \tag{5}$$

where the subcripts $\bar{x}$, x, $\mathring{x}$ denote the backward, forward and central difference quotients, respectivily.

In what follows below it is assumed that the solution u of (1) has in $\bar{Q}$ continous derivatives in t up to order 3, in x_1 and x_2 up to sixth order ($u \in C^{3,6}$), i.e. the coefficients k, r, q are supposed to be sufficiently smooth and therefore we may choose

$$a_1 = k_1\left(x_1 - \frac{h}{2}, x_2, t\right), \quad a_2 = k_2\left(x_1, x_2 - \frac{h}{2}, t\right), \quad b_i = r_i\,, \quad d_i = q_i\,.$$

As a consequence of (2) we then have

$$a_i \geqq c_0 > 0\,, \quad d_i \geqq \frac{1}{c_0} \max_{\omega_\tau \times \omega_h} b_i^2\,, \quad i = 1, 2\,. \tag{6}$$

Finally, we shall use some norms connected with Λ_i,

$$||y||_{1,i}^2 = (a_i y_{\bar{x}_i}, y_{\bar{x}_i}) + (d_i y, y)\,, \quad i = 1, 2\,,$$

$$||y||_1^2 = ||y||_{1,1}^2 + ||y||_{1,2}^2\,.$$

With the notations introduced above the Janenko scheme can now be written as

$$\left.\begin{aligned}
&y^0 = 0\,, \\
&\frac{y^{j+1/2} - y^j}{\tau} = \frac{1}{2} \Lambda_1 (y^{j+1/2} + y^j)\,, \\
&\frac{y^{j+1} - y^{j+1/2}}{\tau} = \frac{1}{2} \Lambda_2 (y^{j+1} + y^{j+1/2})\,, \qquad x \in \omega_h\,, \\
&y^{j+1/2} = \psi^{j+1/2} \qquad \text{(see (3))}\,, \\
&y^{j+1} = \varphi^{j+1}\,, \qquad x \in \gamma\,, \\
&j = 0, 1, \ldots, N-1\,.
\end{aligned}\right\} \tag{7}$$

The coefficients of the operators Λ_i are here taken at the moment $(j + 1/2)\tau$.

Under the assumption that $u \in C^{3,6}$ in [3] it is proved that y approximates u, the solution of (1), with second order, i.e.

$$||y^j - u(x, j\tau)|| = 0(h_1^2 + h_2^2 + \tau^2)\,.$$

The a-priori estimate

We consider now the following difference problem

$$\left.\begin{aligned}
&v^0 = 0\,, \\
&\frac{v^{j+1/2} - v^j}{\tau} = \frac{1}{2} \Lambda_1 (v^{j+1/2} + v^j)\,, \\
&\frac{v^{j+1} - v^{j+1/2}}{\tau} = \frac{1}{2} \Lambda_2 (v^{j+1} + v^{j+1/2})\,, \qquad x \in \omega_h\,, \\
&v^{j+1/2} = \nu^{j+1/2}\,, \quad v^{j+1} = 0\,, \qquad x \in \gamma\,, \\
&j = 0, 1, \ldots, N-1\,,
\end{aligned}\right\} \tag{8}$$

where ν is a function given on γ.

In order to find an estimate of $v = v(\nu)$ by ν we need the following two lemmas, the first of which is well known (a finite analogon of the Grownwall inequality), the second proved easily by applying partial summation, Cauchy inequality and the inequality

$$|ab| \leqq \frac{1}{2\varepsilon} a^2 + \frac{\varepsilon}{2} b^2\,, \qquad \varepsilon > 0\,. \tag{9}$$

Lemma 1. *Let τ, a_k and b_k be nonnegative numbers satisfying*

$$a_1 \leqq b_1\,,$$
$$a_{k+1} \leqq \sum_1^k \tau a_j + b_{k+1}\,, \qquad k = 1, 2, \ldots\,.$$

Then

$$a_{k+1} \leqq b_{k+1} + e^{(k-1)\tau} \sum_1^k \tau b_j\,.$$

Lemma 2. *Suppose* (6) *to be fulfilled. Then for any pair of grid functions y and z with* $y = 0$, $x \in \gamma$, *the following inequality holds:*

$$(\Lambda_i y, y) + (\Lambda_i z, y) \leqq 2\|z\|_{1,i}^2 .$$

Now we proceed to the formulation of the announced estimate.

Theorem 1. *Let be p any grid function with boundary value v, i.e.*

$$p^{j+1/2} = v^{j+1/2}, \qquad x \in \gamma .$$

Then, if (6) *is satisfied, for the solution v of* (8) *the following estimate holds:*

$$\|v^{k+1}\| \leqq M\left(\max_{0\leqq j\leqq k}\|p^{j+1/2}\| + \left(\sum_0^k \tau\|p^{j+1/2}\|_1^2\right)^{1/2} + \left(\sum_1^k \tau\|p_{\bar{t}}^j\|^2\right)^{1/2}\right),$$

where the constant M depends only on T.

Proof: We introduce a new function w, connected with v by

$$w^0 = 0, \quad w^{j+1/2} = v^{j+1/2} - p^{j+1/2}, \quad w^{j+1} = v^{j+1}, \quad j = 1, 2, \dots, N-1 . \tag{10}$$

Then w is the solution of the difference problem

$$w^0 = 0,$$

$$\frac{w^{j+1/2} - w^j}{\tau} = \frac{1}{2}\Lambda_1(p^{j+1/2} + w^{j+1/2} + w^j) - \frac{1}{\tau}p^{j+1/2}, \tag{11}$$

$$(x \in \omega_h),$$

$$\frac{w^{j+1} - w^{j+1/2}}{\tau} = \frac{1}{2}\Lambda_2(p^{j+1/2} + w^{j+1} + w^{j+1/2}) + \frac{1}{\tau}p^{j+1/2}, \tag{12}$$

$$w^{j+1/2} = w^{j+1} = 0, \quad x \in \gamma,$$

$$j = 0, 1, 2, \dots, N-1 .$$

Multiplying equation (11) by $\tau(w^{j+1/2} + w^j)$ and equation (12) by $\tau(w^{j+1} + w^{j+1/2})$, respectively, and summing over ω_h we obtain the identities

$$\|w^{j+1/2}\|^2 - \|w^j\|^2 = -(p^{j+1/2}, w^{j+1/2} + w^j) + \frac{\tau}{2}\left(\Lambda_1(p^{j+1/2} + w^{j+1/2} + w^j), w^{j+1/2} + w^j\right),$$

$$\|w^{j+1}\|^2 - \|w^{j+1/2}\|^2 = (p^{j+1/2}, w^{j+1} + w^{j+1/2}) + \frac{\tau}{2}\left(\Lambda_2(p^{j+1/2} + w^{j+1} + w^{j+1/2}), w^{j+1} + w^{j+1/2}\right).$$

To estimate the second right-hand terms in both identities we now use lemma 2 thus getting

$$\|w^{j+1/2}\|^2 - \|w^j\|^2 \leqq -(p^{j+1/2}, w^{j+1/2} + w^j) + \tau\|p^{j+1/2}\|_{1,1}^2$$

and

$$\|w^{j+1}\|^2 - \|w^{j+1/2}\|^2 \leqq (p^{j+1/2}, w^{j+1} + w^{j+1/2}) + \tau\|p^{j+1/2}\|_{1,2}^2 .$$

By adding we have

$$||w^{j+1}||^2 - ||w^j||^2 \leqq (p^{j+1/2}, w^{j+1} - w^j) + \tau||p^{j+1/2}||_1^2$$

and further, by summing over j from 0 to k,

$$||w^{k+1}||^2 \leqq \sum_0^k (p^{j+1/2}, w^{j+1} - w^j) + \sum_0^k \tau||p^{j+1/2}||_1^2$$

$$= (p^{k+1/2}, w^{k+1}) - \sum_1^k \tau(w^j, p_t^j) + \sum_0^k \tau||p^{j+1/2}||_1^2 .$$

Applying now (9) with $\varepsilon = 1$ to the scalarproducts, gives

$$||w^{k+1}||^2 \leqq \sum_1^k \tau||w^j||^2 + ||p^{k+1/2}||^2 + 2\overline{||p||}_k^2 ,$$

where we have introduced the notation

$$\overline{||p||}_k^2 = \sum_1^k \tau||p_t^j||^2 + \sum_0^k \tau||p^{j+1/2}||_1^2 . \tag{13}$$

Now we set

$$a^{k+1} = ||w^{k+1}||^2 , \qquad b^{k+1} = ||p^{k+1/2}||^2 + 2\overline{||p||}_k^2$$

apply lemma 1 and remark that

$$b_{j+1} \leqq \max_{0\leqq i\leqq k} ||p^{i+1/2}||^2 + 2\overline{||p||}_k^2 , \qquad j \leqq k .$$

This leads to

$$||w^{k+1}||^2 \leqq M_0\left(\max_{0\leqq j\leqq k} ||p^{j+1/2}||^2 + \overline{||p||}_k^2\right),$$

where $M_0 = 2(1 + Te^T)$, and from here, (13) and (10) follows immediately the estimate of theorem 1.

A similar estimate holds for $v^{j+1/2} = w^{j+1/2} + p^{j+1/2}$.

As a corollary of theorem 1 we have now

Theorem 2. *Let the solution of* (1) $u \in C^{3,6}$. *Let z be the solution of* (7), *but, different from* (7) *satisfy the boundary conditions*

$$z^{j+1/2} = \varphi^{j+1/2} + \tau U_1\varphi^{j+1/2} , \quad z^{j+1} = \varphi^{j+1} , \quad x \in \gamma ,$$

where U_1 is the differential operator (4).

Then z approximates u with order of accuracy $O(\tau^2 + h_1^2 + h_2^2)$.

Proof: As mentioned earlier, in [3] was proved the estimate

$$||y^j - u(x, j\tau)|| = O(\tau^2 + h_1^2 + h_2^2)$$

for the solution y of (7). The differentiability conditions satisfied by u are sufficient for the norms

$$||U_2u||, ||U_2u||_1 \quad \text{and} \quad ||(U_2u)_t|| , \qquad t \in [0, T] ,$$

to be bounded. (The operator U_2 contains forth derivatives with respect to the spatial coordinates and second with respect to time [3]). Hence, by setting in (8)

$$\nu = \tau^2 U_2\varphi , \quad x \in \gamma , \quad \text{and} \quad p = \tau^2 U_2 u , \qquad x \in \omega_h ,$$

in (10), we get from theorem 1 for $v = v(\tau^2 U_2 \varphi)$

$$||v|| = O(\tau^2)$$

and theorem 2 follows by noting that $z = y - v(\tau^2 U_2 \varphi)$.

Bibliography

[1] Janenko, N. N., Об одном разностном методе счета многомерного уравнения теплопроводности. ДАН СССР **125**, 6 (1959) 1207—1210.

[2] Janenko, N. N., Die Zwischenschrittmethode zur Lösung mehrdimensionaler Probleme der mathematischen Physik. Springer (1969) (Lecture notes 91).

[3] Stoyan, G., Some economic additive scheme for the solution of boundary value problems of manydimensional parabolic equations. Journ. Num. Math. and Math. Physics **10**, Nr. 3 (1970) 644—653 (russ.).

[4] Samarski, A. A., О разностных схемах для многомерных дифференциальных уравнений математической физики. Aplikace matematiky **10**, 2, 146—163, Praha.

[5] Samarski, A. A., On additive difference schemes. Lecture, delivered at the Intern. Congress of Mathematics, Moscou (1966) (russ.).

[6] Samarski, A. A., Введение в теорию разностных схем. Москва (1971).

[7] Frjasinow, I. W., Экономические симметризованные схемы решения краевых задач для многомерного уравнения параболического типа. Журн. вычисл. мат. и мат. физики **8**, 2 (1968) 436—443.

[8] —, Economic symmetric schemes for the solution of boundary value problems of mathematical physics. Thesis, Institut of Appl. Math. Ac. Sci. SU, Moscou (1968) (russ.).

Anschrift des Autors:

ZIMM der AdW der DDR
108 Berlin, Mohrenstr. 39

Zum Galerkin-Verfahren bei einigen Klassen nichtlinearer Differentialgleichungen im Hilbertraum

KLAUS ZACHARIAS, Berlin

I. Vorbemerkungen

Wir betrachten Anfangswertprobleme für abstrakte Differentialgleichungen der Form

$$\big(A(t)\, u(t)\big)' + B(t)\, u(t) = f(t) \tag{1}$$

und

$$A(t)\, u'(t) + B(t)\, u(t) = f(t)\,. \tag{2}$$

Wir geben Existenz- und Einzigkeitssätze für derartige Probleme an und zeigen, wie man auf der Basis des Galerkinverfahrens konstruktive Näherungsverfahren begründen kann.

Die Lösung u dieser Gleichungen ist dabei auf einem kompakten (Zeit-) Intervall $S = [0, T]$, $T > 0$, definiert und nimmt Werte in einem reellen separablen Hilbertraum H an; wir schreiben dafür $u \in (S \to H)$. Insbesondere soll die Anfangsbedingung

$$u(0) = a \in H \tag{3}$$

erfüllt werden. Das Skalarprodukt und die Norm in H seien mit $(.,.)$ bzw. $\|\,.\,\|$ bezeichnet. Für Funktionen $u \in (S \to H)$ sind Stetigkeit und Differenzierbarkeit in natürlicher Weise erklärt (vgl. [1]). Wir bezeichnen die Menge der stetigen Funktionen auf S mit Werten in H durch $C(S; H)$; mit der Normierung

$$\|u\|_C = \max_{t \in S} \|u(t)\| \tag{4}$$

ist $C(S; H)$ ein reeller Banachraum. Mit $L^2(S; H)$ bezeichnen wir die Menge der (Klassen von) im Bochnerschen Sinne quadratisch summierbaren Funktionen $u \in (S \to H)$; $L^2(S; H)$ ist Banachraum (sogar Hilbertraum) bezüglich der Norm

$$\|u\|_{L^2(S;\, H)} = \left(\int_0^T \|u(t)\|^2\, dt\right)^{1/2}.$$

Näheres über Meßbarkeit, Summierbarkeit usw. bei Funktionen mit Werten in normierten Räumen findet man in [2], [5], [14]; für einige Aussagen über absolut stetige abstrakte Funktionen verweisen wir auf [10], [11].

Es sei noch darauf hingewiesen, daß die hier betrachteten Differentialgleichungen in engem Zusammenhang stehen mit den sogenannten „pseudoparabolischen“ Gleichungen. Wir werden diesen Zusammenhang in Abschnitt V genauer darstellen.

Über die Operatoren $A(t)$, $B(t)$ setzen wir voraus, daß $A(t), B(t) \in (H \to H)$ für jedes $t \in S$, wobei folgende Bedingungen erfüllt seien:

(a) Die Funktion $t \to A(t)\, x$ ist für jedes $x \in H$ stetig auf S.

(b) Die Operatoren $x \to A(t)\, x$ sind (bezüglich $t \in S$ gleichmäßig) Lipschitz-stetig und stark monoton:

$$\|A(t)\, x - A(t)\, y\| \leqq M\|x - y\| \qquad \text{für alle } x, y \in H;$$

$$\big(A(t)\, x - A(t)\, y, x - y\big) \geqq m\|x - y\|^2 \qquad \text{für alle } x, y \in H$$

mit einer Monotoniekonstanten $m > 0$.

(c) Die Funktion $t \to B(t)\, x$ ist für jedes $x \in H$ stetig auf S.

(d) Die Operatoren $x \to B(t)\, x$ sind (bezüglich $t \in S$ gleichmäßig) Lipschitz-stetig:

$$\|B(t)\, x - B(t)\, y\| \leqq L\|x - y\| \qquad \text{für alle } x, y \in H\,.$$

Für die rechten Seiten in den Gleichungen (1), (2) verlangen wir

(e) $$f \in L^2(S; H)\,.$$

II. Existenz- und Einzigkeitssätze

Definition 1. Eine Funktion $u \in (S \to H)$ heißt Lösung des Anfangswertproblems (1,3), falls $u \in C(S; H)$ und die Integralgleichung

$$A(t)\, u(t) = u_0 - \int_0^t \big(B(s)\, u(s) - f(s)\big)\, ds\,, \qquad t \in S\,, \tag{5}$$

mit $u_0 = A(0)\, a$ erfüllt ist.

Definition 2. Eine Funktion $u \in (S \to H)$ heißt Lösung des Anfangswertproblems (2,3), falls $u \in C(S; H)$ und die Integralgleichung

$$u(t) = a + \int_0^t A^{-1}(s)\, \big(- B(s)\, u(s) + f(s)\big)\, ds\,, \qquad t \in S\,, \tag{6}$$

erfüllt ist.

Bemerkung 1. Wegen der absoluten Stetigkeit der auftretenden Bochnerintegrale folgt aus der Gültigkeit von (5) bzw. (6), daß $t \to A(t)\, u(t)$ bzw. $t \to u(t)$ auf S absolut stetig und somit fast überall differenzierbar sind. Der Operator $A^{-1}(s)$ in (6) existiert unter den oben gemachten Voraussetzungen (b) für jedes $s \in S$ (Satz von Browder-Minty [4]).

Wir vereinbaren noch, die Funktionen $t \to A(t)\, u(t)$ usw. abkürzend mit Au zu bezeichnen.

Lemma (Browder-Petryshyn-Weinberg). *Die Operatoren der Familie* $\{A(t)\}$ *mögen für jedes* $t \in S$ *den Bedingungen* (a), (b) *genügen. Dann ist für jedes feste* q *mit* $0 < q < \frac{2m}{M^2}$ *der Operator*

$$x \to x - qA(t)\, x\,, \quad x \in H\,, \quad t \in S\,,$$

kontraktiv mit der Kontraktionskonstanten

$$r = r(q) = \sqrt{1 - 2qm + q^2M^2} < 1 .$$

Zum Beweis des Lemmas verweisen wir auf [3].

Satz 1. *Unter den Voraussetzungen* (a)—(e) *existiert genau eine Lösung u des Anfangswertproblems* (1,3) *mit*

$$u \in C(S; H) \quad \text{und} \quad (Au)' \in L^2(S; H) .$$

Beweis. Wir betrachten für $u \in C(S; H)$ die Abbildung U, definiert durch

$$(Uu)(t) = qu_0 + u(t) - qA(t)\,u(t) - q \int_0^t \big(B(s)\,u(s) - f(s)\big)\,ds \tag{7}$$

für $t \in S$.

Hierbei ist q ein reeller Parameter, der der Einschränkung $0 < q < \frac{2m}{M^2}$ unterliegt. Den Banachraum $C(S; H)$ versehen wir mit der zur C-Norm äquivalenten (C, k)-Norm

$$\|u\|_{C,k} = \max_{t \in S} \{ e^{-kt} \|u(t)\| \} ; \tag{8}$$

k wird später noch geeignet gewählt. Wegen der Lipschitz-stetigkeit von $A(t)$ gilt offenbar

$$U \in \big(C(S; H) \to C(S; H)\big)$$

und für $u, v \in C(S; H)$ ist wegen des Lemmas

$$\|(Uu)(t) - (Uv)(t)\| \leqq r(q)\,\|u(t) - v(t)\| + q \int_0^t \|B(s)\,u(s) - B(s)\,v(s)\|\,ds .$$

Nun ist wegen der Lipschitz-stetigkeit von B

$$\int_0^t \|B(s)\,u(s) - B(s)\,v(s)\|\,ds \leqq L \int_0^t \|u(s) - v(s)\|\,ds = L \int_0^t \|u(s) - v(s)\|\,e^{-ks}e^{ks}ds$$

$$\leqq L\|u - v\|_{C,k} \int_0^t e^{ks}ds \leqq L\|u - v\|_{C,k} \frac{e^{kt}}{k}$$

und somit

$$\|(Uu)(t) - (Uv)(t)\|\,e^{-kt} \leqq r(q)\,\|u(t) - v(t)\|\,e^{-kt} + \frac{qL}{k}\,\|u - v\|_{C,k} .$$

Bildet man die (C, k)-Norm, so folgt

$$\|Uu - Uv\|_{C,k} \leqq \left(r(q) + \frac{qL}{k}\right) \|u - v\|_{C,k} .$$

Wegen $0 < r(q) < 1$ kann man für genügend großes k erreichen, daß

$$0 < r(q) + \frac{qL}{k} < 1 ,$$

d. h., die Abbildung U ist auf dem mit der (C, k)-Norm versehenen Raum $C(S; H)$ für derartige k kontraktiv. Somit existiert genau ein Fixpunkt $u \in C(S; H)$ der Abbildung U:

$$u(t) = (Uu)(t) \qquad \forall\, t \in S$$

oder wegen (7)

$$A(t)\, u(t) = \int_0^t \big(-B(s)\, u(s) + f(s)\big)\, ds\,.$$

Aus der absoluten Stetigkeit des Integrals und der Voraussetzung (e) folgt $(Au)' \in L^2(S; H)$. Damit ist Satz 1 bewiesen.

Satz 2. *Unter den Voraussetzungen* (a)—(e) *existiert genau eine Lösung u des Anfangswertproblems* (2,3) *mit*

$$u \in C(S; H)\,, \qquad u' \in L^2(S; H)\,.$$

Beweis. Wir bezeichnen mit $C_{\text{abs}}(S; H)$ die Menge der auf S definierten absolut stetigen Funktionen, die Werte in H annehmen. Es sei

$$X = \{u\colon u \in C_{\text{abs}}(S; H),\ u' \in L^2(S; H),\ u(0) = a\}\,.$$

Mit der Metrik

$$\varrho_k(u, v) = \left(\int_S \|u'(t) - v'(t)\|^2\, e^{-2kt}\, dt\right)^{1/2}$$

ist X ein vollständiger metrischer Raum. Über die Konstante k werden wir noch geeignet verfügen. Wir definieren für $u \in X$ die Abbildung U durch

$$(Uu)(t) = a + \int_0^t \big(u'(s) - qA(s)\, u'(s) - qB(s)\, u(s) + qf(s)\big)\, ds\,, \tag{9}$$

wobei q (wie im obigen Lemma) der Einschränkung

$$0 < q < \frac{2m}{M^2}$$

unterliegt. Offenbar ist $U \in (X \to X)$ und wegen (9) gilt für $u, v \in X$

$$\big((Uu)(t)\big)' - \big((Uv)(t)\big)' = u'(t) - qA(t)\, u'(t) - \big(v'(t) - qA(t)\, v'(t)\big)$$
$$- q\big(B(t)\, u(t) - B(t)\, v(t)\big)$$

für fast alle $t \in S$. Wegen des Lemmas folgt

$$\varrho_k(Uu, Uv) \leqq r(q)\, \varrho_k(u, v) + qL\left(\int_S \|u(t) - v(t)\|^2\, e^{-2kt}\, dt\right)^{1/2}. \tag{10}$$

Nun ist für $u, v \in X$

$$\|u(t) - v(t)\|^2 \leqq T \int_0^t \|u'(s) - v'(s)\|^2\, ds\,.$$

Damit wird

$$\int_S \|u(t) - v(t)\|^2\, e^{-2kt}\, dt \leqq T \int_S e^{-2kt} \left(\int_0^t \|u'(s) - v'(s)\|^2\, ds\right) dt$$

und nach partieller Integration folgt

$$\int_S \|u(t) - v(t)\|^2 e^{-2kt}\, dt \leqq T\left(-\frac{1}{2k} e^{-2kT} \int_0^T \|u'(s) - v'(s)\|^2\, ds + \frac{1}{2k} \int_0^T \|u'(t) - v'(t)\|^2 e^{-2kt}\, dt\right) \leqq \frac{T}{2k} \varrho_k^2(u, v)\,.$$

Einsetzen in (10) liefert

$$\varrho_k(Uu, Uv) = \left(r(q) + \frac{qL\sqrt{T}}{\sqrt{2k}}\right) \varrho_k(u, v)\,.$$

Wegen $0 < r(q) < 1$ ist für genügend großes k zu erreichen, daß

$$0 < r(q) + \frac{qL\sqrt{T}}{\sqrt{2k}} < 1\,.$$

Die Abbildung U ist mithin in der entsprechenden ϱ_k-Metrik von X kontraktiv, d. h., es existiert ein Fixpunkt $u \in X$ mit

$$u(t) = (Uu)(t) \qquad \forall\, t \in S\,.$$

An Formel (9) liest man ab, daß dieser Fixpunkt die Lösung des Anfangswertproblems (2,3) mit allen behaupteten Eigenschaften ist.

III. Das Galerkinverfahren

Durch das Galerkinverfahren werden wir Näherungslösungen der Gleichungen (1) und (2) konstruieren, die — als Funktionen auf S betrachtet — Werte in endlichdimensionalen Hilberträumen annehmen.

Wir wählen ein in H vollständiges System $\{h_1, h_2, \ldots\}$ linear unabhängiger Elemente und setzen

$$H_n = \operatorname{span}\{h_1, h_2, \ldots, h_n\} \subset H$$

für die lineare Hülle des von den Elementen $h_1, \ldots, h_n$ aufgespannten endlichdimensionalen Teilraumes; H_n ist Hilbertraum bezüglich des von H induzierten Skalarproduktes. Den (orthogonalen) Projektor von H auf H_n bezeichnen wir mit P_n. Wir setzen

$$A_n(t) = P_n A(t)\,, \qquad B_n(t) = P_n B(t)$$

und bemerken, daß diese Operatoren den Bedingungen (a)—(d) genügen, wenn man in den dort angegebenen Formulierungen H durch H_n ersetzt. Als Galerkingleichungen zu den Gleichungen (1), (2) sehen wir

$$\big(A_n(t)\, u_n(t)\big)' + B_n(t)\, u_n(t) = f_n(t) \tag{11}$$

bzw.

$$A_n(t)\, u_n'(t) + B_n(t)\, u_n(t) = f_r(t) \tag{12}$$

mit der Anfangsbedingung

$$u_n(0) = a_n = P_n a \in H_n \tag{13}$$

an; $n = 1, 2, \dots$. Die rechten Seiten f_n seien durch

$$t \to f_n(t) = P_n f(t)\,, \qquad t \in S$$

definiert, wegen (e) ist $f_n \in L^2(S; H_n)$. (Die Räume $L^2(S; H_n)$ bzw. $C(S; H_n)$ sind analog zu $L^2(S; H)$ bzw. $C(S; H)$ erklärt.)

Satz 3. *Unter den Bedingungen* (a)—(e) *besitzen die Galerkingleichungen* (11,13) *für jedes* $n = 1, 2, \dots$ *genau eine Lösung* u_n *mit*

$$u_n \in C(S; H_n) \quad \text{und} \quad (A_n u_n)' \in L^2(S; H_n)\,.$$

Es gilt, wenn u die Lösung des Anfangswertproblems (1,3) *bezeichnet,*

$$\|u_n - u\|_C \to 0 \qquad \text{und} \qquad \|(A_n u_n)' - (Au)'\|_{L^2(S;\,H)} \to 0 \tag{14}$$

für $n \to \infty$.

Beweis. Die Existenz von Lösungen u_n der Galerkingleichungen wird wie in Satz 1 bewiesen; man hat lediglich überall H durch H_n zu ersetzen und von bekannten Eigenschaften von Projektionsoperatoren Gebrauch zu machen. An die Stelle der Abbildung U treten Abbildungen U_n, definiert für $u \in C(S; H_n)$ durch

$$(U_n u)(t) = q u_{0n} + u(t) - q A_n(t)\, u(t) - q \int_0^t \big(B_n(s)\, u(s) - f_n(s)\big)\, ds \tag{15}$$

mit

$$u_{0n} = A_n(0)\, a_n\,.$$

Diese Abbildungen sind in der entsprechenden (C, k)-Normierung von $C(S; H_n)$ kontraktiv mit der gleichen Kontraktionskonstanten wie in Satz 1 — eine Feststellung, die beim Projektions-Iterationsverfahren von Bedeutung sein wird.

Zum Beweis der Konvergenzaussagen (14) zeigen wir zunächst, daß eine Vergleichsfolge $\{w_n\}$ von Funktionen $w_n \in C(S; H_n)$ mit der Eigenschaft

$$\|w_n - u\|_C \to 0 \quad \text{für} \quad n \to \infty \tag{16}$$

existiert. Wählen wir nämlich $w_n(t) = P_n u(t)$ für $n = 1, 2, \dots,\ t \in S$, so ist offenbar $w_n \in C(S; H_n)$ und es gilt wegen der Vollständigkeit des Systems $\{h_1, h_2, \dots\}$ in H

$$\|w_n(t) - u(t)\| \to 0 \quad \text{für} \quad n \to \infty\,, \quad t \in S\,.$$

Da wegen $H_n \subset H_{n+1}$ noch

$$\|w_{n+1}(t) - u(t)\| \leqq \|w_n(t) - u(t)\| \qquad \forall\, t \in S$$

gilt, so konvergiert nach einem Satz von Dini die Folge $\{w_n\}$ auf S gleichmäßig gegen u.

Die Lösungen der Galerkingleichungen (11) und der Gleichung (1,3) genügen den Integralgleichungen

$$A_n(t)\, u_n(t) = u_{0n} - \int_0^t \big(B_n(s)\, u_n(s) - f_n(s)\big)\, ds$$

$$A(t)\, u(t) = u_0 - \int_0^t \big(B(s)\, u(s) - f(s)\big)\, ds\,.$$

Subtraktion dieser Gleichungen und skalare Multiplikation mit $u_n(t) - w_n(t)$ ergibt unter Beachtung der Symmetrie des Projektionsoperators P_n:

$$\big(A(t)\, u_n(t) - A(t)\, w_n(t),\, u_n(t) - w_n(t)\big) + \big(A(t)\, w_n(t) - A(t)\, u(t),\, u_n(t) - w_n(t)\big)$$

$$= \big(u_{0n} - u_0,\, u_n(t) - w_n(t)\big) - \left(\int_0^t \big(B(s)\, u_n(s) - B(s)\, w_n(s)\big)\, ds,\, u_n(t) - w_n(t)\right)$$

$$- \left(\int_0^t \big(B(s)\, w_n(s) - B(s)\, u(s)\big)\, ds,\, u_n(t) - w_n(t)\right).$$

Aus der Monotonie und Lipschitzstetigkeit der Operatoren $A(t)$, $B(t)$ folgt

$$m\|u_n(t) - w_n(t)\| \leqq (M + TL)\, \|w_n - u\|_C + \|u_{0n} - u_0\| + L \int_0^t \|w_n(s) - u_n(s)\|\, ds\,.$$

Das Gronwallsche Lemma liefert nunmehr

$$\|u_n - w_n\|_C \leqq K(\|w_n - u\|_C + \|u_{0n} - u_0\|)$$

mit einer geeigneten Konstanten $K = K(T, m, M, L)$. Mit der Eigenschaft (16) der Folge $\{w_n\}$ folgt hieraus

$$\|u_n - u\|_C \to 0 \quad \text{für} \quad n \to \infty\,.$$

Wegen der absoluten Stetigkeit unbestimmter Integrale gilt bis auf Mengen I_0, I_n $(n = 1, 2, \ldots)$ vom Maß Null auf S

$$\big(A(t)\, u(t)\big)' + B(t)\, u(t) = f(t)$$

bzw.

$$\big(A_n(t)\, u_n(t)\big)' + B_n(t)\, u_n(t) = f_n(t)\,.$$

Es gilt demnach auf $S - \bigcup_{n=0}^{\infty} I_n$, also fast überall auf S

$$\|\big(A(t)\, u(t)\big)' - \big(A_n(t)\, u_n(t)\big)'\| \leqq \|B(t)\, u(t) - P_n B(t)\, u(t)\|$$
$$+ \|B(t)\, u(t) - B(t)\, u_n(t)\| + \|f_n(t) - f(t)\|$$

und weiter

$$\|(Au)' - (A_n u_n)'\|_{L^2(S;\,H)} \leqq K(\|Bu - P_n Bu\|_C + \|u_n - u\|_C + \|f_n - f\|_{L^2(S;\,H)})\,.$$

Mit den gleichen Überlegungen wie beim Beweis von (16) überzeugt man sich, daß

$$\|Bu - P_n Bu\|_C \to 0 \qquad \text{für} \quad n \to \infty\,.$$

Da für $f_n, f \in L^2(S; H)$

$$\|f_n(t) - f(t)\| = \|P_n f(t) - f(t)\| \to 0 \qquad \text{für} \quad n \to \infty$$

fast überall auf S und

$$\|P_n f(t)\| \leqq \|f(t)\|$$

fast überall auf S, so gilt nach dem Konvergenzsatz von Lebesgue

$$\|f_n - f\|_{L^2(S;\,H)} \to 0 \quad \text{für} \quad n \to \infty\,.$$

Damit läßt sich nunmehr auch die zweite Konvergenzaussage beweisen.

Satz 4. *Unter den Bedingungen* (a)—(e) *besitzen die Galerkingleichungen* (12,13) *für jedes* $n = 1, 2, \ldots$ *genau eine Lösung* u_n *mit*

$$u_n \in C(S; H_n), \qquad u_n' \in L^2(S; H_n).$$

Es gilt, wenn u *die Lösung des Anfangswertproblems* (2,3) *bezeichnet,*

$$\|u_n - u\|_C \to 0 \quad \textit{und} \quad \|u_n' - u'\|_{L^2(S;H)} \to 0 \quad \textit{für } n \to \infty.$$

Beweis. Die Existenz der Lösungen der Galerkingleichungen wird wie in Satz 2 bewiesen. Es wird dabei gezeigt, daß die Abbildungen U_n, $n = 1, 2, \ldots$, definiert durch

$$(U_n u)(t) = a_n + \int_0^t \big(u'(s) - qA_n(s)\, u'(s) - qB_n(s)\, u(s) + qf_n(s)\big)\, ds \tag{17}$$

auf $X_n = \{u: u \in C_{\mathrm{abs}}(S; H_n),\ u' \in L^2(S; H_n),\ u(0) = a_n\}$ kontraktiv sind mit einer von n unabhängigen Kontraktionskonstanten. Zur Konvergenzuntersuchung des Galerkinverfahrens geben wir eine Vergleichsfolge $\{w_n\}$ mit den Eigenschaften $w_n \in C(S; H_n)$, $w_n' \in L^2(S; H_n)$ und

$$\|w_n - u\|_C \to 0, \quad \|w_n' - u'\|_{L^2(S;H_n)} \to 0 \quad \text{für} \quad n \to \infty. \tag{18}$$

Wir wählen

$$w_n = P_n u.$$

Die erste der Eigenschaften (18) wurde schon mit (16) bewiesen. In den Differenzierbarkeitspunkten von u, also fast überall, gilt

$$w_n'(t) = P_n u'(t)$$

und deshalb $\|w_n'(t) - u'(t)\| \to 0$ fast überall für $n \to \infty$. Da $\|w_n'(t)\| = \|P_n u'(t)\| \leqq \|u'(t)\|$ fast überall auf S, so folgt aus dem Konvergenzsatz von Lebesgue

$$\|w_n' - u'\|_{L^2(S;H)} \to 0 \qquad \text{für} \quad n \to \infty,$$

womit die zweite der Eigenschaften (18) nachgewiesen ist. Wegen der absoluten Stetigkeit unbestimmter Integrale gilt bis auf Mengen I_0, I_n $(n = 1, 2, \ldots)$ vom Maß Null, also fast überall auf S:

$$A(t)\, u'(t) + B(t)\, u(t) = f(t),$$

$$A_n(t)\, u_n'(t) + B_n(t)\, u_n(t) = f_n(t).$$

Demnach gilt auf $S - \bigcup_{n=0}^{\infty} I_n$, also fast überall auf S, nach Subtraktion und skalarer Multiplikation mit $u_n'(t) - w_n'(t)$ wegen der Symmetrie der Projektoren P_n (das Argument t werde vorübergehend weggelassen):

$$(Au_n' - Aw_n', u_n' - w_n') + (Aw_n' - Au', u_n' - w_n') + (Bu_n - Bw_n, u_n' - w_n')$$
$$+ (Bw_n - Bu, u_n' - w_n') = 0.$$

Abschätzung unter Beachtung von (a)—(d) ergibt

$$m\|u_n'(t) - w_n'(t)\| \leqq M\|w_n'(t) - u'(t)\| + L\|u_n(t) - w_n(t)\| + L\|w_n(t) - u(t)\|.$$

Hieraus folgt mit einer geeigneten, von m, M, L, T abhängenden Konstanten K_1

$$\int_0^s \|u_n'(t) - w_n'(t)\|^2 \, dt$$

$$\leqq K_1 \left(\|w_n' - u'\|^2_{L^2(S;H)} + \|w_n - u\|^2_C + \int_0^s \|w_n(t) - u_n(t)\|^2 \, dt \right). \tag{19}$$

Aus der Darstellung

$$w_n(t) - u_n(t) = \int_0^t \left(w_n'(\tau) - u_n'(\tau) \right) d\tau$$

erhält man

$$\|w_n(t) - u_n(t)\|^2 \leqq T \int_0^t \|w_n'(\tau) - u_n'(\tau)\|^2 \, d\tau \tag{20}$$

und mit der Abkürzung

$$F(s) = \int_0^s \|u_n'(t) - w_n'(t)\|^2 \, dt$$

aus (19)

$$F(s) \leqq K \left(\|w_n' - u'\|^2_{L^2(S;H)} + \|w_n - u\|^2_C + \int_0^s F(t) \, dt \right).$$

Das Gronwallsche Lemma liefert nun

$$\|u_n' - w_n'\|_{L^2(S;H)} \to 0 \qquad \text{für} \quad n \to \infty \, .$$

Mit (20) und den weiteren Eigenschaften der Vergleichsfolge $\{w_n\}$ folgen die übrigen Behauptungen

IV. Das Projektions-Iterationsverfahren

Wir haben im Abschnitt III Galerkinnäherungen als Fixpunkte gewisser (i. a. nichtlinearer) Abbildungen U_n konstruiert. Mit Hilfe der Basiselemente $\{h_1, \ldots, h_n\}$ des endlichdimensionalen Teilraumes $H_n \subset H$ kann man die Galerkinnäherungen in der Form

$$u_n(t) = \sum_{k=1}^{n} c_{nk}(t) \, h_k \, , \qquad t \in S \, , \tag{21}$$

darstellen. Wegen (15) bzw. (17) sind die (reellen) Koeffizienten c_{nk} aus i. a. nichtlinearen Integralgleichungen zu ermitteln. Mit Hilfe des Projektions-Iterationsverfahrens lassen sich Näherungen allein mit Hilfe linearer Verfahren ermitteln. Die Grundlage dafür sind geringfügige Modifikationen eines in [6] angegebenen Satzes.

Satz 5. *Sei u Lösung des Anfangswertproblems* (1,3). *Dann gilt mit* $0 < q < \dfrac{2m}{M^2}$ *für die Iterationsfolge*

$$z_n(t) = q u_{0n} + z_{n-1}(t) - q A_n(t) \, z_{n-1}(t)$$

$$- \, q \int_0^t \left(B_n(s) \, z_{n-1}(s) - f_n(s) \right) ds \, , \qquad t \in S \, , \qquad n = 1, 2, \ldots , \tag{22}$$

bei beliebigem Anfangselement $z_0 \in C(S; H_1)$:

$$||z_n - u||_C \to 0 \qquad \textit{für} \quad n \to \infty .$$

Zum Beweis dieses Satzes verweisen wir auf [6]. Wir geben noch einige Erläuterungen zur Iterationsvorschrift (22). Wegen (15) kann man (22) in der Form

$$z_n(t) = (U_n z_{n-1})(t), \qquad t \in S,$$

schreiben. Macht man für z_n den Ansatz

$$z_n(t) = \sum_{k=1}^{n} a_{nk}(t)\, h_k \tag{23}$$

und entsprechende Ansätze

$$f_n(t) = \sum_{j=1}^{n} s_j(t)\, h_j, \qquad u_{0n} = \sum_{j=1}^{n} b_j h_j$$

mit den Basiselementen $\{h_1, \ldots, h_n\}$, so folgt nach skalarer Multiplikation mit h_k $(k = 1, \ldots, n)$:

$$\begin{aligned}\sum_{j=1}^{n} a_{nj}(t)\,(h_j, h_k) &= q \sum_{j=1}^{n} b_j(h_j, h_k) + \sum_{j=1}^{n-1} a_{n-1j}(t)\,(h_j, h_k) \\ &\quad - q\big(A(t)\, z_{n-1}(t), h_k\big) - q \int_0^t \big(B(s)\, z_{n-1}(s), h_k\big)\, ds \\ &\quad + q \sum_{j=1}^{n} \int_0^t s_j(r)\, dr (h_j, h_k) .\end{aligned}$$

Das ist ein lineares Gleichungssystem für die Koeffizienten $a_{nj} \in C(S)$, $j=1, \ldots, n$. Die Koeffizientenmatrix des Systems (24) ist gerade die Gramsche Matrix G der Basis $\{h_1, \ldots, h_n\}$, d. h.,

$$G = (g_{jk}), \qquad \text{wobei} \quad g_{jk} = (h_j, h_k) \qquad (j, k = 1, \ldots, n) .$$

Man sieht, daß sich die Rechnungen für orthonormierte Basen wesentlich vereinfachen.

Satz 6. *Sei u Lösung des Anfangswertproblems* (2,3). *Dann gilt mit* $0 < q < \frac{2m}{M^2}$ *für die Iterationsfolge*

$$z_n(t) = a_n + \int_0^t \big(z'_{n-1}(s) - qA_n(s)\, z'_{n-1}(s) - qB_n(s)\, z_{n-1}(s) + qf_n(s)\big)\, ds$$

bei beliebigem $z_0 \in X_1$:

$$||z_n - u||_X \to 0 \qquad \text{für} \quad n \to \infty .$$

Zum Beweis vgl. [6]. Wiederum kann man mit einem Ansatz der Form (23) die Berechnung sukzessiver Näherungen z_n auf die Auflösung von linearen Gleichungssystemen und Quadraturen zurückführen.

Bemerkung 2. Wie aus [6] hervorgeht, ist beim Beweis der Sätze 5 und 6 wesentlich, daß die Kontraktionskonstanten der Abbildungen U_n in Satz 3 und Satz 4 von n unabhängig sind.

V. Pseudoparabolische Gleichungen

Die von uns betrachteten Gleichungen (1,2) kann man sich aus pseudoparabolischen Gleichungen hervorgegangen denken. Wir wollen andeuten, wie dieser Übergang vollzogen werden kann.

Es sei H^* der zu H duale Raum, $\|\,.\,\|_*$ die zu $\|\,.\,\|$ duale Norm von H^* und der Wert eines Funktionals $x^* \in H^*$ im Element $x \in H$ werde mit $\langle x^*, x\rangle$ bezeichnet. Wir betrachten die pseudoparabolischen Gleichungen

$$\big(C(t)\,u(t)\big)' + D(t)\,u(t) = g(t)\,, \qquad C(t)\,u'(t) + D(t)\,u(t) = g(t) \tag{25}$$

mit der Anfangsbedingung

$$u(0) = a \in H\,.$$

Die Operatoren $C(t), D(t) \in (H \to H^*)$ seien für jedes $t \in S$ erklärt und mögen folgende Eigenschaften besitzen:

(a_1) Die Funktion $t \to C(t)\,x$ mit Werten in H^* ist für jedes $x \in H$ stetig auf S.

(b_1) Die Operatoren $x \to C(t)\,x$ sind (bezüglich $t \in S$ gleichmäßig) Lipschitz-stetig und stark monoton:

$$\|C(t)\,x - C(t)\,y\|_* \leqq M\|x - y\| \quad \text{für alle} \quad x, y \in H;$$

$$\langle C(t)\,x - C(t)\,y\,x, - y\rangle \geqq m\|x - y\|^2 \quad \text{für alle} \quad x, y \in H \quad \text{mit} \quad m > 0\,.$$

(c_1) Die Funktion $t \to D(t)\,x$ mit Werten in H^* ist für jedes $x \in H$ stetig auf S.

(d_1) Die Operatoren $x \to D(t)\,x$ sind (bezüglich $t \in S$ gleichmäßig) Lipschitz-stetig:

$$\|D(t)\,x - D(t)\,y\|_* \leqq L\|x - y\| \quad \text{für alle} \quad x, y \in H\,.$$

Für die rechten Seiten wird vorausgesetzt:

(e_1) $$g \in L^2(S; H^*)\,.$$

Es sei $R \in (H^* \to H)$ die (lineare) Rieszsche Abbildung von H^* auf H, definiert durch

$$\langle x^*, y\rangle = (Rx^*, y) \qquad \forall\, y \in H\,. \tag{26}$$

Diese Abbildung existiert wegen des Darstellungssatzes von Riesz für lineare Funktionale im Hilbertraum und vermittelt einen Isomorphismus zwischen H und H^*. Es gilt außerdem

$$\|R\|_{H^* \to H} = 1\,, \qquad \|Rx^*\| = \|x^*\|_*\,.$$

Wir setzen nun für $t \in S$

$$A(t) = RC(t)\,, \quad B(t) = RD(t)\,, \quad f(t) = Rg(t)\,.$$

Aus den Eigenschaften des Rieszschen Operators ergibt sich sofort, daß aus den Bedingungen (a_1)—(e_1) die Bedingungen (a)—(e) von Abschnitt I folgen. Man kann nun zeigen, daß die Gleichungen (1,2) den Gleichungen (25) (mit entsprechend definiertem Lösungsbegriff) äquivalent sind.

Bemerkung 3. Die Bezeichnung „pseudoparabolische Gleichungen" für (25) geschieht in Anlehnung an T. W. Ting und R. E. Showalter (vgl. [12], [13]), die derartige Gleichungen für den Fall betrachtet haben, daß $C(t)$, $D(t)$ lineare elliptische Differentialoperatoren sind.

Bemerkung 4. Eine Darstellung der hier skizzierten Theorie für stetige rechte Seiten ist in [8], [9] durchgeführt. Dort wird auch auf die physikalische Interpretation pseudoparabolischer Gleichungen eingegangen, vgl. auch [7].

Man kann die oben gemachten Voraussetzungen an verschiedenen Stellen etwas abschwächen; insbesondere kann man die Operatoren B als die „Vorgeschichte" berücksichtigende sog. Volterraoperatoren wählen.

Literatur

[1] Bourbaki, N., Fonctions d'une variable réelle. Paris 1960.
[2] Bourbaki, N., Intégration. Paris 1952.
[3] Browder, F. E., and Petryshyn, W. V., Construction of fixed points of nonlinear mappings in Hilbert space. J. Math. Analysis Appl. **20** (1967) 197–228.
[4] Browder, F. E., The solvability of nonlinear functional equations. Duke Math. J. **30** (1963) 557–566.
[5] Dinculeanu, N., Vector measures. Berlin 1966.
[6] Gajewski, H., Iterationsverfahren bei Folgen kontraktiver Operatoren. Monatsber. Dt. Akad. Wiss. **11** (1969) 818–826.
[7] Gajewski, H., und Zacharias, K., Zur starken Konvergenz des Galerkinverfahrens bei einer Klasse pseudoparabolischer partieller Differentialgleichungen. Math. Nachr. **47** (1970) 365–376.
[8] Gajewski, H., und Zacharias, K., Über eine Klasse nichtlinearer Differentialgleichungen im Hilbertraum. Ersch. J. Math. Anal. Appl. **44** (1973) 71–87.
[9] Gajewski, H., und Zacharias, K., Über eine weitere Klasse nichtlinearer Differentialgleichungen im Hilbertraum. Ersch. Math. Nachr. **57** (1973) 127–140.
[10] Gelfand, I. M., Abstrakte Funktionen und lineare Operatoren. Mat. Sbornik **4** (1938) 235–286.
[11] Komura, Y., Nonlinear semigroups in Hilbert space. J. Math. Soc. Japan **19** (1967) 493–507.
[12] Showalter, R. E., Partial differential equations of Sobolev-Galpern type. Pacific J. Math. **31** (1969) 787–793.
[13] Ting, T. W., Parabolic and pseudo-parabolic differential equations. J. Math. Soc. Japan **21** (1969) 440–453.
[14] Yosida, K., Functional analysis. Berlin 1965.

Anschrift des Autors:

ZIMM der AdW der DDR
108 Berlin, Mohrenstraße 39

Existenz, Eindeutigkeit, Eigenschaften und Anwendungen des Abbildungsgrades im R^n

EBERHARD ZEIDLER, Leipzig

Herrn Prof. Joachim Focke zum 50. Geburtstag

Einleitung

Der Abbildungsgrad, der ursprünglich in analytischer Form von Kronecker 1869 [61] eingeführt wurde und in seiner topologischen Fassung auf Brouwer 1912 [13] zurückgeht, besitzt grundlegende Bedeutung für die Behandlung nichtlinearer Probleme. Das Ziel dieser Arbeit ist es, einen elementaren und sehr kurzen analytischen Zugang zum Begriff des Abbildungsgrades im R^n zu beschreiben und daraus in möglichst rascher Weise alle für den Analytiker bedeutsamen Resultate zu gewinnen. Elementar soll dabei heißen, daß lediglich Grundkenntnisse aus der Differential- und Integralrechnung sowie aus der linearen Algebra benutzt werden. Den Ausgangspunkt für diese Arbeit bildete der Wunsch des Verfassers, einen Satz über Vektorfelder auf Sphären mit Gruppeninvarianz von Krasnosjelski 1955 [56] und Browder 1970 [25] (Verallgemeinerung des Antipodensatzes von Borsuk) auf beliebige beschränkte Gebiete zu übertragen (§ 12) und einen möglichst einfachen analytischen Beweis für den Satz von Hopf über Abbildungsklassen auf Sphären zu erhalten (§ 10).

In der Literatur findet man heute sehr verschiedenartige Definitionen des Abbildungsgrades. Mit Methoden der algebraischen Topologie arbeiten: Brouwer 1912 [13], Lefschetz 1949 [64] (simpliziale Approximation, Überdeckungszahlen), Alexandrow, Hopf 1935 [2], Cronin 1964 [33] (Schnittzahltheorie), Pontrjagin 1947 [80], Eilenberg, Steenrod 1952 [39], Dold 1965 [36], Spanier 1966 [94] (Homomorphismen der Homologiegruppen), Rado, Reichelderfer 1956 [81] (Kohomologietheorie), Krasnosjelski 1956 [57] (Spernersches Lemma, Numerierung simplizialer Komplexe im Anschluß an Knaster, Kuratowski, Mazurkiewicz 1929 [54]), Whitney 1937 [102] („Standardabbildung"). Rein analytische Definitionen sind enthalten in Kronecker 1869 [61], 1878 [62] (Charakteristik von $n + 1$ Funktionen in n Variablen, Zurückführung auf ein Oberflächenintegral, vgl. auch Picard 1892 [78]), Hadamard 1910 [44] (Kroneckersches Oberflächenintegral), Nagumo 1951 [73] (Lemma von Sard), Heinz 1959 [45] (Gebietsintegral, vgl. auch Ortega, Rheinboldt 1970 [75]), J. T. Schwartz 1964 [91] (Verschmelzung von Nagumo [73] mit Heinz [45], vgl. auch Berger 1968 [8]), Krasnosjelski 1966 [59] (Argumentprinzip für ebene Vektorfelder).

Damit entsteht die Frage, ob alle diese verschiedenartigen Definitionen zum gleichen Abbildungsgrad führen. In § 2 geben wir vier sehr naturgemäße Axiome an und zeigen in § 4, daß damit der Abbildungsgrad eindeutig festgelegt ist. Durch geeignete Kombination von Ideen aus Nagumo [73], Heinz [45] und J. T. Schwartz [91] beweisen wir in § 5 die Existenz des Abbildungsgrades.

Der hier vorgestellte Zugang führt zu Vereinfachungen gegenüber jeder der drei soeben genannten Arbeiten. Den Schlüssel bildet wie bei Nagumo [73] eine von A. B. Brown 1935 [27] stammende vereinfachte Version des Lemmas von Sard 1942 [86] (Approximationssatz in § 3) und eine Darstellungsformel von Heinz [45] (vgl. (16)), die es gestattet, den Abbildungsgrad durch Auswertung eines Gebietsintegrals auch numerisch zu berechnen.

In den Arbeiten von Nagumo [73], Heinz [45] und J. T. Schwartz [91] werden eine Reihe von wichtigen Eigenschaften des Abbildungsgrades bewiesen. Unter anderem wird jedoch nicht auf folgende Fragen eingegangen: Verhältnis zwischen unwesentlichen Abbildungen und Abbildungsgrad, Satz von Hopf, topologische Invarianz des Abbildungsgrades, Verallgemeinerung des Antipodensatzes von Borsuk mit den sich daraus ergebenden Folgerungen, analytische Abbildungen und Bifurkation, nichtlineare Eigenwertprobleme, verallgemeinerte Fredholmsche Alternativen, Existenzaussagen für monotone, quasimonotone und koerzitive Operatoren im R^n. Mit der vorliegenden geschlossenen Darstellung ohne Benutzung von Hilfsmitteln der algebraischen Topologie hofft der Verfasser, eine Lücke in der Literatur schließen zu helfen. Der Kenner wird auch bemerken, daß für eine Reihe bekannter Resultate sehr kurze Beweise angegeben werden. Die Arbeit ist wie auch die Monographie von Granas 1962 [42] im Geist der Erweiterungstheorie geschrieben, d. h. der Erweiterungssatz für stetige Funktionen von Tietze 1915 [95] und der Homotopieerweiterungssatz von Borsuk 1936 [10] spielen eine wichtige Rolle.

Die grundlegenden Eigenschaften des Abbildungsgrades sind in zehn Theoremen zusammengefaßt. Anwendungen dieser Theoreme werden in Sätzen formuliert. Diese Anwendungen beziehen sich jedoch nicht auf konkrete Probleme, sondern stellen ihrerseits wichtige Existenzaussagen der nichtlinearen Analysis dar.

Wir beschränken uns bewußt auf den R^n, alle Resultate sind jedoch im Hinblick auf funktionalanalytische Verallgemeinerungen formuliert. Über Verallgemeinerungen des Abbildungsgrades auf unendlichdimensionale Räume und über den damit eng zusammenhängenden Begriff des Fixpunktindex kann man sich orientieren in den Übersichtsartikeln von van der Walt 1963 [99], Eells 1966 [38], Fadell 1970 [40], Sadowski 1972 [85] sowie in der Monographie von F. B. Brown 1970 [28]. Für differentialgeometrische Verallgemeinerungen verweisen wir gesondert auf de Rham 1955 [34] und J. T. Schwartz 1968 [90]. Tiefliegende topologische und analytische Verallgemeinerungen beinhalten im Sinne der globalen Analysis das Periodizitätstheorem von Bott 1959 [12] über die Homotopien von Abbildungen in die lineare Gruppe, der Indexsatz von Atiyah, Singer 1963 [5], 1968 [6], Atiyah, Segal 1968 [7] sowie seine Verallgemeinerung durch Hörmander 1971 [50] und die verallgemeinerte Lefschetzsche Fixpunktformel von Atiyah, Bott 1967 [4] (vgl. Kotake 1969 [55]). Eine elementare Einführung in diesen Problemkreis gibt der Vortrag von Atiyah 1967 [3] (vgl. auch die Seminarausarbeitung von Palais 1965 [76]). Der Indexsatz besagt grob, daß der analytische Index $i_a(D)$ eines elliptischen Operators D auf einer Mannigfaltigkeit mit gewissen Eigenschaften ($i_a(D) = \dim \ker D - \dim \operatorname{coker} D$) gleich ist einem topologischen Index i_t, der sich aus den topologischen Eigenschaften der Abbildung ergibt, die durch das Symbol von D beschrieben wird.

§ 1. Bezeichnungen

In der gesamten Arbeit sei $\mathfrak{G}$ stets eine offene, beschränkte und nichtleere Teilmenge des R^n, $n = 1, 2, \ldots,$ mit den Punkten $x = (\xi_1, \ldots, \xi_n)$. $F \in C(\overline{\mathfrak{G}})$ bedeute abweichend vom üblichen Gebrauch

$$F: \overline{\mathfrak{G}} \subset R^n \xrightarrow{\text{stetig}} R^n$$

mit der zusätzlichen Forderung, daß auf dem Rand $\partial\mathfrak{G}$ keine Nullstellen liegen, also $\partial\mathfrak{G}: F(x) \neq 0$. Ferner benötigen wir zwei weitere engere Abbildungsklassen

$$C(\overline{\mathfrak{G}}) \supset C^{(1)}(\overline{\mathfrak{G}}) \supset C_e^{(1)}(\overline{\mathfrak{G}}) .$$

$C^{(1)}(\overline{\mathfrak{G}})$ umfaßt die Abbildungen $F = (f_1, \ldots, f_n) \in C(\overline{\mathfrak{G}})$, deren Komponenten f_i zusätzlich auf einer offenen Menge $\mathfrak{G}_1 \supset \overline{\mathfrak{G}}$ stetig differenzierbar sind, $\mathfrak{G}_1$ hängt von F ab. Der Fréchet-Ableitung $F'(x)$ entspricht eine Matrix (a_{ij}) mit $a_{ij} = \partial f_i / \partial \xi_j$, det $F'(x)$ ist gleich der Jacobischen Funktionaldeterminante.

$C_e^{(1)}(\overline{\mathfrak{G}})$ enthält alle $F \in C^{(1)}(\overline{\mathfrak{G}})$, die auf $\overline{\mathfrak{G}}$ höchstens endlich viele Nullstellen $x_1, \ldots, x_k \in \mathfrak{G}$ besitzen, wobei diese außerdem nicht entartet sind, also $F(x_i) = 0$, det $F'(x_i) \neq 0$.

Das Skalarprodukt in R^n bezeichnen wir mit $\langle x, y \rangle$. Außerdem verwenden wir folgende Normen

$$||x|| = \sqrt{\sum_{i=1}^{n} \xi_i^2}\,, \qquad ||F||_{C(\overline{\mathfrak{G}})} = \max_{x \in \overline{\mathfrak{G}}} ||F(x)||\,,$$

$$||F||_{C^{(1)}(\overline{\mathfrak{G}})} = ||F||_{C(\overline{\mathfrak{G}})} + \sum_{i,j} \max_{x \in \overline{\mathfrak{G}}} \left|\frac{\partial f_i}{\partial \xi_j}\right| .$$

Ist A ein linearer Operator im R^n, dem eine (n, n)-Matrix entspricht, dann verstehen wir unter $||A||$ die Operatornorm. Zur identischen Abbildung $I(x) := x$ gehört der Einheitsoperator I.

Zwei Abbildungen $F, G \in C(\overline{\mathfrak{G}})$ heißen homotop auf $\mathfrak{M} \subseteq \overline{\mathfrak{G}}$ (in Zeichen $\mathfrak{M}: F \cong G$) genau dann, wenn es eine für $x \in \mathfrak{M}$, $0 \leqq t \leqq 1$ erklärte Homotopieabbildung

$$H: \mathfrak{M} \times \langle 0, 1 \rangle \xrightarrow{\text{stetig}} R^n \tag{1}$$

gibt mit

$$H(x, 0) = F(x)\,, \quad H(x, 1) = G(x)\,, \quad H(x, t) \neq 0; \quad x \in \mathfrak{M}\,, \quad 0 \leqq t \leqq 1\,. \tag{2}$$

$\mathfrak{M}: F \cong G$ ist eine Äquivalenzrelation.

Für offene Kugeln vereinbaren wir die Bezeichnung

$$\mathfrak{K}_\varrho^{(n)}(x_0) := \{x \in R^n; ||x - x_0|| < \varrho\}\,. \tag{3}$$

Sehr häufig werden wir den Erweiterungssatz von Tietze 1915 [95] benutzen. Er besagt, daß sich jede Abbildung

$$F: \mathfrak{A} \subset R^n \xrightarrow{\text{stetig}} R^m\,, \quad \mathfrak{A} \text{ kompakt} \tag{4}$$

stetig auf R^n erweitern läßt (vgl. Dieudonné [35] S. 95). Wegen der Retrakteigenschaft kompakter und konvexer Mengen im R^m läßt sich obendrein stets erreichen, daß das Bild der erweiterten Abbildung $\tilde{F}$ in der abgeschlossenen konvexen Hülle des Bildes $F(\mathfrak{A})$ der Ausgangsabbildung F liegt, also

$$\tilde{F}\colon R^n \xrightarrow{\text{stetig}} \overline{\operatorname{konv} F(\mathfrak{A})} \subset R^m\,. \tag{4'}$$

Einen kurzen und elementaren Beweis hierfür findet man in Granas [42] S. 29. Für hier nicht benötigte Verallgemeinerungen vergleiche man Dugundij 1951 [37] und Borsuk [11] S. 77.

Eine wichtige Rolle werden ebenfalls der klassische Satz über implizite Funktionen und der Approximationssatz von Weierstraß spielen (vgl. Dieudonné [35] S. 268 bzw. S. 141).

§ 2. Axiome für den Abbildungsgrad

Der axiomatisch einzuführende Abbildungsgrad soll ein Maß für die Anzahl der Nullstellen von $F \in C(\overline{\mathfrak{G}})$ in $\mathfrak{G}$ sein.

Axiom 1 (*Ganzzahligkeit*). *Jeder offenen, beschränkten und nichtleeren Menge $\mathfrak{G}$ des R^n, $n = 1, 2, \ldots$, und jeder Abbildung $F \in C(\overline{\mathfrak{G}})$ werde eine ganze Zahl $d_{\mathfrak{G}}(F)$ zugeordnet* (*Abbildungsgrad*).

Axiom 2 (*Normierung*).

a) *Aus* $\mathfrak{G}\colon F \neq 0$ *folgt* $d_{\mathfrak{G}}(F) = 0$.

b) *Für* $F(x) := x - x_0$, $x_0 \in \mathfrak{G}$ *ist* $d_{\mathfrak{G}}(F) = 1$.

Axiom 3 (*Additivität*). *Zerfällt $\mathfrak{G}$ in der Form $\overline{\mathfrak{G}} = \bigcup_{i=1}^{m} \overline{\mathfrak{G}}_i$, wobei die $\mathfrak{G}_i \neq \emptyset$ paarweise disjunkte, offene Mengen sind, und besitzt $F \in C(\overline{\mathfrak{G}})$ neben $\partial\mathfrak{G}$ auch auf $\partial\mathfrak{G}_i$ keine Nullstellen, also $F \in C(\overline{\mathfrak{G}}_i)$, $i = 1, \ldots, m$, dann ist*

$$d_{\mathfrak{G}}(F) = \sum_{i=1}^{m} d_{\mathfrak{G}_i}(F)\,.$$

Axiom 4 (*Homotopie*).

Aus $\partial\mathfrak{G}\colon F \simeq G$ *folgt* $d_{\mathfrak{G}}(F) = d_{\mathfrak{G}}(G)$.

Die Axiome A_1, A_2, A_3 sind naturgemäße Forderungen an eine „Anzahl der Nullstellen". Um A_4 zu motivieren, formulieren wir zunächst

Satz 1. (*Bedingungen, die zu A_4 äquivalent sind*). *Für ein ganzzahliges Funktional $d_{\mathfrak{G}}(F)$ auf $C(\overline{\mathfrak{G}})$ ist A_4 zu jeder der beiden folgenden Bedingungen äquivalent:*

a) $d_{\mathfrak{G}}(F)$ *ist stetig auf* $C(\overline{\mathfrak{G}})$.

b) *Aus* $F, G \in C(\overline{\mathfrak{G}})$ *und*

$$\partial\mathfrak{G}\colon ||F(x) - G(x)|| < \gamma(F)\,, \qquad \gamma(F) := \min_{x \in \partial\mathfrak{G}} ||F(x)|| > 0 \tag{5}$$

folgt $d_{\mathfrak{G}}(F) = d_{\mathfrak{G}}(G)$.

Beweis. $A_4 \Rightarrow$ b). Aus (5) folgt für $H(x, t) := F + t(G - F)$

$$\partial\mathfrak{G}\colon ||H(x, t)|| \geqq ||F(x)|| - ||G(x) - F(x)|| > 0\,, \quad 0 \leqq t \leqq 1\,;$$

also $\partial\mathfrak{G}\colon F \cong G$, und nach A_4 ist $d_{\mathfrak{G}}(F) = d_{\mathfrak{G}}(G)$.

b) $\Rightarrow$ a). Man beachte die Normdefinition in $C(\overline{\mathfrak{G}})$.

a) $\Rightarrow A_4$). Sei $\partial\mathfrak{G}\colon F \cong G$ mit $H(x, t)$ gemäß (1), (2).

Wir erweitern $H(x, t)$ zunächst stetig durch $\tilde{H}(x, 0) = F(x)$, $\tilde{H}(x, 1) = G(x)$ für $x \in \overline{\mathfrak{G}}$ und gehen anschließend nach (4) zu einer stetigen Erweiterung $\tilde{H}(x, t)$ auf $\overline{\mathfrak{G}} \times \langle 0, 1\rangle$ über. Die Funktion $f(t) := d_{\mathfrak{G}}(\tilde{H}(x, t))$ ist nach a) stetig und nach Voraussetzung ganzzahlig, also $f(0) = f(1)$, d. h. $d_{\mathfrak{G}}(F) = d_{\mathfrak{G}}(G)$, q. e. d.

Die gewünschte Motivierung für A_4 ergibt sich nun durch Betrachtung der beiden Gleichungen

$$\mathfrak{G}\colon F(x) = 0\,, \quad \mathfrak{G}\colon F_1(x) = 0\,, \quad \max_{x \in \overline{\mathfrak{G}}} ||F(x) - F_1(x)|| < \varepsilon$$

mit $F, F_1 \in C(\overline{\mathfrak{G}})$. Ist $d_{\mathfrak{G}}(F) \neq 0$, dann besitzt $\mathfrak{G}\colon F(x) = 0$ nach A_2 a) eine Lösung (Kroneckersches Existenzprinzip). Für hinreichend kleine $\varepsilon > 0$ gilt nach Satz 1 b) $d_{\mathfrak{G}}(F_1) = d_{\mathfrak{G}}(F) \neq 0$, so daß auch $\mathfrak{G}\colon F_1(x) = 0$ eine Lösung hat. A_4 beinhaltet somit die Stabilität der Existenzaussage.

Um die Bedeutung dieser Axiome zu illustrieren, wollen wir zeigen, daß allein aus A_2 und A_4 der Fixpunktsatz von Brouwer 1912 [13] folgt. Dieser Satz besagt, daß die stetige Abbildung $F\colon \overline{\mathfrak{K}}_\varrho^{(n)}(0) \to \overline{\mathfrak{K}}_\varrho^{(n)}(0)$ einer abgeschlossenen Kugel in sich stets einen Fixpunkt besitzt. Hätte F keinen Fixpunkt, dann könnten wir die stetige Abbildung $G\colon x \in \overline{\mathfrak{K}}_\varrho^{(n)}(0) \to x^* \in \partial\mathfrak{K}_\varrho^{(n)}(0)$ konstruieren, die jedem $x \in \overline{\mathfrak{K}}_\varrho^{(n)}(0)$ den auf der Geraden $y = F(x) + t(x - F(x))$, $-\infty < t < \infty$, liegenden Randpunkt $x^* \in \partial\mathfrak{K}_\varrho^{(n)}(0)$ mit $t \geqq 1$ zuordnet. Wegen $\partial\mathfrak{K}_\varrho^{(n)}(0)$: $G(x) = x$ gilt nach A_4, A_2 b) $d_{\mathfrak{K}_\varrho^{(n)}(0)}(G) = 1$, also besitzt $G(x)$ nach A_2 a) eine Nullstelle in $\mathfrak{K}_\varrho^{(n)}(0)$ im Widerspruch zu $G(\overline{\mathfrak{K}}_\varrho^{(n)}(0)) = \partial\mathfrak{K}_\varrho^{(n)}(0)$ (vgl. Eilenberg, Steenrod [39]).

Gilt $F(x_0) = 0$, und ist x_0 eine isolierte Nullstelle, dann definieren wir den Index dieser Nullstelle durch

$$d_{x_0}(F) := d_{\mathfrak{K}_\varrho^{(n)}(x_0)}(F)\,, \qquad 0 < \varrho \leqq \varrho_0\,. \tag{6}$$

Nach A_3, A_2 a) ist dieser Wert bei hinreichend kleinem festem ϱ_0 für alle ϱ mit $0 < \varrho \leqq \varrho_0$ konstant (vgl. (3)).

Ist $\partial\mathfrak{G}\colon F(x) \neq y$, also $F - y \in C(\overline{\mathfrak{G}})$, dann kann man den Abbildungsgrad bezüglich y durch

$$\deg(F, \mathfrak{G}, y) := d_{\mathfrak{G}}(F - y) \tag{7}$$

definieren.

Besitzt F auf $\partial\mathfrak{G}$ keine Fixpunkte, also $\partial\mathfrak{G}\colon F(x) \neq x$, $F - I \in C(\overline{\mathfrak{G}})$, dann läßt sich durch $i_{\mathfrak{G}}(F) := d_{\mathfrak{G}}(F - I)$ ein Fixpunktindex einführen (Lefschetzsche Zahl).

§ 3. Ein Approximationssatz

Die Abbildungen $G \in C_e^{(1)}(\overline{\mathfrak{G}})$ besitzen im Hinblick auf $d_{\mathfrak{G}}(G)$ besonders günstige Eigenschaften. Unser Ziel ist es zunächst, alle $F \in C(\overline{\mathfrak{G}})$ durch derartige Abbildungen beliebig genau zu approximieren.

Satz 2. *(Approximationssatz).*

a) *Zu $F \in C(\overline{\mathfrak{G}})$ und beliebig vorgegebenem $\varepsilon > 0$ existiert stets ein $G \in C_e^{(1)}(\overline{\mathfrak{G}})$ mit*

$$||F - G||_{C(\overline{\mathfrak{G}})} < \varepsilon . \tag{8}$$

b) *Ist $F \in C^{(1)}(\overline{\mathfrak{G}})$, dann kann man G in der Form $G(x) = F(x) - y_s$ wählen, also gilt* (8) *auch in der Norm von $C^{(1)}(\overline{\mathfrak{G}})$. Genauer hat man, daß die Menge*

$$\mathfrak{M}' := \{y \in F(\overline{\mathfrak{G}}) - F(\partial\mathfrak{G}); \quad F - y \notin C_e^{(1)}(\overline{\mathfrak{G}})\} \tag{8'}$$

das n-dimensionale Maß Null besitzt.

Beweis. 1. Schritt: Approximationssatz von Weierstraß ([35] S. 141).

Wir approximieren $F \in C(\overline{\mathfrak{G}})$ zunächst durch $P \in C^{(1)}(\overline{\mathfrak{G}})$ mit $||F-P||_{C(\overline{\mathfrak{G}})} < \varepsilon/2$ und konstruieren

$$\mathfrak{M} := \{y \in R^n;\ y = P(x),\ \det P'(x) = 0,\ x \in \overline{\mathfrak{G}}\} .$$

2. Schritt: $\mathfrak{M}$ besitzt keine inneren Punkte in R^n und hat das n-dimensionale Maß Null (A. B. Brown 1935 [27] S. 382). Es genügt zu zeigen, daß sich $\mathfrak{M}$ durch endlich viele „Kreiszylinder" mit beliebig kleinem Gesamtvolumen überdecken läßt.

Wir schließen $\overline{\mathfrak{G}}$ in einen Würfel $\overline{\mathfrak{W}}$ vom Volumen V ein, zerlegen diesen in N abgeschlossene Würfel $\overline{\mathfrak{W}}_i$ der Kantenlänge $\delta = \sqrt[n]{V/N}$ und betrachten nur solche $\overline{\mathfrak{W}}_i$, die einen Punkt $x_i \in \overline{\mathfrak{G}}$ mit $\det P'(x_i) = 0$ enthalten. Nach dem Taylorschen Lehrsatz ist dann

$$x \in \overline{\mathfrak{W}}_i \cap \overline{\mathfrak{G}}: P(x) = P(x_i) + P'(x_i)\,(x - x_i) + r := L(x) + r .$$

Aus $||r|| = o(||x - x_i||)$ folgt

$$x \in \overline{\mathfrak{W}}_i \cap \overline{\mathfrak{G}}: ||P(x) - L(x)|| \leqq s(\delta)\,\delta , \qquad \lim_{\delta \to 0} s(\delta) = 0 , \tag{9}$$

$$x \in \overline{\mathfrak{W}}_i \cap \overline{\mathfrak{G}}: ||L(x) - L(x_i)|| \leqq ||P'(x_i)||\,||x - x_i|| \leqq M\delta . \tag{10}$$

M hängt wegen $P \in C^{(1)}(\overline{\mathfrak{G}})$ nicht von $\overline{\mathfrak{W}}_i$ ab.

Auf Grund von $\det P'(x_i) = 0$ und (10) bildet L den „Würfel" $\overline{\mathfrak{W}}_i \cap \overline{\mathfrak{G}}$ in den Durchschnitt einer Hyperebene $\mathfrak{H}_i$ mit einer Kugel vom Radius $M\delta$ ab. Nach (9) ist somit $P(\overline{\mathfrak{W}}_i \cap \overline{\mathfrak{G}})$ enthalten in einem „Kreiszylinder" der Höhe $2s(\delta)\,\delta$ mit „Deckflächen" vom Radius $\big(M\delta + s(\delta)\,\delta\big)$ „parallel" zu $\mathfrak{H}_i$. Da die Anzahl der $\overline{\mathfrak{W}}_i$ höchstens gleich $N = V/\delta^n$ ist, wird $\mathfrak{M}$ durch N derartige Kreiszylinder mit einem Gesamtvolumen

$$V_\delta = (V/\delta^n) \cdot E_{n-1}\big(M\delta + s(\delta)\,\delta\big)^{n-1} \cdot 2s(\delta)\,\delta \to 0 \quad \text{für} \quad \delta \to 0$$

überdeckt. E_{n-1} ist das Volumen der $(n-1)$-dimensionalen Einheitskugel mit $E_0 = 1$.

3. Schritt: Konstruktion von $G \in C_e^{(1)}$. Wegen $0 \notin \operatorname{int} \mathfrak{M}$ existiert ein Punkt $y_s \notin \mathfrak{M}$, $||y_s|| < \varepsilon/2$ mit $G: = P - y_s \in C^{(1)}(\overline{\mathfrak{G}})$, $||F - G||_{C(\overline{\mathfrak{G}})} < \varepsilon$. Nach Konstruktion von $\mathfrak{M}$ folgt aus $\mathfrak{G}: G(x_N) = 0$ stets $\det G'(x_N) \neq 0$. Der Satz über implizite Funktionen ([35] S. 268) ergibt dann wegen $\partial\mathfrak{G}: G(x) \neq 0$ sofort die Isoliertheit dieser Nullstelle, also ist auf Grund der Kompaktheit von $\overline{\mathfrak{G}}$ die Anzahl der Nullstellen von G endlich, d. h. $G \in C_e^{(1)}(\overline{\mathfrak{G}})$. q. e. d.

§ 4. Eindeutigkeit des Abbildungsgrades

Satz 3. *(Eindeutigkeitssatz). Es gibt höchstens einen Abbildungsgrad $d_{\mathfrak{G}}(F)$, der den Axiomen A_1, A_2, A_3, A_4 genügt. Für $F \in C_e^{(1)}(\overline{\mathfrak{G}})$ muß notwendigerweise gelten*

$$d_{\mathfrak{G}}(F) = \sum_{i=1}^{k} \operatorname{sgn} \det F'(x_i) , \qquad F(x_i) = 0 , \tag{11}$$

wenn $x_1, \ldots, x_k \in \mathfrak{G}$ die Nullstellen von $F(x)$ sind. Nach Definition von $C_e^{(1)}(\overline{\mathfrak{G}})$ ist $\det F'(x_i) \neq 0$.

Beweis. Nach der zu A_4 äquivalenten Aussage b) von Satz 1 und nach (8) genügt es, die Eindeutigkeit für $F \in C_e^{(1)}(\overline{\mathfrak{G}})$ zu beweisen. Nach (6) und A_3, A_2 a) ist ferner

$$F \in C_e^{(1)}(\overline{\mathfrak{G}}): d_{\mathfrak{G}}(F) = \sum_{i=1}^{k} d_{x_i}(F) , \qquad F(x_i) = 0 .$$

Wir brauchen also lediglich $d_{x_i}(F) = \operatorname{sgn} \det F'(x_i)$ nachzuweisen.

1. Schritt: $d_{x_i}(F) = d_{x_i}\big(F'(x_i)(x - x_i)\big)$. Wegen $F(x_i) = 0$ ergibt der Taylorsche Lehrsatz

$$||x - x_i|| = \varrho: F(x) = F'(x_i)(x - x_i) + r , \qquad ||r|| = o(\varrho) .$$

Wir konstruieren $H(x, t): = F'(x_i)(x - x_i) + t\, r$, $0 \leqq t \leqq 1$, dann gilt

$$||x - x_i|| = \varrho: ||H(x, t)|| \geqq ||\big(F'(x_i)\big)^{-1}||^{-1}\, ||x_i - x|| - ||r|| > 0$$

für hinreichend kleines ϱ, also $\partial\mathfrak{K}_\varrho^{(n)}(x_i): F'(x_i)(x - x_i) \cong F(x)$. Aus A_4 folgt die Behauptung.

2. Schritt: $d_{x_i}\big(F'(x_i)(x - x_i)\big)$ hängt nur von $s := \operatorname{sgn} \det F'(x_i)$ ab. Wir betrachten die beiden (n, n)-Diagonalmatrizen

$$I_{\pm 1} = \begin{pmatrix} \pm 1 & & 0 \\ & 1 & \\ & & \ddots \\ 0 & & & 1 \end{pmatrix}$$

und zeigen: Ist $s = \pm 1$, dann existiert eine stetige Matrizenfunktion H_t, $0 \leqq t \leqq 1$ mit

$$H_0 = F'(x_i) , \quad H_1 = I_s , \quad \det H_t \neq 0 , \quad 0 \leqq t \leqq 1 . \tag{12}$$

Um diese wohlbekannte Tatsache einzusehen, schreiben wir die zu $F'(x_i)$ gehörige Matrix als ein Produkt von Additionsmatrizen A (man ersetze in

I_{+1} eine Null durch $a \neq 0$), Vertauschungsmatrizen V (man vertausche in I_{+1} zwei Zeilen), Spiegelungsmatrizen S (man ersetze in I_{+1} eine Eins durch $a = -1$) mit einer Diagonalmatrix $D = (d_i)$, $d_i > 0$. Dieses Produkt beschreibt den üblichen Reduktionsprozeß, der $F'(x_i)$ in Diagonalgestalt überführt. Für A bzw. D konstruiere man A_t bzw. D_t mit $A_1 = D_1 = I_{+1}$, indem man a durch $(1-t)\,a$ bzw. d_i durch $(1-t)\,d_i + t$ ersetzt. Für V, S wähle man V_t, S_t mit $V_1 = S_1 = I_{-1}$ in der Form von Matrizen, die Drehungen in geeigneten zweidimensionalen Unterräumen beschreiben. H_t ist dann das Produkt der A_t, V_t, S_t, D_t. Aus $\det H_t \neq 0$, $0 \leqq t \leqq 1$ folgt $\operatorname{sgn}\det H_1 = \operatorname{sgn}\det H_0 = s$, also $H_1 = I_s$.

Da H_t^{-1} für $0 \leqq t \leqq 1$ ebenfalls stetig in t ist, gilt $||H_t^{-1}||^{-1} \geqq \alpha > 0$. Setzen wir $H(x, t) := H_t(x - x_i)$, $0 \leqq t \leqq 1$, dann folgt aus

$$||x - x_i|| = \varrho\colon ||H(x, t)|| \geqq ||H_t^{-1}||^{-1}\,||x - x_i|| \geqq \alpha\varrho > 0$$

nach A_4 und (12) sofort

$$d_{x_i}\big(F'(x_i)\,(x - x_i)\big) = d_{x_i}\big(I_s(x - x_i)\big)\,, \qquad s := \operatorname{sgn}\det F'(x_i)\,. \tag{13}$$

3. Schritt: $d_{x_i}\big(F'(x_i)\,(x - x_i)\big) = 1$ für $s = 1$. Das folgt aus (13) und A_2 b).

4. Schritt: $d_{x_i}\big(F'(x_i)\,(x - x_i)\big) = -1$ für $s = -1$. Wegen (13) genügt es, ein spezielles Beispiel zu betrachten. Sei $x_\pm = (\pm 1, 0, \ldots, 0)$. Wir wählen zunächst $x_i = x_-$ und konstruieren

$$H(x, t) := (\xi_1^2 + 1 - 2t, \xi_2, \ldots, \xi_n)\,, \qquad 0 \leqq t \leqq 1\,. \tag{14}$$

Offensichtlich gilt $\partial\mathfrak{K}\colon H(x, t) \neq 0$ mit $\mathfrak{K} = \{x\colon ||x|| \leqq 2\}$, also $\partial\mathfrak{K}\colon H(x, 0) \cong H(x, 1)$. $H(x, 0)$ besitzt keine Nullstelle, somit ist nach A_4 und A_2 a): $d_{\mathfrak{K}}\big(H(x, 1)\big) = d_{\mathfrak{K}}\big(H(x, 0)\big) = 0$. $H(x, 1)$ besitzt genau die zwei isolierten Nullstellen $x_\pm$ mit $\operatorname{sgn}\det H'_x(x_\pm, 1) = \pm 1$. Nach A_3, A_2 a) und nach dem 3. Schritt gilt somit

$$0 = d_{\mathfrak{K}}\big(H(x, 1)\big) = d_{x_+}\big(H(x, 1)\big) + d_{x_-}\big(H(x, 1)\big) = 1 + d_{x_-}\big(H(x, 1)\big)\,,$$

folglich $d_{x_-}\big(H(x, 1)\big) = d_{x_-}\big(I_{-1}(x - x_-)\big) = -1$. Für beliebige x_i betrachte man $H(x + x_- - x_i, t)$. q. e. d.

§ 5. Existenz des Abbildungsgrades

Beim Beweis der Existenz des Abbildungsgrades benutzen wir lediglich Satz 1 und Satz 2 (Approximationssatz).

Definition 1. *(Abbildungsgrad)*

a) Für $F \in C_e^{(1)}(\overline{\mathfrak{G}})$ definieren wir $d_{\mathfrak{G}}(F)$ durch (11), also

$$d_{\mathfrak{G}}(F) = \sum_{i=1}^{k} \operatorname{sgn}\det F'(x_i)\,, \qquad F(x_i) = 0\,.$$

Besitzt F auf $\mathfrak{G}$ keine Nullstellen, dann sei $d_{\mathfrak{G}}(F) = 0$.

b) Ist $F \in C(\overline{\mathfrak{G}})$, so wähle man eine nach Satz 2 existierende Abbildung $G \in C_e^{(1)}(\overline{\mathfrak{G}})$ mit

$$||F - G||_{C(\overline{\mathfrak{G}})} < \frac{\gamma(F)}{3}, \quad \gamma(F) := \min_{x \in \partial \mathfrak{G}} ||F(x)|| > 0 \tag{15}$$

und setze $d_{\mathfrak{G}}(F) := d_{\mathfrak{G}}(G)$.

Für $F \in C_e^{(1)}(\overline{\mathfrak{G}})$ ist somit die Anzahl der Nullstellen in $\mathfrak{G}$ mindestens gleich $|d_{\mathfrak{G}}(F)|$. Diese Definition legt es nahe, Aussagen über den Abbildungsgrad nach folgendem Schema zu beweisen: a) Beweis für $F \in C_e^{(1)}(\overline{\mathfrak{G}})$, b) Ausdehnung auf $F \in C(\overline{\mathfrak{G}})$ durch Approximationsprozesse.

Satz 4. *(Rechtfertigung der Definition und Darstellungssatz)*

a) *Alle Abbildungen* $G \in C_e^{(1)}(\overline{\mathfrak{G}})$ *mit* (15) *besitzen bei festem* $F \in C(\overline{\mathfrak{G}})$ *den gleichen Abbildungsgrad* $d_{\mathfrak{G}}(G)$, *deshalb ist die Definition von* $d_{\mathfrak{G}}(F)$ *sinnvoll.*

b) *Für* $F \in C^{(1)}(\overline{\mathfrak{G}})$ *gilt folgende Integraldarstellung*

$$d_{\mathfrak{G}}(F) = \int_{\mathfrak{G}} \varphi_\varepsilon(||F(x)||) \det F'(x)\, dx := J_\varepsilon(F) \tag{16}$$

für alle ε *mit* $0 < \varepsilon < \gamma(F)$. *Dabei ist*

$$\varphi_\varepsilon(r) = A_{\varepsilon,n} \max\left(0, (\varepsilon - r)\left(r - \frac{\varepsilon}{2}\right)\right) \tag{17}$$

mit ($O_n = 2\pi^{n/2}/\Gamma(n/2)$, $n > 1$ *Oberfläche der Einheitskugel im* R^n, $O_1 = 1$, n *Dimension von* $\mathfrak{G}$)

$$A_{\varepsilon,n} = \frac{2^{n+2}(n+2)(n+1)\,n}{\varepsilon^{n+2} O_n\left(2^{n+1}(n-2) + (n+4)\right)}.$$

Zusatz 1. Das Funktional $d_{\mathfrak{G}}(F)$ ist stetig auf $C(\overline{\mathfrak{G}})$, denn für zwei in $C(\overline{\mathfrak{G}})$ hinreichend benachbarte Abbildungen $F, F_1 \in C(\overline{\mathfrak{G}})$ kann man stets ein gemeinsames $G \in C_e^{(1)}(\overline{\mathfrak{G}})$ finden, das (15) gleichzeitig für F und F_1 befriedigt. Nach Satz 1 ist somit $d_{\mathfrak{G}}(F) = d_{\mathfrak{G}}(G)$ für alle $F, G \in C(\overline{\mathfrak{G}})$ mit

$$\partial\mathfrak{G}: ||F(x) - G(x)|| < \gamma(F), \quad \gamma(F) := \min_{x \in \partial\mathfrak{G}} ||F(x)|| > 0. \tag{18}$$

Daraus resultiert eine bequeme Methode zur numerischen Auswertung von $d_{\mathfrak{G}}(F)$: Man approximiere $F \in C(\overline{\mathfrak{G}})$ durch $G \in C^{(1)}(\overline{\mathfrak{G}})$ gemäß (18), berechne $d_{\mathfrak{G}}(G)$ durch (16) und berücksichtige $d_{\mathfrak{G}}(F) = d_{\mathfrak{G}}(G)$. Da $d_{\mathfrak{G}}(G)$ ganzzahlig ist, braucht die numerische Berechnung von (16) nicht mit allzu großer Genauigkeit zu erfolgen.

Zusatz 2. Der Kern $\varphi_\varepsilon(r)$ in (16) besitzt folgende Eigenschaften:

$$\begin{aligned} &\alpha)\ \varphi_\varepsilon(r) \quad \text{stetig für } 0 \leqq r < \infty, \\ &\beta)\ \varphi_\varepsilon(r) = 0 \text{ in einer Umgebung von } r = 0 \text{ und für } r \geqq \varepsilon, \\ &\gamma)\ \int_{R^n} \varphi_\varepsilon(||x||)\, dx \equiv O_n \int_0^\infty \varphi_\varepsilon(r)\, r^{n-1}\, dr = 1. \end{aligned} \tag{19}$$

Eine geringfügige Modifikation des nachfolgenden Beweises zeigt, daß (16) für jede Funktion $\varphi(r)$ gilt, die den Bedingungen (19) für ein festes ε mit $0 < \varepsilon < \gamma(F)$ genügt (vgl. Heinz [45]). Die konkrete Gestalt (17) ist lediglich für numerische Zwecke angegeben worden.

Beweis. 1. Schritt: $F \in C^{(1)}(\overline{\mathfrak{G}})$, $J_\varepsilon(F) = \text{const.}$, $0 < \varepsilon < \gamma(F)$ (Heinz [45]). Nach dem verallgemeinerten Approximationssatz von Weierstraß (vgl. Whitney 1934 [101]) existiert eine Folge $F_n \in C^{(2)}(\overline{\mathfrak{G}})$, die in der Norm von $C^{(1)}(\overline{\mathfrak{G}})$ gegen F konvergiert. Wir können deshalb $F \in C^{(2)}(\overline{\mathfrak{G}})$ voraussetzen.

Wir setzen $\varphi := \varphi_\varepsilon - \varphi_{\varepsilon_1}$, $0 < \varepsilon_1 < \varepsilon$ und werden

$$J_\varepsilon(F) - J_{\varepsilon_1}(F) = \int_{\mathfrak{G}} \varphi(||F(x)||) \det F'(x)\, dx = 0 \tag{20}$$

über den Satz von Gauß nachweisen.

Nach (19) gilt: $\varphi(r)$ stetig für $0 \leqq r < \infty$, $\varphi(r) = 0$ in einer Umgebung von $r = 0$ und für $r \geqq \varepsilon$ sowie

$$\int_0^\infty \varphi(r)\, r^{n-1}\, dr = \int_{\varepsilon_1/2}^{\varepsilon} \varphi(r)\, r^{n-1}\, dr = 0\,.$$

Für die Funktion

$$r > 0: \psi(r) := r^{-n} \int_0^r \varphi(t)\, t^{n-1}\, dt\,, \qquad \psi(0) := 0$$

ergibt sich dann: $\psi(r)$ stetig differenzierbar für $r \geqq 0$, $\psi(r) = 0$ in einer Umgebung von $r = 0$ und für $r \geqq \varepsilon$ sowie

$$r\psi'(r) + n\psi(r) = \varphi(r)\,.$$

Daraus erhält man für $g_i(y) := \psi(||y||)\, \eta_i$, $y = (\eta_1, \ldots, \eta_n)$ sofort

$$\sum_{i=1}^n \frac{\partial g_i}{\partial \eta_i} = \psi'(||y||)\, ||y|| + n\psi(||y||) = \varphi(||y||)\,.$$

Sei ferner

$$A_{ij} = \frac{1}{(n-1)!} \sum_{i_\alpha, j_\beta = 1}^n \operatorname{sgn} \begin{pmatrix} j\; j_1 \cdots j_{n-1} \\ i\; i_1 \cdots i_{n-1} \end{pmatrix} \frac{\partial f_{i_1}}{\partial \xi_{j_1}} \cdots \frac{\partial f_{i_{n-1}}}{\partial \xi_{j_{n-1}}}$$

die Adjunkte von $\partial f_i / \partial \xi_j$ in $\det F'(x)$, dann gilt

$$\sum_{j=1}^n \frac{\partial A_{ij}}{\partial \xi_j} = \frac{n-1}{(n-1)!} \sum_{j, i_\alpha, j_\beta = 1}^n \operatorname{sgn} \begin{pmatrix} j\; j_1 \cdots j_{n-1} \\ i\; i_1 \cdots i_{n-1} \end{pmatrix} \frac{\partial^2 f_{i_1}}{\partial \xi_{j_1} \partial \xi_j} \frac{\partial f_{i_2}}{\partial \xi_{j_2}} \cdots \frac{\partial f_{i_{n-1}}}{\partial \xi_{j_{n-1}}} = 0\,,$$

was man sogleich durch Vertauschung von j mit j_1 erkennt.

Aus den bereitgestellten Formeln folgt mit $F(x) = (f_1, \ldots, f_n)$ und $x = (\xi_1, \ldots, \xi_n)$ nun der gewünschte Divergenzausdruck

$$\begin{aligned} \varphi(||F(x)||) \det F'(x) &= \det F'(x) \sum_{i=1}^n \frac{\partial g_i}{\partial \eta_i}\bigg|_{y=F(x)} \\ &= \sum_{i,k=1}^n \det F'(x)\, \delta_{ik} \frac{\partial g_i}{\partial \eta_k} = \sum_{i,j,k=1}^n \left(A_{ij} \frac{\partial f_k}{\partial \xi_j} \right) \frac{\partial g_i}{\partial \eta_k} \\ &= \sum_{i,j=1}^n A_{ij} \frac{\partial g_i}{\partial \xi_j} = \sum_{j=1}^n \frac{\partial}{\partial \xi_j} \left(\sum_{i=1}^n A_{ij} g_i(F(x)) \right). \end{aligned} \tag{21}$$

Diese Formel gilt mit $A_{11} = 1$ auch im R^1. Wegen $0 < \varepsilon < \gamma(F)$ (vgl. (15)) gilt $||F(x)|| \geqq \varepsilon$ in einem Randstreifen von $\overline{\mathfrak{G}}$, also ist dort auch $A_{ij} g_i(F(x)) = 0$. (20) folgt dann aus (21) durch Anwendung des Gaußschen Satzes.

2. Schritt: Für $F \in C_e^{(1)}(\overline{\mathfrak{G}})$ gilt (16). Besitzt F zunächst auf $\overline{\mathfrak{G}}$ keine Nullstellen, dann ist $\overline{\mathfrak{G}}$: $||F(x)|| \geqq \alpha > 0$. Wählt man in (16) ein festes ε mit $0 < \varepsilon < \alpha$, dann ist $\varphi_\varepsilon(||F(x)||) \equiv 0$ nach (19 β), also $J_\varepsilon(F) = 0 = d_{\mathfrak{G}}(F)$. Aus dem 1. Schritt folgt dann $J_\varepsilon(F) = d_{\mathfrak{G}}(F)$ für alle ε mit $0 < \varepsilon < \gamma(F)$.

Seien jetzt $x_1, \ldots, x_k \in \mathfrak{G}$ die Nullstellen von F auf $\overline{\mathfrak{G}}$, dann gibt es wegen $\det F'(x_i) \neq 0$ nach dem Satz über implizite Funktionen ein $\varepsilon_0 > 0$ und Umgebungen $\overline{\mathfrak{W}}_i$ von x_i, die durch $y = F(x)$ homöomorph auf die Kugel $\overline{\mathfrak{K}}_{\varepsilon_0}^{(n)}(0)$ abgebildet werden. Ferner ist

$$\overline{\mathfrak{G}} - \bigcup_{i=1}^{k} \overline{\mathfrak{W}}_i\colon ||F(x)|| \geqq \gamma_0 > 0 \,.$$

Wir wählen nun ε mit $0 < \varepsilon < \min(\gamma_0, \varepsilon_0)$, dann wird $\overline{\mathfrak{U}}_i := F^{-1}(\overline{\mathfrak{K}}_\varepsilon^{(n)}(0)) \subset \overline{\mathfrak{W}}_i$ homöomorph auf $\overline{\mathfrak{K}}_\varepsilon^{(n)}(0)$ abgebildet mit

$$\overline{\mathfrak{G}} - \bigcup_{i=1}^{k} \overline{\mathfrak{U}}_i\colon ||F(x)|| \geqq \varepsilon > 0 \,, \qquad \varphi_\varepsilon(||F(x)||) = 0 \,. \tag{22}$$

Gleichzeitig kann ε so klein gewählt werden, daß $\operatorname{sgn} \det F'(x_i)$ auf $\overline{\mathfrak{U}}_i$ konstant ist. Variablentransformation führt nun auf

$$\begin{aligned} J_\varepsilon(F) &\overset{(22)}{=} \sum_{i=1}^{k} \int_{\mathfrak{U}_k} \varphi_\varepsilon(||F(x)||) \det F'(x)\, dx \\ &= \sum_{i=1}^{k} \operatorname{sgn} \det F'(x_i) \int_{\mathfrak{K}_\varepsilon^{(n)}(0)} \varphi_\varepsilon(||y||)\, dy \\ &\overset{(19)}{=} \sum_{i=1}^{k} \operatorname{sgn} \det F'(x_i) = d_{\mathfrak{G}}(F) \,. \end{aligned}$$

3. Schritt: $J_\varepsilon(F)$ ist ganzzahlig für $F \in C^{(1)}(\overline{\mathfrak{G}})$, $0 < \varepsilon < \gamma(F)$.

Nach Satz 2b) existiert bei festem ε mit $0 < \varepsilon < \gamma(F)$ eine Folge

$$F_n \in C_e^{(1)}(\overline{\mathfrak{G}})\colon ||F_n - F||_{C^{(1)}(\overline{\mathfrak{G}})} \to 0 \qquad \text{für } n \to \infty \tag{23}$$

$$||F - F_n||_{C(\overline{\mathfrak{G}})} < \frac{\gamma(F)}{3}\,, \qquad \min_{x \in \partial\mathfrak{G}} ||F_n(x)|| > \varepsilon\,, \qquad n = 1, 2, \ldots \,.$$

Der 2. Schritt ergibt $d_{\mathfrak{G}}(F_n) = J_\varepsilon(F_n)$, andererseits folgt aus (16) offensichtlich

$$d_{\mathfrak{G}}(F_n) = J_\varepsilon(F_n) \to J_\varepsilon(F) \qquad \text{für} \quad n \to \infty \,. \tag{24}$$

Als Grenzwert einer Folge ganzer Zahlen ist $J_\varepsilon(F)$ selbst ganzzahlig.

4. Schritt: Beweis von Behauptung a). Die Abbildungen $G_1, G_2 \in C_e^{(1)}(\overline{\mathfrak{G}})$ mögen beide (15) genügen. Wir betrachten $F_t := (1 - t)\, G_1 + t G_2$, $0 \leqq t \leqq 1$ mit

$$x \in \partial\mathfrak{G}\colon ||F_t(x)|| = ||F(x) + (1 - t)\,(G_1(x) - F(x)) + t(G_2(x) - F(x))||$$

$$\geqq ||F(x)|| - ||G_1(x) - F(x)|| - ||G_2(x) - F(x)|| \geqq \gamma - \frac{\gamma}{3} - \frac{\gamma}{3} = \frac{\gamma(F)}{3}.$$

Die Funktion $f(t) := J_\varepsilon(F_t(x))$, $0 < \varepsilon < \gamma(F)/3$ ist für $0 \leqq t \leqq 1$ stetig (vgl. (16)) und nach dem 3. Schritt ganzzahlig, also $f(0) = f(1)$. Folglich gilt nach dem 2. Schritt: $d_{\mathfrak{G}}(G_1) = d_{\mathfrak{G}}(G_2)$.

5. Schritt: Beweis von Behauptung b). Sei $F \in C^{(1)}(\overline{\mathfrak{G}})$. Für die Folge (23) gilt nach Satz 4 a) $d_{\mathfrak{G}}(F_n) = d_{\mathfrak{G}}(F)$, also ist nach (24) auch $J_\varepsilon(F) = d_{\mathfrak{G}}(F)$. q. e. d.

Theorem 1. *(Existenz- und Eindeutigkeitssatz). Es existiert genau ein Abbildungsgrad $d_{\mathfrak{G}}(F)$, der den Axiomen A_1, A_2, A_3, A_4 genügt.*

Beweis. Die Eindeutigkeit wurde bereits in Satz 3 bewiesen. A_1, A_2, A_3 sind nach Definition 1 für $F \in C_e^{(1)}(\overline{\mathfrak{G}})$ offensichtlich erfüllt, im Fall $F \in C(\overline{\mathfrak{G}})$ benutze man geeignete Approximationen. A_4 folgt aus Zusatz 1 und Satz 1 a). q. e. d.

§ 6. Folgerungen aus den Axiomen

Die folgenden Aussagen ergeben sich unmittelbar aus den Axiomen ohne Benutzung der konkreten Gestalt von $d_{\mathfrak{G}}(F)$.

Theorem 2. *(Eigenschaften von $d_{\mathfrak{G}}(F)$, die aus A_i folgen).*

a) *Randwerteigenschaft: Der Abbildungsgrad hängt nur von den Randwerten ab, d. h. aus $\partial\mathfrak{G}$: $F = G$, $F, G \in C(\overline{\mathfrak{G}})$ folgt nach A_4 auch $d_{\mathfrak{G}}(F) = d_{\mathfrak{G}}(G)$.*

b) *Indexsumme: Besitzt $F(x) \in C(\overline{\mathfrak{G}})$ auf $\overline{\mathfrak{G}}$ nur die endlich vielen Nullstellen $x_1, \ldots, x_k \in \mathfrak{G}$, dann gilt nach A_3, A_2 a) und* (6)

$$d_{\mathfrak{G}}(F) = \sum_{i=1}^{k} d_{x_i}(F)\,, \qquad F(x_i) = 0\,. \tag{25}$$

c) *Komponentenabhängigkeit: Ist $\mathfrak{U}$ eine Komponente (maximale zusammenhängende Menge) von $R^n - F(\partial\mathfrak{G})$ und ist F auf $\overline{\mathfrak{G}}$ stetig, dann gilt*

$$d_{\mathfrak{G}}(F - y_1) = d_{\mathfrak{G}}(F - y_2)\,, \qquad y_1, y_2 \in \mathfrak{U}\,. \tag{26}$$

d) *Kroneckersches Existenzprinzip: Gilt $F - y_0 \in C(\overline{\mathfrak{G}})$, also $\partial\mathfrak{G}$: $F(x) \neq y_0$, und ist $d_{\mathfrak{G}}(F - y_0) \neq 0$, dann besitzt die Gleichung*

$$\mathfrak{G}: F(x) = y\,, \qquad ||y - y_0|| < \gamma\,, \qquad \gamma := \min_{x \in \partial\mathfrak{G}} ||F(x) - y_0|| > 0 \tag{27}$$

eine Lösung. Genauer existiert eine Lösung für alle y, die sich mit y_0 durch eine stetige Kurve $y(t)$ verbinden lassen, die das Bild des Randes $F(\partial\mathfrak{G})$ nicht schneidet. Die Menge dieser y bezeichnen wir mit $\mathfrak{U}(y_0)$.

e) *Normsatz: Für $F \in C(\overline{\mathfrak{G}})$ gilt $\partial\mathfrak{G}$: $F(x) \simeq F(x)/||F(x)||$ mit $H(x, t) = F(x)/ \big((1 - t) + t||F(x)||\big)$, also $d_{\mathfrak{G}}(F(x)) = d_{\mathfrak{G}}(F(x)/||F(x)||)$.*

Benutzt man die konkrete Gestalt von $d_{\mathfrak{G}}(F)$, dann kann man das Existenzprinzip d) *wesentlich verschärfen.*

Korollar zum Kroneckerschen Existenzprinzip. *Gilt in d) stärker $F \in C^{(1)}(\overline{\mathfrak{G}})$, dann besitzt $\mathfrak{G}$: $F(x) = y$ wegen* (8') *und Definition* 1 *für fast alle $y \in \mathfrak{U}(y_0)$, d. h. bis auf eine Menge vom n-dimensionalen Maß Null mindestens $m = |d_{\mathfrak{G}}(F - y_0)|$ verschiedene Lösungen $x_1(y), \ldots, x_m(y) \in \mathfrak{G}$ mit* $\det F'(x_i(y)) \neq 0$.

Hängt $F(x, p)$ zusätzlich von einem Parameter $p \in R^l$ ab, wobei $F(x) := F(x, 0)$ den oben genannten Bedingungen genügt, und ist $F(x, p)$, $F'_x(x, p)$ stetig für $(x, p) \in \overline{\mathfrak{G}} \times \{p: ||p|| < p_0\}$, dann ergibt sich nach dem Satz über implizite Funktionen unter Berücksichtigung von $\det F'_x(x_i(y), 0) \neq 0$ *aus jedem $x_i(y)$, $i = 1, \ldots, m$ ein bezüglich p: $||p|| < \varepsilon(y)$ stetiger Lösungszweig $x_i(y, p), x_i(y, 0) = x_i(y)$ der Gleichung $\mathfrak{G}$: $F(x, p) = y$ für fast alle $y \in \mathfrak{U}(y_0)$ (vgl. Cronin* 1954 [30], 1964 [33]).

Beweis. Zu c). Es existiert eine stetige Kurve $y(t) \in \mathfrak{U}$, $0 \leqq t \leqq 1$ mit $y(0) = y_1$, $y(1) = y_2$. Wegen $y(t) \notin F(\partial\mathfrak{G})$ ist $\partial\mathfrak{G}$: $H(x, t) := F(x) - y(t) \neq 0$, also $\partial\mathfrak{G}$: $F - y_1 \cong F - y_2$, und (26) folgt aus A_4.

Zu d). Die in d) angegebenen y-Werte liegen alle in der gleichen Komponente von $R^n - F(\partial\mathfrak{G})$, somit gilt nach c): $d_{\mathfrak{G}}(F - y) = d_{\mathfrak{G}}(F - y_0) \neq 0$. Die Behauptung folgt dann aus A_2 a). q. e. d.

Die folgenden Existenzaussagen ergeben sich alle bis auf g) aus dem „Prinzip des spitzen Winkels", das in der Theorie der monotonen Operatoren eine weitreichende Verallgemeinerung erfahren hat (vgl. die Monographie von Wainberg 1972 [97]).

Satz 5. *(Existenzsätze, die aus A_2, A_4 folgen.)* a) *Prinzip vom spitzen Winkel: Genügt $F \in C(\overline{\mathfrak{G}})$, $0 \in \mathfrak{G}$ der zusätzlichen Bedingung $\partial\mathfrak{G}$: $\langle F(x), x\rangle \geqq 0$, dann besitzt entweder die Gleichung $\partial\mathfrak{G}$: $F(x) = 0$ eine Lösung oder es ist $\partial\mathfrak{G}$: $F(x) \cong I(x)$, also $d_{\mathfrak{G}}(F) = 1$, und für $\mathfrak{G}$: $F(x) = y$ gelten alle in Theorem* 2d) *gemachten Existenzaussagen mit $y_0 = 0$. Offensichtlich liegt für $\partial\mathfrak{G}$: $\langle F(x), x\rangle > 0$ (echter spitzer Winkel) stets der zweite Fall vor (vgl. Browder* 1963 [21], *Minty* 1963 [72]).

b) *Ein Fixpunktprinzip: $\mathfrak{G}$ sei eine offene, beschränkte Menge des R^n mit $0 \in \mathfrak{G}$. Jede Abbildung der Form*

$$F: \overline{\mathfrak{G}} \subset R^n \xrightarrow{\text{stetig}} R^n\,, \qquad \partial\mathfrak{G}: \langle F(x), x\rangle \leqq ||x||^2 \tag{28}$$

besitzt auf $\overline{\mathfrak{G}}$ einen Fixpunkt (vgl. Krasnosjelski 1956 [57], *Shinbrot* 1964 [92]). *Ist $\mathfrak{G}$ speziell eine Kugel mit dem Mittelpunkt in* 0, *dann bedeutet* (28) *genau $\partial\mathfrak{G}$: $||F(x)|| \cos\left(F(x), x\right) \leqq ||x||$, d. h. das Bild $F(x)$ eines Randpunktes x liegt stets auf der Seite der Tangentialebene durch x, auf der auch* 0 *liegt. Speziell existiert also ein Fixpunkt, wenn die Kugel $\overline{\mathfrak{G}}$ stetig in sich abgebildet wird (Brouwer* 1912 [13]) *oder wenigstens das Bild des Randes $F(\partial\mathfrak{G})$ wieder auf $\overline{\mathfrak{G}}$ liegt (Rothe* 1937 [83]).

c) *Allgemeiner Brouwerscher Fixpunktsatz: Jede stetige Abbildung einer kompakten und konvexen Menge $\mathfrak{K} \subset R^n$ in sich besitzt einen Fixpunkt.*

d) *Koerzitive Abbildungen: Gilt*

$$F: R^n \xrightarrow{\text{stetig}} R^n\,, \qquad \lim_{||x|| \to \infty} \frac{\langle F(x), x\rangle}{||x||} = +\infty\,, \tag{29}$$

dann ist die Gleichung $F(x) = y$ für jedes $y \in R^n$ lösbar. Ist F noch zusätzlich streng monoton, d. h.

$$\langle F(x) - F(y), x - y\rangle > 0\,, \qquad x \neq y\,, \quad x, y \in R^n\,, \tag{30}$$

dann ist die Lösung auch eindeutig (vgl. Browder 1966 [23]).

e) *Stark monotone Abbildungen: Für jede stark monotone, stetige Abbildung*

$$F\colon\ R^n \xrightarrow{\text{stetig}} R^n\,,\qquad \langle F(x) - F(y), x - y\rangle \geqq \gamma(\|x - y\|)\,\|x - y\|$$

$$\lim_{t\to+\infty} \gamma(t) = +\infty\,,\qquad \gamma(t) > 0 \quad \text{für} \quad t > 0, \gamma(0) = 0\,,$$

ist die Gleichung $F(x) = y$ *für alle* $y \in R^n$ *eindeutig lösbar (vgl. Minty* 1962 [71], *Browder* 1963 [20]*).*

f) *Retraktsatz: Der Rand* $\partial\mathfrak{G}$ *einer offenen, beschränkten Menge* $\mathfrak{G}$ *des* R^n *ist niemals Retrakt von* $\overline{\mathfrak{G}}$, *d. h., es gibt keine stetige Abbildung* $R\colon \overline{\mathfrak{G}} \to \partial\mathfrak{G}$, $\partial\mathfrak{G}$: $R(x) = x$.

g) *Prinzip der a-priori-Abschätzungen: Existiert für die stetige Abbildung* $F\colon R^n \to R^n$ *eine a-priori-Abschätzung, d. h., es gibt Zahlen* λ_0, $r_0 > 0$ *mit*

$$0 < \lambda < \lambda_0\,,\qquad \lambda F(x) = x \Rightarrow \|x\| \leqq r_0\,, \tag{31}$$

dann besitzt die Gleichung $\lambda F(x) = x$ *für alle* λ *mit* $0 \leqq \lambda \leqq \lambda_0$ *eine Lösung (vgl. Leray, Schauder* 1934 [67])*)*

Zusatz. Aus dem Satz über die Gebietsinvarianz, den wir in § 11 in einfacher Weise aus dem Antipodensatz von Borsuk herleiten, ergibt sich, daß die eineindeutigen und stetigen Abbildungen des R^n auf den R^n in d), e) sogar Homöomorphismen darstellen.

Beweis. Zu a). Sei $\partial\mathfrak{G}$: $F(x) \neq 0$, dann ist mit $H(x, t) := (1 - t)\,x + tF(x)$ wegen

$$\partial\mathfrak{G}\colon \|H(x, t)\|^2 = (1 - t)^2\,\|x\|^2 + 2t(1 - t)\,\langle F(x), x\rangle + t^2\|F(x)\|^2 > 0\,,\qquad 0 \leqq t \leqq 1$$

stets $\partial\mathfrak{G}$: $F(x) \cong I(x)$.

Zu b). Aus (28) folgt $\langle x - F(x), x\rangle \geqq 0$, somit besitzt $\overline{\mathfrak{G}}$: $x - F(x) = 0$ nach a) eine Lösung.

Zu c). Ist $\mathfrak{K}$ eine Kugel, dann folgt die Behauptung aus b). Im allgemeinen Fall wählen wir eine umfassendere Kugel $\mathfrak{K}_\varrho^{(n)}(0) \supset \mathfrak{K}$ und erweitern $F\colon \mathfrak{K} \to \mathfrak{K}$ nach (4′) stetig zu $\tilde{F}\colon \mathfrak{K}_\varrho^{(n)}(0) \to \mathfrak{K}$. Der Fixpunkt $x = \tilde{F}(x) \in \mathfrak{K}$ ist dann wegen $\mathfrak{K}$: $F(x) = \tilde{F}(x)$ zugleich Fixpunkt von F.

Zu d). Für $\|x\| = \varrho$ und ϱ hinreichend groß gilt bei festem y

$$\langle F(x) - y, x\rangle \geqq \|x\|\left(\frac{\langle F(x), x\rangle}{\|x\|} - \|y\|\right) > 0\,,$$

somit ist $\|x\| \leqq \varrho$: $F(x) - y = 0$ nach a) lösbar. Die Richtigkeit der Eindeutigkeitsaussage ist offensichtlich.

Zu e). Wegen

$$\langle F(x), x\rangle = \langle F(x) - F(0), x\rangle + \langle F(0), x\rangle \geqq (\gamma(\|x\|) - \|F(0)\|)\,\|x\|$$

sind die Voraussetzungen von d) erfüllt.

Zu f). Gäbe es eine solche Abbildung R, dann dürften wir o. E. d. A. $0 \in \mathfrak{G}$ voraussetzen. Aus $\partial\mathfrak{G}: \langle R(x), x\rangle = ||x||^2 > 0$ folgt dann jedoch nach a) $0 \in F(\mathfrak{G})$ im Widerspruch zu $0 \notin \partial\mathfrak{G} = F(\overline{\mathfrak{G}})$.

Zu g). Aus (31) folgt bei festem λ: $0 \leqq \lambda < \lambda_0$

$$||x|| = 2r_0: H_\lambda(x, t) := x - t\lambda F(x) \neq 0\,, \qquad 0 \leqq t \leqq 1\,,$$

also $d_{\mathfrak{K}^{(n)}_{2r_0}(0)}\big(I(x) - \lambda F(x)\big) = d_{\mathfrak{K}^{(n)}_{2r_0}(0)}\big(I(x)\big) = 1$, und Theorem 2d) ergibt die Existenz einer Lösung. Für $\lambda = \lambda_0$ gilt entweder $||x|| = 2r_0: \lambda_0 F(x) \neq x$, dann liefert der vorangegangene Schluß eine Lösung x mit $||x|| < 2r_0$ oder es existiert wegen $||x|| = 2r_0: \lambda_0 F(x) = x$ trivialerweise eine Lösung x mit $||x|| = 2r_0$. q. e. d.

§ 7. Lineare Abbildungen und nichtlineare Eigenwertprobleme

Theorem 3. *(Lineare Abbildungen)* a) *Gegeben sei*

$$L:\ R^n \xrightarrow{\text{linear}} R^n\,.$$

Existiert L^{-1}, dann gilt für $F(x) := L(x) - x_0$ und $\mathfrak{G}$ mit $\partial\mathfrak{G}$: $L(x) \neq x_0$ stets

$$d_{\mathfrak{G}}(F) = \begin{cases} \operatorname{sgn}\det L = \operatorname{sgn}\prod_{i=1}^{n}\lambda_i\,, & L^{-1}(x_0) \in \mathfrak{G}\,,\\ 0\,, & L^{-1}(x_0) \notin \overline{\mathfrak{G}}\,. \end{cases} \tag{32}$$

Dabei sind λ_i die Eigenwerte von L entsprechend ihrer algebraischen Vielfachheit, d.h. entsprechend der Anzahl, in der sie in der Jordanschen Normalform von L vorkommen.

b) *Ist μ keine charakteristische Zahl von L, d.h. es existiert $(I-\mu L)^{-1}$, dann ist nach* (6) *und* (32)

$$\begin{aligned} d_{x=0}\big(I(x) - \mu L(x)\big) &= \operatorname{sgn}\prod_{i=1}^{n}(1 - \mu\lambda_i)\\ &= (\operatorname{sgn}\det L)\prod_{i=1}^{n}(\mu_i - \mu)\,, \qquad \mu_i = \frac{1}{\lambda_i}. \end{aligned} \tag{33}$$

Beweis. Man beachte $F'(x) = L$ und Definition 1. Die Beziehung $\det L = \prod_{i=1}^{n}\lambda_i$ ergibt sich aus der Jordanschen Normalform, q. e. d.

Theorem 3 kann man heranziehen, um den Nullstellenindex zu berechnen. Besitzt nämlich die Abbildung

$$F: \mathfrak{K}^{(n)}_\varrho(x_0) \subset R^n \xrightarrow{\text{stetig}} R^n\,, \qquad F(x_0) = 0$$

in x_0 eine Fréchet-Ableitung $F'(x_0)$ mit $\det F'(x_0) \neq 0$, dann zeigen die gleichen Überlegungen wie im 1. Schritt des Beweises zum Satz 3, daß x_0 eine isolierte Nullstelle ist und

$$d_{x_0}(F) = d_{x_0}\big(F'(x_0)\,(x - x_0)\big) = \operatorname{sgn}\prod_{i=1}^{n}\lambda_i$$

gilt, wenn λ_i die Eigenwerte von $F'(x_0)$ sind.

Wir wollen jetzt Theorem 3 anwenden, um Existenzaussagen für das nichtlineare Eigenwertproblem

$$\partial\mathfrak{G}\colon F(x) = \lambda x\,,\ \lambda \neq 0\,,\ 0 \subset \mathfrak{G}\,,\ \mathfrak{G}\ \text{offen, beschränkt im Raum}\ R^n \tag{34}$$

zu erhalten. Zu diesem Zweck stellen wir einige Definitionen bereit. Unter einem Kegel $\mathfrak{K}$ verstehen wir eine abgeschlossene, konvexe Menge ungleich $\{0\}$ mit $t\mathfrak{K} = \mathfrak{K}$ für $t \geqq 0$ und $\mathfrak{K} \cap -\mathfrak{K} = \{0\}$. Jeder solche Kegel induziert vermöge

$$x \geqq_{\mathfrak{K}} y \Leftrightarrow x - y \in \mathfrak{K}; \quad x >_{\mathfrak{K}} y \Leftrightarrow x \geqq_{\mathfrak{K}} y\,, \quad x \neq y \tag{35}$$

eine Halbordnung im R^n. Wählt man speziell

$$\mathfrak{K}_+ = \{(\xi_1, \dots, \xi_n) \in R^n; \quad \xi_i \geqq 0\}\,,$$

dann schreiben wir statt $\geqq_{\mathfrak{K}_+}$ kurz $\geqq$; $x \geqq y$ ist gleichbedeutend mit $\xi_i \geqq \eta_i$, $i = 1, \dots, n$; $x = (\xi_1, \dots, \xi_n)$, $y = (\eta_1, \dots, \eta_n)$.

Die folgenden Existenzaussagen ergeben sich alle bis auf h) in einfacher Weise aus dem Eigenwertprinzip a).

Satz 6. *(Nichtlineare Eigenwertprobleme)* a) *Eigenwertprinzip: Aus* $F, G \in C(\overline{\mathfrak{G}})$ *und* $d_{\mathfrak{G}}(F) \neq d_{\mathfrak{G}}(G)$ *folgt die Existenz einer Lösung des Eigenwertproblems* $\partial\mathfrak{G}$: $F(x) = \lambda G(x), \lambda < 0$.

b) *Igeltheorem: Für* $F \in C(\overline{\mathfrak{G}})$ *und n ungerade besitzt* (34) *stets eine Lösung. Ist n gerade, dann existiert eine Lösung von* (34) *im Fall* $d_{\mathfrak{G}}(F) \neq 1$.

c) *Lückensatz: Läßt* $F \in C(\overline{\mathfrak{G}})$ *auf* $\partial\mathfrak{G}$ *eine Richtung* $x_0 \neq 0$ *aus, also* $\partial\mathfrak{G}$: $F(x) \neq \tau x_0$, $0 < \tau < \infty$, *dann ist* $d_{\mathfrak{G}}(F) = 0$, *und* (34) *besitzt zwei Eigenlösungen* $x_1, x_2 \in \partial\mathfrak{G}$, $\lambda_1 > 0$, $\lambda_2 < 0$.

d) *Tangentenvektorfelder (Poincaré-Brouwer): Auf der Oberfläche einer Kugel* $\mathfrak{K}_\varrho^{(n)}(x_0)$ *des* R^n *existiert genau dann ein stetiges, nichtverschwindendes Tangentenvektorfeld* $F(x)$, *wenn n gerade ist. Gilt* $x_0 = 0$, *dann ist nach* b): $d_{\mathfrak{K}_\varrho^{(n)}(0)}(F) = 1$.

e) *Positive Eigenlösungen: Gilt*

$$F\colon \partial\mathfrak{G} \cap \mathfrak{K} \subset R^n \xrightarrow{\text{stetig}} \mathfrak{K} \subset R^n\,, \qquad F(x) \neq 0\,, \tag{36}$$

und ist $\mathfrak{K}$ *ein Kegel, dann besitzt* (34) *eine Eigenlösung der Form* $x \in \partial\mathfrak{G} \cap \mathfrak{K}$, $\lambda > 0$, *also* $x >_{\mathfrak{K}} 0$, $\lambda > 0$ *(vgl. Rothe* 1944 [84], *Krasnosjelski* 1956 [57]).

f) *Minorantenprinzip: Gilt* (36) *ohne die Bedingung* $\partial\mathfrak{G} \cap \mathfrak{K}$: $F(x) \neq 0$, *dann besitzt* (34) *eine Eigenlösung* $x >_{\mathfrak{K}} 0, \lambda > 0$, *wenn es feste Zahlen* $\gamma_0 > 0$, $0 < s \leqq 1$ *und ein* $x_0 >_{\mathfrak{K}} 0$ *gibt mit*

$$\partial\mathfrak{G} \cap \mathfrak{K}\colon F(x) \geqq_{\mathfrak{K}} \alpha^s(x)\,\gamma_0 x_0\,. \tag{37}$$

Dabei ist das nichtlineare Funktional $\alpha(x)$ *durch*

$$\alpha(x) := \max\,\{\alpha; \alpha \geqq 0,\ x \geqq_{\mathfrak{K}} \alpha x_0\} \tag{38}$$

erklärt (vgl. Krasnoselski 1956 [57], 1962 [58], *Zeidler* 1971 [105]).

g) *Satz von Perron* 1907 [77]: *Ist* $A \neq 0$ *ein linearer Operator des* R^n *mit der zugehörigen Matrix* (a_{ij}), *und gilt* $a_{ij} \geqq 0$, *dann besitzt* $A(x) = \lambda x$ *eine Eigenlösung* $\lambda > 0$, $x = \{\xi_1, \dots, \xi_n\} > 0$, *d. h.* $\xi_i > 0$, $i = 1, \dots, n$, *wenn eine der beiden Bedingungen erfüllt ist:* α) $\det(a_{ij}) \neq 0$, β) $a_{i_0 i_0} > 0$ *für ein Diagonalelement* $a_{i_0 i_0}$.

h) *Positive, monotone und asymptotisch verschwindende Operatoren: Genügt die Abbildung*

$$F: \mathfrak{K} \subset R^n \xrightarrow{\text{stetig}} R^n\,, \qquad \mathfrak{K} \text{ Kegel}$$

zusätzlich den drei Bedingungen: 1) $F(y) \geqq_{\mathfrak{K}} F(x) \geqq_{\mathfrak{K}} 0$ *für* $y \geqq_{\mathfrak{K}} x \geqq_{\mathfrak{K}} 0$, 2) $||x|| \leqq \varrho$; $\max ||F(x)|| = o(\varrho)$ *für* $\varrho \to \infty$, 3) $\mu_1 F(x_1) \geqq x_1$ *für festes* $\mu_1 > 0$, $x_1 >_{\mathfrak{K}} 0$, *dann besitzt das nichtlineare Eigenwertproblem*

$$\mu F(x) = x\,, \quad \mu > 0\,, \quad x >_{\mathfrak{K}} 0 \tag{38'}$$

für alle $\mu: \mu \geqq \mu_1$ *eine Lösung* $x: x \geqq_{\mathfrak{K}} x_1$. *Folglich gilt für* (38′) *ohne* 3) *folgende Alternative: Entweder besitzt* (38′) *überhaupt keine Lösung, oder mit jeder Lösung* $\mu_1 > 0$, $x_1 >_{\mathfrak{K}} 0$ *existiert auch für jedes* $\mu: \mu \geqq \mu_1$ *eine Lösung* $x >_{\mathfrak{K}} 0$ *mit* $x \geqq_{\mathfrak{K}} x_1$ *(vgl. Zeidler* 1970 [107]*).*

Zusatz. Die Bedingung (37) ist z. B. erfüllt, wenn $F(x)$ auf $\partial\mathfrak{G} \cap \mathfrak{K}$ eine Minorante $M(x)$ besitzt, also $\partial\mathfrak{G} \cap \mathfrak{K}: F(x) \geqq_{\mathfrak{K}} M(x)$, wobei gilt: $M(x_2) \geqq_{\mathfrak{K}} M(x_1) \geqq_{\mathfrak{K}} 0$ für $x_2 \geqq_{\mathfrak{K}} x_1 \geqq_{\mathfrak{K}} 0$, $M(tx) = t^s M(x)$ für $t \geqq 0$, $x \geqq u\, 0$, $M(x_0) \geqq_{\mathfrak{K}} \gamma_0 x_0$, denn dann ist

$$\partial\mathfrak{G} \cap \mathfrak{K}: F(x) \geqq_{\mathfrak{K}} M(x) \geqq_{\mathfrak{K}} M\big(\alpha(x)\, x_0\big) \geqq_{\mathfrak{K}} \alpha^s(x)\, \gamma_0 x_0 \tag{37'}$$

(Prinzip der monotonen Minoranten von Krasnosjelski). Die vom Verfasser stammende Bedingung (37) ist jedoch auch anwendbar, wenn keine monotonen Minoranten vorliegen (vgl. [105]).

Beweis. Zu a). Wegen $d_{\mathfrak{G}}(F) \neq d_{\mathfrak{G}}(G)$, $\partial\mathfrak{G}: F(x), G(x) \neq 0$ muß $\partial\mathfrak{G}: H(x, t) := (1 - t)\, F(x) + tG(x) = 0$ eine Lösung $x_1 \in \partial\mathfrak{G}$, $0 < t_1 < 1$ besitzen.

Zu b). Man setze in a) $G = \pm I$ und beachte $d_{\mathfrak{G}}\big(I(x)\big) = 1$, $d_{\mathfrak{G}}\big(-I(x)\big) = (-1)^n$ nach (32).

Zu c). Nach Voraussetzung ist $\partial\mathfrak{G}: H(x, t) := (1 - t)\, F(x) - tx_0 \neq 0$, $0 \leqq t \leqq 1$, also $\partial\mathfrak{G}: F(x) \simeq G(x)$, $G(x) := -x_0$, folglich $d_{\mathfrak{G}}(F) = d_{\mathfrak{G}}(G) = 0$, und wegen $d_{\mathfrak{G}}(F) \neq d_{\mathfrak{G}}\big(\pm I(x)\big)$ ergibt a) die Behauptung.

Zu d). Wir können o. E. d. A. annehmen, daß der Nullpunkt der Mittelpunkt der Kugel ist.

Ist $n = 2m$, dann bildet $F(x) = (\xi_2, -\xi_1; \ldots; \xi_{2m}, -\xi_{2m-1})$, $x = \{\xi_1, \ldots, \xi_{2m})$ wegen $||x|| = \varrho: F(x) \neq 0$, $\langle F(x), x\rangle = 0$ ein nichtverschwindendes, stetiges Tangentenvektorfeld auf der Kugeloberfläche $\partial\mathfrak{K}_\varrho^{(n)}(0)$.

Existiert umgekehrt ein Feld $F(x)$ mit $||x|| = \varrho: F(x) \neq 0$, $\langle F(x), x\rangle = 0$, dann ist

$$||x|| = \varrho: H(x, t) = x \cos \pi t + \frac{\varrho F(x)}{||F(x)||} \sin \pi t \neq 0$$

wegen $||H(x, t)||^2 = \varrho^2$, also $||x|| = \varrho: I(x) = H(x, 0) \simeq H(x, 1) = -I(x)$, folglich $d_{x=0}\big(I(x)\big) = d_{x=0}\big(-I(x)\big)$, d. h. $1 = (-1)^n$, somit muß n gerade sein.

Zu e). Wir erweitern F nach (4′) zu einer auf $\overline{\mathfrak{G}}$ stetigen Abbildung $\tilde{F}$: $\overline{\mathfrak{G}} \to \mathfrak{A} := \overline{\text{konv}\, F(\partial\mathfrak{G} \cap \mathfrak{K})}$. Aus $0 \notin F(\partial\mathfrak{G} \cap \mathfrak{K}) \subset \mathfrak{K}$ folgt $0 \notin \mathfrak{A}$, also $\tilde{F} \in C(\overline{\mathfrak{G}})$. Wählen wir ferner ein $x_0 \neq 0$ mit $x_0 \notin \mathfrak{K}$, dann läßt $F(x)$ die Richtung x_0 aus. Folglich existiert nach c) eine Eigenlösung $\partial\mathfrak{G}: \tilde{F}(x) = \lambda x, \lambda > 0$. Wegen $\tilde{F}(\overline{\mathfrak{G}}) \subseteq \mathfrak{A} \subseteq \mathfrak{K}$ ist dann auch $x \in \mathfrak{K}$ und somit $\tilde{F}(x) = F(x)$.

Zu f). Die Hilfsprobleme

$$\partial\mathfrak{G}: \lambda_n x_n = F(x_n) + \frac{x_0}{n}, \quad \lambda_n > 0, \quad x_n >_{\mathfrak{K}} 0 \tag{40}$$

besitzen nach e) Lösungen. Übergang zur Norm zeigt sofort $0 < \lambda_n \leqq b$. Geeignete Teilfolgen ergeben somit durch $\lambda_{n'} \to \lambda \geqq 0$, $x_{n'} \to x >_{\mathfrak{K}} 0$ für $n' \to \infty$ Eigenlösungen von (34).

Um $\lambda > 0$ nachzuweisen, genügt es, $0 < a \leqq \lambda_n$ zu zeigen. Offensichtlich gilt $\alpha(x_n) \leqq c$, sonst würden Teilfolgen nach (38) auf $0 \geqq_{\mathfrak{K}} x_0$ führen. Aus (40) und (37) folgt nun aber

$$\lambda_n x_n \geqq_{\mathfrak{K}} \alpha^s(x_n)\, \gamma_0 x_0, \qquad \lambda_n \alpha(x_n) \geqq \frac{1}{n},$$

also nach (38)

$$\lambda_n \alpha(x_n) \geqq \alpha^s(x_n)\, \gamma_0, \qquad \lambda_n \geqq \alpha^{s-1}(x_n)\, \gamma_0 \geqq c^{s-1}\gamma_0 .$$

Zu g). Im Fall α) ist (36) mit $\mathfrak{K} = \mathfrak{K}_+$ erfüllt, im Fall β) gibt es ein $x_0 > 0$ mit $A(x_0) \geqq \gamma_0 x_0$, und (37′) mit $A = M$, $s = 1$, liefert (37).

Zu h). Bei festem $\mu: \mu \geqq \mu_1$ (μ_1 aus 3)) und hinreichend großem ϱ wird

$$\mathfrak{M} := \{x; x \geqq_{\mathfrak{K}} x_1, \|x\| \leqq \varrho\}, \qquad \mathfrak{M} \text{ konvex, kompakt}$$

wegen 1), 2) durch $A(x) := \mu F(x)$ stetig in sich abgebildet. Der vorhandene Fixpunkt $x \in \mathfrak{M}$ ist der gesuchte Eigenvektor, q. e. d.

§ 8. Produkteigenschaften des Abbildungsgrades

Theorem 4. *(Produkt- und Reduktionseigenschaften).* a) *Kartesisches Produkt: Aus $\mathfrak{G}_1 \subset R^n$, $\mathfrak{G}_2 \subset R^m$, $F_i \in C(\overline{\mathfrak{G}}_i)$, $i = 1, 2$, folgt für die Produktabbildung*

$$F := (F_1, F_2): \overline{\mathfrak{G}}_1 \times \overline{\mathfrak{G}}_2 \to R^n \times R^m$$

stets $F \in C(\overline{\mathfrak{G}}_1 \times \overline{\mathfrak{G}}_2)$ und

$$d_{\mathfrak{G}_1 \times \mathfrak{G}_2}(F) = d_{\mathfrak{G}_1}(F_1)\, d_{\mathfrak{G}_2}(F_2) . \tag{41}$$

b) *Reduktionssatz: Sei*

$$f(x): \overline{\mathfrak{G}} \subset R^n \xrightarrow{\text{stetig}} R^m, \qquad m < n, \qquad f = (f_1, \dots, f_m, 0 \dots, 0)$$

und $\partial\mathfrak{G}$: $F(x) := x + f(x) \neq 0$. Ist dann $\overline{F}$ die durch Einschränkung auf $\overline{\mathfrak{G}} \cap R^m$ aus F entstehende Abbildung, dann gilt

$$d_{\mathfrak{G}}(F) = d_{\mathfrak{G} \cap R^m}(\overline{F}) . \tag{42}$$

c) *Abbildungen mit Dimensionserniedrigung: Gilt für $F \in C(\overline{\mathfrak{G}})$, $\mathfrak{G}$ offen, beschränkt in R^n zusätzlich*

$$F: \overline{\mathfrak{G}} \subset R^n \xrightarrow{\text{stetig}} R^m, \qquad m < n,$$

dann ist $d_{\mathfrak{G}}(F) = 0$.

d) *Produktsatz von Leray* 1935 [65], 1950 [66]*: Sei*

$$\overline{\mathfrak{G}} \subset R^n \xrightarrow[\text{stetig}]{F} F(\overline{\mathfrak{G}}) \subset R^n \xrightarrow[\text{stetig}]{G} R^n$$

mit $\partial\mathfrak{G}$: $G(F(x)) \neq 0$, *also* $G(F(x)) \in C(\overline{\mathfrak{G}})$, *dann gilt*

$$d_{\mathfrak{G}}(G(F(x))) = \sum_{\mathfrak{U}_i \subseteq F(\mathfrak{G})} d_{\mathfrak{U}_i}(G)\, d_{\mathfrak{G}}(F - y_i)\,, \qquad y_i \in \mathfrak{U}_i\,. \tag{43}$$

Summiert wird dabei über alle Komponenten $\mathfrak{U}_i$ *von* $R^n - F(\partial\mathfrak{G})$ *mit* $\mathfrak{U}_i \subseteq F(\mathfrak{G})$. *Die* y_i *können beliebig aus* $\mathfrak{U}_i$ *gewählt werden. In* (43) *sind stets nur endlich viele Summanden ungleich Null.* G *sei auf* $F(\overline{\mathfrak{G}})$ *erklärt.*

e) *Exzisionseigenschaft: Sei* $F \in C(\overline{\mathfrak{G}})$, $\mathfrak{A} \subset \overline{\mathfrak{G}}$, $\mathfrak{A}$ *kompakt und* $\mathfrak{A}$: $F(x) \neq 0$, *dann gilt*

$$d_{\mathfrak{G}-\mathfrak{A}}(F) = d_{\mathfrak{G}}(F)\,.$$

Der Produktsatz von Leray spielt beim Beweis vieler tieferliegender Eigenschaften des Abbildungsgrades eine zentrale Rolle.

Der Reduktionssatz ist Ausgangspunkt für die Erweiterung des Abbildungsgrades auf „kompakte" Abbildungen in lokalkonvexen Räumen (vgl. Leray-Schauder 1934 [67], Nagumo 1951 [74], J. T. Schwartz 1969 [91]).

Beweis. Zu a). Sei $F_i \in C_e^{(1)}(\overline{\mathfrak{G}}_i)$, $i = 1, 2$, mit den Nullstellen $x_k^{(1)}, x_l^{(2)}$, dann besitzt F die Nullstellen $y_{kl} = (x_k^{(1)}, x_l^{(2)})$, also

$$\begin{aligned} d_{\mathfrak{G}_1\times\mathfrak{G}_2}(F) &= \sum_{k,l} \operatorname{sgn}\det F'(y_{kl}) \\ &= \sum_{k,l} \operatorname{sgn}\det F_1'(x_k^{(1)}) \operatorname{sgn}\det F_2'(x_l^{(2)}) = d_{\mathfrak{G}_1}(F_1)\, d_{\mathfrak{G}_2}(F_2)\,. \end{aligned}$$

Der allgemeine Fall ergibt sich durch hinreichend feine Approximation.

Zu b). Wegen Definition 1 genügt es, $F \in C_e^{(1)}(\overline{\mathfrak{G}})$ zu betrachten. Aus $\mathfrak{G}$: $F(x_i) = 0$ folgt $x_i = -f(x_i) \in R^m$, somit besitzen F und $\overline{F}$ die gleichen Nullstellen. Die Behauptung folgt dann aus $\det F'(x_i) = \det \overline{F}'(x_i)$.

Zu c). Es gilt $F = (f_1, \dots, f_m, 0, \dots, 0)$. Nach dem Approximationssatz von Weierstraß ist $\partial\mathfrak{G}$: $F \cong \overline{F}$ mit $\overline{F} = (\bar{f}_1, \dots, \bar{f}_m, 0, \dots, 0)$, $\overline{F} \in C^{(1)}(\overline{\mathfrak{G}})$, also $\det \overline{F}'(x) \equiv 0$, und nach (16) gilt $d_{\mathfrak{G}}(F) = d_{\mathfrak{G}}(\overline{F}) = 0$.

Zu e). Für $F \in C_e^{(1)}(\overline{\mathfrak{G}})$ folgt die Behauptung unmittelbar aus Definition 1. Im allgemeinen Fall benutze man eine hinreichende feine Approximation $G \in C_e^{(1)}(\overline{\mathfrak{G}})$ von F.

Zu d). Gilt $G(F(x))$, $G, F \in C_e^{(1)}$, dann ergibt sich (43) durch elementare Berechnungen (vgl. 5. Schritt). Den allgemeinen Fall führen wir durch geeignete Approximationen, die allerdings einer gewissen Sorgfalt bedürfen, auf den oben erwähnten Spezialfall zurück.

1. Schritt: Erweiterung von G. Sei $\mathfrak{W}$ ein Würfel mit $F(\overline{\mathfrak{G}}) \subset \mathfrak{W}$, der alle beschränkten $\mathfrak{U}_i$ enthalte. Wir erweitern G stetig auf $\overline{\mathfrak{W}}$. Wegen

$$0 < \gamma := \min_{x \in \partial\mathfrak{G}} \|G(F(x))\| = \min_{y \in F(\partial\mathfrak{G})} \|G(y)\| \tag{44}$$

und $\partial\mathfrak{U}_i \subseteq F(\partial\mathfrak{G})$ ist gleichzeitig

$$0 < \gamma \leqq \min_{y \in \partial\mathfrak{U}} \|G(y)\|\,. \tag{45}$$

Außerdem folgt aus der gleichmäßigen Stetigkeit von G auf $\overline{\mathfrak{W}}$

$$y_i \in \overline{\mathfrak{W}}\colon ||G(y_1) - G(y_2)|| < \frac{\gamma}{2} \quad \text{für} \quad ||y_1 - y_2|| < \alpha \,. \tag{46}$$

2. Schritt: Approximation von G. Die gleichen Überlegungen wie im Beweis zu Satz 2 ergeben die Existenz einer Abbildung $P = (p_1, \ldots, p_n)$ (p_i Polynome) mit

$$||G - P||_{C(\overline{\mathfrak{W}})} < \frac{\gamma}{2}, \tag{47}$$

wobei alle Nullstellen von P auf $\overline{\mathfrak{W}}$ nichtentartet sind. Da sich diese nach dem Satz über implizite Funktionen nur am Rand häufen können, läßt sich durch geringfügige Verkleinerung von $\mathfrak{W}$ stets $P \in C_e^{(1)}(\overline{\mathfrak{W}})$ erreichen mit den endlich vielen Nullstellen ($\mathfrak{U}_\infty$ unbeschränkt)

$$y_j^{(i)} \in \mathfrak{U}_i\,, \quad i = 1, \ldots, N, \infty\colon P(y_j^{(i)}) = 0\,, \quad \det P'(y_j^{(i)}) \neq 0\,, \tag{48}$$

wenn wir die $\mathfrak{U}_i$ geeignet numerieren. Wegen (45), (47) können keine Nullstellen auf $\partial\mathfrak{U}_i$ liegen. Außerdem ist nach (45), (47)

$$d_{\mathfrak{U}_i}(G) = d_{\mathfrak{U}_i}(P) = \sum_{j=1}^{j_i} \operatorname{sgn} \det P'(y_j^{(i)})\,, \qquad i = 1, \ldots, N\,. \tag{49}$$

$$d_{\mathfrak{U}_i}(G) = d_{\mathfrak{U}_i}(P) = 0\,, \qquad \mathfrak{U}_i \subseteq F(\mathfrak{G})\,, \qquad i > N\,. \tag{49'}$$

3. Schritt: Approximation von F. Der Beweis von Satz 2 (3. Schritt) lehrt, daß $H \in C_e^{(1)}(\overline{\mathfrak{G}})$ genau dann gilt, wenn $H \in C^{(1)}(\overline{\mathfrak{G}})$ ist und H in $\mathfrak{G}$ nur nichtentartete Nullstellen besitzt. Somit ist $C_e^{(1)}(\overline{\mathfrak{G}})$ in $C^{(1)}(\overline{\mathfrak{G}})$ bezüglich der Normtopologie offen, denn das Komplement ist wegen der Kompaktheit von $\overline{\mathfrak{G}}$ abgeschlossen.

Aus $\partial\mathfrak{G}\colon F(x) \neq y_j^{(i)}$ ergibt sich für alle $y_j^{(i)}$

$$0 < \beta := \min_{x \in \partial\mathfrak{G}} ||F(x) - y_j^{(i)}|| \tag{50}$$

und $F - y_j^{(i)} \in C(\overline{\mathfrak{G}})$. Dann existiert einheitlich für alle $y_j^{(i)}$ eine Abbildung Q mit

$$Q - y_j^{(i)} \in C_e^{(1)}(\overline{\mathfrak{G}})\,, \quad \overline{\mathfrak{G}}\colon ||F(x) - Q(x)|| < \varepsilon\,, \quad 0 < \varepsilon < \min(\alpha, \beta, \gamma)\,, \tag{51}$$

wobei ε weiterhin so klein sei, daß $Q(\overline{\mathfrak{G}}) \subset \mathfrak{W}$ gilt. Die Existenz von Q mit $Q - y_1^{(1)} \in C_e^{(1)}(\overline{\mathfrak{G}})$ folgt sofort aus Satz 2b). Unter Berücksichtigung der Offenheit von $C_e^{(1)}(\overline{\mathfrak{G}})$ in $C^{(1)}(\overline{\mathfrak{G}})$ kann man durch schrittweise Abänderung um hinreichend kleine Konstanten (Satz 2b)) die Beziehung (51) für alle $y_j^{(i)}$ erreichen. Nach (50), (51), (5) ist obendrein

$$d_{\mathfrak{G}}\big(F(x) - y_j^{(i)}\big) = d_{\mathfrak{G}}\big(Q(x) - y_j^{(i)}\big)\,. \tag{52}$$

4. Schritt: Approximation von $G(F(x))$. Wegen (46), (47), (51) gilt

$$\overline{\mathfrak{G}}\colon ||G(F(x)) - P(Q(x))|| \leq ||G(F(x)) - G(Q(x))||$$
$$+ ||G(Q(x)) - P(Q(x))|| < \frac{\gamma}{2} + \frac{\gamma}{2} = \gamma\,,$$

also nach (44), (5)

$$d_{\mathfrak{G}}\big(G(F(x))\big) = d_{\mathfrak{G}}\big(P(Q(x))\big) . \tag{53}$$

5. Schritt: Berechne $d_{\mathfrak{G}}\big(P(Q(x))\big)$. Nach (51) nimmt Q nur in endlich vielen Punkten $x_{jk}^{(i)}$ die Werte $y_j^{(i)}$ an, d. h.

$$x_{jk}^{(i)} \in \mathfrak{G}:\ Q(x_{jk}^{(i)}) = y_j^{(i)} , \qquad \det Q'(x_{jk}^{(i)}) \neq 0 ,$$

also $T(x) := P\big(Q(x)\big) \in C_e^{(1)}(\overline{\mathfrak{G}})$. Somit ist nach (11)

$$d_{\mathfrak{G}}\big(P(Q(x))\big) = \sum_{i,j,k} \operatorname{sgn} \det T'(x_{jk}^{(i)})$$

$$= \sum_{i,j,k} \operatorname{sgn} \det P'(y_j^{(i)}) \operatorname{sgn} \det Q'(x_{jk}^{(i)}) \overset{(11)}{=} \sum_{i,j} \operatorname{sgn} \det P'(y_j^{(i)})\, d_{\mathfrak{G}}(Q - y_j^{(i)})$$

$$\overset{(52)}{=} \sum_{i,j} \operatorname{sgn} \det P'(y_j^{(i)})\, d_{\mathfrak{G}}(F - y_j^{(i)})$$

$$\overset{(49)\,(26)}{=} \sum_{i,j} \operatorname{sgn} \det P'(y_j^{(i)})\, d_{\mathfrak{G}}(F - y_i) \overset{(49)}{=} \sum_{i=1}^{N} d_{\mathfrak{U}_i}(G)\, d_{\mathfrak{G}}(F - y_i),\ y_i \in \mathfrak{U}_i .$$

Besitzt die Komponente $\mathfrak{U}_i$ einen Punkt $\tilde{y}_i \in \mathfrak{U}_i$ mit $\tilde{y}_i \notin F(\mathfrak{G})$, dann ist nach A_2 a) und (26) $d_{\mathfrak{G}}(F - y_i) = d_{\mathfrak{G}}(F - \tilde{y}_i) = 0$ für alle $y_i \in \mathfrak{U}_i$. Wegen dieser Bemerkung, und wegen (49′) gilt dann tatsächlich (43), q.e.d.

§ 9. Homöomorphismen und die topologische Invarianz des Abbildungsgrades

Der Produktsatz von Leray (Theorem 4d) stellt das geeignete Mittel zur Untersuchung von Homöomorphismen dar.

Theorem 5. *(Abbildungsgrad von Homöomorphismen).*

a) *Signum eines Homöomorphismus: Gegeben sei der Homöomorphismus* $U: \overline{\mathfrak{G}} \xrightarrow[\text{auf}]{\text{homöomorph}} U\overline{\mathfrak{G}}$, $\mathfrak{G}$ *zusammenhängend.*

Für einen beliebigen Punkt $x_i \in \mathfrak{G}$ *gilt*

$$d_{x_i}\big(U(x) - U(x_i)\big) = d_{\mathfrak{G}}\big(U(x) - U(x_i)\big) = \text{const} . \tag{54}$$

Diese von x_i *unabhängige Konstante bezeichnen wir als Signum* $\operatorname{sgn}_{\mathfrak{G}} U$ *des Homöomorphismus.*

Das Signum besitzt folgende Eigenschaften. Für

$$\overline{\mathfrak{G}} \xrightarrow[\text{homöomorph}]{U} U\overline{\mathfrak{G}} \xrightarrow[\text{homöomorph}]{V} VU\overline{\mathfrak{G}}$$

gilt

$$\operatorname{sgn}_{\mathfrak{G}} V\big(U(x)\big) = \operatorname{sgn}_{\mathfrak{G}} U \operatorname{sgn}_{U\mathfrak{G}} V . \tag{55}$$

Speziell ist nach (32) *für einen linearen Operator* L *mit* $\det L \neq 0$

$$\operatorname{sgn}_{\mathfrak{G}} L = \operatorname{sgn} \det L , \tag{56}$$

also auch

$$\operatorname{sgn}_{\mathfrak{G}} U \cdot \operatorname{sgn}_{U\mathfrak{G}} U^{-1} = \operatorname{sgn}_{\mathfrak{G}} I = 1 . \tag{57}$$

Folglich gilt nach A_1: $\operatorname{sgn}_{\mathfrak{G}} U = \pm 1$.

b) *Verpflanzungssatz: Gegeben seien $F \in C(\overline{\mathfrak{G}})$ ($\mathfrak{G}$ zusammenhängend) und die Homöomorphismen*

$$U\colon \overline{\mathfrak{G}} \subset R^n \xrightarrow[\text{in}]{\text{homöomorph}} R^n\,, \qquad V\colon R^n \xrightarrow[\text{in}]{\text{homöomorph}} R^n\,, \qquad V(0) = 0\,.$$

Dann gilt

$$d_{U\mathfrak{G}}\big(VF(U^{-1}(x))\big) = \operatorname{sgn}_{\mathfrak{G}} U \operatorname{sgn}_{R^n} V d_{\mathfrak{G}}\big(F(x)\big)\,, \tag{58}$$

wobei $F_1(x) := F\big(U^{-1}(x)\big)$ die aus $F(x)$ durch Verpflanzung entstehende Abbildung ist, d. h., das folgende Diagramm ist kommutativ

$$\begin{array}{ccccc} \overline{\mathfrak{G}} & \xrightarrow{F} & F(\overline{\mathfrak{G}}) & \xrightarrow{V} & VF(\overline{\mathfrak{G}})\,. \\ {\scriptstyle U}\searrow & & \nearrow{\scriptstyle F_1} & & \\ & U\overline{\mathfrak{G}} & & & \end{array} \tag{59}$$

Aus (58) *folgt insbesondere, daß der Abbildungsgrad bis auf das Vorzeichen, das von der Orientierung abhängt, eine topologische Invariante darstellt (Brouwer* 1912 [16]).

Beweis. Zu a). Da $U\mathfrak{G}$ zusammenhängend ist, folgt (54) aus Theorem 2c). Nach Theorem 4d) gilt

$$d_{\mathfrak{G}}\big(V(U(x)) - V(U(x_1))\big) = d_{\mathfrak{G}}\big(U(x) - U(x_1)\big)\, d_{U\mathfrak{G}}\big(V(y) - V(U(x_1))\big)\,,$$

woraus sich unmittelbar (55) ergibt.

Zu b). Wegen (59) folgt aus Theorem 4d)

$$d_{\mathfrak{G}}(F) = d_{\mathfrak{G}}\big(U - U(x_1)\big)\, d_{U\mathfrak{G}}(F_1) = \operatorname{sgn}_{\mathfrak{G}} U\, d_{U\mathfrak{G}}(F_1)\,.$$

Damit ist zunächst (58) für $V = I$ bewiesen. Um den allgemeinen Fall zu erledigen, nehmen wir zunächst $0 \in F(\mathfrak{G})$ an und wählen diejenige Komponente $\mathfrak{U}$ von $R^n - F(\partial\mathfrak{G})$ mit $0 \in \mathfrak{U}$. Dann gilt nach dem schon bewiesenen Spezialfall unter Beachtung von Theorem 4d) und $V(0) = 0$

$$\begin{aligned} d_{U\mathfrak{G}}\big(VF(U^{-1}(x))\big) &= \operatorname{sgn}_{\mathfrak{G}} U d_{\mathfrak{G}}\big(VF(x)\big) \\ &= \operatorname{sgn}_{\mathfrak{G}} U d_{\mathfrak{G}}(F)\, d_{\mathfrak{U}}(V) \\ &= \operatorname{sgn}_{\mathfrak{G}} U d_{\mathfrak{G}}(F) \operatorname{sgn}_{R^n} V\,. \end{aligned}$$

Im Fall $0 \notin F(\overline{\mathfrak{G}})$ ist auch $0 \notin VF(\overline{\mathfrak{G}}) = VF_1(U\overline{\mathfrak{G}})$, also nach A_2 a) $d_{\mathfrak{G}}(F) = d_{U\mathfrak{G}}\big(VF(U^{-1}(x))\big) = 0$, q. e. d.

Aus dem Produktsatz von Leray folgt nach Leray 1950 [66] (vgl. auch J. T. Schwartz [91]) ferner der

Satz 7. *(Trennungssatz von Brouwer* 1912 [15]*). Sind die beiden kompakten Mengen $\mathfrak{M}, \mathfrak{N} \subseteq R^n$ homöomorph mit*

$$U\colon \mathfrak{M} \subset R^n \xrightarrow[\text{auf}]{\text{homöomorph}} \mathfrak{N} \subset R^n\,,$$

dann besitzen $R^n - \mathfrak{M}$ und $R^n - \mathfrak{N}$ die gleiche (möglicherweise unendliche) Anzahl von Komponenten.

Für den R^2 ist der von Jordan 1893 [53] stammende Jordansche Kurvensatz ein Spezialfall des Trennungssatzes.

Beweis. Wir erweitern U zu F und U^{-1} zu G, also

$$F\colon R^n \to R^n\,, \qquad \mathfrak{M}\colon F(x) = U(x)$$
$$G\colon R^n \leftarrow R^n\,, \qquad \mathfrak{N}\colon G(y) = U^{-1}(y)$$

und bezeichnen mit $\mathfrak{M}_i$, $\mathfrak{N}_k$, $\mathfrak{N}_i^{(l)}$ die beschränkten Komponenten der folgenden Mengen

$$R^n - \mathfrak{M}\colon \mathfrak{M}_i\,; \quad R^n - \mathfrak{N}\colon \mathfrak{N}_k\,; \qquad R^n - F(\partial\mathfrak{M}_i)\colon \mathfrak{N}_i^{(l)}\,.$$
$$i = 1, \ldots, n'\,, \qquad k = 1, \ldots, m'\,.$$

Dabei sind auch abzählbare Indexmengen zugelassen, kurz „$n' = \infty$" oder „$m' = \infty$". Die eindeutig festgelegten unbeschränkten Komponenten bezeichnen wir durch $\tilde{\mathfrak{M}}$, $\tilde{\mathfrak{N}}$, $\tilde{\mathfrak{N}}_i$.

Aus $\partial\mathfrak{M}_i \subseteq \mathfrak{M}$ folgt $F(\partial\mathfrak{M}_i) \subseteq F(\mathfrak{M}) = \mathfrak{N}$, also

$$R^n - \mathfrak{N} \subseteq R^n - F(\partial\mathfrak{M}_i)\,,$$

somit ist auf Grund des Zusammenhangs jede Menge $\mathfrak{N}_k$ Teilmenge einer Komponente $\mathfrak{N}_i^{(l)}$ oder $\tilde{\mathfrak{N}}_i$. Für die Mengen

$$\mathfrak{A}_i^{(l)} := \overline{\mathfrak{N}}_i^{(l)} - \bigcup_{\mathfrak{N}_k \subseteq \mathfrak{N}_i^{(l)}} \mathfrak{N}_k$$

gilt $\mathfrak{A}_i^{(l)} \subseteq \mathfrak{N}$, also $G(\mathfrak{A}_i^{(l)}) \subseteq \mathfrak{M}$, folglich ist $\mathfrak{A}_i^{(l)}\colon G(y) \neq x_j$, $x_j \in \mathfrak{M}_j$.

Wegen der Kompaktheit von $\overline{\mathfrak{N}}_i^{(l)}$ besitzt die Gleichung $\mathfrak{N}_k\colon G(y) = x_j$ nur für endliche viele $\mathfrak{N}_k \subseteq \mathfrak{N}_i^{(l)}$ eine Lösung, sonst könnte man eine Teilfolge $y_{k'} \to y$ für $k' \to \infty$ konstruieren mit $\mathfrak{N}_{k'}\colon G(y_{k'}) = x_j$, also $G(y) = x_j$, während andererseits auf Grund der paarweisen Disjunktheit der offenen Mengen $\mathfrak{N}_{k'}$ im Widerspruch dazu $y \in \mathfrak{A}_i^{(l)}$ sein müßte. Nach Theorem 4e) und A_2 a) gilt dann

$$\sum_{\mathfrak{N}_k \subseteq \mathfrak{N}_i^{(l)}} d_{\mathfrak{N}_k}\bigl(G(y) - x_j\bigr) = d_{\mathfrak{N}_i^{(l)}}\bigl(G(y) - x_j\bigr)\,. \tag{60}$$

Aus $\partial\mathfrak{M}_i \subseteq \mathfrak{M}$ ergibt sich ferner

$$\partial\mathfrak{M}_i\colon G\bigl(F(x)\bigr) - x_j = I(x) - x_j\,, \qquad x_j \in \mathfrak{M}_j\,.$$

Nach A_2 b) und nach dem Produktsatz (Theorem 4d)) ist somit

$$\begin{aligned}\delta_{ij} &= d_{\mathfrak{M}_i}\bigl(G(F(x)) - x_j\bigr)\\ &= \sum_l d_{\mathfrak{M}_i}\bigl(F(x) - y_i^{(l)}\bigr)\, d_{\mathfrak{N}_i^{(l)}}\bigl(G(y) - x_j\bigr)\,.\end{aligned}$$

Daraus folgt mit $y_k \in \mathfrak{N}_k$, $x_j \in \mathfrak{M}_j$, $y_i^{(l)} \in \mathfrak{N}_i^{(l)}$, $\tilde{y}_i \in \tilde{\mathfrak{N}}_i$ unter Beachtung von

$$\mathfrak{N}_k \subseteq \tilde{\mathfrak{N}}_i\colon d_{\mathfrak{M}_i}\bigl(F(x) - y_k\bigr) = d_{\mathfrak{M}_i}\bigl(F(x) - \tilde{y}_i\bigr) = 0$$

(vgl. Theorem 2c), A_2 a)) sofort

$$\begin{aligned}
&\sum_k d_{\mathfrak{M}_i}(F(x) - y_k)\, d_{\mathfrak{N}_k}(G(y)) - x_j) \\
&= \sum_l{}' \sum_{\mathfrak{N}_k \subseteq \mathfrak{N}_i^{(l)}} d_{\mathfrak{M}_i}(F(x) - y_k)\, d_{\mathfrak{N}_k}(G(y) \quad x_j) \\
&\overset{(26)}{=} \sum_l d_{\mathfrak{M}_i}(F(x) - y_i^{(l)}) \sum_{\mathfrak{N}_k \subseteq \mathfrak{N}_i^{(l)}} d_{\mathfrak{N}_k}(G(y) - x_j) \\
&\overset{(60)}{=} \sum_l d_{\mathfrak{M}_i}(F(x) - y_i^{(l)})\, d_{\mathfrak{N}_i^{(l)}}(G(y) - x_j) = \delta_{ij} \,. \qquad (61)
\end{aligned}$$

Durch Vertauschen von F mit G entsteht andererseits

$$\sum_k d_{\mathfrak{N}_i}(G(y) - x_k)\, d_{\mathfrak{M}_k}(F(x) - y_j) = \delta_{ij} \,. \qquad (62)$$

Wir betrachten nunmehr die beiden Matrizen

$$A := (d_{\mathfrak{M}_i}(F - y_k)) \,, \qquad B: = (d_{\mathfrak{N}_k}(G(y) - x_j))$$
$$i, j = 1, \dots, n' \,, \qquad k = 1, \dots, m' \,.$$

Aus (61), (62) erhält man dann

$$AB = I_{n'} \,, \qquad BA = I_{m'}$$

mit

$$A: R^{m'} \to R^{n'} \,, \quad B: R^{n'} \to R^{m'} \,, \quad I_{n'}: R^{n'} \xrightarrow{\text{Identität}} R^{n'} \,.$$

Aus

$$\begin{aligned}
n' &= \dim I_{n'}(R^{n'}) = \dim AB(R^{n'}) \leqq m' \\
m' &= \dim I_{m'}(R^{m'}) = \dim BA(R^{m'}) \leqq n'
\end{aligned}$$

folgt nun aber $m' = n'$. Das gilt auch für „$m' = \infty$", „$n' = \infty$", q. e. d.

§ 10. Unwesentliche Abbildungen und der Satz von Hopf

Definition 2. (Unwesentliche Abbildung). Eine Abbildung

$$F: \mathfrak{M} \subset R^n \xrightarrow{\text{stetig}} R^m \,, \qquad \mathfrak{M}: F(x) \neq 0 \,, \qquad \mathfrak{M} \text{ kompakt} \,,$$

heißt genau dann unwesentlich bezüglich $\tilde{\mathfrak{M}} \supset \mathfrak{M}$, wenn es eine nullstellenfreie, stetige Erweiterung $\tilde{F}$ von F auf $\tilde{\mathfrak{M}}$ gibt mit

$$\tilde{F}: \tilde{\mathfrak{M}} \subset R^n \xrightarrow{\text{stetig}} R^m \,, \qquad \tilde{\mathfrak{M}}: \tilde{F}(x) \neq 0 \,, \qquad \tilde{\mathfrak{M}} \text{ kompakt} \,.$$

Existiert keine derartige Erweiterung, dann heißt die Abbildung wesentlich bezüglich $\tilde{\mathfrak{M}}$.

Satz 8. *(Unwesentliche Abbildungen).* a) *Homotopieerweiterungssatz von Borsuk* 1936 [11]*: Gegeben seien zwei homotope Abbildungen*

$$F, G: \mathfrak{M} \subset R^n \xrightarrow{\text{stetig}} R^m \,, \qquad \mathfrak{M}: F \simeq G \,,\; F(x), G(x) \neq 0 \,, \qquad \mathfrak{M} \textit{ kompakt.}$$

Ist dann G bezüglich der kompakten Menge $\tilde{\mathfrak{M}} \supset \mathfrak{M}$ unwesentlich, dann gilt das auch für F, und die Erweiterungen sind ebenfalls homotop, d. h. $\tilde{\mathfrak{M}}: \tilde{F} \simeq \tilde{G}$.

b) *Dimensionsdefekt: Jede Abbildung*

$$F: \mathfrak{M} \subset R^n \xrightarrow{\text{stetig}} R^m, \qquad \mathfrak{M}: F(x) \neq 0, \qquad \mathfrak{M} \text{ kompakt}$$

ist für $n < m$ bezüglich jeder kompakten Menge $\tilde{\mathfrak{M}} \subset R^n$ unwesentlich.

Beweis. Zu a). Sei $H(x, t)$ die zu $\mathfrak{M}: F \simeq G$ zugehörige Homotopieabbildung mit $H(x, 1) = F$, $H(x, 0) = G$, $\mathfrak{M}: H(x, t) \neq 0$, $0 \leqq t \leqq 1$. Wir konstruieren

$$H_1(x, t) = \begin{cases} \tilde{G}(x), & x \in \tilde{\mathfrak{M}}, \quad t = 0 \\ H(x, t), & x \in \mathfrak{M}, \quad 0 \leqq t \leqq 1 \end{cases}$$

und erweitern $H_1(x, t)$ stetig auf $\tilde{\mathfrak{M}} \times \langle 0, 1 \rangle$ zu $\tilde{H}_1(x, t)$. Mit Hilfe der kompakten Menge

$$\tilde{\mathfrak{M}}_1 := \{x \in \tilde{\mathfrak{M}}: \tilde{H}_1(x, t) = 0 \quad \text{für ein } t \text{ mit } 0 \leqq t \leqq 1\}$$

konstruieren wir ferner

$$\lambda(x) = \begin{cases} 0, & x \in \tilde{\mathfrak{M}}_1 \\ 1, & x \in \mathfrak{M} \end{cases}$$

und erweitern $\lambda(x)$ nach (4′) stetig auf $\tilde{\mathfrak{M}}$ zu $\tilde{\lambda}(x)$ mit $0 \leqq \tilde{\lambda}(x) \leqq 1$. Setzen wir nun

$$\tilde{F}(x) := \tilde{H}(x, 1), \quad \tilde{H}(x, t) := H_1(x, \tilde{\lambda}(x)\, t), \quad x \in \tilde{\mathfrak{M}}, \quad 0 \leqq t \leqq 1,$$

dann gilt $\tilde{\mathfrak{M}}: \tilde{F}(x) \neq 0$, und wegen $\tilde{H}(x, 0) = \tilde{G}(x)$, $\tilde{\mathfrak{M}}: \tilde{H}(x, t) \neq 0$ ist $\tilde{\mathfrak{M}}: \tilde{F} \simeq \tilde{G}$.

Zu b). Wir erweitern zunächst F in beliebiger stetiger Weise auf $\tilde{\mathfrak{M}}$. Die Bildmenge $F(\tilde{\mathfrak{M}})$ besitzt keine inneren Punkte (vgl. 2. Schritt im Beweis zu Satz 2). Folglich gibt es ein x_1 mit hinreichend kleiner Norm $\|x_1\|$, so daß gilt $\tilde{\mathfrak{M}}: F(x) - x_1 \neq 0$, $\mathfrak{M}: F \simeq F - x_1$. Aus a) folgt die Existenz einer nullstellenfreien Erweiterung $\tilde{F}$ von F. q. e. d.

Theorem 6. *(Abbildungsklassen).* a) *Charakterisierung unwesentlicher Abbildungen: Sei $\mathfrak{G}$ ein beschränktes Gebiet des R^n, dann ist die Abbildung*

$$F: \partial\mathfrak{G} \subset R^n \xrightarrow{\text{stetig}} R^n, \qquad \partial\mathfrak{G}: F(x) \neq 0$$

genau dann unwesentlich bezüglich $\overline{\mathfrak{G}}$, wenn $d_{\mathfrak{G}}(F) = 0$ gilt.

b) *Satz von Hopf* 1927 [48]: *Im R^n, $n \geqq 2$ gilt für eine kompakte Menge $\overline{\mathfrak{K}}$, die zu einer Kugel homöomorph ist:*
$d_{\mathfrak{K}}(F) = d_{\mathfrak{K}}(G)$ ist äquivalent zu $\partial\mathfrak{K}: F \simeq G$.
Zu jeder ganzen Zahl m existiert eine Abbildung $F \in C(\overline{\mathfrak{K}})$ mit $d_{\mathfrak{K}}(F) = m$. Folglich charakterisiert m die Abbildungsklassen (Homotopieklassen).

c) *Abbildungsklassen im R^1: Sei $\overline{\mathfrak{K}}$ im R^1 das abgeschlossene Intervall $\langle a, b \rangle$, und sei $F(x)$ eine für $a \leqq x \leqq b$ stetige Funktion mit $F(a)$, $F(b) \neq 0$, also $F \in C(\overline{\mathfrak{K}})$,* *dann gilt*

$$d_{\mathfrak{R}}(F) = \begin{cases} 0\,, & F(a)F(b) > 0\,, \\ 1\,, & F(a) < 0\,, \quad F(b) > 0 \\ -1\,, & F(a) > 0\,, \quad F(b) < 0\,. \end{cases}$$

Wie die beiden Funktionen $F_{\pm}(x) = \text{const.} = \pm 1$ *zeigen, gilt* b) *nicht im* R^1.

Für $n = 2$ geht Aussage b) auf Brouwer 1912 [17], 1913 [18] zurück.

Wir werden a) durch ähnliche Überlegungen wie beim Beweis des Eindeutigkeitssatzes (Satz 3) unter Heranziehung von Satz 8 erhalten; b) ergibt sich dann nach Hopf [48] unmittelbar aus a).

Beweis. Zu a). Nach A_2 a) und Theorem 2a) (Randwerteigenschaft) ist $d_{\mathfrak{G}}(F) = 0$ notwendig für eine bezüglich $\overline{\mathfrak{G}}$ unwesentliche Abbildung. Die Hinlänglichkeit dieser Bedingung beweisen wir in mehreren Schritten.

1. Schritt: Approximation von $F \in C(\overline{\mathfrak{G}})$, $d_{\mathfrak{G}}(F) = 0$. Nach Satz 2a) existiert ein $G \in C_e^{(1)}(\overline{\mathfrak{G}})$ mit

$$\partial\mathfrak{G}\colon F \cong G\,, \qquad 0 = d_{\mathfrak{G}}(G) = \sum_{i=1}^{k} \operatorname{sgn}\det G'(x_i)\,, \qquad G(x_i) = 0\,. \tag{64}$$

Folglich gibt es zu jeder „positiven" Nullstelle x_+ eine dazugehörige „negative" Nullstelle x_- mit $G(x_{\pm}) = 0$, $\operatorname{sgn}\det G'(x_{\pm}) = \pm 1$. Unser Ziel ist es, eine Abbildung $\tilde{G} \in C(\overline{\mathfrak{G}})$ zu konstruieren mit

$$\partial\mathfrak{G}\colon G \cong \tilde{G}\,, \qquad \overline{\mathfrak{G}}\colon \tilde{G}(x) \neq 0\,. \tag{65}$$

Die Unwesentlichkeit von F bezüglich $\overline{\mathfrak{G}}$ ergibt sich dann sofort aus Satz 8a). Sei vorerst $n \geqq 3$. Wir werden $\tilde{G}$ aus G durch Beseitigung der Nullstellenpaare $\{x_+, x_-\}$ konstruieren („Zerstrahlung der Elektron-Positron-Paare").

2. Schritt: Hantelkonstruktion. Wir nehmen zunächst an, daß G in $\mathfrak{G}$ nur zwei Nullstellen $x_{\pm} \in \mathfrak{G}$ besitzt, die sich überdies in $\mathfrak{G}$ durch ein Geradenstück verbinden lassen, o. E. d. A. sei $x_{\pm} = (\pm 1, 0, \ldots, 0)$.

Wir konstruieren die abgeschlossene „Hantel" $\overline{\mathfrak{H}} = \overline{\mathfrak{Z}} \cup \overline{\mathfrak{K}}_+ \cup \overline{\mathfrak{K}}_-$, die aus dem „Kreiszylinder" $\overline{\mathfrak{Z}}$ und den „gekappten" Kugeln $\overline{\mathfrak{K}}_{\pm}$ mit $x_{\pm} \in \mathfrak{K}_{\pm}$ besteht (Abb. 1a)).

3. Schritt: Eine spezielle Abbildung. Parallel zu (14) wählen wir die speziellen Abbildungen

$$G_0(x) := H_0(x, 0)\,, \quad G_1(x) = H_0(x, 1)\,, \quad H_0(x, t) := (\xi_1^2 + 1 - 2t, \xi_2, \ldots, \xi_n)\,.$$

Offensichtlich gilt

$$\partial\mathfrak{H}\colon G_0 \cong G_1\,, \qquad \overline{\mathfrak{H}}\colon G_0(x) \neq 0\,. \tag{66}$$

$G_1(x)$ besitzt genau in $x_{\pm}$ Nullstellen und zwar mit $\operatorname{sgn}\det G_1'(x_{\pm}) = \pm 1$. Für hinreichend kleine Kugeln $\mathfrak{K}_{\pm}$ gilt (vgl. 1. Schritt und 2. Schritt des Beweises zu Satz 3)

$$\partial\mathfrak{K}_{\pm}\colon G \cong G_1, \quad H_{\pm}(x, \tfrac{1}{3}) = G(x)\,, \quad H_{\pm}(x, \tfrac{2}{3}) = G_1(x)\,, \tag{67}$$

$H_{\pm}(x, t)$ sei die zugehörige Homotopieabbildung mit veränderter t-Skala.

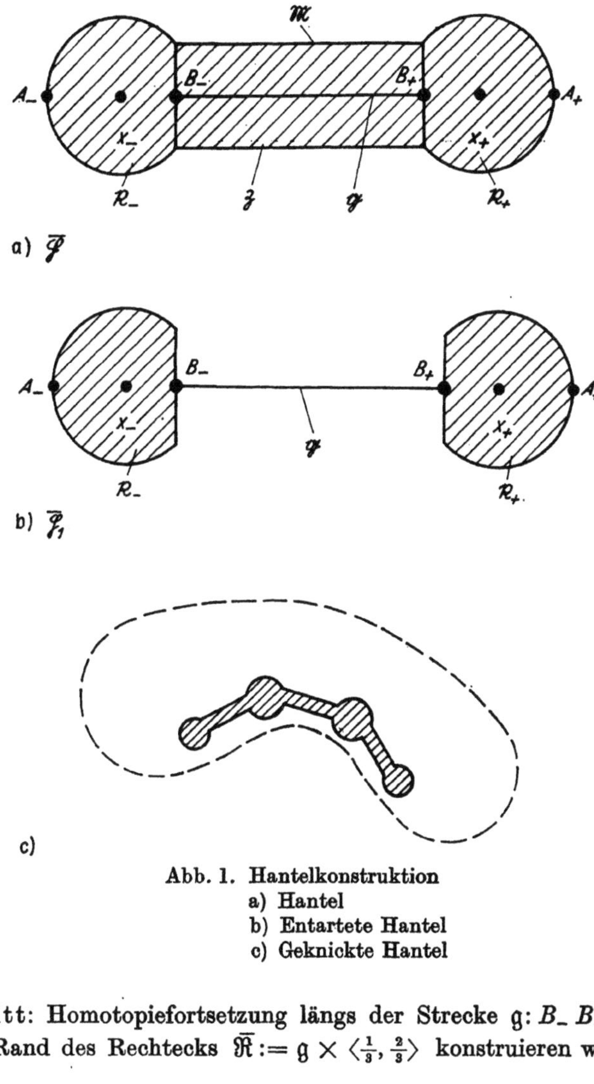

Abb. 1. Hantelkonstruktion
a) Hantel
b) Entartete Hantel
c) Geknickte Hantel

4. Schritt: Homotopiefortsetzung längs der Strecke $\mathfrak{g}: B_- B_+$ (Abb. 1a)). Auf dem Rand des Rechtecks $\overline{\mathfrak{R}} := \mathfrak{g} \times \langle \frac{1}{3}, \frac{2}{3} \rangle$ konstruieren wir die stetige Abbildung

$$H_1(x, t) = \begin{cases} H_\pm(x, t)\,, & x = x_{B_\pm}\,, \quad \frac{1}{3} \leqq t \leqq \frac{2}{3} \\ G(x)\,, & x \in \mathfrak{g}\,, \quad t = \frac{1}{3} \\ G_1(x)\,, & x \in \mathfrak{g}\,, \quad t = \frac{2}{3} \end{cases}$$

und erweitern $H_1(x, t)$ nach Satz 8b) zu einer nullstellenfreien, stetigen Abbildung $\tilde{H}_1(x, t)$ von $\overline{\mathfrak{R}}$ in R^n. Das ist wegen $n \geqq 3$ möglich.

5. Schritt: Homotopiefortsetzung von (67) auf $\partial\mathfrak{H}$. Wir betrachten nunmehr im Zeitintervall $0 \leqq t \leqq \frac{1}{3}$ eine stetige Deformation des Hantelrandes $\partial\mathfrak{H}$

in den Rand $\partial \mathfrak{H}_1$ der entarteten Hantel $\overline{\mathfrak{H}}_1 = \mathfrak{g} \cup \overline{\mathfrak{K}}_+ \cup \overline{\mathfrak{K}}_-$ (Abb. 1b)) mit den Bahnkurven x_t, also $x_0 \in \partial \mathfrak{H}$, $x_{1/3} \in \partial \mathfrak{H}_1$, wobei die Punkte $A_\pm$ festbleiben (Einschnürung). Diese Deformation werde so konstruiert, daß sich die Punkte der Mantelfläche $\mathfrak{M}$ von $\mathfrak{Z}$ radial auf $\mathfrak{g}$ zubewegen und die Punkte von $\partial \mathfrak{K}_+ \cap \partial \mathfrak{H}$ Großkreisbögen von $\partial \mathfrak{K}_\pm$ und radiale Strahlen der Deckflächen von $\mathfrak{Z}$ durchlaufen.

Die gesuchte Homotopie ergibt sich nun aus

$$\partial \mathfrak{H}\colon G \cong G_1\,, \quad H(x, t) := \begin{cases} G(x_t)\,, & x \in \partial \mathfrak{H}\,, \quad 0 \leqq t \leqq \frac{1}{3} \\ H_\pm(x_{1/3}, t)\,, & x \in \partial \mathfrak{K}_\pm \cap \partial \mathfrak{H}\,, \quad \frac{1}{3} \leqq t \leqq \frac{2}{3} \\ \tilde{H}_1(x_{1/3}, t)\,, & x \in \mathfrak{M}\,, \quad \frac{1}{3} \leqq t \leqq \frac{2}{3} \\ G_1(x_{1-t})\,, & x \in \partial \mathfrak{H}\,, \quad \frac{2}{3} \leqq t \leqq 1\,. \end{cases} \tag{68}$$

Dabei besitzt $H(x, t)$ folgende anschauliche Bedeutung:

α) Zeitraum $0 \leqq t \leqq \frac{1}{3}$; stetige Deformation von $\partial \mathfrak{H}$ in $\partial \mathfrak{H}_1$ längs der Bahnkurven x_t, wobei die entsprechenden Bildpunkte von G durchlaufen werden,

β) Zeitraum $\frac{1}{3} \leqq t \leqq \frac{2}{3}$; G wird stetig in G_1 überführt vermöge der Homotopieabbildungen $H_\pm(x, t)$, $\tilde{H}_1(x, t)$,

γ) Zeitraum $\frac{2}{3} \leqq t \leqq 1$; Deformation von $\partial \mathfrak{H}$ in $\partial \mathfrak{H}_1$ wird rückgängig gemacht, wobei die entsprechenden Bildpunkte von G_1 durchlaufen werden.

Nach (68), (66) gilt $\partial \mathfrak{H}\colon G \cong G_0$, $\overline{\mathfrak{H}}\colon G_0(x) \neq 0$, folglich existiert nach Satz 8a) eine stetige Abbildung $\tilde{G}$ mit $\partial \mathfrak{H}\colon G = \tilde{G}$, $\overline{\mathfrak{H}}\colon \tilde{G}(x) \neq 0$. Somit sind die Nullstellen $x_\pm$ beseitigt worden und durch $\overline{\mathfrak{G}} - \mathfrak{H}\colon \tilde{G}(x) := G(x)$ entsteht die gesuchte Abbildung $\tilde{G}$ in (65).

Kann man das Nullstellenpaar $\{x_+, x_-\}$ nur durch einen Streckenzug in $\mathfrak{G}$ verbinden, dann benutze man eine „geknickte Hantel" mit Kugelgelenken (Abb. 1c)). Da diese homöomorph zu einer gewöhnlichen Hantel ist, verlaufen die Überlegungen analog zu den vorangegangenen. Sind mehrere Nullstellenpaare vorhanden, dann werden diese getrennt beseitigt.

6. Schritt: Konstruktion von $\tilde{G}$ in (65) für $n = 1$. Da $\mathfrak{G}$ zusammenhängend ist, gilt $\mathfrak{G} = \{x; a < x < b\}$. Sei $G \in C_e^{(1)}(\mathfrak{G})$. Wegen des paarweisen Auftretens der Nullstellen ist $G(a)\, G(b) > 0$, und die Konstruktion von $\tilde{G}$ liegt auf der Hand.

7. Schritt: Konstruktion von $\tilde{G}$ in (65) für $n = 2$. Sei $\mathfrak{G}$ zunächst der Einheitskreis mit dem Mittelpunkt in Null und $\partial \mathfrak{G} = \{(\xi_1, \xi_2);\ \xi_1 = \cos\varphi,\ \xi_2 = \sin\varphi\}$. Dann besitzt $G \in C(\overline{\mathfrak{G}})$, $d_{\mathfrak{G}}(G) = 0$ die Gestalt

$$\partial \mathfrak{G}\colon G = ||G(x)|| \left(\cos\psi(\varphi), \sin\psi(\varphi)\right) := ||G(x)||\; G_1(x)\,.$$

$\psi(\varphi)$ ist eine stetige Funktion von φ, $0 \leqq \varphi \leqq 2\pi$. Wegen

$$\cos\psi(0) = \cos\psi(2\pi),\ \sin\psi(0) = \sin\psi(2\pi) \quad \text{gilt} \quad \psi(2\pi) = 2\pi k + \psi(0)$$

(k ganze Zahl). Aus

$$\partial \mathfrak{G}\colon H_k(x, t) := \left(\cos\{(1-t)\,\psi(\varphi) + tk\varphi\},\quad \sin\{(1-t)\,\psi(\varphi) + tk\varphi\}\right) \neq 0$$

(man beachte die Eindeutigkeit von $H_k(x, t)$ auf $\partial\mathfrak{G}$) und Theorem 2e) folgt somit $\partial\mathfrak{G}$: $G \cong G_1 \cong H_k(x, 1)$, also $0 = d_{\mathfrak{G}}(G) = d_{\mathfrak{G}}(H_k(x, 1))$. Nach Theorem 10a) ist $d_{\mathfrak{G}}(H_k(x, 1)) = k$ im Falle $k \neq 0$, folglich muß $k = 0$ gelten. Die in (65) gesuchte Abbildung $\tilde{G}$ ergibt sich somit durch $\tilde{G} := H_0(x, 1) = (1, 0)$.

Ist $\mathfrak{G}$ kein Kreis, dann betrachten wir wie im 5. Schritt geknickte Hanteln $\mathfrak{H}$ (Abb. 1c)). Wegen des Vorhandenseins genau eines Nullstellenpaares $(x_+, x_-\}$ in $\mathfrak{H}$ gilt $d_{\mathfrak{G}}(G) = 0$. $\mathfrak{H}$ läßt sich homöomorph auf den Einheitskreis $\mathfrak{K}$ abbilden. Nach Theorem 5b) gilt dann für die auf $\mathfrak{K}$ verpflanzte Abbildung $G_{\mathfrak{K}}$ ebenfalls $d_{\mathfrak{K}}(G_{\mathfrak{K}}) = 0$, deshalb kann die oben durchgeführte Überlegung benutzt werden, um das Nullstellenpaar $\{x_+, x_-\}$ zu beseitigen.

Zu b). Nach Theorem 5b) dürfen wir beim Beweis von b) stets annehmen, daß $\mathfrak{K}$ eine Kugel ist.

1. Schritt: Konstruktion der Homotopie. Wegen A_4 folgt aus $\partial\mathfrak{K}$: $F \cong G$ sofort $d_{\mathfrak{K}}(F) = d_{\mathfrak{K}}(G)$.

Um die umgekehrte Behauptung zu beweisen, betrachten wir neben $\mathfrak{K}$ noch eine kleinere konzentrische Kugel $\mathfrak{K}_1 \subset \mathfrak{K}$, die aus $\mathfrak{K}$ durch die Ähnlichkeitstransformation $A: \mathfrak{K} \to \mathfrak{K}_1$ hervorgeht. Durch Verpflanzung entsteht $G_1(x) := G(A^{-1}x)$. Nach Theorem 5b) gilt $d_{\mathfrak{K}_1}(G_1) = d_{\mathfrak{K}}(G)$.

Sei weiter $G_2 \in C(\overline{\mathfrak{K} - \mathfrak{K}_1})$ eine nach (4′) existierende stetige Abbildung mit $\partial\mathfrak{K}_1$: $G_2(x) = G_1(x)$, $\partial\mathfrak{K}$: $G_2(x) = F(x)$. Nach A_3 und Theorem 2a) ist

$$d_{\mathfrak{K}}(G) = d_{\mathfrak{K}}(F) = d_{\mathfrak{K}_1}(G_1) + d_{\mathfrak{K}-\overline{\mathfrak{K}}_1}(G_2)\,, \qquad \text{also } d_{\mathfrak{K}-\overline{\mathfrak{K}}_1}(G_2) = 0\,.$$

Folglich existiert nach Theorem 6a) eine nullstellenfreie, stetige Erweiterung $\tilde{G}_2$ der Randwerte von G_2 auf $\mathfrak{K} - \overline{\mathfrak{K}}_1$.

Nunmehr ergibt sich die gesuchte Homotopie durch

$$\partial\mathfrak{K}\colon F \cong G\,,\ H(x, t) := \tilde{G}_2(x_t),\, H(x, 0) = F(x)\,,\ H(x, 1) = \tilde{G}_2(x_1) = G(x)\,,$$

wenn wir mit x_t die Bahnkurven bezeichnen, die ein Randpunkt $x \in \partial\mathfrak{K}$ durchläuft, der sich im Zeitraum $0 \leqq t \leqq 1$ gleichförmig und radial nach $x_1 \in \partial\mathfrak{K}_1$ bewegt.

2. Schritt: Existenz von Abbildungsklassen für $n \geqq 2$.

Die Abbildung $F_m = (\varrho^m \cos m\varphi, \varrho^m \sin m\varphi, \xi_3, \ldots, \xi_n)$ mit $\xi_1 = \varrho \cos\varphi$, $\xi_2 = \varrho \sin\varphi$, $0 \leqq \varphi \leqq 2\pi$, $0 \leqq \varrho \leqq 1$ im abgeschlossenen Gebiet $\overline{\mathfrak{G}} = \{(\xi_1, \xi_2);\ \xi_1^2 + \xi_2^2 \leqq 1, -1 \leqq \xi_3, \ldots, \xi_n \leqq 1\}$ besitzt nach Theorem 4a), c) und Theorem 10a) sowie nach A_2 b) den Abbildungsgrad $d_{\mathfrak{G}}(F_m) = m$. Da $\overline{\mathfrak{G}}$ homöomorph zur Kugel und somit homöomorph zu $\overline{\mathfrak{K}}$ ist, existiert nach Theorem 5b) zu jedem m eine Abbildung F auf $\overline{\mathfrak{K}}$ mit $d_{\mathfrak{K}}(F) = m$.

Beweis. Zu c). Man approximiere $F \in C(\overline{\mathfrak{G}})$ durch $G \in C_e^{(1)}(\overline{\mathfrak{G}})$, dann folgt (63) aus der Tatsache, daß die Ableitungen zweier benachbarter Nullstellen von G unterschiedliches Vorzeichen besitzen, q. e. d.

Andersgeartete Beweise des Satzes von Hopf findet man in Alexandrow-Hopf 1935 [2], Alexandrow 1947 [1], Hurewicz-Wallmann 1948 [52], Lefschetz 1949 [64] (simpliziale Approximation), Whitney 1937 [102] („Standardabbildungen"), Spanier 1966 [94] (höhere Homotopiegruppen). Der vorliegende Beweis zeichnet sich durch besondere Einfachheit aus.

Für Abbildungen von n-dimensionalen Polyedern in n-dimensionale Sphären gab Hopf 1933 [49] ein vollständiges Invariantensystem an. Whitney 1937 [103] konnte zeigen, daß eine eineindeutige Korrespondenz zwischen den Abbildungsklassen und den Elementen der n-ten Kohomologiegruppe besteht. Eine ausführliche Darstellung hierüber findet man in Hu 1959 [51]. Eine tiefliegende Verallgemeinerung stellt das Periodizitätstheorem von Bott 1959 [12] dar, das Aussagen über die Homotopieklassen von Abbildungen $F: \partial\mathfrak{K}_\varrho^{(n)}(0) \to GL(N, C)$, $N \geqq 2n$ aus Sphären in die lineare Gruppe über dem Körper C der komplexen Zahlen macht (vgl. Atiyah 1967 [3]).

§ 11. Der Antipodensatz von Borsuk und seine Folgerungen

Theorem 7. *(Antipodensatz von Borsuk* 1933 [9]). a) *Spezielle Fassung: Sei* $\mathfrak{K}_\varrho^{(n)}(0)$ *eine Kugel des* R^n *mit dem Mittelpunkt im Nullpunkt und dem Radius* ϱ, *dann gilt für*

$$F \in C\big(\overline{\mathfrak{K}}_\varrho^{(n)}(0)\big)\,, \qquad \|x\| = \varrho\colon F(x) = -F(-x) \tag{69}$$

stets $d_{\mathfrak{K}_\varrho^{(n)}(0)}(F) = 2m + 1$, *$m$ ganze Zahl.*

b) *Allgemeinere Fassung: Die Aussage bleibt richtig, wenn man* (69) *durch*

$$F \in C\big(\overline{\mathfrak{K}}_\varrho^{(n)}(0)\big)\,, \qquad \|x\| = \varrho\colon F(x) \neq tF(-x)\,, \qquad 0 \leqq t \leqq 1 \tag{69'}$$

ersetzt.

Beim Beweis werden wir wesentlich den Verpflanzungssatz (Theorem 5b) und Satz 8b) benutzen.

Beweis. Zu a). Sei etwa $n = 2$. Wir betrachten eine Zerlegung von $\mathfrak{K}_\varrho^{(n)}(0)$ in der Form

$$\overline{\mathfrak{K}}_\varrho^{(n)}(0) = \overline{\mathfrak{K}}_1 \bigcup_i \pm \overline{\mathfrak{A}}_i \qquad \text{(vgl. Abb. 2).} \tag{70}$$

Unser Ziel ist es, die Randwerte von F auf $\partial\mathfrak{K}_\varrho^{(n)}(0)$ so stetig zu $\tilde{F}$ in das Innere fortzusetzen, daß gilt

$$\overline{\mathfrak{K}}_1\colon \tilde{F}(x) = x\,, \quad \partial(\pm\mathfrak{A}_i)\colon \tilde{F}(x) \neq 0\,, \quad d_{\mathfrak{A}_i}(\tilde{F}) = d_{-\mathfrak{A}_i}(\tilde{F})\,. \tag{71}$$

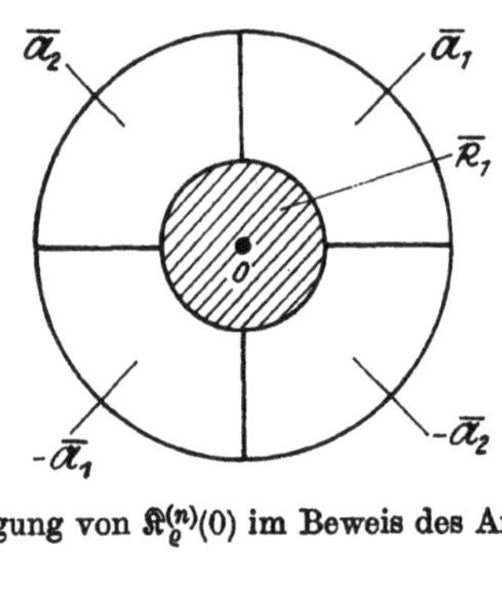

Abb. 2. Zerlegung von $\mathfrak{K}_\varrho^{(n)}(0)$ im Beweis des Antipodensatzes

Dann folgt aus (70), A_3, A_2 b) und Theorem 2a) sofort die Behauptung

$$d_{\mathfrak{K}_\varrho^{(n)}(0)}(F) = d_{\mathfrak{K}_\varrho^{(n)}(0)}(\tilde{F}) = 1 + \sum_i 2d_{\mathfrak{A}_i}(\tilde{F}) = 1 + 2m .$$

Um (71) zu konstruieren, setzen wir zunächst $\overline{\mathfrak{K}_1}$: $\tilde{F}(x) = x$. Damit ist $\tilde{F}$ zugleich auf $\overline{\mathfrak{K}}_1 \cap \partial\mathfrak{A}_1$, $\partial\mathfrak{K}_\varrho^{(n)}(0) \cap \partial\mathfrak{A}_1$ festgelegt. Nach Satz 8b) können wir dann $\tilde{F}$ nullstellenfrei und stetig auf den restlichen Teil von $\partial\mathfrak{A}_1$ fortsetzen; $\tilde{F}(x)$ auf $\overline{\mathfrak{A}}_1$ ergibt sich durch weitere stetige Fortsetzung der Randwerte.

Auf $-\overline{\mathfrak{A}}_1$ erklären wir $\tilde{F}$ durch Spiegelung, also $-\overline{\mathfrak{A}}_1$: $\tilde{F}(x) = -\tilde{F}(-x)$. Nach Theorem 5b) ist

$$d_{-\mathfrak{A}_i}(\tilde{F}) = \operatorname{sgn}_{R^n}\big(-I(x)\big)\operatorname{sgn}_{R^n}\big(-I(x)\big)\, d_{\mathfrak{A}_i}(\tilde{F}) = d_{\mathfrak{A}_i}(\tilde{F}) .$$

Die Fortsetzung auf $\pm\overline{\mathfrak{A}}_2$ geschieht analog.

Offensichtlich ist diese Konstruktion für alle $n \geqq 1$ durchführbar.

Zu b). Gilt (69′), dann ist

$$\partial\mathfrak{K}_\varrho^{(n)}(0)\colon F(x) \cong F(x) - F(-x) , \quad H(x, t) = F(x) - tF(-x) , \quad 0 \leqq t \leqq 1 ,$$

somit nach a): $d_{\mathfrak{K}_\varrho^{(n)}(0)}(F) = d_{\mathfrak{K}_\varrho^{(n)}(0)}\big(F(x) - F(-x)\big) = 2m + 1$. q. e. d.

Satz 9. *(Folgerungen aus dem Antipodensatz).* a) *Satz über die Gebietsinvarianz von Brouwer* 1912 [14]: *Bei einer Abbildung der Form*

$$F\colon \mathfrak{A} \subseteq R^n \xrightarrow[\text{eineindeutig}]{\text{stetig}} F(\mathfrak{A}) \subseteq R^n$$

gehen innere Punkte wieder in innere Punkte über. Ist $\mathfrak{A}$ offen, dann stellt F einen Homöomorphismus dar (vgl. Schauder 1932 [87], *Nagumo* 1951 [74]).

b) *Existenzsatz für quasimonotone Operatoren: Ist die Abbildung*

$$F\colon R^n \xrightarrow{\text{stetig}} R^n$$

quasimonoton, d. h.

$$|\langle F(x) - F(y), x - y\rangle| \geqq \gamma(||x - y||)\, ||x - y|| , \quad x, y \in R^n , \tag{72}$$

$$\lim_{t \to +\infty} \gamma(t) = +\infty , \quad \gamma(t) > 0 \quad \textit{für} \quad t > 0 , \quad \gamma(0) = 0 ,$$

dann ist F ein Homöomorphismus von R^n auf R^n. Die Gleichung $F(x) = y$ ist somit für alle $y \in R^n$ eindeutig lösbar, und die Lösungen hängen stetig von y ab (vgl. Browder 1964 [22], *Wainberg* 1972 [97]).

c) *Borsuk-Ulam Theorem (Borsuk* 1933 [9]*): Zu jeder Abbildung*

$$F\colon \partial\mathfrak{K}_\varrho^{(n+1)}(0) \subset R^{n+1} \xrightarrow{\text{stetig}} R^n$$

existiert ein Punkt $x_0 \in \partial\mathfrak{K}_\varrho^{(n+1)}(0)$ mit $F(x_0) = F(-x_0)$.

d) *Überdeckungssatz von Ljusternik-Schnirelman* 1930 [68] *und Borsuk* 1933 [9]. *Wird die Oberfläche der n-dimensionalen Kugel $\mathfrak{K}_\varrho^{(n)}(x_0)$ durch p abgeschlossene Mengen $\mathfrak{A}_1, \ldots, \mathfrak{A}_p$ überdeckt mit $1 \leqq p \leqq n$, dann enthält wenigstens eines der $\mathfrak{A}_i$ ein Antipodenpaar $\{x, -x\}$. Für $p = n + 1$ gibt es eine Überdeckung, die nicht diese Eigenschaft besitzt.*

e) *Verallgemeinerte Fredholmsche Alternative I: Die Gleichung*

$$L(x) + N(x) = y\,, \qquad R^n \xrightarrow[\text{stetig}]{L,\,N} R^n\,, \qquad L(x) \not\equiv 0 \tag{72'}$$

ist für alle $y \in R^n$ *lösbar, wenn bei festem* $k = 1, 2, \ldots$ *gilt:*

$$L(tx) = t^k L(x)\,, \quad 0 < t < \infty\,, \quad L(-x) = -L(x)\,, \quad L(x) = 0 \Rightarrow x = 0$$

$$||x|| = \varrho\colon \max ||N(x)|| = o(\varrho^k)\,, \qquad \varrho \to \infty$$

(vgl. Hess 1972 [46]*).*

f) *Verallgemeinerte Fredholmsche Alternative II: Der Operator L in* (72') *sei linear, und für die nichtlineare Störung* $N(x)$ *gelte*

$$||x|| = \varrho\colon \max ||N(x)|| = o(\varrho)\,, \quad \varrho \to \infty\,; \qquad N(R^n) \subset (\ker L^*)^\perp$$

mit dem zu L adjungierten Operator L^* *und*

$$\ker L^* = \{x;\ L^*(x) = 0\}\,,\ (\ker L^*)^\perp = \{y;\ \langle x, y\rangle = 0,\ L^*(x) = 0\}\,.$$

Unter diesen Voraussetzungen ist (72') *genau dann lösbar, wenn* $y \in (\ker L^*)^\perp$ *erfüllt ist (vgl. Hess* 1972 [47]*).*

Satz d) stellt eines der Kernstücke der Theorie von Ljusternik-Schnirelman 1930 [68] dar (vgl. Krasnosjelski 1956 [57], Wainberg 1956 [96], Berger 1968 [8], Browder 1970 [24]).

Beweis. Zu a) (vgl. Granas [42]). Sei o. E. d. A. $0 \in \mathfrak{A}$ ein innerer Punkt von $\mathfrak{A}$ mit $F(0) = y_0$. Wegen der Eineindeutigkeit der Abbildung ist

$$||x|| = \varrho\colon F(x) - y_0 \neq 0\,,\ H(x, t) := F\left(\frac{x}{1+t}\right) - F\left(-\frac{tx}{1+t}\right) \neq 0\,,$$
$$0 \leqq t \leqq 1\,,$$

denn aus $H(x_1, t) = 0$ folgt sofort $x_1/(1+t) = -tx_1/(1+t)$, $0 \leqq t \leqq 1$, also $x_1 = 0$. Somit gilt $||x|| = \varrho\colon F(x) - y_0 \cong H(x, 1)$. Wegen $H(x, 1) = -H(-x, 1)$ ist nach dem Antipodensatz $d_{\mathfrak{K}_\varrho^{(n)}(0)}\big(F(x) - y_0\big) = d_{\mathfrak{K}_\varrho^{(n)}(0)}\big(H(x, 1)\big) \neq 0$. Das Kroneckersche Existenzprinzip (Theorem 2d)) ergibt dann sofort die Behauptung.

Zu b). Gilt $F(x) = F(y)$, dann folgt aus (72) sofort $x = y$, die Abbildung ist also eineindeutig. Aus a) ergibt sich dann die Offenheit von $F(R^n)$. Gleichzeitig ist $F(R^n)$ abgeschlossen, also $F(R^n) = R^n$, und nach a) handelt es sich um einen Homöomorphismus. Die Abgeschlossenheit von $F(R^n)$ folgt unmittelbar aus (72), denn sei $F(x_n) \to y_0$ für $n \to \infty$, dann ist nach (72)

$$||F(x_n) - F(x_m)||\ ||x_n - x_m|| \geqq |\langle F(x_n) - F(x_m), x_n - x_m\rangle| \geqq \gamma(||x_n - x_m||)\ ||x_n - x_m||\,,$$

also $||F(x_n) - F(x_m)|| \geqq \gamma(||x_n - x_m||)$, folglich ist wegen $\lim\limits_{t \to +\infty} \gamma(t) = +\infty$ die Menge $\{x_n\}$ beschränkt. Die konvergente Teilfolge $x_{n'} \to x_0$ für $n' \to \infty$ ergibt $F(x_0) = y_0$, also $y_0 \in F(R^n)$.

Zu c). Wir setzen $U := -I$, d. h. $U(x) := -x$. Ist die Behauptung falsch, dann gilt

$$\partial\mathfrak{K}_\varrho^{(n+1)}(0)\colon F_1(x) := \big(f_1(x) - f_1(U(x)), \ldots, f_n(x) - f_n(U(x)), 0\big) \neq 0\,,$$

also nach Theorem 4c) $d_{\mathfrak{K}_\varrho^{(n+1)}(0)}(F_1) = 0$, während der Antipodensatz im Widerspruch dazu $d_{\mathfrak{K}_\varrho^{(n+1)}(0)}(F_1) \neq 0$ liefert.

Zu d). Wir dürfen o. E. d. A. voraussetzen, daß der Mittelpunkt der Kugel im Nullpunkt liegt.

Sei zunächst $1 \leqq p \leqq n$. Durch Hinzufügen von $n - p$ Punkten, die kein Antipodenpaar enthalten, erhält man eine Überdeckung aus n abgeschlossenen Mengen $\mathfrak{A}_1, \dots, \mathfrak{A}_n$, und es genügt somit, die Behauptung für $p = n$ zu beweisen. Angenommen, die Behauptung ist falsch, d. h. $\mathfrak{A}_i \cap -\mathfrak{A}_i = \emptyset$, $i = 1, \dots, n$. Wir setzen $U := -I$, also $U(x) := -x$ und konstruieren durch stetige Erweiterung von

$$\partial\mathfrak{K}_\varrho^{(n)}(0)\colon f_i(x) = \begin{cases} 1\,, & x \in \mathfrak{A}_i\,, \quad i = 1, \dots, n \\ -1\,, & x \in U\mathfrak{A}_i\,, \end{cases}$$

auf $\overline{\mathfrak{K}}_\varrho^{(n)}(0)$ eine stetige Abbildung $F = (f_1, \dots, f_n)$. Unter Berücksichtigung der Überdeckungseigenschaften der $\mathfrak{A}_i$ und $U\mathfrak{A}_i$ folgt aus der Konstruktion der $f_i(x)$

$$\partial\mathfrak{K}_\varrho^{(n)}(0)\colon F(x) \neq (\tau, 0, \dots, 0), \tau \geqq 0\,, \quad F(x) \neq tF(Ux)\,, \quad 0 \leqq t \leqq 1\,.$$

Denn wäre etwa $F(x_1) = (\tau, 0, \dots, 0)$, $\tau \geqq 0$, dann müßte nach Konstruktion $x_1 \notin \mathfrak{A}_i$, $x_1 \notin U\mathfrak{A}_i$, $i = 2, \dots, n$, gelten, also $x_1 \in \mathfrak{A}_1 \cap U\mathfrak{A}_1$ im Widerspruch zu $\mathfrak{A}_1 \cap U\mathfrak{A}_1 = \emptyset$.

Nach dem Lückensatz Satz 6c) gilt dann $d_{\mathfrak{K}_\varrho^{(n)}(0)}(F) = 0$, während aus dem Antipodensatz (69′) die dazu widersprüchliche Aussage $d_{\mathfrak{K}_\varrho^{(n)}(0)}(F) \neq 0$ folgt.

Wir konstruieren nun durch Induktion nach n eine Überdeckung von $\partial\mathfrak{K}_\varrho^{(n)}(0)$ aus abgeschlossenen Mengen $\mathfrak{A}_1, \dots, \mathfrak{A}_{n+1}$ mit $\mathfrak{A}_i \cap -\mathfrak{A}_i = \emptyset$. Für $n = 1$, $n = 2$ existiert offensichtlich eine derartige Überdeckung. Sei nun die Existenz einer solchen Überdeckung bereits für $n - 1$ mit $n \geqq 3$ bewiesen. Wir überdecken den Äquator von $\partial\mathfrak{K}_\varrho^{(n)}(0)$ mit n abgeschlossenen Mengen $\tilde{\mathfrak{A}}_1, \dots, \tilde{\mathfrak{A}}_n$, $\tilde{\mathfrak{A}}_i \cap -\tilde{\mathfrak{A}}_i = \emptyset$ und wählen einen Breitenkreis $\mathfrak{b}$ auf der Südhalbkugel. $\mathfrak{A}_i$ bestehe dann aus allen Großkreisbögen, die von $\mathfrak{b}$ zum Nordpol führen und einen Punkt von $\tilde{\mathfrak{A}}_i$ enthalten; $\mathfrak{A}_{n+1}$ sei die Menge aller Punkte, die (einschließlich $\mathfrak{b}$) südlich von $\mathfrak{b}$ liegen.

Zu e). Aus der Homogenität von L folgt

$$\gamma(L) := \min_{||x|| = \varrho} ||L(x)|| \geqq a\varrho^k\,, \qquad a > 0\,.$$

Bei festem y ist deshalb für hinreichend große ϱ stets

$$||x|| = \varrho\colon ||L(x) - \big(L(x) + N(x) - y\big)|| \leqq o(\varrho^k) + ||y|| < \gamma(L)\,,$$

also gilt nach Satz 1 unter Berücksichtigung des Antipodensatzes

$$d_{\mathfrak{K}_\varrho^{(n)}(0)}\big(L(x) + N(x) - y\big) = d_{\mathfrak{K}_\varrho^{(n)}(0)}\big(L(x)\big) \neq 0\,,$$

woraus sich nach dem Kroneckerschen Existenzprinzip sofort die Behauptung ergibt.

Zu f): Die Bedingung $y \in (\ker L^*)^\perp$ ist wegen $L(R^n) = (\ker L^*)^\perp$ offensichtlich notwendig für die Lösbarkeit von (72′).

Um die Existenz einer Lösung von (72') nachzuweisen, führen wir f) auf e) durch Einführung einer Pseudoresolvente zurück und konstruieren hierzu

$$\overline{L}(x) := L(x) + \sum_{i=1}^{m} \langle x_i, x\rangle\, x_i^* , \quad x_i \in \ker L , \quad x_i^* \in \ker L^* ,$$

wobei $\{x_i\}$ bzw. $\{x_i^*\}$ Basen in ker L bzw. ker L^* darstellen. Aus $\overline{L}(x) = 0$ folgt $x = 0$, denn aus $L(R^n) = (\ker L^*)^\perp$ ergibt sich $\langle x_i, x\rangle = 0$, $i = 1, \dots, m$, also $x \in (\ker L)^\perp$ und $L(x) = 0$, folglich $x = 0$. Somit existiert nach e) eine Lösung des modifizierten Problems $\overline{L}(x) + N(x) = y$. Diese Lösung ist jedoch wegen

$$-\sum_{i=1}^{m} \langle x_i, x\rangle\, x_i^* = L(x) + N(x) - y \in (\ker L^*)^\perp ,$$

d. h. $\langle x_i, x\rangle = 0$, $i = 1, \dots, m$ auch Lösung von (72') $L(x) + N(x) = y$, q. e. d.

§ 12. Der verallgemeinerte Antipodensatz

Theorem 8. *(Verallgemeinerter Antipodensatz).* a) *Spezielle Fassung: $\mathfrak{G}$ sei ein beschränktes Gebiet des R^n, und $\overline{\mathfrak{G}}$ sei invariant bezüglich des Homöomorphismus p-ter Ordnung ($p = 2, 3, \dots$)*

$$U\colon \overline{\mathfrak{G}} \subset R^n \xrightarrow[\text{auf}]{\text{homöomorph}} \overline{\mathfrak{G}} , \qquad U^p(x) = I(x) . \tag{73}$$

Daneben sei noch ein Homöomorphismus q-ter Ordnung gegeben (q Teiler von p, $q = 1, 2, \dots$)

$$V\colon R^n \xrightarrow[\text{auf}]{\text{homöomorph}} R^n , \qquad V^q(x) = I(x) , \qquad V(0) = 0 , \tag{74}$$

und die Abbildungen

$$F_j\colon \partial\mathfrak{G} \subset R^n \xrightarrow{\text{stetig}} R^n , \qquad \partial\mathfrak{G}\colon F_j(x) \neq 0 , \qquad j = 1, 2 \tag{75}$$

mögen die für die folgenden Aussagen grundlegende Invarianzeigenschaft

$$\partial\mathfrak{G}\colon F_j(x) = VF_j(U^{-1}(x)) , \qquad j = 1, 2 \tag{76}$$

besitzen.

Wir setzen $\gamma := \operatorname{sgn}_{\mathfrak{G}} U \operatorname{sgn}_{R_n} V$ (vgl. § 9)*, dann gilt $\gamma = \pm 1$, und aus $\gamma = -1$ folgt, daß p gerade ist. Wir unterscheiden nun zwei Fälle.*

Fall 1: $U, U^2, \dots, U^{p-1}$ besitzen auf $\overline{\mathfrak{G}}$ keine Fixpunkte. Dann gilt

$$\gamma = 1\colon \qquad d_{\mathfrak{G}}(F_j) \equiv 0 \pmod p \tag{77}$$

$$\gamma = -1\colon \qquad d_{\mathfrak{G}}(F_j) = 0 . \tag{78}$$

Außerdem existiert eine stetige Erweiterung $\tilde{F}_j$ von F_j auf $\overline{\mathfrak{G}}$ mit $\overline{\mathfrak{G}}\colon \tilde{F}_j(x) = V\tilde{F}_j(U^{-1}(x))$.

Fall 2: U besitzt auf $\partial\mathfrak{G}$ keine Fixpunkte, in $\mathfrak{G}$ seien genau die m Fixpunkte $x_1, \dots, x_m \in \mathfrak{G}$ vorhanden, und $U^2, \dots, U^{p-1}$ haben auf $\overline{\mathfrak{G}}$ keine weiteren Fixpunkte als diese $x_i \in \mathfrak{G}$. Dann gilt

$$\gamma = 1\colon \qquad d_{\mathfrak{G}}(F_1) \equiv d_{\mathfrak{G}}(F_2) \pmod p \tag{79}$$

$$\gamma = -1\colon \qquad d_{\mathfrak{G}}(F_1) = d_{\mathfrak{G}}(F_2) . \tag{80}$$

Gibt es ferner stetige Abbildungen G_i, $i = 1, \ldots, m$, auf kleinen Kugeln um x_i mit den isolierten Nullstellen x_i und

$$G_i\colon \mathfrak{K}_\varrho^{(n)}(x_i) \xrightarrow{\text{stetig}} R^n\,, \qquad G_i(x) = VG_i(U^{-1}(x))\,,$$

dann ist

$$\gamma = 1: \qquad d_{\mathfrak{G}}(F_j) \equiv \sum_{i=1}^{m} d_{x_i}(G_i) \qquad (\mathrm{mod}\ p) \tag{81}$$

$$\gamma = -1: \qquad d_{\mathfrak{G}}(F_j) = \sum_{i=1}^{m} d_{x_i}(G_i)\,. \tag{82}$$

Ferner existieren stetige Erweiterungen $\tilde{F}_j$ von F_j auf $\overline{\mathfrak{G}}$ mit $\overline{\mathfrak{G}}$: $\tilde{F}_j(x) = V\tilde{F}_j(U^{-1}(x))$.

b) *Allgemeinere Fassung: Besitzen die F_j, $j = 1, 2$ anstelle von* (76) *die Eigenschaft*

$$\partial\mathfrak{G}\colon H_j(x, t) := F_j(x) + \sum_{k=1}^{p-1} tV^kF_j(U^{-k}(x)) \neq 0\,, \qquad 0 \leqq t \leqq 1\,, \tag{83}$$

dann ist $\partial\mathfrak{G}$: $F_j(x) \cong H_j(x, 1)$. Wegen $\partial\mathfrak{G}$: $H_j(x, 1) = VH_j(U^{-1}(x), 1)$ gelten dann für F_j ebenfalls die Aussagen (77) *bis* (82).

Dieses Theorem verallgemeinert Sätze von Fet 1954 [41], Krasnosjelski 1955 [56], Browder 1970 [25] (vgl. Satz 10).

Beim Beweis benutzen wir durch Modifizierung des Beweisgedankens von Krasnosjelski [56] die gleiche einfache Idee wie beim Beweis des Antipodensatzes. Wiederum spielen der Verpflanzungssatz (Theorem 5b)) und Satz 8b) die entscheidende Rolle.

Beweis. Zu a). 1. Schritt: Ausschließung der Fixpunkte $x_1, \ldots, x_m \in \mathfrak{G}$. Wir umgeben jeden Fixpunkt x_i mit einer hinreichend kleinen offenen Kugel $\mathfrak{K}(x_i)$ und setzen

$$\mathfrak{G}_0^{(i)} := \mathfrak{K}(x_i) \cup U\mathfrak{K}(x_i) \cup \cdots \cup U^{p-1}\mathfrak{K}(x_i) \tag{84}$$

sowie

$$\mathfrak{G}_0 := \bigcup_{i=1}^{m} \mathfrak{G}_0^{(i)} \subset \mathfrak{G}\,, \qquad \mathfrak{G}_0^{(i)} \cap \mathfrak{G}_0^{(j)} = \emptyset\,, \qquad i \neq j\,.$$

Offensichtlich gilt wegen $U^p(x) = I(x)$

$$U\mathfrak{G}_0^{(i)} = \mathfrak{G}_0^{(i)} \tag{85}$$

und

$$\partial\mathfrak{G}_0^{(i)} \subseteq \partial\mathfrak{K}(x_i) \cup U(\partial\mathfrak{K}(x_i)) \cup \cdots \cup U^{p-1}(\partial\mathfrak{K}(x_i))\,. \tag{86}$$

2. Schritt: Zerlegung von $\mathfrak{G} - \overline{\mathfrak{G}}_0$. Nach (85) ist $U(\overline{\mathfrak{G}} - \mathfrak{G}_0) = \overline{\mathfrak{G}} - \mathfrak{G}_0$. Unser Ziel besteht darin, $F_1(x)$ stetig zu $\tilde{F}_1$ auf $\overline{\mathfrak{G}} - \mathfrak{G}_0$ fortzusetzen mit $\overline{\mathfrak{G}} - \mathfrak{G}_0$: $\tilde{F}_1(x) = V\tilde{F}_1(U^{-1}(x))$.

Da $U, U^2, \ldots, U^{p-1}$ auf $\overline{\mathfrak{G}} - \mathfrak{G}_0$ keine Fixpunkte besitzen, gilt

$$x \in \overline{\mathfrak{G}} - \mathfrak{G}_0\colon ||U^k(x) - x|| \geqq 2\varrho_0 > 0\,, \qquad k = 1, \ldots, p-1\,. \tag{87}$$

Wir überdecken $\overline{\mathfrak{G}} - \mathfrak{G}_0$ mit einem festen Netz abgeschlossener Würfel $\overline{\mathfrak{W}}_j$, also

$$\overline{\mathfrak{G}} - \mathfrak{G}_0 \subset \bigcup_{j=1}^{w} \overline{\mathfrak{W}}_j\,, \qquad \mathfrak{W}_i \cap \mathfrak{W}_j = \emptyset\,, \qquad i \neq j$$

mit

$$x, y \in \overline{\mathfrak{W}}_j\colon \|x - y\| \leqq \varrho_0 \,. \tag{88}$$

Wir greifen zunächst $\mathfrak{W}_1$ heraus und setzen

$$\mathfrak{G}_1^{(0)} := \mathfrak{W}_1 \cap (\mathfrak{G} - \overline{\mathfrak{G}}_0)\,, \qquad \partial \mathfrak{G}_1^{(0)} \subseteq \partial \mathfrak{W}_1 \cup \partial \mathfrak{G} \bigcup_{i=1}^{m} \partial \mathfrak{G}_0^{(i)} \,.$$

Auf $\overline{\mathfrak{G}}_1^{(0)} \cap \partial \mathfrak{G}$ ist F_1 bereits bekannt. Nach Satz 8b) unter Berücksichtigung von (86) können wir F_1 stetig und nullstellenfrei auf $\partial \mathfrak{G}_1^{(0)}$ ausdehnen. Schließlich erweitern wir F_1 in irgendeiner Form stetig auf die gesamte Menge $\overline{\mathfrak{G}}_1^{(0)}$ und erhalten

$$\tilde{F}_1\colon \overline{\mathfrak{G}}_1^{(0)} \subset R^n \xrightarrow{\text{stetig}} R^n\,, \qquad \partial \mathfrak{G}_1^{(0)}\colon \tilde{F}_1(x) \neq 0 \,.$$

Nunmehr konstruieren wir sukzessiv für $j = 1, 2, \ldots, p-1$

$$\mathfrak{G}_1^{(j)} := U \mathfrak{G}_1^{(j-1)}\,, \qquad \overline{\mathfrak{G}}_1^{(j)} : \tilde{F}_1(x) = V \tilde{F}_1(U^{-1}(x)) \,.$$

Wegen (76) entsteht auf $\overline{\mathfrak{G}}_1^{(j)} \cap \partial \mathfrak{G}$ wieder F_1. Auf Grund von (87), (88) sind die $\overline{\mathfrak{G}}_1^{(j)}$, $j = 1, \ldots, p-1$ paarweise disjunkt. Nach dem Verpflanzungssatz (58) gilt

$$d_{\mathfrak{G}_1^{(j)}}(\tilde{F}_1) = \operatorname{sgn}_{\mathfrak{G}} U \operatorname{sgn}_{R^n} V \, d_{\mathfrak{G}_1^{(j-1)}}(\tilde{F}_1)\,, \qquad j = 1, \ldots, p-1 \,. \tag{89}$$

Schließlich setzen wir

$$\mathfrak{G}_1 := \bigcup_{j=0}^{p-1} \mathfrak{G}_1^{(j)} \subseteq \mathfrak{G} - \overline{\mathfrak{G}}_0\,, \qquad \overline{\mathfrak{G}}_1^{(j)} \cap \overline{\mathfrak{G}}_1^{(k)} = \emptyset\,, \qquad j \neq k \,.$$

3. Schritt: Zerlegung von $\mathfrak{G} - \overline{\mathfrak{G}}_0 - \overline{\mathfrak{G}}_1$.

Wir greifen $\mathfrak{W}_2$ heraus und betrachten

$$\mathfrak{G}_2^{(0)} := \mathfrak{W}_2 \cap (\mathfrak{G} - \overline{\mathfrak{G}}_0 - \overline{\mathfrak{G}}_1)\,, \qquad \partial \mathfrak{G}_2^{(0)} \subseteq \partial \mathfrak{W}_2 \cup \partial \mathfrak{G} \bigcup_{i=1}^{m} \partial \mathfrak{G}_0^{(i)} \bigcup_{j=0}^{p-1} \partial \mathfrak{G}_1^{(j)} \,.$$

Auf $\overline{\mathfrak{G}}_2^{(0)} \cap (\partial \mathfrak{G} \cup \overline{\mathfrak{G}}_1)$ ist $\tilde{F}_1$ bereits bekannt. Wir setzen $\tilde{F}_1$ stetig und nullstellenfrei auf den restlichen Teil von $\partial \mathfrak{G}_2^{(0)}$ fort, erweitern anschließend $\tilde{F}_1$ stetig auf ganz $\overline{\mathfrak{G}}_2^{(0)}$ und konstruieren

$$\mathfrak{G}_2^{(j)} := U \mathfrak{G}_2^{(j-1)}\,, \quad \overline{\mathfrak{G}}_2^{(j)}\colon \tilde{F}_1(x) = V \tilde{F}_1(U^{-1}(x))\,, \qquad j = 1, \ldots, p-1 \,.$$

Nach spätestens w Schritten ist $\overline{\mathfrak{G}}$ völlig zerlegt

$$\overline{\mathfrak{G}} = \bigcup_{i=1}^{m} \overline{\mathfrak{G}}_0^{(i)} \bigcup_{j=0}^{p-1} \overline{\mathfrak{G}}_1^{(j)} \cup \cdots \bigcup_{j=0}^{p-1} \overline{\mathfrak{G}}_w^{(j)}\,,$$

wobei die zugehörigen offenen Mengen paarweise disjunkt sind. Ferner existiert auf $\overline{\mathfrak{G}}$ eine stetige Abbildung $\tilde{F}_1$ mit

$$\partial \mathfrak{G}_k^{(l)}\colon \tilde{F}_1(x) \neq 0\,, \qquad d_{\mathfrak{G}_r^{(j)}}(\tilde{F}_1) = \gamma d_{\mathfrak{G}_r^{(j-1)}}(\tilde{F}_1)$$

$$k = 0, \ldots, w\,, \qquad r = 1, \ldots, w\,, \qquad j = 1, \ldots, p-1\,, \qquad \gamma = \operatorname{sgn}_{\mathfrak{G}} U \operatorname{sgn}_{R^n} V \,.$$

Nach A_3 gilt somit

$$d_{\mathfrak{G}}(F_1) = d_{\mathfrak{G}}(\tilde{F}_1) = \sum_{i=1}^{m} d_{\mathfrak{G}_0^{(i)}}(\tilde{F}_1) + (1 + \gamma + \cdots + \gamma^{p-1}) \sum_{r=1}^{w} d_{\mathfrak{G}_r^{(0)}}(\tilde{F}_1) . \qquad (90)$$

4. Schritt: Beweis von (77), (78). Besitzen $U, U^2, \ldots, U^{p-1}$ auf $\overline{\mathfrak{G}}$ keine Fixpunkte, dann kann man die vorangegangene Konstruktion mit $\mathfrak{G}_0^{(i)} = \emptyset$ durchführen. Es entsteht

$$d_{\mathfrak{G}}(F_1) = d_{\mathfrak{G}}(\tilde{F}_1) = (1 + \gamma + \cdots + \gamma^{p-1}) \sum_{r=1}^{w} d_{\mathfrak{G}_r^{(0)}}(\tilde{F}_1) . \qquad (91)$$

Gilt $\gamma = 1$, dann ergibt sich sofort (77).

Ist $\gamma = -1$, dann muß wegen

$$1 = \operatorname{sgn}_{\mathfrak{G}} I = \operatorname{sgn}_{\mathfrak{G}} V^p U^p = (\operatorname{sgn}_{R^n} V \operatorname{sgn}_{\mathfrak{G}} U)^p = (-1)^p$$

(vgl. (55)) p gerade sein, also $d_{\mathfrak{G}}(F_1) = 0$.

5. Schritt: Beweis von (79), (80). Wir führen für F_2 die gleiche Konstruktion wie für F_1 durch mit dem Unterschied, daß wir auf $\partial\mathfrak{G}_0^{(i)}$ von vornherein die Werte $\tilde{F}_1(x)$ vorgeben. Dann entsteht

$$d_{\mathfrak{G}}(F_2) = \sum_{i=1}^{m} d_{\mathfrak{G}_0^{(i)}}(\tilde{F}_1) + (1 + \gamma + \cdots + \gamma^{p-1}) \sum_{r=1}^{w} d_{\mathfrak{G}_r^{(0)}}(\tilde{F}_2) , \qquad (92)$$

und (79), (80) ergeben sich aus (90), (92).

6. Schritt: Beweis von (81), (82). Wir führen für F_j die soeben beschriebene Konstruktion durch mit dem Unterschied, daß wir auf $\partial\mathfrak{G}_0^{(i)}$ die Werte $G_i(x)$ vorgeben. Dann ergibt sich

$$d_{\mathfrak{G}}(F_j) = \sum_{i=1}^{m} d_{x_i}(G_i) + (1 + \gamma + \cdots + \gamma^{p-1}) \sum_{r=1}^{w} d_{\mathfrak{G}_r^{(0)}}(\tilde{F}_j) ,$$

woraus wir (81), (82) erhalten, q. e. d.

Satz 10. *(Folgerungen aus dem verallgemeinerten Antipodensatz).* a) *Periodische Felder auf Sphären: Auf der Kugel $\mathfrak{K}_\varrho^{(n)}(0)$ des R^n mit dem Mittelpunkt* 0 *sei ein Homöomorphismus p-ter Ordnung* ($p = 2, 3, \ldots$)

$$U\colon \partial\mathfrak{K}_\varrho^{(n)}(0) \xrightarrow[\text{auf}]{\text{homöomorph}} \partial\mathfrak{K}_\varrho^{(n)}(0) , \qquad U^p(x) = I(x) \qquad (93)$$

gegeben, wobei $U, U^2, \ldots, U^{p-1}$ keine Fixpunkte auf $\partial\mathfrak{K}_\varrho^{(n)}(0)$ besitzen sollen. Durch Verpflanzung auf konzentrische Sphären läßt sich U zu

$$\tilde{U}\colon R^n \xrightarrow[\text{auf}]{\text{homöomorph}} R^n , \qquad \tilde{U}(0) = 0$$

erweitern, wobei $\tilde{U}, \tilde{U}^2, \ldots, \tilde{U}^{p-1}$ genau $x = 0$ als einzigen Fixpunkt besitzen. Genügt

$$F\colon \partial\mathfrak{K}_\varrho^{(n)}(0) \xrightarrow{\text{stetig}} R^n \qquad (94)$$

dann der Bedingung

$$\partial\mathfrak{K}_\varrho^{(n)}(0)\colon F(x) = \tilde{U}F(\tilde{U}^{-1}(x)) , \qquad (95)$$

dann gilt $d_{\mathfrak{K}_\varrho^{(n)}(0)}(F) = 1 \pmod p$ *(vgl. Krasnosjelski* 1955 [56], *Browder* 1970 [25]*).*

b) *Involutionen auf Sphären: Gegeben sei eine Involution U auf* $\partial\mathfrak{K}_\varrho^{(n)}(0)$, *d. h. ein fixpunktfreier Homöomorphismus der Form* (93) *mit* $p = 2$. *Genügt dann F in* (94) *statt* (95) *der Bedingung*

$$\partial\mathfrak{K}_\varrho^{(n)}(0)\colon F(x) \neq tF(U(x))\,, \qquad 0 \leqq t \leqq 1\,, \tag{96}$$

dann ist $d_{\mathfrak{K}_\varrho^{(n)}(0)}(F) \equiv 1 \pmod 2$ *(vgl. Fet* 1954 [41]*). Genauer gilt*

$$\operatorname{sgn}_{R^n} \tilde{U} = (-1)^n\colon \qquad d_{\mathfrak{K}_\varrho^{(n)}(0)}(F) \equiv 1 \pmod 2$$

$$\operatorname{sgn}_{R^n} \tilde{U} = (-1)^{n+1}\colon \qquad d_{\mathfrak{K}_\varrho^{(n)}(0)}(F) = 1\,.$$

c) *Verallgemeinertes Borsuk-Ulam-Theorem von Fet* 1954 [41]:
Zu jeder Abbildung

$$F\colon \partial\mathfrak{K}_\varrho^{(n+1)}(0) \subset R^{n+1} \xrightarrow{\text{stetig}} R^n$$

und zu jeder Involution U von $\partial\mathfrak{K}_\varrho^{(n+1)}(0)$ *existiert ein Punkt* $x_0 \in \partial\mathfrak{K}_\varrho^{(n+1)}(0)$ *mit* $F(x_0) = F(U(x_0))$.

d) *Verallgemeinerter Überdeckungssatz von Ljusternik-Schirelman-Borsuk: Ist U eine Involution von* $\partial\mathfrak{K}_\varrho^{(n)}(0)$ *und wird* $\partial\mathfrak{K}_\varrho^{(n)}(0)$ *durch p abgeschlossene Mengen* $\mathfrak{A}_1, \ldots, \mathfrak{A}_p$ *überdeckt, dann enthält für* $1 \leqq p \leqq n$ *wenigstens eines der* $\mathfrak{A}_i$ *ein involutorisches Paar* $\{x, U(x)\}$ *(Fet* 1954 [41]*).*

Beweis. Zu a). Wir wenden Theorem 8 (79) an mit $U = \tilde{U}$, $V = \tilde{U}$, $F_1 = F$, $F_2 = I$.

Zu b). Wir benutzen (81), (82), (83) in Theorem 8 mit $U = \tilde{U}$, $V = -I$, $G_1 = I - \tilde{U}$. Offensichtlich ist wegen $\tilde{U}^2 = I$

$$G_1(x) = -(\tilde{U}^{-1} - I)(x) = VG_1(\tilde{U}^{-1}(x))\,, \qquad \tilde{U}^{-1} = \tilde{U}\,.$$

$G_1(x)$ besitzt 0 als einzige Nullstelle, und es gilt

$$\partial\mathfrak{K}_\varrho^{(n)}(0)\colon I(x) \cong G_1(x)\,, \quad H(x, t) = I(x) - t\tilde{U}(x)\,, \qquad 0 \leqq t \leqq 1\,,$$

denn aus $\|x_1\| = \varrho$: $H(x_1, t) = 0$ folgt wegen $\|\tilde{U}(x_1)\| = \varrho$ sofort $t = 1$ im Widerspruch zu $\|x\| = \varrho$: $H(x, 1) \neq 0$.

Zu c), d). Man vergleiche die Beweise von Satz 9c), d) und ersetze (69′) durch (96). q. e. d.

§ 13. Analytische Abbildungen und Bifurkationsprobleme

Definition 3. (Abbildungen vom analytischen Typ; Browder 1970 [26]) $F \in C(\overline{\mathfrak{G}})$ heißt vom analytischen Typ genau dann, wenn es eine Folge $F_k \in C^{(1)}(\overline{\mathfrak{G}})$ gibt mit den beiden Eigenschaften:

a) F_k konvergiert für $k \to \infty$ gleichmäßig auf $\overline{\mathfrak{G}}$ gegen F.

b) $\mathfrak{M}_k := \{x \in \mathfrak{G};\ \det F_k'(x) > 0\}$ ist dicht in $\mathfrak{G}$.

Theorem 9. *(Abbildungen vom analytischen Typ; Browder 1970 [26]). Ist $F \in C(\overline{\mathfrak{G}})$ vom analytischen Typ, dann gilt:*

a) $d_{\mathfrak{G}}(F) \geqq 0$,

b) $d_{\mathfrak{G}}(F) \geqq 1$ *genau dann, wenn* $\mathfrak{G}\colon F(x) = 0$ *lösbar,*

c) $d_{\mathfrak{G}}(F) = 1$ *impliziert, daß die abgeschlossene Menge der Nullstellen* $\mathfrak{N} = \{x \in \mathfrak{G}$ $F(x) = 0\}$ *nicht in zwei disjunkte, abgeschlossene und nichtleere Teilmengen zerfällt.*

d) $\mathfrak{U}$ *sei Komponente von* $R^n - F(\partial\mathfrak{G})$. *Besitzt* $\mathfrak{G}\colon F(x) = y$ *für festes* $y = y_0 \in \mathfrak{U}$ *eine Lösung, dann auch für jedes* $y \in \mathfrak{U}$. *Für fast alle* $y \in \mathfrak{U}$ *besitzt* $\mathfrak{G}\colon F(x) = y$ *obendrein genau* $m = d_{\mathfrak{G}}(F(x) - y_0)$ *Lösungen im Fall* $F \in C^{(1)}(\overline{\mathfrak{G}})$.

Beweis. Zu a). Nach (5) ist $d_{\mathfrak{G}}(F) = d_{\mathfrak{G}}(F_k)$ für $k \geqq k_0$, und aus (16) folgt $d_{\mathfrak{G}}(F_k) \geqq 0$.

Zu b). Gilt $d_{\mathfrak{G}}(F) \geqq 1$, dann ist $\mathfrak{G}\colon F(x) = 0$ nach Theorem 2d) lösbar. Sei umgekehrt $F(x_0) = 0$, $x_0 \in \mathfrak{G}$. Wir wählen zu $\gamma(F) := \min ||F(x)||$, $x \in \partial\mathfrak{G}$ zwei Zahlen $k_0, \delta_0 > 0$ mit

$$\overline{\mathfrak{G}}\colon ||F_{k_0}(x) - F(x)|| < \frac{\gamma(F)}{2}, \quad ||F_{k_0}(x)|| < \frac{\gamma(F)}{2} \quad \text{für} \quad ||x - x_0|| < \delta_0 .$$

Wegen Definition 3b) gibt es ein x_1 mit

$$||x_1 - x_0|| < \delta_0, \qquad \det F'_{k_0}(x_1) > 0, \qquad ||F_{k_0}(x_1)|| < \frac{\gamma(F)}{2}.$$

Die Abbildung $G(x) := F_{k_0}(x) - F_{k_0}(x_1) \in C(\overline{\mathfrak{G}})$ ist vom analytischen Typ und besitzt in x_1 eine isolierte Nullstelle mit $d_{x_1}(G) = \operatorname{sgn} \det F'_{k_0}(x_1) = 1$. Nach Theorem 9a) und A_3 ist $d_{\mathfrak{G}}(G) \geqq 1$, und nach (5) folgt aus $\overline{\mathfrak{G}}\colon ||G(x) - F(x)|| < \gamma(F)$ auch $d_{\mathfrak{G}}(F) = d_{\mathfrak{G}}(G) \geqq 1$.

Zu c). Ist die Behauptung nicht richtig, dann gibt es zwei disjunkte, offene Teilmengen $\mathfrak{G}_1, \mathfrak{G}_2 \subset \mathfrak{G}$, die alle Nullstellen enthalten und nach A_3, A_2 a) und Theorem 9b) gilt

$$d_{\mathfrak{G}}(F) = d_{\mathfrak{G}_1}(F) + d_{\mathfrak{G}_2}(F) \geqq 2$$

im Widerspruch zur Voraussetzung $d_{\mathfrak{G}}(F) = 1$.

Zu d). Da $F(x) - y_0 \in C(\overline{\mathfrak{G}})$ vom analytischen Typ ist, gilt nach b) $d_{\mathfrak{G}}(F - y_0) \geqq 1$, und die Behauptung folgt aus Theorem 2d) und dem dazugehörigen Korollar. q. e. d.

Definiton 4. (Analytische Abbildungen). $\mathfrak{U}^* \subset C^n$ sei ein Gebiet des Raumes C^n von n komplexen Veränderlichen $z = (z_1, \ldots, z_n)$, und die n komplexwertigen Funktionen $w_i = f_i^*(z_1, \ldots, z_n)$, $i = 1, \ldots, n$ seien holomorph in $\mathfrak{U}^*$. Sie beschreiben die Abbildung

$$F^*(z) = (f_1^*, \ldots, f_n^*)\colon \mathfrak{U}^* \subset C^n \to C^n . \tag{96'}$$

Trennt man Real- und Imaginärteil und identifiziert man $\mathfrak{U}^* \subset C^n$ mit $\mathfrak{U} \subset R^{2n}$, so entsteht die reelle Abbildung

$$F = (f_1, \ldots, f_{2n}) : \mathfrak{U} \subset R^{2n} \xrightarrow{\text{stetig}} R^{2n} . \tag{96''}$$

F heißt dann analytisch in $\mathfrak{U}$.

Satz 11. *(Abbildungsklassen vom analytischen Typ; Browder* 1970 [26]*). $F \in C(\overline{\mathfrak{G}})$ sei stetig auf einer umfassenden offenen Menge $\mathfrak{U} \supset \overline{\mathfrak{G}}$. Jede der beiden folgenden Bedingungen garantiert, daß F vom analytischen Typ ist.*

a) *F ist analytisch in $\mathfrak{U}$.*

b) *F ist monoton, d. h.*

$$\langle F(x_1) - F(x_2),\ x_1 - x_2 \rangle \geqq 0\,, \qquad x_1, x_2 \in \mathfrak{U}\,.$$

Beweis. Zu a). Nach (98′) ist $\det F'(x) = |\det F^{*\prime}(z)|^2$. Für die Abbildung $F_\varepsilon^*(z) := F^*(z) + \varepsilon z$ ist $\det F_\varepsilon^{*\prime}(z)$ ein Polynom n-ten Grades in ε. Zu vorgegebenem k und festem $z^{(0)}$ läßt sich ein ε_k mit $0 < \varepsilon_k < 1/k$ finden, so daß $\det F_{\varepsilon_k}^{*\prime}(z^{(0)}) \neq 0$ gilt. F_k in Definition 3a) entstehe aus $F_{\varepsilon_k}^*(z)$. Für F_k ist $\mathfrak{N} = \{x \in \mathfrak{U};\ \det F'(x) > 0\}$ dicht in $\mathfrak{G}$, anderenfalls wäre $\det F'(x) = 0$ in einer Umgebung von $x^{(1)} \in \mathfrak{G}$, also $\det F_{\varepsilon_k}^{\prime *}(z^{(1)}) = 0$ in einer Umgebung von $z^{(1)}$. Da $\mathfrak{U}^*$ zusammenhängend ist, würde daraus unter Berücksichtigung der Holomorphie von $F_{\varepsilon_k}^*(z)$ auch $\det F_{\varepsilon_k}^*(z^{(0)}) = 0$ folgen, was im Widerspruch zur Konstruktion der ε_k steht.

Zu b). Sei $M_k(y) \geqq 0$, $y \in R^n$ eine beliebig oft differenzierbare Funktion mit

$$\int_{R^n} M_k(y)\,dy = 1\,, \quad M_k(y) = 0 \qquad \text{für } \|y\| \geqq \frac{1}{k}.$$

Wir konstruieren die Mittelfunktionen

$$x \in \overline{\mathfrak{G}}\colon F_k(x) := \int_{R^n} \left(F(x-y) + \frac{1}{k}(x-y)\right) M_k(y)\,dy$$

$$= \int_{R^n} \left(F(y) + \frac{1}{k}y\right) M_k(x-y)\,dy\,. \tag{97}$$

F_k ist beliebig oft differenzierbar und konvergiert für $k \to \infty$ gleichmäßig auf $\overline{\mathfrak{G}}$ gegen F (vgl. Smirnow [93] S. 180). Aus (97) und der Monotonie von F folgt

$$\langle F_k(x_1) - F_k(x_2), x_1 - x_2 \rangle \geqq \frac{1}{k}\|x_1 - x_2\|^2\,,$$

also

$$\langle F_k'(x_1)\,(x_1 - x_2), x_1 - x_2 \rangle \geqq \frac{1}{k}\|x_1 - x_2\|^2\,.$$

Die letzte Aussage beweist man indirekt. Der Widerspruch ergibt sich leicht aus der Bilinearität. $F_k'(x_1)$ besitzt somit nur Eigenwerte $\lambda_i \geqq 1/k$, also $\det F_k'(x_1) > 0$, q. e. d.

Für analytische Abbildungen (vgl. Definition 4) ergeben sich im Vergleich zu Theorem 9 schärfere Aussagen. Im folgenden Theorem benutzen wir die Bezeichnungen (96′), (96″) von Definition 4.

Theorem 10. *(Analytische Abbildungen).* a) *Analytische Abbildungen im R^2: Die zu*

$$w = z^k \quad \text{bzw.} \quad w = \bar{z}^k, \quad k = 1, 2, \ldots,$$

gehörigen reellen Abbildungen F bzw. $\bar{F}$ im R^2 besitzen den Index $d_{x=0}(F) = k$ bzw. $d_{x=0}(\bar{F}) = -k$.

b) *Lineare Abbildungen: Ist $f_i^* = \sum_{j=1}^{n} a_{ij} z_j$, dann gilt*

$$\det F'(x) = |\det a_{ij}|^2 , \tag{98}$$

also $d_{x=0}(F) = 1$ im Fall $\det a_{ij} \neq 0$. Für allgemeine analytische Abbildungen ist folglich

$$\det F'(x) = \left|\det \frac{\partial f_i^*(z)}{\partial z_j}\right|^2 . \tag{98'}$$

c) *Polynomabbildungen: Sind die $f_i^*(z_1, \ldots, z_n)$ homogene Polynome $P_i^{(k_i)}$ vom Grade $k_i \geqq 1$, dann ist $x = 0$ genau dann die einzige Nullstelle von $F(x)$, wenn $R \neq 0$ gilt. Dabei ist die Resultante $R(a_1, \ldots, a_q)$ ein Polynom in den Koeffizienten der f_i^* (vgl. van der Waerden* [100] *S. 15). Im Fall $R \neq 0$ ist*

$$d_{x=0}(F) = \prod_{i=1}^{n} k_i . \tag{99}$$

d) *Allgemeine analytische Abbildungen: Besitzen die holomorphen Funktionen $f_i^*(z_1, \ldots, z_n)$ die Gestalt*

$$f_i^*(z_1, \ldots, z_n) = \sum_{m=k_i}^{\infty} P_i^{(m)}(z_1, \ldots, z_n) , \qquad k_i \geqq 1 ,$$

wobei die $P_i^{(m)}$ homogene Polynome m-ten Grades sind, und ist F auf dem Rand eines Gebietes $\mathfrak{G}$ mit $0 \in \mathfrak{G}$, $\bar{\mathfrak{G}} \subset \mathfrak{U}$ ungleich Null, d. h. $\partial\mathfrak{G}$: $F(x) \neq 0$, dann gilt

$$d_{\mathfrak{G}}(F) \geqq \prod_{i=1}^{n} k_i . \tag{100}$$

Außerdem hat F (vgl. Remmert, Stein 1953 [82] *Satz* 7*) nur endlich viele Nullstellen x_i, $i = 1, \ldots, l$ in $\mathfrak{G}$, also*

$$d_{\mathfrak{G}}(F) = \sum_{i=1}^{l} d_{x_i}(F) , \qquad d_{x_i}(F) \geqq 1 , \tag{100'}$$

und es ist

$$d_{x=0}(F) = 1 \text{ genau dann, wenn } \det F'(0) \neq 0 . \tag{101}$$

Nach b) *folgt aus* $\det F'(0) \neq 0$ *stets* $\det F'(0) > 0$.

Besitzt das zugehörige reduzierte Gleichungssystem der Glieder niedrigsten Grades $P_i^{(k_i)}(z_1, \ldots, z_n) = 0$, $i = 1, \ldots, n$, nur die triviale Lösung $z = 0$, d. h. $R \neq 0$, dann gilt

$$d_{x=0}(F) = \prod_{i=1}^{n} k_i . \tag{100''}$$

Allgemeine analytische Abbildungen sind vom analytischen Typ, deshalb gelten für sie alle Aussagen von Theorem 9.

Ist eine der Funktionen f_i^* *identisch Null, dann folgt nach Theorem* 4c) *aus* $\partial\mathfrak{G}: F(x) \neq 0$ *stets* $d_{\mathfrak{G}}(F) = 0$.

e) *Zusammenhang mit reellen Abbildungen: Besitzen die* $f_i^*(z_1, \ldots, z_n)$ *in* d) *nur reelle Koeffizienten, dann induzieren sie neben der Abbildung* F *im* R^{2n} *noch eine Abbildung* F_r *im* R^n. *Ist nun* $\mathfrak{K}_\varrho^{(2n)}(0): F(x) \neq 0$, *so gilt*

$$|d_{\mathfrak{K}_\varrho^{(n)}(0)}(F_r)| \leqq d_{\mathfrak{K}_\varrho^{(2n)}(0)}(F)$$
$$d_{\mathfrak{K}_\varrho^{(n)}(0)}(F_r) \equiv d_{\mathfrak{K}_\varrho^{(2n)}(0)}(F) \pmod 2 .$$

Ist speziell $d_{\mathfrak{K}_\varrho^{(2n)}(0)}(F)$ *eine ungerade Zahl, dann hat man* $d_{\mathfrak{K}_\varrho^{(n)}(0)}(F_r) \neq 0$, *folglich besitzt nach Theorem* 2d) *das Gleichungssystem*

$$\mathfrak{K}_\varrho^{(n)}(0): f_i^*(\xi_1, \ldots, \xi_n) = \eta_i , \qquad i = 1, \ldots, n$$

für reelle η_i *mit* $|\eta_i| < \delta$ *eine „reelle" Lösung* $x = (\xi_1, \ldots, \xi_n) \in \mathfrak{K}_\varrho^{(n)}(0)$.

Diese Aussagen findet man im wesentlichen bei Cronin 1950 [29], 1953 [31], 1954 [32]. Verallgemeinerungen enthält J. T. Schwartz 1963 [89].

Beweis. Zu b). Durch eine Ähnlichkeitstransformation in C^n, die die Determinante nicht ändert und der im R^{2n} ebenfalls eine Ähnlichkeitstransformation entspricht, geht (a_{ij}) in Jordansche Normalform über. Dabei entsteht beim Übergang von C^n zu R^{2n} aus dem Komplex

$$\begin{pmatrix} \lambda_k & 1 & \\ & \ddots & \ddots \\ & & \ddots \end{pmatrix}, \quad \lambda_k = a_k + ib_k$$

der neue Komplex

$$\begin{pmatrix} a_k & -b_k & 1 & 0 & \\ b_k & a_k & 0 & 1 & \\ \ddots & \ddots & \ddots & \ddots & \\ \end{pmatrix},$$

somit ist $\det F'(x) = \prod_{k=1}^{n} (a_k^2 + b_k^2) = \prod_{k=1}^{n} |\lambda_k|^2 = |\det a_{ij}|^2$.

Zu a). Anstelle von $w = F^*(z) := z^k$ betrachten wir

$$|z| = \varrho: w = F_\varepsilon^*(z) := z^k - \varepsilon = \prod_{j=1}^{k} (z - \zeta_j) , \quad \zeta_i \neq \zeta_j , \quad i \neq j .$$

Für hinreichend kleines $\varepsilon > 0$ gilt $\mathfrak{K}_\varrho^{(2)}(0): F_\varepsilon \cong F$ (vgl. Definition 4), also

$$d_{\mathfrak{K}_\varrho^{(2)}(0)}(F) = \sum_{j=1}^{k} \operatorname{sgn} \det F_\varepsilon'(x_j) \overset{(98')}{=} \sum_{j=1}^{k} \operatorname{sgn} \left|\frac{dF_\varepsilon^*(\zeta_j)}{dz}\right|^2 = k .$$

Die zu $w = \bar{z}^k$ gehörige Abbildung $\overline{F}$ entsteht aus F durch eine Spiegelung S der unabhängigen Variablen mit $\operatorname{sgn}_{R^2} S = -1$, also nach (58) $d_{\mathfrak{K}_\varrho^{(2)}(0)}(\overline{F}) = - d_{\mathfrak{K}_\varrho^{(2)}(0)}(F)$.

Zu c). Die Koeffizienten der homogenen Polynome $P_i^{(k_i)}$ seien $a_1, \ldots, a_q$, und die entsprechenden Koeffizienten der Normalform $\tilde{P}_i^{(k_i)}(z) := z_i^{k_i}$ bezeichnen wir mit $b_1, \ldots, b_q$. Für

$$g(\lambda) = R((1 - \lambda)\, a_1 + \lambda b_1, \ldots, (1 - \lambda)\, a_q + \lambda b_q)$$

gilt wegen der vorausgesetzten Einzigkeit der Nullstelle $z = 0$ des Systems $w_i = P_i^{(k_i)}(z)$, $i = 1, \ldots, n$, und wegen der offensichtlichen Einzigkeit der Nullstelle $z = 0$ von

$$w_i = z_i^{k_i}, \qquad i = 1, \ldots, n, \tag{102}$$

nach den Eigenschaften der Resolvente: $g(0), g(1) \neq 0$. Da $g(\lambda)$ ein Polynom in λ darstellt, existiert ein Weg $\lambda = \lambda(t)$, $0 \leqq t \leqq 1$ in der komplexen λ-Ebene mit $g(\lambda(t)) \neq 0$, $0 \leqq t \leqq 1$. Die Abbildung $w_i = P_i^{(k_i)}(z_1, \ldots, z_n; t)$ mit den Koeffizienten $(1 - \lambda(t))\, a_j + \lambda(t)\, b_j$ besitzt $z = 0$ als einzige Nullstelle, folglich gilt $\mathfrak{K}_\varrho^{(2n)}(0)$: $F \cong F_N$, wobei F_N der Abbildung (102) entspricht. Nach Theorem 10a) und nach dem Satz über das kartesische Produkt (Theorem 4a)) gilt

$d_{\mathfrak{K}_\varrho^{(2n)}(0)}(F) = d_{\mathfrak{K}_\varrho^{(2n)}(0)}(F_N) = \prod_{i=1}^{n} k_i$.

Zu (100). Seien wiederum $a_1, \ldots, a_q$ die Koeffizienten der homogenen Polynome $P_i^{(k_i)}$. Ist $R(a_1, \ldots, a_q) = 0$, dann kann man durch eine kleine Abänderung der $a_1, \ldots, a_q$ stets $R(a_1, \ldots, a_q) \neq 0$ erreichen, und die so entstehende Abbildung ist auf $\partial\mathfrak{G}$ homotop zur Ausgangsabbildung. Wir dürfen also von vornherein $R(a_1, \ldots, a_q) \neq 0$ annehmen.

Neben $w_i = f_i^*(z_1, \ldots, z_n)$, $i = 1, \ldots, n$, betrachten wir noch $w_i = P_i^{(k_i)}(z_1, \ldots, z_n)$, $i = 1, \ldots, n$. Die zugehörigen reellen Abbildungen seien F bzw. F_P. Wegen der Einzigkeit der Nullstelle $x = 0$ von R^{2n}: $F_P(x) = 0$ gilt in jedem Oberflächenpunkt Q der Einheitskugel des R^{2n} für einen gewissen Index $P_i^{(k_i)}(Q) \neq 0$. Nach dem Überdeckungssatz von Heine-Borel kann die Oberfläche der Einheitskugel durch endlich viele abgeschlossene Mengen $\mathfrak{A}_j$ überdeckt werden mit $\mathfrak{A}_j$: $P_{i_j}^{(k_{i_j})} \neq 0$. Wegen der Homogenität der $P_i^{(k_i)}$ gilt eine analoge Aussage zugleich auf jeder konzentrischen Kugeloberfläche $\partial\mathfrak{K}_\varrho^{(2n)}(0)$. Wählt man ϱ hinreichend klein, dann ist (5) erfüllt, weil die Glieder $P_i^{(m)}(z_1, \ldots, z_n)$ in $f_i^*(z_1, \ldots, z_n)$ mit $m > k_i$ von höherer Ordnung klein werden als $P_i^{(k_i)}(z_1, \ldots, z_n)$, folglich gilt nach Satz 1 $d_{\mathfrak{K}_\varrho^{(2n)}(0)}(F) = d_{\mathfrak{K}_\varrho^{(2n)}(0)}(F_P)$. Nach (99) ist somit $d_{\mathfrak{K}_\varrho^{(2n)}(0)}(F) = \prod_{i=1}^{n} k_i$, und aus Theorem 9a) folgt

$$d_{\mathfrak{G}}(F) = d_{\mathfrak{K}_\varrho^{(2n)}(0)}(F) + d_{\mathfrak{G} - \bar{\mathfrak{K}}_\varrho^{(2n)}(0)}(F) \geqq \prod_{i=1}^{n} k_i,$$

womit (100) bewiesen ist.

Zu (101). Ist $\det F'(0) > 0$, dann gilt nach Definition 1 $d_{x=0}(F) = 1$.

Sei umgekehrt $x = 0$ eine isolierte Nullstelle von F, und es gelte $d_{x=0}(F) = 1$. Wir nehmen $\det F'(0) = 0$ an und führen diese Annahme zum Widerspruch.

Nach (100) müssen die zugehörigen holomorphen Funktionen die Gestalt

$$w_i = \sum_{j=1}^{n} a_{ij} z_j + \sum_{m=m_i \geqq 2}^{\infty} P_i^{(m)}(z_1, \ldots, z_n) =: f_i^*(z_1, \ldots, z_n), \qquad i = 1, \ldots, n \tag{103}$$

haben. Keine der Funktionen $f_i^*(z_1, \ldots, z_n)$ kann identisch verschwinden, sonst wäre $d_{x=0}(F) = 0$. Außerdem dürfen die a_{ij} nicht alle gleich Null sein, sonst wäre nach (100) $d_{x=0}(F) \geqq 2$.

Durch eine eineindeutige Variablentransformation der unabhängigen Variablen $z_1, \dots, z_n$ und der abhängigen Variablen $w_1, \dots, w_n$ läßt sich (103) auf die Form bringen $w' = \Theta(z')$:

$$w'_i = z'_i + \sum_{m=m'_i \geq 2}^{\infty} Q_i^{(m)}(z'_1, \dots, z'_n), \qquad i = 1, \dots, r \tag{104}$$

$$w'_j = \sum_{m=m'_j \geq 2}^{\infty} Q_j^{(m)}(z'_1, \dots, z'_n), \qquad j = r+1, \dots, n,$$

wobei r mit $0 < r < n$ der Rang von (a_{ij}) ist und die $Q_l^{(m)}(z'_1, \dots, z'_n)$ homogene Polynome m-ten Grades sind. Bezeichnen wir die zu (104) gehörige reelle Abbildung mit F_Θ, dann gilt nach dem Verpflanzungssatz (58) $d_{x=0}(F) = d_{x'=0}(F_\Theta)$, denn aus Theorem 10b) folgt $\operatorname{sgn}_{R^{2n}} U = +1$, wenn der Homöomorphismus U aus einer eineindeutigen linearen Abbildung in C^n entsteht.

Die ersten r Gleichungen in (104) kann man bei festen z'_j, $j = r+1, \dots, n$, mit $|z'_j| < \delta$ in einer Umgebung von $w'_i = 0$, $i = 1, \dots, r$, eindeutig nach z'_i auflösen

$$z'_i = g_i(w'_1, \dots, w'_r, z'_{r+1}, \dots, z'_n), \qquad i = 1, \dots, r. \tag{105}$$

Wir konstruieren nun die beiden Abbildungen $v = \varphi(z')$:

$$v_i = z'_i + \sum_{m=m'_i}^{\infty} Q_i^{(m)}(z'_1, \dots, z'_n), \qquad i = 1, \dots, r, \tag{106}$$

$$v_j = z'_j \qquad j = r+1, \dots, n,$$

und $w' = \psi(v)$:

$$w'_i = v_i \qquad i = 1, \dots, r,$$

$$w'_j = v_j + \left\{ -v_j + \sum_{m=m'_j}^{\infty} Q_j^{(m)}(g_1, \dots, g_r;\ v_{r+1}, \dots, v_n) \right\} \tag{107}$$

$$:= q_j(v_1, \dots, v_n), \qquad j = r+1, \dots, n,$$

mit $g_i = g_i(v_1, \dots, v_r;\ v_{r+1}, \dots, v_n)$.

Offensichtlich gilt (vgl. (104), (105))

$$w' = \Theta(z') = \psi(\varphi(z')).$$

Bezeichnen wir mit F_ψ, F_φ die zugehörigen reellen Abbildungen im R^{2n}, dann ist

$$x''' = F_\Theta(x') = F_\psi(x''),\ x'' = F_\varphi(x'). \tag{108}$$

Die Abbildung F_φ ist nach (106) ein Homöomorphismus in einer kleinen Umgebung von $x' = 0$ mit $d_{x'=0}(F_\varphi(x')) = d_{x'=0}(I) = 1$.

Nach dem Reduktionssatz (Theorem 4b)) ist $d_{x''=0}(F_\psi)$ gleich dem Index der durch

$$w'_j = q_j(0, \dots, 0, v_{r+1}, \dots, v_n) = \sum_{m=\alpha_j}^{\infty} T_j^{(m)}(v_{r+1}, \dots, v_n), \qquad j = r+1, \dots, n,$$

im R^{2n-2r} induzierten Abbildungen F_q. Nach Konstruktion von g_i in (105) gilt $\alpha_j \geq 2$, also nach Theorem 10d)

$$d_{x''=0}(F_\psi) = d_{x''=0}(F_q) \geq \Pi \alpha_j \geq 2.$$

Aus (108) entsteht nun nach dem Produktsatz von Leray (Theorem 4d))

$$d_{x=0}(F) = d_{x'=0}(F_\Theta) = d_{x'=0}(F_\varphi)\, d_{x''=0}(F_\psi) \geqq 2$$

im Widerspruch zur Voraussetzung $d_{x=0}(F) = 1$, also gilt $\det F'(0) \neq 0$.

Zu e). Ändern wir die Funktionen $f_i^*(z)$, $z = (z_1, \ldots, z_n)$ um hinreichend kleine reelle Konstanten, d. h. $g_i^*(z) := f_i^*(z) + \alpha_i$, $\bar{\alpha}_i = \alpha_i$, dann gilt für die zugehörigen Abbildungen G_r bzw. G im R^n bzw. R^{2n} nach Satz 2b)

$$d_{\mathfrak{K}_\varrho^{(n)}(0)}(G_r) = d_{\mathfrak{K}_\varrho^{(n)}(0)}(F_r)\,, \quad d_{\mathfrak{K}_\varrho^{(2n)}(0)}(G) = d_{\mathfrak{K}_\varrho^{(2n)}(0)}(F)$$

und

$$G_r \in C_\varrho^{(1)}\big(\overline{\mathfrak{K}}_\varrho^{(n)}(0)\big)\,, \quad d_{\mathfrak{K}_\varrho^{(n)}(0)}(G_r) = \sum_{i=1}^{k} \operatorname{sgn} \det G_r'(x_i)\,, \quad G_r(x_i) = 0\,.$$

Um die Behauptung aus (100′) zu erhalten, erklären wir die beiden Homöomorphismen U bzw. V, die durch den Übergang zu konjugiert komplexen Größen $z \to \bar{z}$ bzw. $g_i^*(z) \to \bar{g}_i^*(z)$, $i = 1, \ldots, n$, im R^{2n} induziert werden. Es ist $\operatorname{sgn}_{R^{2n}} U = \operatorname{sgn}_{R^{2n}} V = (-1)^n$ (vgl. Theorem 5a)). Da die Koeffizienten der $g_i^*(z)$ reell sind, gilt $\bar{g}_i^*(\bar{z}) = g_i^*(z)$, d. h. $VG(U^{-1}(x)) = G(x)$, und das Gleichungssystem $g_i^*(z) = 0$, $i = 1, \ldots, n$, besitzt nach Theorem 10d) die endlich vielen Nullstellen

$$z^{(i)} = \bar{z}^{(i)}\,, \quad i = 1, \ldots, k; \qquad z^{(j)}, \bar{z}^{(j)}\,, \quad j = k+1, \ldots, s\,,$$

denen $x_i \in R^n$, $x_j \in R^{2n}$, $U(x_j) \in R^{2n}$ entsprechen.

In den Punkten $x_i \in R^n$ gilt wegen $\det G_r'(x_i) \neq 0$ auch $\det G'(x_i) \neq 0$, also $\det G'(x_i) > 0$ und $d_{x_i}(G) = 1$.

Für $x_j \in R^{2n}$ hat man nach Theorem 5 (58)

$$d_{U(x_j)}\big(VG(U^{-1}(x))\big) = d_{U(x_j)}\big(G(x)\big) = d_{x_j}\big(G(x)\big) \geqq 1\,.$$

Aus (100′) ergibt sich somit

$$d_{\mathfrak{K}_\varrho^{(2n)}(0)}(G) = \sum_{i=1}^{k} d_{x_i}(G) + \sum_{j=k+1}^{s} d_{x_j}(G) + \sum_{j=k+1}^{s} d_{U(x_j)}(G)$$

$$= k + 2 \sum_{j=k+1}^{s} d_{x_j}(G)\,. \quad \text{q. e. d.}$$

Die folgenden Sätze beinhalten weitreichende Folgerungen aus Theorem 10.

Satz 12. *(Verallgemeinerter Fundamentalsatz der Algebra). Vorgelegt sei das Gleichungssystem*

$$f_i^* := \sum_{m=1}^{k_i} P_i^{(m)}(z_1, \ldots, z_n) + a_i = 0\,, \quad k_i \geqq 1,\ i = 1, \ldots, n\,, \tag{109}$$

wobei die a_i feste komplexe Zahlen und die $P_i^{(m)}$ homogene Polynome m-ten Grades sind.

Besitzt dann das reduzierte Problem

$$g_i^* := P_i^{(k_i)}(z_1, \ldots, z_n) = 0\,, \quad i = 1, \ldots, n\,, \tag{110}$$

$z = 0$ *als einzige Lösung, so hat* (109) *eine Lösung.*

Die Anzahl der Lösungen ist höchstens gleich $m = \prod_{i=1}^{n} k_i$. *Sind die Koeffizienten der* f_i^* *alle reell, dann existiert in dem Fall, daß* m *eine ungerade Zahl ist, auch eine reelle Lösung.*

Beweis. F, G seien die zu (109), (110) gehörigen Abbildungen im R^{2n}. Angenommen, $R^{2n} : F(x) = 0$ besitzt keine Lösung, dann ergeben die gleichen Überlegungen wie im Beweis von Theorem 10d), daß auf der Oberfläche einer hinreichend großen Kugel $\mathfrak{K}_\varrho^{(2n)}(0)$ die Abschätzung (5) gilt, also $d_{\mathfrak{K}_\varrho^{(2n)}(0)}(F) = d_{\mathfrak{K}_\varrho^{(2n)}(0)}(G) = \prod_{i=1}^{n} k_i \geqq 1$ nach (99). Das Kroneckersche Existenzprinzip (Theorem 2d)) ergibt im Widerspruch zu unserer Annahme die Existenz einer Lösung. Die übrigen Aussagen folgen aus Theorem 10d), e). q. e. d.

Wir betrachten jetzt das Gleichungssystem

$$f_i^*(p, z) := \sum_{m'=m_i}^{\infty} P_i^{(m')}(p; z_1, \dots, z_n) = 0\,, \qquad i = 1, \dots, n \tag{111}$$

mit $z = (z_1, \dots, z_n)$, $p = (p_1, \dots, p_m)$. Die $P_i^{(m')}$ seien homogene Polynome in $z_1, \dots, z_n$ vom Grade m', $m_i \geqq 0$, es gelte $f_i^*(0, 0) = 0$, $i = 1, \dots, n$, und es sei

$$p = 0 : f_i^*(0, z) = \sum_{m'=k_i}^{\infty} P_i^{(m')}(0, z_1, \dots, z_n)\,, \qquad k_i \geqq 1\,, \tag{111'}$$

wobei die Koeffizienten $a_l = a_l(p_1, \dots, p_m)$ stetig von p_j abhängen. Für $|p_j| < \delta$, $j = 1, \dots, m$, sei gleichmäßig $|a_l(p_1, \dots, p_m)| \leqq \alpha_l$, und die entsprechenden Reihen mit den Koeffizienten α_l mögen für $|z_i| \leqq \varrho_0$, $\varrho_0 > 0$, $i = 1, \dots, n$, absolut konvergieren.

Neben der trivialen Lösung $z = 0$, $p = 0$ werden weitere Lösungen gesucht.

Satz 13. *(Verallgemeinerter Satz über implizite Funktionen). Für* (111) *gelten folgende Existenzaussagen.*

Fall 1*:* $z = 0$ *ist für* $p = 0$ *keine isolierte Lösung von* (111).

Dann existiert trivialerweise eine Folge von Lösungen $z^{(k)} \neq 0$, $p = 0$ *mit* $z^{(k)} \to 0$ *für* $k \to \infty$.

Fall 2*:* $z = 0$ *ist für* $p = 0$ *isolierte Lösung von* (111).

Fall 2.1. $\det \dfrac{\partial f_i^*}{\partial z_j}(0, 0) \neq 0$ *(regulärer Fall).*

Es gibt dann Zahlen $\varepsilon_0, \delta_0 > 0$, *so daß für festes* p *mit* $|p_j| < \delta_0$, $j = 1, \dots, m$, *genau eine Lösung* z *von* (111) *existiert mit* $|z_i| < \varepsilon_0$, $i = 1, \dots, n$. *Die Lösung* $z_i = z_i(p_1, \dots, p_m)$ *hängt stetig von* $p_1, \dots, p_m$ *ab.*

Fall 2.2. $\det \dfrac{\partial f_i^*}{\partial z_j}(0, 0) = 0$ *(singulärer Fall).*

2.2.1. *Für* $p \neq 0$, $|p_j| < \delta_0$, $j = 1, \dots, n$, *ist* $(p, 0)$ keine Lösung von (111): *Zu beliebig kleinem* $\varepsilon > 0$ *gibt es ein* $\delta(\varepsilon) > 0$, *so daß zu fest vorgegebenem* $p \neq 0$, $|p_j| < \delta(\varepsilon)$, $j = 1, \dots, m$, *eine Lösung* $z \neq 0$ *von* (111) *existiert mit* $|z_i| < \varepsilon$, $i = 1, \dots, n$.

Besitzt für $p = 0$ *das reduzierte System der Glieder niedrigsten Grades* $P_i^{(k_i)}(0, z_1, \dots, z_n) = 0$, $i = 1, \dots, n$, *(vgl.* (111')*) nur die triviale Lösung* $z = 0$, *dann gilt für* $0 < \varepsilon < \varepsilon_0$

$$d_{\mathfrak{K}_\varepsilon^{(2n)}(0)}(F(p, x)) = \prod_{i=1}^{n} k_i \,, \qquad |p_j| < \delta(\varepsilon) \,, \qquad j = 1, \dots, m,$$

wenn $F(p, x)$ *die zu* (111) *gehörige Abbildung im* R^{2n} *beschreibt. Ist* $\prod_{i=1}^{n} k_i$ *eine ungerade Zahl, und haben die* $f_i^*(p, z)$ *in* (111) *für reelle* p_j *auch reelle Koeffizienten, dann existiert zu jedem* $p \neq 0$: $|p_j| < \delta(\varepsilon)$, $j = 1, \dots, m$: *eine reelle Lösung* $z \neq 0$ *von* (111) *mit* $z_i = \bar{z}_i$, $|z_i| < \varepsilon$, $i = 1, \dots, n$.

2.2.2. det $\dfrac{f_i^*(p, z)}{\partial z_j} \neq 0$ *für* $p \neq 0$, $|p_j| < \delta_0$, $j = 1, \dots, m$: *Zu beliebig kleinem* $\varepsilon > 0$ *gibt es ein* $\delta(\varepsilon) > 0$, *so daß zu fest vorgegebenem* $p \neq 0$: $|p_j| < \delta(\varepsilon)$ *eine Lösung* $z \neq 0$ *von* (111) *existiert mit* $|z_i| < \varepsilon$, $i = 1, \dots, n$.

Speziell existiert im Fall 2.2.1. *bzw. im Fall* 2.2.2. *eine Folge von Lösungen* $z^{(k)} \neq 0$, $p^{(k)} \neq 0$ *mit* $z^{(k)} \to 0$, $p^{(k)} \to 0$ *für* $k \to \infty$.

Unter weiter einschränkenden Voraussetzungen kann man im Fall 2.2.1. eine stärkere Aussage beweisen (vgl. den Rangsatz in Dieudonné [35] S. 276).

Verallgemeinerungen findet man in Cronin 1953 [31], Melamed 1962 [69]. Methoden zur tatsächlichen Berechnung der Lösungen unter Benutzung des Kroneckerschen Ausschließungsverfahrens sind enthalten in Wainberg, Trenogin 1969 [98], Kranosjelski 1969 [60].

Beweis. Zu Fall 2.1. Die Behauptung ergibt sich, indem man auf

$$P_i^{(1)}(0, z) = -\sum_{m' \neq 1} P_i^{(m')}(p, z) + \{P_i^{(1)}(0, z) - P_i^{(1)}(p, z)\}$$

sukzessive Approximation anwendet (vgl. z. B. Zeidler [106] S. 43).

Zu Fall 2.2.1. Da $x = 0$ isolierte Lösung von R^{2n}: $F(0, x) = 0$ ist, gilt nach Theorem 10d) für hinreichend kleines $\varepsilon > 0$ und $p \neq 0$: $|p_j| < \delta(\varepsilon)$, $j = 1, \dots, m$, unter besonderer Beachtung von det $F'_x(0, 0) = 0$ (vgl. (101))

$$d_{\mathfrak{K}_\varepsilon^{(2n)}(0)}(F(p, x)) = d_{\mathfrak{K}_\varepsilon^{(2n)}(0)}(F(0, x)) \geqq \max\left\{\prod_{i=1}^{n} k_i,\, 2\right\}.$$

Das Kroneckersche Existenzprinzip (Theorem 2d)) liefert dann sofort die Existenzaussage. Die weiteren Behauptungen ergeben sich aus (100″) und Theorem 10e).

Zu Fall 2.2.2. Ist $x = 0$ keine Lösung von $\mathfrak{K}_\varepsilon^{(2n)}(0)$: $F(p, x) = 0$, $p \neq 0$: $|p_j| < \delta(\varepsilon)$, $j = 1, \dots, m$, dann schließt man wie im vorangegangenen Beweis. Anderenfalls muß neben der Lösung $x = 0$ wegen det $F'(p, 0) > 0$, also $d_{x=0}(F(p, x) = 1$ und $d_{\mathfrak{K}_\varepsilon^{(2n)}(0)}(F(p, x)) \geqq 2$, eine weitere Lösung $x \neq 0$ existieren, q. e. d.

Wir wollen Satz 13 auf nichtlineare Eigenwertprobleme anwenden und betrachten hierzu die beiden Aufgaben

$$\mu F^*(z) = z \colon \mu f_i^*(z_1, \dots, z_n) = z_i \,, \quad f_i^*(0, \dots, 0) = 0 \,, \quad i = 1, \dots, n \,, \tag{112}$$

und

$$\mu F(x) = x \colon \mu f_i(\xi_1, \dots, \xi_n) = \xi_i \,, \quad f_i(0, \dots, 0) = 0 \,, \qquad i = 1, \dots, n \,. \tag{113}$$

In (112) sei μ ein komplexer Parameter, und die $f_i^*(z_1, \ldots, z_n)$ seien in einer Umgebung von $(0, \ldots, 0)$ in C^n holomorph (komplexes Eigenwertproblem). Die komplexe Zahl $\mu_0 \neq 0$ heißt komplexer Bifurkationspunkt von (112) genau dann, wenn es eine Folge von komplexen Lösungen $z^{(k)} \in C^n$, $\mu^{(k)} \in C^1$ mit $z^{(k)} \neq 0$ und $z^{(k)} \to 0$, $\mu^{(k)} \to \mu_0$ für $k \to \infty$ gibt.

In (113) sei μ ein reeller Parameter, und die $f_i(\xi_1, \ldots, \xi_n)$ seien in einer Umgebung von $(0, \ldots, 0)$ im R^n stetig differenzierbare reelle Funktionen (reelles Eigenwertproblem). Die reelle Zahl $\mu_0 \neq 0$ heißt reeller Bifurkationspunkt von (113) genau dann, wenn es eine Folge von reellen Lösungen $x^{(k)} \in R^n$, $\mu^{(k)} \in R^1$ gibt mit $x^{(k)} \neq 0$ und $x^{(k)} \to 0$, $\mu^{(k)} \to \mu_0$ für $k \to \infty$.

Entscheidend für derartige Bifurkationsprobleme sind die Eigenschaften der linearisierten Aufgaben

$$\mu F^{*\prime}(0)\, z = z\colon \mu \sum_{j=1}^{n} \frac{\partial f_i^*}{\partial z_j}(0, \ldots, 0)\, z_j = z_i\,, \qquad i = 1, \ldots, n\,, \tag{112'}$$

$$\mu F'(0)\, x = x\colon \mu \sum_{j=1}^{n} \frac{\partial f_i}{\partial \xi_j}(0, \ldots, 0)\, \xi_j = \xi_i\,, \qquad i = 1, \ldots, n\,. \tag{113'}$$

Besitzen diese Gleichungen für $\mu = \mu_0 \neq 0$ eine nichttriviale Lösung, dann heißt μ_0 charakteristische Zahl. Unter der algebraischen Zahl Vielfachheit von μ_0 verstehen wir die Vielfachheit des Auftretens des Eigenwertes $\lambda_0 = 1/\mu_0$ in der Jordanschen Normalform von $F^{*\prime}(0)$ bzw. $F'(0)$. Es sei stets $F^{*\prime}(0) \neq 0$, $F'(0) \neq 0$.

Satz 14. *(Nichtlineare Eigenwertprobleme).* a) *Komplexes Bifurkationsproblem* (112)*: Die komplexe Zahl $\mu_0 \neq 0$ ist genau dann Bifurkationspunkt von* (112), *wenn μ_0 charakteristische Zahl von* (112') *ist, d. h.* $\det(\mu_0 F^{*\prime}(0) - I) = 0$ (*Melamed* 1963 [70]).

b) *Reelles Bifurkationsproblem* (113): *Ist die reelle Zahl $\mu_0 \neq 0$ Bifurkationspunkt von* (113), *dann ist μ_0 charakteristische Zahl von* (113'), *d. h.* $\det(\mu_0 F'(0) - I) = 0$.

Umgekehrt ist jedoch nur jede charakteristische Zahl μ_0 von (113') *mit ungerader algebraischer Vielfachheit Bifurkationspunkt von* (113). *Besitzt zusätzlich* (113) *für $\mu = \mu_0$, $\|x\| = \varrho_1 > 0$ keine Lösung, dann existiert ein $\varepsilon_1 > 0$, so daß* (113) *für alle $\mu\colon \mu_0 < \mu < \mu_0 + \varepsilon_1$ oder alle $\mu\colon \mu_0 - \varepsilon_1 < \mu < \mu_0$ eine Lösung $x \neq 0$ hat (Krasnoselski* 1956 [57]*)*.

Beweis. Zu a). Ist $\det(\mu_0 F^{*\prime}(0) - I) \neq 0$, dann besitzt (112) nach dem Satz über implizite Funktionen zu vorgegebenem μ mit $|\mu - \mu_0| < \delta_0$ genau eine Lösung $z = z(\mu)$ mit $|z(\mu)| < \varepsilon_0$. Da $z = 0$ stets Lösung ist, gilt $z(\mu) = 0$, folglich kann μ_0 nicht Bifurkationspunkt sein. Analog schließt man im reellen Fall.

Gilt $\det(\mu_0 F^{*\prime}(0) - I) = 0$, dann liegt in (112) entweder der Fall 1 von Satz 13 oder wegen der Isoliertheit der charakteristischen Zahlen handelt es sich mit $p = \mu - \mu_0$ um Fall 2.2.2. von Satz 13. In jedem Fall ist μ_0 Bifurkationspunkt.

Zu b). Wir setzen $G_\mu(x) := \mu F(x) - x$. Sei μ_0 charakteristische Zahl von (113') mit ungerader algebraischer Vielfachheit, dann ist wegen $\det((\mu_0 + \varepsilon)$

$F'(0) - I) \neq 0$ für hinreichend kleines $\varepsilon > 0$ der Punkt $x = 0$ isolierte Lösung von $G_{\mu_0 \pm \varepsilon}(x) = 0$, also $\partial \mathfrak{K}_\varrho^{(n)}\colon G_{\mu_0 \pm \varepsilon}(x) \neq 0$. Nach (33) gilt

$$d_{\mathfrak{K}_\varrho^{(n)}(0)}(G_{\mu_0+\varepsilon}) = - d_{\mathfrak{K}_\varrho^{(n)}(0)}(G_{\mu_0-\varepsilon}) ,$$

d. h.

$$\partial \mathfrak{K}_\varrho^{(n)}(0)\colon H(x, t) := tG_{\mu_0+\varepsilon}(x) + (1 - t)\, G_{\mu_0-\varepsilon}(x) = 0$$

für $t = t_1$ $x = x_1$ mit $0 < t_1 < 1$, $||x_1|| = \varrho$, folglich

$$||x_1|| = \varrho\colon (\mu_0 + 2t_1\varepsilon - \varepsilon)\, F(x_1) = x_1 .$$

Da ϱ, ε beliebig klein gewählt werden können, ist μ_0 Bifurkationspunkt.

Gilt zusätzlich $\partial \mathfrak{K}_{\varrho_1}^{(n)}\colon G_{\mu_0}(x) \neq 0$, dann ist auch $\partial \mathfrak{K}_{\varrho_1}^{(n)}(0)\colon G_{\mu_0 \pm \varepsilon}(x) \neq 0$ und

$$\partial \mathfrak{K}_{\varrho_1}^{(n)}(0)\colon G_{\mu_0+\varepsilon} \cong G_{\mu_0-\varepsilon} \qquad \text{für} \quad 0 < \varepsilon < \varepsilon_1 .$$

Wegen $\det G'_{\mu_0 \pm \varepsilon}(0) \neq 0$ ist $x = 0$ isolierte Lösung von $G_{\mu_0 \pm \varepsilon}(x) = 0$. Nach (33) gilt

$$d_{x=0}(G_{\mu_0 \pm \varepsilon}(x)) = \pm 1 , \qquad 0 < \varepsilon < \varepsilon_1$$

oder

$$d_{x=0}(G_{\mu_0 \pm \varepsilon}(x)) = \mp 1 ,$$

somit muß wegen

$$d_{\mathfrak{K}_{\varrho_1}^{(n)}(0)}(G_{\mu_0+\varepsilon}) = d_{\mathfrak{K}_{\varrho_1}^{(n)}(0)}(G_{\mu_0-\varepsilon}) , \qquad 0 < \varepsilon < \varepsilon_1$$

entweder $G_{\mu_0+\varepsilon}(x) = 0$ oder $G_{\mu_0-\varepsilon}(x) = 0$ in $\mathfrak{K}_{\varrho_1}^{(n)}(0)$ eine weitere von $x = 0$ verschiedene Lösung besitzen. q. e. d.

Literatur

[1] Alexandrow, P. S., Комбинаторная топология, Москва (1947).

[2] Alexandrow, P. S. und Hopf, H., Topologie, Berlin: Springer-Verlag (1935).

[3] Atiyah, M. F., Algebraic topology and elliptic operators. Comm. Pure Appl. Math. **20** (1967) 237–249.

[4] Atiyah, M. F., and Bott, R., A Lefschetz fixed point formula for elliptic complexes. Ann. of Math. **86** (1967) 374–407.

[5] Atiyah, M. F., and Singer, I. M., The index of elliptic operators on compact manifolds. Bull. of AMS **69** (1963) 422–433.

[6] Atiyah, M. F., and Singer, I. M., The index of elliptic operators I. Ann. of Math. **87** (1968) 484–530; III, Ann. of Math. **87** (1968) 546–603.

[7] Atiyah, M. F., and Segal, G. B., The index of elliptic operators II. Ann. of Math. **87** (1968) 531–545.

[8] Berger, M., and Berger, M., Perspectives in Nonlinearity. New York, Amsterdam: W. A. Benjamin (1968).

[9] Borsuk, K., Drei Sätze über die n-dimensionale Euklidische Sphäre. Fund. Math. **21** (1933) 177–190.

[10] Borsuk, K., Sur les prolongements des transformations continues. Fund. Math. **28** (1936) 239–243.

[11] —, Theory of retracts. Warszawa: Polish Scientific Publishers (1966).

[12] Bott, R., Stable homotopy of the classical groups. Ann. of Math. **70** (1959) 313–337.

[13] Brouwer, L. E. J., Über Abbildungen von Mannigfaltigkeiten, Math. Ann. **71** (1912) 97–115.
[14] –, Beweis der Invarianz des n-dimensionalen Gebiets, Math. Ann. **71** (1912) 305–313.
[15] –, Beweis des Jordanschen Satzes für den n-dimensionalen Raum, Math. Ann. **71** (1912) 314–327.
[16] –, Über Jordansche Mannigfaltigkeiten, Math. Ann. **71** (1912) 320–327.
[17] –, Sur la notion de „classe" de transformations d'une multiplicité, Proc. V. Intern. Congress of Math. Vol. II, Cambridge (1912) 9–10.
[18] –, Over één-éénduidige continue transformaties van oppervlakken in zichzelf, Akad. Amsterdam, Versl. **21** (1913) 300–309.
[19] –, Aufzählung der Abbildungsklassen endlichfach zusammenhängender Flächen, Math. Ann. **82** (1921) 208–286.
[20] Browder, F. E., On the solvability of nonlinear functional equations, Duke Math. J. **30** (1963) 557–566.
[21] –, Nonlinear elliptic boundary value problems, Bull. of AMS **69**, 6 (1963) 862–874.
[22] –, Remarks on nonlinear functional equations, Proc. Nat. Acad. Sci. USA **51** (1964).
[23] –, Problèmes non-linéaires, Montreal (1966).
[24] –, Nonlinear eigenvalue problems and group invariance, Functional Analysis and Related Fields, 1–58, Berlin-Heidelberg-New York: Springer-Verlag (1970).
[25] –, Group invariance in nonlinear functional analysis, Bull of AMS **76** (1970) 986–992.
[26] –, Nonlinear mappings of analytic type in Banach space, Math. Ann. **185** (1970) 259–278.
[27] Brown, A. B., Functional dependence. Transact. of AMS **38** (1935) 379–394.
[28] Brown, F. B., The Lefschetz fixed point theorem, Glenview, Illinois, London: Scott, Foresman and Co. (1970).
[29] Cronin, J., Branch points of solutions of equations in Banach spaces I, Transact. of AMS **69**, 2 (1950) 208–281.
[30] –, Branch points of solutions of equations in Banach spaces II, Transact. of AMS **74** (1954) 207–222.
[31] –, Analytical functional mappings. Ann. of Math. **58**, 1 (1953) 175–181.
[32] –, Topological degree of some mappings, Proc. AMS **5** (1954) 175–178.
[33] –, Fixed points and topological degree in nonlinear analysis. Amer. Math. Soc. (1964).
[34] de Rham, G., Variétés différentiables, Paris: Hermann (1955).
[35] Dieudonné, J., Grundzüge der modernen Analysis, Berlin: Deutscher Verlag der Wissenschaften (1972) (Übersetz. aus dem Engl.).
[36] Dold, A., Fixed point index and fixed point theorem for euclidean neighbourhood retracts, Topology **4** (1965) 1–8.
[37] Dugundij, J., An extension of Tietze's Theorem. Pac. Journ. Math. **1** (1951) 353–367.
[38] Eells, J., A setting for global analysis, Bull. of AMS **72** (1966) 751–807.
[39] Eilenberg, S., and Steenrod, N., Foundations of algebraic topology. Princeton (1952).
[40] Fadell, E., Recent results in the fixed point theory of continuous maps, Bull. of AMS **76** (1970) 10–29.
[41] Fet, A. I., Обощение теоремы Люстерника-Шнирельмана о покрытиях сфер и некоторых связанных с ней теорем. ДАН СССР **95**, 6 (1954) 1149—1151.
[42] Granas, A., The theory of compact vector fields and some of its applications to topology of functional spaces. Warszawa: Rozprawy Matematyczne (1962).
[43] –, Leray-Schauder Index and the fixed point theory for arbitrary ANR's. Bull. Soc. math. France **100** (1972) 209–228.
[44] Hadamard, J., Note sur quelques applications de l'indice de Kronecker: in Tannery, Introduction a la théorie des fonctions d'une variable, Tome II, Paris: Herman (1910).

[45] Heinz, E., An elementary theory of the degree of a mapping in n-dimensional space, Journal Math. Mech. **8** (1959) 231–247.
[46] Hess, P., On nonlinear mappings of monotone type homotopic to odd operators, J. Funct. Anal. **11** (1972) 138–167.
[47] –, On the Fredholm alternative for nonlinear functional equations in Banach spaces, Proc. AMS **33** (1972) 55–61.
[48] Hopf, H., Abbildungsklassen n-dimensionaler Mannigfaltigkeiten, Math. Ann. **96** (1927) 209–224.
[49] –, Die Klassen der Abbildungen der n-dimensionalen Polyeder auf die n-dimensionale Sphäre, Comm. Math. Helvetici **5** (1933) 365–384.
[50] Hörmander, L., On the Index of Pseudodifferential operators, Ellipt. Differentialgl. II, 127–146, Berlin: Akademie-Verlag (1971).
[51] Hu, S. T., Homotopy theory, New York, London: Academic Press (1959).
[52] Hurewicz, W., and Wallman, H., Dimension theory, Princeton (1948).
[53] Jordan, C., Cours d'Analyse I, Paris (1893).
[54] Knaster, B., Kuratowski, C., und Mazurkiewicz, S., Ein Beweis des Fixpunktsatzes für n-dimensionale Simplexe, Fund. Math. **14** (1929) 132–137.
[55] Kotake, T., The fixed point theorem of Atiyah-Bott via parabolic operators, Comm. Pure Appl. Math. **22** (1969) 789–806.
[56] Krasnoselski, M. A., О вычислении вращения векторного поля на конечномерной сфере. ДАН СССР **101**, 3 (1955) 401–404.
[57] –, Топологические методы в теории нелинейных интегральных уравнений, Москва (1956).
[58] –, Положительные решения операторных уравнений, Москва (1962).
[59] Krasnosjelski, M. A., u. a., Vektorfelder in der Ebene, Berlin: Akademie-Verlag (1966) (Übersetz. aus dem Russ.).
[60] –, Приближенное решение операторных уравнений, Москва (1969).
[61] Kronecker, L., Über Systeme von Funktionen mehrerer Variabler I, II, Gesammelte Werke Bd. I, 177–234, Leipzig: B. G. Teubner (1895).
[62] –, Über die Charakteristik von Funktionensystemen, Ges. Werke Bd. II, 73–82, Leipzig: B. G. Teubner (1897).
[63] –, Sur le nombre des racines communes à plusieurs équations simultanées, Ges. Werke Bd. III, 2, 209–215, Leipzig, Berlin: B. G. Teubner (1931).
[64] Lefschetz, S., Introduction to topology, Princeton (1949).
[65] Leray, J., Topologie des espaces abstrait de M. Banach, C. R. Acad. Sci. Paris **200** (1935) 1082–1084.
[66] –, La théorie des points fixes et ses applications en analyse, Proc. Int. Congress of Math., Vol. II, 202–208, Cambridge (1950).
[67] Leray, J., et Schauder, J., Topologie et équations fonctionelles, Ann. Sci. Ecole Norm. Sup. **51** (1943) 45–78.
[68] Ljusternik, L. A., u. Schnirelman, L. T., Топологические методы в вариационных задачах, Москва (1930). (Übersetzung aus dem Russ.: Méthodes topologiques dans les problèmes variationnels, Paris: Hermann 1934).
[69] Melamed, W. B., К задаче о ветвлении решений нелинейного аналитического уравнения, ДАН СССР **145**, 3 (1962) 531–533.
[70] –, О точках бифуркации одного класса уравнений, ДАН СССР **152** (1963) 821–822.
[71] Minty, G. J., Monotone non-linear operators in Hilbert space, Duke Math. J. **29**, 3 (1962) 341–346.
[72] –, On a monotonicity method for the solutions of non-linear equations in Banach spaces. Proc. Nat. Acad. Sci. USA **50**, 6 (1963).
[73] Nagumo, M., A theory of degree of mapping based on infinitesimal analysis, Amer. J. Math. **73** (1951) 485–496.

[74] —, Degree of mapping in convex linear topological spaces, Amer. J. Math. 73 (1951) 497–511.
[75] Ortega, J. M., and Rheinholdt, W. C., Iterative solution of nonlinear equations in several variables, New York, London: Academic Press (1970).
[76] Palais, R., Seminar on the Atiyah-Singer Index theorem, Princeton (1965).
[77] Perron, O., Zur Theorie der Matrices, Math. Ann. 64 (1907) 248–263.
[78] Picard, É., Sur le nombre des racines communes à plusieurs équations simultanées, J. Mathém. pures et appl. (1892) 5–24.
[79] —, Traité d'Analyse I, Paris: Gauthier-Villars (1901).
[80] Pontrjagin, L. S., Grundzüge der kombinatorischen Topologie, Berlin: Deutscher Verlag der Wissenschaften 1956 (Übersetz. aus dem Russ.).
[81] Rado, T., and Reichelderfer, P. V., Continuous transformations in analysis, Berlin-Göttingen-Heidelberg: Springer-Verlag (1955).
[82] Remmert, R. und Stein, K., Über die wesentlichen Singularitäten analytischer Mengen. Math. Ann. 126 (1953) 263–306.
[83] Rothe, E., Theorie der topologischen Ordnung und der Vektorfelder in Banachschen Räumen, Comp. Math. 5 (1937) 177–197.
[84] —, On non-negative functional transformations. Amer. J. Math. 66 (1944) 245–254.
[85] Sadowski, B. N., Предельно компактные и уплотняющие операторы, УМН 27, 1 (163), 1972, 883—897.
[86] Sard, A., The measure of the critical values of differentiable maps, Bull. of AMS 48 (1942) 883–897.
[87] Schauder, J., Über den Zusammenhang zwischen der Eindeutigkeit und Lösbarkeit partieller Differentialgleichungen zweiter Ordnung vom elliptischen Typ, Math. Ann. 106 (1932) 661–721.
[88] Schnirelmann, L., Über eine neue kombinatorische Invariante, Monatshefte Math. Phys. 37 (1930) 131–134.
[89] Schwartz, J. T., Compact analytical mappings of B-spaces and a theorem of Jane Cronin, Comm. Pure Appl. Math. 16 (1963) 253–260.
[90] —, Differential geometry and topology, Notes on Math. and its Appl., Courant Institut of Math. Sciences, New York (1968).
[91] —, Nonlinear functional analysis, Notes on Math. and its Appl., New York: Gordon and Breach Science Publ. Inc. (1969).
[92] Shinbrot, M., A fixed point theorem and some applications. Arch. Rat. Mech. Anal. 17 (1964) 255–271.
[93] Smirnow, W. I., Lehrgang der höheren Mathematik V, Berlin: Verlag der Wissenschaften (1962).
[94] Spanier, E. H., Algebraic Topology, New York: McGraw Hill Book Company (1966).
[95] Tietze, H., Über Funktionen, die auf einer abgeschlossenen Menge stetig sind, Journal für reine und angew. Math. 145 (1915) 9–14.
[96] Wainberg, M. M., Вариационные методы исследования нелинейных операторов, Москва (1) (1956).
[97] —, Вариационный метод и метод монотонных операторов в теории нелинейных уравнений, Москва (1972).
[98] Wainberg, M. M., u. Trenogin, W. A., Теория ветвления решений нелинейных уравнений, Москва (1969).
[99] van der Walt, T., Fixed and almost fixed points, Amsterdam: Mathem. Centre Tracts (1963).
[100] van der Waerden, B. L., Moderne Algebra II, Berlin: Springer-Verlag (1940).
[101] Whitney, H., Extensions of differentiable functions. Transact. of AMS 26 (1934) 63–89.

[102] —, On the maps of an n-sphere into another n-sphere, Duke Math. J. **3**, 1 (1937) 46—50.
[103] —, The maps of an n-complex into an n-sphere, Duke Math. J. **3**, 1 (1937) 51—55.
[104] Zarantonello, E. H., Solving functional equation by contractiv averaging, Univ. Wisconsin: Math. Res. Center Techn. Rep. **160** (1960).
[105] Zeidler, E., Zur Methode der monotonen Minoranten von M. A. Krasnosjelski, Math. Nachr. **49** (1971) 69—83.
[106] —, Beiträge zur Theorie und Praxis freier Randwertaufgaben, Berlin: Akademie-Verlag (1971).
[107] —, Über eine Klasse nichtlinearer singulärer Randwertaufgaben, Math. Nachr. **43** (1970) 321—351.

Anschrift des Autors:

Sektion Mathematik
Karl-Marx-Universität
701 Leipzig
Karl-Marx-Platz

On a class of nonlinear eigenvalue problems arising in reactor physics

ROSTISLAV ZEZULA, Prague

Introduction

Let us study a class of optimization problems which is a generalization of the Goertzel's problem of critical homogenized thermal reactor [1], [2], [3] with the least amount of fuel, because there is some technical interest in extremalizing some others reacter parameters, e.g., the specific output per volumen unity [4]. We shall give some necessary conditions for the existence of a solution for this class of problems.

§ 1. Formulation of the problem of OCHTRR — Optimal Critical Homogenized Thermal Reactor with Reflector

Let Ω_1, $\Omega_2 \subset \Omega_1$ be two nonempty domains in the Euclidean space E_i, $i = 1, 2, 3$, filled by a medium characterized by the spatial distribution of fuel $\psi = \psi(x)$, $x \in \Omega_1$, which we shall consider as regulation, and by the given kernel of the (linear) integral neutron transport operator $H = H(x, x')$; $x, x' \in \Omega_1$, or some of its approximations. Both ψ and H describe the scattering, absorption and fission properties of the medium.

Let the medium filling Ω_1 form a critical homogenized thermal reactor with thermal neutron flux $\phi = \phi(x)$, $x \in \Omega_1$ and let us consider the following mathematical model for optimal regulation of this reactor with help of the properly chosen spatial distribution of the fuel $\psi = \psi_{\text{opt.}}$:

$$\phi = H\psi\phi ; \quad \psi = \psi_{\text{opt.}} \quad \text{(criticality condition)} \tag{1.1}$$

$$g(\psi_{\text{opt.}}) = \underset{\psi \in U}{\text{extr.}}\, g(\psi) \qquad \text{(optimization condition)} \tag{1.2}$$

where

$$g(\psi) = \int_{\Omega_1} G(\phi_\psi, \psi)\, dx \,. \tag{1.2a}$$

In the expression (1.2a) for the optimization functional denotes $G = G(\phi, \psi)$ a given real function of two real variables, $\phi_\psi \not\equiv 0$, $\phi_\psi = H\psi\phi_\psi$ denotes the nontrivial solution of the criticality condition (1.1), i.e., the eigenfunction of the operator $H\psi$ (properly normed), corresponding to the eigenvalue 1, and the class $U \subset L_2(\Omega_1)$ of admissible regulations ψ is defined as follows:

Definition 1.1. $U \subset L_2(\Omega_1)$ contains these bounded piecewise continuous regulations $\psi = \psi(x)$ which have the support Ω_2 and for which the reactor is critical in Ω_1 with bounded ϕ_ψ, having the following regularity property: there exists a function $\varphi = \varphi(\psi) \in C_1(E_1)$, for which the equation

$$\phi_\psi(x) = \varphi[\psi(x)] \tag{1.2b}$$

holds for all $x \in \Omega_2$.

Physical assumptions used in the construction of the reactor model (1.1), (1.2) and their mathematical equivalents are the following:

(1.i) The reactor considered is a thermal homogenized reactor with reflector formed by moderator, i.e.,

a) the thermal neutron flux $\phi = \phi(x) \geqq 0$, the absorption characteristics $\Sigma_a^U = \Sigma_a^U(x) \geqq 0$, $x \in \Omega_1$ of fuel and $\Gamma_a^M > 0$ of moderator and the kernel of the neutron transport operator $H = H(x, x')$ do not depend neither on the neutron energy E, nor on the velocity direction Ω of the neutron nor on the time t,

b) $0 < \Sigma_a^U(x) \in C(\Omega_2)$, by homogeneity of the reactor medium,

c) the reactor core Ω_2 has a positive distance from the extrapolated outer reactor boundary $\partial\Omega_1^e$ which is the same for all neutron energies, so that the thermal neutron flux ϕ (for all $\psi\phi$ in the domain D_H of the linear operator H) has to satisfy the following boundary condition

$$\phi|_{\partial\Omega_1^e} = 0\,,\ D_H \subset\subset L_2(\Omega_1) \tag{1.3}$$

(1.ii) There is neglected the moderator expelling by the fuel, i.e., $H = H(x, x')$ does not depend on ψ and $\Sigma_a^M = \text{const.}$ in Ω_1, so that

$$0 < \psi(x) \equiv \frac{\Sigma_a^U(x)}{\Sigma_a^M} \in C(\Omega_2) \quad \text{if} \quad x \in \Omega_2 \tag{1.3a}$$

is proportional to the fuel concentration.

(1.iii) The reactor core $\emptyset \neq \Omega_2$ and Ω_1 are both convex and the medium in Ω_1 has some symmetry, so that the kernel of the integral transport operator H is symmetric, i.e., we have

$$v, w \in D_H \Rightarrow (v, Hw) = (Hv, w) = (w, Hv) \tag{1.4}$$

where $(.,.)$ denotes the usual scalar product in the real Hilbert space $L_2(\Omega_1)$.

(1.iv) The reactor is critical, i.e., for each $\psi \in U \subset L_2(\Omega_1)$ the operator $H\psi$ has the value 1 as simple dominant characteristic value, to which there corresponds a unique (properly normed) eigenfunction ϕ_ψ satisfying the criticality condition (1.1) and the boundary condition (1.3).

Remark 1.1: For validity of the assumption (1.iv) one has to verify that the transport operator H is positive in the sense of Krejn-Rutman, leaving invariant the cone of nonnegative functions in $D_H \subset\subset L_2(\Omega_1)$.

§ 2. Necessary conditions for the existence of a solution for the OCHTRR

We will now seek for necessary conditions for the existence of a nontrivial solution $\psi, 1, \phi_\psi \not\equiv 0$ of the problem (1.1), (1.2), (1.2a), (1.3), satisfying the physical assumptions (1.i), (1.ii), (1.iii), (1.iv). Because the fixed critical eigenvalue 1 in this problem gives not insight enough for our optimization purposes, we will, using the relation

$$\frac{\psi(x)}{\int\limits_{\Omega_1} G(\phi, \psi)\, dx'} = \frac{u(x)}{\int\limits_{\Omega_1} G(\phi, u)\, dx'}, \tag{2.1}$$

$$\phi,\ G = G(\phi, w) \text{ fixed}, \qquad u \geqq 0, \quad \psi \in U$$

transform the problem (1.1), (1.2), (1.2a), (1.3), (1.i)—(1.iv) to the following nonlinear eigenvalue problem

$$\phi = gR_u^G \phi \tag{2.2}$$

with the eigenvalue

$$g = \int\limits_{\Omega_1} G(\phi, \psi)\, dx' \tag{2.3}$$

which is a functional depending on ϕ, ψ.

The nonlinear operator R_u^G: $C(\Omega_1) \to C(\Omega_1)$ in (2.2) depending on u and G is given by the relation

$$u\phi \in D_H \subset\subset L_2(\Omega_1) \Rightarrow R_u^G \phi = \frac{Hu\phi}{\int\limits_{\Omega_1} G[\phi(x'), u(x')]\, dx'} \tag{2.3a}$$

so that the transformed criticality condition (1.1) takes the form

$$\phi \int\limits_{\Omega_1} G(\phi, u)\, dx' = gHu\phi, \qquad g = \int\limits_{\Omega_1} G(\phi, \psi)\, dx' \tag{2.4}$$

For given $u \geqq 0$ there shall exist in virtue of (1.1) and of the criticality assumption (1.iv) a simple positive least eigenvalue

$$g_u = \int\limits_{\Omega_1} G(\phi_g, \psi)\, dx. \tag{2.4a}$$

of (2.2), depending on u, G, to which there exist the corresponding unique (if properly normed) nonnegative eigenfunction $\phi_g = \phi_g(u, G)$, satisfying the criticality condition (1.1) and the boundary condition (1.3). For fixed $G = G(\phi, u)$ we will denote the solution of the criticality condition (2.4) u, g_u, ϕ_g.

Our task will be, considering $G = G(\phi, u)$ as fixed, to determine for which u satisfying the transformation (2.1), this simple least eigenvalue $g_u = \int\limits_{\Omega_1} G(\phi_g, \psi)\, dx' > 0$, with eigenfunction $\phi_g \geqq 0$, will be extremal, i.e., will fulfill the transformed optimization condition

$$g_u = \int\limits_{\Omega_1} G(\phi_g, \psi)\, dx' = \operatorname*{extr}_{U_1} \int\limits_{\Omega_1} G\left(\phi_g, \frac{g_u}{\int\limits_{\Omega_1} G(\phi_g, u)} u\right) dx' \tag{2.5}$$

where U_1 denotes the set of those u for which (2.1) holds with $\phi = \phi_g$.

Lemma 2.1. *Let the function $G = G(\phi, \psi)$ in the optimization functional* (1.2a) *satisfy the following conditions:*

$$G(\phi, \psi) > 0 \quad \textit{for} \quad \phi \geqq 0, \quad \psi > 0, \tag{2.6}$$

ϕ not zero identically, (positivity condition)

$$G(\phi, \psi) \in C_1(E_2), \qquad G[\phi(x), \psi(x)] \in C_1(\Omega_2) \tag{2.6a}$$

and bounded (regularity conditions)

$$G(\phi, \beta\psi) = \beta G(\phi, \psi) \quad \textit{for each} \quad \beta \in E_1 \quad \textit{(homogeneity condition)} \tag{2.6b}$$

For

$$\alpha = \alpha(\psi, u) = \frac{\int_{\Omega_1} G(\phi_g, \psi)\, dx}{\int_{\Omega_1} G(\phi_g, u)\, dx} \text{ there exists the mean value } \bar{\alpha} = \frac{\int_{\Omega_1} \alpha(\psi, u)\, G(\phi_g, u)\, dx}{\int_{\Omega_1} G(\phi_g, u)\, dx}. \tag{2.6c}$$

Then both reactor optimization problems (1.1)—(1.3), (1.i)—(1.iv) *and* (2.4), (2.5) *are equivalent, i.e.,*

$$\exists\, \{\psi = \psi_{\text{opt.}}, 1, \phi_\psi\} : \phi_\psi = 1 \cdot H\psi\phi_\psi \wedge g(\psi_{\text{opt.}}) = \underset{\psi \in U}{\text{extr.}}\, g(\psi)$$

$$\Leftrightarrow \exists\, \{u, g_u, \phi_g\} : \phi_g \int_{\Omega_1} G(\phi_g, u)\, dx' = g_u H u \phi_g \wedge g_u = \text{extr.} \int_{\Omega_1} G(\phi_g, \psi)\, dx'$$

(where extremum is taken over ψ, u fulfilling 2.1, *i.e., $\psi \in U$, $u \in U_1$) and $\phi_\psi = \phi_g$, $\psi_{\text{opt.}} = \bar{\alpha} u_{\text{opt.}}$.*

Proof: From (2.1) and (2.6), (2.6a), (2.6b), it follows by substituting into (2.4) (see remark 2.1), that (1.1) must hold for ψ, 1, ϕ_g iff (2.4) holds for u, g_u, ϕ_g. Using (1.iv), it follows from the simplicity of the eigenvalue 1 that g_u is also a simple eigenvalue, with eigenfunction $\phi_g = \phi_\psi$. But $0 \neq \alpha \in E_1$ by (2.6), (2.6a) and we have therefore (see remark 2.1):

$$\psi = \alpha u$$

so that it follows from (2.6b), (2.6c) and (1.2a), (2.5)

$$g(\psi) = \int_{\Omega_1} G(\phi_\psi, \psi)\, dx = \int_{\Omega_1} G(\phi_g, \alpha u)\, dx = \int_{\Omega_1} \alpha G(\phi_g, u)\, dx = \bar{\alpha} \int_{\Omega_1} G(\phi_g, u)\, dx$$
$$= \int_{\Omega_1} G(\phi_g, \bar{\alpha} u)\, dx\,.$$

Therefore, the extremum values of the functionals $g(\psi)$, g_u resp. will be attained for $\psi_{\text{opt.}} = \bar{\alpha} u_{\text{opt.}}$, q.e.d.

Remark 2.1. Algorithmical aspect of the transformation (2.1): For given u and $G(\phi, u)$ let ϕ be in the domain D_{Hu} of the operator Hu, so that $R_u^G \phi$ can be computed. Let $g = g_u$ be the least positive eigenvalue of the operator R_u^G and let $\phi = \phi_g$ be the corresponding eigenfunction. Then, for $\psi = \psi(x)$ the relation (2.1) takes the form (with right side independent on ψ)

$$\psi(x) = \frac{\int_{\Omega_1} G(\phi_g, \psi)\, dx}{\int_{\Omega_1} G(\phi_g, u)\, dx}\, u(x) = \frac{g}{\int_{\Omega_1} G(\phi_g, u)\, dx}\, u(x)$$

where

$\int_{\Omega_1} G(\phi_g, \psi)\, dx = g$, $\quad \phi_g = g R_u^G \phi_g$ and g, $\int_{\Omega_1} G(\phi_g, u)\, dx$ depend only on G, u.

Theorem 2.1. *Let us make the following assumptions:*

(2.i) *Let there exist for $\forall\, u \in U_1$, $\psi \in U$ a nontrivial (properly normed) solution u, $g_u = \int_{\Omega_1} G(\phi_g, \psi)\, dx$, $\phi_g = \phi_\psi$ of the transformed criticality condition* (2.4), *so that*

$$\phi_g \int_{\Omega_1} G(\phi_g, u)\, dx' = g_u H u \phi_g\,.$$

(2.ii) *Let the optimization function* $G(\phi, \psi)$ *fulfil the conditions* (2.6), (2.6a), (2.6b), (2.6c) *of lemma* 2.1 *and the further condition*

$$\exists\, u, g_u, \phi_g = \varphi(u) \in C_1(E_1) \text{ satisfying (2.i) such that } G(\phi_g, u) = \int \varphi^2(u)\, du + \text{const}. \tag{2.7}$$

Then the following assertion is valid: If some u, g_u, ϕ_g *satisfying* (2.i), (2.ii) *extremalizes the functional*

$$g_u = \int_{\Omega_1} G(\phi_g, \psi)\, dx$$

then the integral condition

$$\int_{\Omega_1} u(x)\, [\phi_g(x)]^2\, dx = \int_{\Omega_1} G[\phi_g(x), u(x)]\, dx \tag{2.8}$$

is fulfilled.

Proof: We can eliminate the unknown functional dependence

$$\phi_g = \phi_g(u, G),\ (\phi_g, \phi_g) \text{ arbitrary} \tag{2.9}$$

by considering, for fixed function G, the functional

$$I = I(\phi, u, g) = (u\phi, \phi \int_{\Omega_1} G(\phi, u)\, dx' - gHu\phi) \tag{2.9a}$$

which, in virtue of the transformed criticality condition (2.4) takes identically the zero value for each, u, g_u, ϕ_g satisfying (2.4), so that for the first variation of the functional

$$I_g = I_g(u) = I(\phi_g, u, g_u) \equiv 0 \tag{2.10}$$

depending only on u, the condition

$$\delta I_g = 0 \tag{2.11}$$

must hold.

Using this condition, we can calculate the variation δg of the functional $g = g_u$ and, by substituting it in the well known necessary condition for the existence of an extremum of the functional g_u

$$\delta g_u = 0 \tag{2.12}$$

we shall obtain necessary conditions which have to fulfill the functions $\phi_g = \phi_\psi$ and ψ realizing $\underset{\psi \in U}{\text{extr.}} \int_{\Omega_1} G(\phi_\psi, \psi)\, dx$. We have

$$\begin{aligned} 0 = \delta I_g &= \delta[(u\phi_g, \phi_g) \int_{\Omega_1} G(\phi_g, u)\, dx' - g_u(u\phi_g, Hu\phi_g)] \\ &= (\delta u\phi_g, \phi_g) \int_{\Omega_1} G(\phi_g, u)\, dx' + (u\phi_g, \phi_g) \int_{\Omega_1} \delta G(\phi_g, u)\, dx' \\ &\quad - g_u(\delta u\phi_g, Hu\phi_g) - g_u(u\phi_g, H\delta u\phi_g) - \delta g_u(u\phi_g, Hu\phi_g)\,. \end{aligned} \tag{2.13}$$

By help of the relation (2.4), written for $g = g_u$ and $\phi = \phi_g$, it can be easily seen the validity of the relation

$$(\delta u\phi_g, \phi_g) \int_{\Omega_1} G(\phi_g, u)\, dx' = g_u(\delta u\phi_g, Hu\phi_g) \tag{2.14}$$

so that we have, because the transport operator H is symmetric by (1.iii)

$$g_u(u\phi_g, H\delta u\phi_g) = (\delta u\phi_g, g_u Hu\phi_g) = (\delta u\phi_g, \phi_g) \int_{\Omega_1} G(\phi_g, u)\, dx'\,. \tag{2.15}$$

Substituting (2.14) and (2.15) into (2.13) we get the relation

$$\delta g_u(u\phi_g, Hu\phi_g) = (u\phi_g, \phi_g) \int_{\Omega_1} \delta G(\phi_g, u)\, dx' - (\delta u\phi_g, \phi_g) \int_{\Omega_1} G(\phi_g, u)\, dx' \quad (2.16)$$

Using the change of integration variables and by help of the definition of the scalar product in the real Hilbert space $L_2(\Omega_1)$

$$v, w \in L_2(\Omega_1) \Rightarrow (v, w) \equiv \int_{\Omega_1} v(x)\, w(x)\, dx \quad (2.17)$$

we can write the second term on the right side of the relation (2.16) as follows:

$$\begin{aligned}(\delta u\, \phi_g, \phi_g) \int_{\Omega_1} G[\phi_g(x'), u(x')]\, dx' &= \int_{\Omega_1} \delta u(x)\, \phi_g^2(x)\, dx \int_{\Omega_1} G[\phi_g(x'), u(x')]\, dx' \\ &= \int_{\Omega_1} \delta u(x')\, \phi_g^2(x')\, dx' \int_{\Omega_1} G[\phi_g(x), u(x)]\, dx\,. \quad (2.18)\end{aligned}$$

We can further, using the regularity of G and ϕ_g assumed in (2.6a) and (2.ii) and in lemma 2.1, calculate the integral in the first term on the right side of (2.16) in the following manner [5]:

$$\begin{aligned}&\int_{\Omega_1} \delta G[\phi_g(u), u]\, dx' = \delta \int_{\Omega_1} G[\phi_g(u), u]\, dx' = \frac{d}{d\alpha} \int_{\Omega_1} G(\phi_g + \alpha\, \delta\phi_g, u + \alpha\, \delta u)\, dx' \Big|_{\alpha=0} \\ &= \int_{\Omega_1} \frac{d}{d\alpha} G(\phi_g + \alpha\, \delta\phi_g, u + \alpha\, \delta u)\, dx' \Big|_{\alpha=0} \\ &= \int_{\Omega_1} \left(\frac{\partial G}{\partial u}\, \delta u\right) dx' + \int_{\Omega_1} \left(\frac{\partial G}{\partial \phi_g}\, \delta\phi_g\right) dx' = \int_{\Omega_1} \left(\frac{\partial G}{\partial u}\, \delta u\right) dx' + \int_{\Omega_1} \left(\frac{\partial G}{\partial \phi_g} \cdot \frac{d\phi_g}{du}\, \delta u\right) dx' \\ &= \int_{\Omega_1} \delta u \left[\frac{\partial G}{\partial u} + \frac{\partial G}{\partial \phi_g} \frac{d\phi_g}{du}\right] dx'\,, \quad (2.19)\end{aligned}$$

so that we have, in virtue of the assumption (2.7) on the function G, for some u, $g_u, \phi_g = \varphi(u)$ satisfying the transformed criticality equation (2.4)

$$\int_{\Omega_1} \delta G[\phi_g(u), u]\, dx' = \int_{\Omega_1} \delta u(x')\, \varphi^2[u(x')]\, dx\,. \quad (2.20)$$

Using (2.20) and the definition (2.17) of the scalar product we can write the first term on the right side of (2.16) in the following form

$$(u\phi_g, \phi_g) \int_{\Omega_1} \delta G(\phi_g, u)\, dx' = \int_{\Omega_1} \phi_g u\phi_g\, dx \int_{\Omega_1} \delta u(x')\, \phi_g^2(x')\, dx'\,. \quad (2.21)$$

Further, by help of (2.17) and (2.4a) we can rewrite the left side of (2.16) in the following way

$$\begin{aligned}\delta g_u(u\phi_g, Hu\phi_g) &= \delta g_u \frac{1}{g_u} (u\phi_g, \phi_g) \int_{\Omega_1} G[\phi_g(x'), u(x')]\, dx' \\ &= \delta g_u \frac{1}{g_u} \int_{\Omega_1} \phi_g u\phi_g\, dx \int_{\Omega_1} G[\phi_g(x')\, u(x')]\, dx'\,. \quad (2.22)\end{aligned}$$

By substituting the relations (2.18), (2.21) and (2.22) in the equation (2.16) for δg_u we obtain

$$\delta g_u \left\{ \frac{1}{g_u} \int_{\Omega_1} \phi_g u \phi_g \, dx \int_{\Omega_1} G[\phi_g(x'), u(x')] \, dx' \right\}$$

$$= \int_{\Omega_2} \delta u(x') \, \varphi^2[u(x')] \, dx' \left\{ \int_{\Omega_2} \phi_g u \phi_g \, dx - \int_{\Omega_2} G[\phi_g(x), u(x)] \, dx \right\}. \tag{2.23}$$

Because the continuous eigenfunction $\varphi[u(x)] = \phi_g(x)$ cannot be zero identically, and because ϕ_g, u, $G(\phi_g, u)$ are supposed to be bounded and positive, it follows from the equations (2.23) and (2.12) by help of the fundamental lemma of the calculus of variations [5] that we have

$$\delta g_u = 0 \Leftrightarrow \int_{\Omega_2} u(x) \, [\phi_g(x)]^2 \, dx = \int_{\Omega_2} G[\phi_g(x), u(x)] \, dx \,, \tag{2.24}$$

for each u, g_u, ϕ_g satisfying the assumptions (2.i), (2.ii) of theorem (2.1), q.e.d.

Theorem 2.2. *Under the assumptions of theorem* 2.1, *the solution* ψ, 1, $\phi_\psi = \phi_g$ *of the problem* (1.1)—(1.3), (1.i)—(1.iv) *has to fulfil the following inequality:*

$$\overline{[\phi_g^2]} \cdot \frac{\int_{\Omega_1} u \, dx}{|\Omega_1|} \geqq \overline{G} \,, \quad (|\Omega_1| = \text{measure } \Omega_1) \tag{2.24a}$$

where $\overline{[\phi_g^2]}$ *denotes the mean value of* ϕ_g^2 *in the reactor core*

$$\overline{[\phi_g^2]} = \frac{\int_{\Omega_2} \phi_g^2 u \, dx}{\int_{\Omega_2} u \, dx} = \frac{\int_{\Omega_2} \varphi^2(u) \, u \, dx}{\int_{\Omega_2} u \, dx} \tag{2.24b}$$

and $\overline{G}$ *denotes the mean value of* $G = G(\phi_g, u)$ *in the whole reactor*

$$\overline{G} = \frac{1}{|\Omega_1|} \int_{\Omega_1} G[\phi_g(x), u(x)] \, dx \,. \tag{2.24c}$$

Proof: Because there is no fuel in the reflector (by definition 1.1), we have

$$u \in U_1 \wedge x \in \Omega_1 \backslash \Omega_2 \Rightarrow \delta u(x) = 0$$

and therefore, using the assumption (2.7) of theorem 2.1

$$\delta g_u = \delta \int_{\Omega_1} G(\phi_g, \overline{\alpha} u) \, dx = \overline{\alpha} \int_{\Omega_2} \delta G(\phi_g, u) \, dx = \overline{\alpha} \int_{\Omega_2} \delta u \left(\frac{\partial G}{\partial u} + \frac{\partial G}{\partial \phi_g} \frac{d\phi_g}{du} \right) dx$$

$$= \overline{\alpha} \int_{\Omega_2} \delta u [\varphi^2(u)] \, dx \geqq 0 \,.$$

From (2.23) it follows therefore, that for u, ϕ_g, $G(\phi_g, u)$ the relation

$$\int_{\Omega_2} \phi_g^2 u \, dx - \int_{\Omega_2} G(\phi_g, u) \, dx \geqq 0 \tag{2.24d}$$

must hold, which can be written in the form ($|\Omega|$ = measure Ω)

$$\frac{\int\limits_{\Omega_1} \phi_g^2 u \, dx}{\int\limits_{\Omega_1} u \, dx} \int\limits_{\Omega_1} u \, dx - |\Omega_1| \frac{\int\limits_{\Omega_1} G(\phi_g, u) \, dx}{|\Omega_1|} \geqq 0 . \tag{2.24e}$$

Because $|\Omega_1| > 0$, (2.24e) is equivalent to (2.24a), (2.24b), Q.E.D.

Corollary 2.1. Theorem 2.1 can be used for algorithmization in two ways: 1) We take a solution u, g_u, $\phi_g = \varphi(u)$ of (2.4), construct the function

$$h(u) = \int \varphi^2(u) \, du = \text{const.}$$

put $h(u)$ in the form $h(u) = G[\varphi(u), u]$ and verify if the conditions (2.6), (2.6a), (2.6b), (2.6c), (2.8) are fulfilled by $G[\varphi(u), u]$.

2) For given $G(\phi_g, u)$ we take $\frac{\partial G}{\partial u} + \frac{\partial G}{\partial \phi_g} \frac{d\phi_g}{du} = \phi_g^2$ and compute $\phi_g = \varphi(u, C)$ from the Riccati differential equation

$$\frac{\partial G}{\partial \phi_g} \cdot \frac{d\phi_g}{du} = \phi_g^2 - \frac{\partial G}{\partial u} . \tag{2.25}$$

Now, we shall try seek for $u = u(x)$ and $g = g_u$ such, that the criticality condition $\phi_g = g_u R_u^G \phi_g$ and the necessary condition (2.24) would be fulfilled. We have from the criticality condition

$$\int\limits_{\Omega_1} u(x) \, [\phi_g(x)]^2 \, dx = g_u^2 \int\limits_{\Omega_1} u(x) \, [R_u^G \phi_g]^2 \, dx \tag{2.26}$$

so that, using the condition (2.24), we obtain by comparison

$$g_u^2 = g_u^2(G) = \frac{\int\limits_{\Omega_1} G[\phi_g(x), u(x)] \, dx}{\int\limits_{\Omega_1} u(x) \, [R_u^G \phi_g]^2 \, dx} = \frac{\left[\int\limits_{\Omega_1} G(\phi_g, u) \, dx\right]^3}{\int\limits_{\Omega_1} u(H u \phi_g)^2 \, dx} \quad \text{for} \quad \forall u(x) \in U_1 . \tag{2.27}$$

Therefore, for g_u given by (2.27) the condition (2.24) will be fulfilled for all $u(x) \in U_1$ and we have only to determine $u = u(x)$ from the criticality condition, i.e., to find the eigenvector of R_u^G corresponding to its least positive eigenvalue g_u, which can be done by well known methods, e.g., by searching for $u(x)$ minimizing (2.27).

Let us consider some examples.

Example 1. Let us consider the case $G(\phi, u) = u$ (Goertzel's problem, [1]). Then, by (2.25), we have for ϕ_g the equation

$$0 = 0 \cdot \frac{d\phi_g}{du} = \phi_g^2 - 1 \Leftrightarrow \phi_g = 1 = \text{const.}$$

Putting $u_1(x) = \psi_1(x)$ as the corresponding fuel distribution, we shall have for it the criticality condition

$$1 = H\psi_1 .$$

The least positive eigenvalue $g_u > 0$ of the operator R_u^G will fulfil, by (2.27), the relation

$$g_u^2 = \frac{\int\limits_{\Omega_1} \psi_1(u)\,dx}{\int\limits_{\Omega_1} \psi_1 \cdot (H\psi_1)^2 \frac{1}{\left[\int\limits_{\Omega_1} \psi_1(x)\,dx\right]^2}\,dx} = \frac{\int\limits_{\Omega_1} \psi_1(x)\,dx}{\int\limits_{\Omega_1} \psi_1\,dx \cdot \frac{1}{\left(\int\limits_{\Omega_1} \psi_1\,dx\right)^2}} = \left[\int\limits_{\Omega_1} \psi_1\,dx\right]^2$$

so that we have

$$g_u = \int\limits_{\Omega_1} \psi_1\,dx\,.$$

The necessary condition (2.24) for optimality of the reactor will be in this case trivially fulfilled:

$$\int\limits_{\Omega_1} u(x) \cdot 1^2\,dx = \int\limits_{\Omega_1} u(x)\,dx\,.$$

The criticality condition (2.4) takes in this case the form

$$\phi_g = g_u R_u^G \phi_g \Rightarrow 1 = \left[\int\limits_{\Omega_1} \psi_1\,dx\right] \cdot \frac{H\psi_1 \cdot 1}{\int\limits_{\Omega_1} \psi_1\,dx} = 1 \cdot 1 = 1$$

so that it has the solution $u = \psi_1$, $g_u = \int\limits_{\Omega_1} \psi_1\,dx$, $\phi_g \equiv 1$. The inequality (2.24) in the reflector $\Omega_1 \setminus \Omega_2$ will also be fulfilled trivially in this case, because we have for $G(\psi, u) = u$ by (2.24c) and (2.24b)

$$\overline{G} = \frac{\int\limits_{\Omega_1} u\,dx}{|\Omega_1|}\,, \qquad \overline{[\phi_g^2]} = 1$$

so that the inequality (2.24a) takes the form

$$1 \cdot \frac{\int\limits_{\Omega_1} u\,dx}{|\Omega_1|} \geqq \frac{\int\limits_{\Omega_1} u\,dx}{|\Omega_1|}\,.$$

Example 2. Let us consider the case $\phi_g = \varphi(u) \equiv 1$, $H\psi_1 = Hu_1 = 1$ (Wilkins [6]). We have in this case, by (2.7)

$$G(\phi_g, u) = G[\varphi(u), u] = \int \varphi^2(u)\,du + \text{const.} = u + \text{const.} = u_1 = \psi_1\,.$$

Clearly, conditions (2.6), (2.6a), (2.6b), (2.6c) on $G(\phi, \psi)$ are fulfilled in this case, and we can again verify, that the sufficient condition (2.8) for optimality of the reactor is fulfilled.

Example 3. Let $G = G(u)$ be a given function of the fuel spatial distribution $u \in U_1$ only. Then, by (2.25), we have for ϕ_g the equation

$$0 \equiv 0 \cdot \frac{d\phi_g}{du} = \phi_g^2 - \frac{dG(u)}{du}$$

so that we have, because $\phi_g^2 = [\varphi(u)]^2$ by (1.2b):

$$G(u) = \int [\varphi(u)]^2\,du + \text{const.} = \varkappa(u)\,.$$

Let us compute, using (2.27), for $u = u(x) \in U_1$ the value

$$g_u = \left\{ \frac{\left[\int_{\Omega_1} \varkappa(u)\, dx \right]^3}{\int_{\Omega_1} u\,(Hu\,\varphi(u))^2\, dx} \right\}^{1/2}.$$

For these u, g_u, ϕ_g both the condition (2.7) and the necessary optimality condition (2.8) of Theorem 2.1 will be fulfilled, if $u \in U_1$ will be so that u, g_u, ϕ_g fulfil the criticality condition (2.26). Therefore, we must search for $u(x) \in U_1$ giving the extremal value to g_u, because this is the sufficient condition for these u, g_u, ϕ_g to fulfil also the criticality condition (2.4).

If we take $u = u(x) \in U_1$ having the inverse function $x = x(u) \in C_1(E_1)$, then the inequality (2.24d) in the reflector takes the form

$$\int_{u(\Omega_2)} u\varphi^2(u)\, x'(u)\, du \geqq \int_{u(\Omega_1)} \varkappa(u)\, x'(u)\, du$$

from which it follows, that for $\phi_{\max}^2 = \max\limits_{u \in u(\Omega_2)} \varphi^2(u)$ we have the inequality

$$\phi_{\max}^2 \geqq \frac{\int_{u(\Omega_1)} \varkappa(u)\, x'(u)\, du}{\int_{u(\Omega_2)} u x'(u)\, du},$$ because $\varkappa(u)$ is increasing.

Example 4. Let us consider the case $G(\phi_g, u) = \phi_g \cdot u$ (reactor with extremal total output). By (2.25), we have in this case for ϕ_g the equation

$$u \cdot \frac{d\phi_g}{du} = \phi_g^2 - \phi_g$$

which is a homogeneous equation of Riccati type, so that it can be by substitution

$$\varrho = \frac{1}{\phi_g}, \qquad \frac{d\varrho}{du} = -\frac{1}{\phi_g^2} \frac{d\phi_g}{du}$$

transformed to the linear equation

$$\frac{d\varrho}{du} - \frac{1}{u}\varrho + 1 = 0$$

which has the general solution (with C as integration constant)

$$\varrho(u) = e^{\lg u} \left(C - \int 1 \cdot e^{-\lg u}\, du\right) = u(C - \lg u)$$

so that we have

$$\phi_g = \frac{1}{u(C - \lg u)}.$$

For the least positive eigenvalue g_u we obtain from (2.27) the relation

$$0 < g_u = \left[\frac{\left(\int_{\Omega_1} \phi_g u\, dx \right)^3}{\int_{\Omega_1} u(Hu\phi_g)^2\, dx} \right]^{1/2}.$$

For these u, g_u, ϕ_g will be fulfilled both the conditions (2.7) and (2.8) of theorem 2.1, if $u \in U_1$ will be so that u, g_u, ϕ_g fulfil the criticality condition (2.26). Therefore, we must construct $u = u(x)$ which extremalizes g_u, because for this u the criticality condition (2.26) will also be fulfilled.

Inequality (2.24d) in the reflector $\Omega_1 \setminus \Omega_2$ takes in this case the form (if again $u \in U_1$ is taken so that it has the inverse function $x = x(u) \in C_1(E_1)$)

$$\int_{u(\Omega_3)} \phi_g \cdot \frac{1}{(C - \lg u)} x'(u)\, du \geqq \int_{u(\Omega_3)} \frac{1}{(C - \lg u)} x'(u)\, du\,,$$

so that, if we denote $\phi_{g\max} = \max\limits_{u \in u(\Omega_3)} \dfrac{1}{u(C - \lg u)}$ we have for $\phi_{g_{\max}}$ the conditions

$$\phi_{g\max} \geqq 1 \qquad \text{if} \qquad \int_{u(\Omega_3)} \frac{1}{(C - \ln u)}\, x'(u)\, du > 0\,,$$

$$\phi_{g\max} \leqq 1 \qquad \text{if} \qquad \int_{u(\Omega_3)} \frac{1}{(C - \lg u)}\, x'(u)\, du < 0$$

resp.

§ 3. Applications to engineering regulation problems

In practice of reactor physics, the engineer often does not know the optimizing functional g, i.e., the optimization function $G(\phi, \psi)$. Instead of this, it is desirable to him to keep some physical quantity, i.e., some function $F(\phi, \psi)$ on constant level for obtain a technically preferable regulation, i.e., he postulates the following mathematical model for OCHTRR:

$$\phi = H\psi\phi \qquad \text{(criticality condition)}\,, \tag{3.1}$$

$$F(\phi, \psi) = \text{const.} \qquad \text{(regulation condition)}\,. \tag{3.2}$$

Usually, the condition (3.2) is taken in one of the following forms (3.2a), (3.2b) respective, containing the implicite functions ϕ, ψ resp.

$$\phi = \phi(\psi, C_1); \qquad C_1 = \text{const.}, \qquad F[\phi(\psi, C_1), \psi] = C_1 \tag{3.2a}$$

$$\psi = \psi(\phi, C_2); \qquad C_2 = \text{const.}, \qquad F[\phi, \psi(\phi, C_2)] = C_2\,. \tag{3.2b}$$

In this case, we can substitute in (3.1) and obtain the following two nonlinear reactor regulation models:

$$\phi(\psi, C_1) = H\psi\phi(\psi, C_1)\,, \tag{3.3a}$$

$$\phi = H\psi(\phi, C_2)\,\phi \tag{3.3b}$$

resp.

One can try to obtain the approximate solution ψ, ϕ resp. of those nonlinear equations by solving them with the aids of nonlinear functional analysis and numerical analysis. But there is also another interesting question (due to Goertzel [1], Wilkins [6]):

Which is the relation between the variational optimal regulation problems (1.1)—(1.3), (1.i)—(1.iv) and the technical regulation problems (3.1), (3.2)?
A partial answer to this question is given by the following

Theorem 3.1. *Let for the optimal critical reactor problem* (2.4), (2.5) *with* $\psi = \psi(u)$, $\phi = \phi_g$ *the assumptions* (2.i), (2.ii) *of Theorem* 2.1 *be valid with given* $G = G(\phi, \psi)$, *and moreover, let us have in the condition* (2.7)

$$\int \varphi^2(u)\, du + \text{const.} = G_1(\phi_g, \psi) + \text{const.} \tag{3.4}$$

$$G(\phi_g, u) = G_0(\phi_g, \psi)\,. \tag{3.4a}$$

Then for the existence of a solution ϕ, ψ *of the regulation problem* (3.1), (3.2) *with*

$$F(\phi_g, \psi) \equiv G_0(\phi_g, \psi) - G_1(\phi_g, \psi) \tag{3.5}$$

the condition (2.8) *is necessary.*

Proof: Follows immediately from theorem 2.1, because $G_0(\phi_g, \psi) - G_1(\phi_g, \psi) = \text{const.}$ by (2.7) and (3.4), (3.4a). In virtue of theorem 2.1, relation (2.8) is a necessary condition for the existence of a solution u, g_u, ϕ_g of the optimal critical reactor problem (2.4), (2.5) and, by construction of $F(\phi, \psi)$ in (3.5), this $\phi = \phi_g$ and $\psi = \psi(u)$ will be also a solution of the regulation problem (3.1), (3.2), q.e.d.

Remark 3.1: The inverse problem: Find $G(\phi, \psi)$ in (1.2), if $F(\phi, \psi)$ in (3.1) is given, is of great technical interest, but much more difficult.

References

[1] Goertzel, G., Minimum critical mass and flat flux. J. Nuclear Energy, Vol. 2, (1956) 193—201.
[2] Bartošek, V., and Zezula, R., Flat Flux in a slab reactor with natural uranium. Journal of Nuclear Energy Parts A/B, Vol. 20 (1966) 129—134.
[3] Zezula, R., A sufficient condition for flattening of the thermal neutron flux and some related problems (in onedimensional geometries). Apl. Mat. 14, Nr. 2 (1969) 134—145.
[4] —, Onedimensional homogenized reactor with natural Uranium and with flattened specific output. Apl. Mat. 14, Nr. 4 (1969), 323—336.
[5] Elsgolz, Calculus of variations (in Russian). 2nd ed., Moscow, GITTL, 1958.
[6] Wilkins, J. E., Nucl. Sci. Eng. 1 (1956) 68.

Anschrift des Autors:

Mathematical Institute of Charles University
18600 Prague 8, Sokolovská 83, ČSSR

Bei Fragen zur Produktsicherheit wenden Sie sich bitte an:
If you have any questions regarding product safety,
please contact:

Walter de Gruyter GmbH
Genthiner Straße 13
10785 Berlin
productsafety@degruyterbrill.com